ASPHALT MATERIALS SCIENCE AND TECHNOLOGY

ASPHALT MATERIALS SCIENCE AND TECHNOLOGY

JAMES G. SPEIGHT PhD, DSC

ELSEVIER

Amsterdam • Boston • Heidelberg • London
New York • Oxford • Paris • San Diego
San Francisco • Singapore • Sydney • Tokyo
Butterworth-Heinemann is an imprint of Elsevier

B H

Butterworth-Heinemann is an imprint of Elsevier
The Boulevard, Langford Lane, Kidlington, Oxford OX5 1GB, UK
225 Wyman Street, Waltham, MA 02451, USA

Notices
Knowledge and best practice in this field are constantly changing. As new research and experience broaden our understanding, changes in research methods, professional practices, or medical treatment may become necessary.

Practitioners and researchers must always rely on their own experience and knowledge in evaluating and using any information, methods, compounds, or experiments described herein. In using such information or methods they should be mindful of their own safety and the safety of others, including parties for whom they have a professional responsibility.

To the fullest extent of the law, neither the Publisher nor the authors, contributors, or editors, assume any liability for any injury and/or damage to persons or property as a matter of products liability, negligence or otherwise, or from any use or operation of any methods, products, instructions, or ideas contained in the material herein.

ISBN: 978-0-12-800273-5

Library of Congress Cataloging-in-Publication Data
A catalog record for this book is available from the Library of Congress.

British Library Cataloguing-in-Publication Data
A catalogue record for this book is available from the British Library.

For Information on all Butterworth-Heinemann publications visit our website at http://store.elsevier.com/

CONTENTS

PART II ASPHALT TECHNOLOGY

PREFACE

The use of petroleum is an old art but the petroleum industry and the use of asphalt are essentially a twentieth century industry. The first use of petroleum came over 5000 years ago when it was recognized that the heavier derivatives of petroleum (asphalt) could be used for caulking, as adhesives for jewelry, for construction purposes, as well as for medicinal and other purposes.

The facile production of asphalt makes it an ideal choice for the rehabilitation of busy highways and aviation facilities. Whether on the highway or airport, asphalt pavements give the best value for the taxpayer dollar, with the lowest life cycle cost and the highest residual value. Asphalt is one of the indispensable materials of life in America. In fact, 94% of the paved roads in America are surfaced with asphalt. However, asphalt is a complex product that is used with aggregate, as one example, to build roads. To many, asphalt production is an art. This book will close the gap between the wide distribution and use of asphalt and knowledge of its chemical characterization for research and development.

It is the purpose of this book to present the production and applications of asphalt, as well as the various standard test methods used to assess the properties and potential uses of the manufactured products. In addition, the chemistry of asphalt is also presented with a discussion of how asphalt can change in service, which can influence the properties of the asphalt and the means by which asphalt is used for roadway construction. However, first and foremost, an important aspect of asphalt science and technology is an understanding of the terminology being used.

This book will assist the reader to understand the mix design procedures, the relevant test methods, and also presents an introduction to paving technology. It will serve as an information source to individuals who are experienced in the some aspects of asphalt science and technology but require background knowledge in other aspects. The book will also provide an easy-to-read reference source for individuals who are relatively new to asphalt pavement technology who will find information on materials, asphalt pavements, asphalt concrete mixtures, and mix design methods.

This book provides the linkages to provide an understanding of the issues associated with the technology surrounding the production and utilization of asphalt, including loading and transporting quarried materials, production of the bituminous product, and the associated health, safety, and environmental issues.

The intended audience for this book is wide-ranging within the scientific and engineering disciplines and includes the newer inductees into asphalt science and technology petroleum science and technology, as well as those already involved in this important field of science and engineering who need a broader coverage of the subject area.

This book is an easy-to-read update on asphalt science and technology that is suitable for scientific and engineering disciplines. The book also explains why asphalt lasts or fails in service. The book is also a one-of-a-kind reference that offers detailed coverage of the chemistry, technology, and engineering of asphaltic products that are used for construction of asphaltic roadways.

James G. Speight, Phd, DSc
Laramie, WY, USA

CONVERSION FACTORS

WEIGHT

1 ounce (1 oz) = 28.3495 grams (18.2495 g)
1 pound (1 lb) = 0.454 kilogram
1 pound (1 lb) = 454 grams (454 g)
1 kilogram (1 kg) = 2.20462 pounds (2.20462 lb)
1 stone (English, 1 st) = 14 pounds (14 lb)
1 ton (US; 1 short ton) = 2000 lbs
1 ton (English; 1 long ton) = 2240 lbs
1 metric ton = 2204.62262 pounds
1 tonne = 2204.62262 pounds

TEMPERATURE CONVERSIONS

°F = (°C × 1.8) + 32
°C = (°F − 32)/1.8
(°F − 32) × 0.555 = °C
Absolute zero = −273.15°C
Absolute zero = −459.67°F

AREA

1 square centimeter (1 cm^2) = 0.1550 square inches
1 square meter 1 (m^2) = 1.1960 square yards
1 hectare = 2.4711 acres
1 square kilometer (1 km^2) = 0.3861 square miles
1 square inch (1 $inch^2$) = 6.4516 square centimeters
1 square foot (1 ft^2) = 0.0929 square meters
1 square yard (1 yd^2) = 0.8361 square meters
1 acre = 4046.9 square meters = 43,560 sq ft
1 acre foot = 7758.0 bbl
1 square mile (1 mi^2) = 2.59 square kilometers

CONCENTRATION

1 part per million (1 ppm) = 1 microgram per liter (1 μg/L)
1 microgram per liter (1 μg/L) = 1 milligram per kilogram (1 mg/kg)
1 microgram per liter (1 μg/L) × 6.243 × 10^8 = 1 lb per cubic foot (1 lb/ft^3)
1 microgram per liter (1 μg/L) × 10^{-3} = 1 milligram per liter (1 mg/L)
1 milligram per liter (1 mg/L) × 6.243 × 10^5 = 1 pound per cubic foot (1 lb/ft^3)
1 gram mole per cubic meter (1 g mol/m^3) × 6.243 × 10^5 = 1 pound per cubic foot (1 lb/ft^3)
10,000 ppm = 1% w/w
1 ppm hydrocarbon in soil × 0.002 = 1 lb of hydrocarbons per ton of contaminated soil

OTHER APPROXIMATIONS

1 acre-inch of liquid = 27,150 gallons = 3.630 ft^3
1 atmosphere = 14.7 pounds per square inch (14.7 psi)
1 atmosphere = 760 mm Hg = 14.696 psia = 29.91 in. Hg
1 atmosphere = 1.0133 bars = 33.899 ft. H_2O
1 barrel (oil) = 42 gal = 5.6146 cu ft
1 barrel (water) = 350 lb at 60°F
1 barrel per day = 1.84 cu cm per second
1 Btu = 778.26 ft-lb
1 centipoise × 2.42 = lb mass/(ft) (h), viscosity
1 centipoise × 0.000672 = lb mass/(ft) (s), viscosity
1 cubic foot = 28,317 cu cm = 7.4805 gal
1 cubic yard (1 yd^3) = 27 ft^3 = 0.765 m^3
1 cubic yard (clayey soils-excavated) = 1.1–1.2 tons (US)
1 cubic yard (sandy soils-excavated) = 1.2–1.3 tons (US)
Density of water at 60°F = 0.999 g/cu cm = 62.367 lb/cu ft = 8.337 lb/gal
1 foot depth in 1 acre (*in situ*) = 1613 × (20–25% excavation factor) = ~2000 yd^3
1 gallon = 231 cu in. = 3785.4 cu cm = 0.13368 cu ft
1 gallon of water (US gallon) = 8.34 lbs
1 gallon of water (imperial gallon, UK gallon) = 10 lbs
1 inch = 2.54 cm
1 kilopascal (kPa) × 9.8692 × 10^{-3} = 14.7 pounds per square inch (14.7 psi)
1 meter = 100 cm = 1000 mm = 10 microns = 10 angstroms (A)
1 ounce = 28.35 grams
1 pound = 453.59 grams = 7000 grains
1 square mile = 640 acres

PART I

Asphalt Science

CHAPTER 1

Nomenclature and Terminology

1.1 INTRODUCTION

In the modern sense of the word, *asphalt* is a derivative of petroleum and was first used more than 5000 years ago when it was recognized that the higher boiling (non-volatile under the conditions of use) derivatives of petroleum (asphalt) could be used for caulking, water proofing, adhesives for jewelry, and as a mastic for construction purposes. There is also documented use of asphalt for medicinal purposes or for punishment—it is recorded in the Code of Hammurabi a miscreant could be punished by having hot bitumen (sometimes called *natural asphalt*) from one of the nearby seepages poured over his (or her) head. Approximately 2000 years ago, Arabian scientists developed methods for the distillation of petroleum that were introduced into Europe by way of Spain. Interest in petroleum was also documented in China, where petroleum was encountered when drilling for salt, and appears in documents of the third century AD. The Baku region of northern Persia was also reported (by Marco Polo in 1271–1273) to have a commercial petroleum industry. Of course, it must be realized that there may be vast differences between commercial operations in the thirteenth century and commercial operations in the twentieth century!

Interest in petroleum continued up to modern times with an increasing interest in nafta (naphtha) when it was discovered that this material could be used as an illuminant and as a supplement to asphalt incendiaries in warfare. The nafta of that time was obtained from shallow wells or by the destructive distillation of asphalt. This may perhaps be equated to modern-day coking operations (Speight, 2014), in which the overall objective of the process is to convert the non-volatile residua into liquid fuels. The modern oil industry began in 1859 with the discovery and subsequent commercialization of petroleum in Pennsylvania (Bell, 1945). It is from this time that the modern petroleum industry evolved into its present form and the use of asphalt expanded phenomenally.

Asphalt Materials Science and Technology.
DOI: http://dx.doi.org/10.1016/B978-0-12-800273-5.00001-5

Asphalt is the product of a refinery operation whereby a residuum is treated by air blowing or by a solvent method to produce a product that meets specifications for a variety of road/highway construction and other uses. However, the term *asphalt* is a term that had been maligned and used loosely to describe a variety of naturally occurring and manufactured substances or in some cases the end product of refining that was considered to be of limited use. In some cases, coal tar was mixed with asphalt for waterproofing and road paving purposes, which was also a means of removing coal tar and converting it into a product of commercial importance. However, there have been many terms (some favorable, some unfavorable) that have been used historically to describe asphalt or bitumen (Table 1.1) and while there have been valiant attempts to define bitumen and asphalt (Abraham, 1945; Barth, 1962; Zakar, 1971; Speight, 1992, 2014), confusion still exists because of the lack of understanding of the nature of the origin and/or the production of the various materials. It is the purpose of this chapter to bring common sense to the confusion by offering explanations of definitions that are in keeping with the origin and nature of the various types of bitumen, asphalt, tar, and pitch. For the purposes of this text, the term *asphalt* is used to describe the refinery product and the term *bitumen* is used to describe the naturally occurring material.

Asphalt production from crude petroleum as a product of refineries in the early twentieth century and the increasing popularity of the automobile served to greatly expand the asphalt industry. Typically, asphalt is produced from a petroleum residuum. A *residuum* or *resid* (pl. *residua*, *resids*) is the non-distillable fraction of petroleum (Table 1.2) (Speight, 2014). Distillation temperatures are ordinarily limited to temperatures below the onset of thermal degradation (<350°C, <660°F) but it is possible to recover gas oils with true boiling points up to 565°C (1050°F) by vacuum distillation but maintaining the actual distillation temperature below the limits given above. The shifting product demand to lighter products coupled with the influx of heavy oil (which has high proportions of residua) into refineries has caused attention to be focused on the conversion of residua to distillate products. In the past, residua were used primarily as industrial fuels or asphalt precursors (Speight, 2011, 2014).

Asphalt may be residual (straight-run) asphalt, which is made up of the non-volatile hydrocarbons in the feedstock, along with similar materials produced by thermal alteration during the distillation sequences, or they may be produced by air blowing residua. Alternatively, asphalt may be the residuum from a vacuum distillation unit. In either case, the properties

Table 1.1 Linguistic Origins of Words Related to the Various Aspects of Petroleum Science and Technology

Language	Word	Possible Meaning
Sumerian	esir	Petroleum, bitumen
	esir-lah	Hard/glossy asphalt
	esir-harsag	Rock asphalt
	esir-é-a	Mastic asphalt
	esir-ud-du-a	Pitch
	kupru	Slime, pitch
Sanskrit	jatu	Bitumen, pitch
	śilā-jatu	Rock asphalt
	aśmajātam-jatu	Rock asphalt
Assyrian/Akkadian	idd, ittû, it-tû-u	Bitumen
	amaru	Bitumen
	sippatu	Pitch
Hebrew	zephet	Bitumen
	kopher or kofer	Pitch
	hêmâr	Pitch
Arabic and Turkish	seyali	Bitumen
	zift or zipht	Bitumen or pitch
	chemal	rock asphalt
	humar (houmar)	Rock asphalt
	gasat (qasat)	Rock asphalt
	ghir or gir	Asphalt mastic
	kir or kafr	Asphalt mastic or pitch
	mûmûia	Bitumen
	neftgil	Petroleum wax, mineral wax
Greek	maltha	Soft asphalt
	asphaltos	Bitumen
	pissasphaltos	Rock asphalt
	pittasphaltos	Rock asphalt
	pittolium	Rock asphalt
	pissa or pitta	Pitch
	ampelitis	Mineral wax and asphaltites
Latin	maltha	Soft asphalt
	bitumen liquidum	Soft Asphalt
	pix	Pitch

of the asphalt are, essentially, the properties of the residuum (Speight and Ozum, 2002; Parkash, 2003; Hsu and Robinson, 2006; Gary et al., 2007; Speight, 2014). If the properties are not suitable for the asphalt product to meet specifications, changing the properties by, for example, blowing is necessary.

Table 1.2 Properties of Tia Juana Crude Oil and the 650°F, 950°F, and 1050°F Residua

		Residua		
	Whole Crude	**650°F+**	**950°F+**	**1050°F+**
Yield, vol. %	100.0	48.9	23.8	17.9
Sulfur, wt. %	1.08	1.78	2.35	2.59
Nitrogen, wt. %		0.33	0.52	0.60
API gravity	31.6	17.3	9.9	7.1
Carbon residue, wt. %				
Conradson		9.3	17.2	21.6
Metals				
Vanadium, ppm		185		450
Nickel, ppm		25		64
Viscosity				
Kinematic				
at 100°F	10.2	890		
at 210°F		35	1010	7959
Furol				
at 122°F		172		
at 210°F			484	3760
Pour point, °F	−5	45	95	120

With the development of gasoline-powered and diesel-powered engines there arose a need for better roadways—the wagon and stage-coach tracks of pervious eras were totally unsuitable for vehicular travel. As a result of the need for better roadways, the fundamentals of a good roadway design evolved and the requirements were (i) proper design and (ii) proper construction so that the roadway would remain in good usable condition for several years with only minimal maintenance. In addition, as the twentieth century evolved, the roadway had to be suitable for traffic loads under all weather conditions for a specified design life and so the achievements of the early roadway design and construction led to more reliable roadways and, hence, more comfortable means of transportation.

This was achieved by: (i) choice of a suitable asphalt binder, (ii) choice of a suitable mineral aggregate, (iii) stabilizing and strengthening of the sub-grade to provide a stable base and sub-base layers above the sub-grade, (iv) providing adequate water drainage, because water can weaken soils and asphalt pavements, and (v) construction of a roadway surface and structure that has sufficient depth to structurally carry all expected traffic loads for a period of time. In addition, the full strength and prevention of water penetration into the pavement and its base must also be assured.

Finally, the roadway surface (the wearing course) must resist wear, deformation, and weather, and remain skid-resistant.

In relation to asphalt, the definitions of petroleum and associated materials are varied, to the point of being archaic, and can lead to much confusion. The terms *slime* and *tar* in ancient texts are used to identify asphalt and bitumen. However, the definitions evolved and many in current use came into being during the latter part of the nineteenth century and the early part of the twentieth century at a time when the nomenclature of organic chemistry was not properly formalized. In summary, these definitions are descriptions of convenience and not necessarily accurate in a scientific sense. The increasing commercial importance of asphaltic products has been manifested by greater efforts to perfect and standardize methods of testing (see Chapters 4 and 5) through various government agencies in many countries, as well as agencies that specialize in the development of standard test methods (e.g., American Association of State Highway and Transportation Officials, AASHTO, and the American Society for Testing and Materials, ASTM). This is not the case with the issues surrounding the definitions of petroleum and related materials.

Thus, it is the purpose of this book to present the production and application of asphalt, as well as the various standard test methods used to assess the properties and potential uses of the manufactured products. In addition, the chemistry of asphalt is also presented with a discussion of how asphalt can change in service, which can influence the properties of the asphalt and the means by which asphalt is used for construction of asphaltic roadways.

However, first and foremost, an important aspect of asphalt science and technology is an understanding of the terminology being used. The purpose of this chapter is to provide descriptions of the terminology that is used when asphalt roadways are planned and constructed.

1.2 ASPHALT

Asphalt is produced in a variety of types and grades ranging from hard-brittle solids to near water-thin liquids. The semi-solid form, known as asphalt cement, is the basic material used in asphalt concrete pavements. Liquid asphalt is produced when asphalt cement is blended (cut back) with petroleum-derived distillates or emulsified with water and an emulsifying agent. Liquid asphalt products may be produced for various uses and applications.

By definition, asphalt is a bituminous material because it is a hydrocarbonaceous material which is soluble in carbon disulfide and many aromatic solvents. Asphalt is a thermoplastic material insofar as it gradually liquefies when heated. At ambient temperatures asphalt is typically a semi-solid material, but at high temperatures (120–165°C, 250–325°F) asphalt is in liquid form and can be pumped through pipes, sprayed through nozzles, or mixed with aggregate, provided the high temperature is maintained. When asphalt cools, it returns to a semi-solid form and can be used as a cementing and waterproofing material—two properties that confer upon asphalt strength and flexibility.

The tar obtained from the destructive distillation of coal is also (incorrectly) often referred to as a bituminous material. However, because the properties of both petroleum-derived asphalt and coal tar differ significantly, one should not be confused with the other and the two materials should be treated as separate chemical entities (Puzinauskas and Corbett, 1978; Speight, 2013d, 2014).

Asphalt production from crude petroleum as a product of refineries in the early twentieth century and the increasing popularity of the automobile served to greatly expand the asphalt industry. Asphalt may be residual (straight-run) asphalt, which is made up of the non-volatile hydrocarbons in the feedstock, along with similar materials produced by thermal alteration during the distillation sequences, or they may be produced by air blowing residua. Alternatively, asphalt may be the residuum from a vacuum distillation unit. In either case, the properties of the asphalts are, essentially, the properties of the residuum (see Speight, 2014). If the properties are not suitable for the asphalt product to meet specifications, changing the properties by, for example, blowing is necessary.

The major source of asphalt is petroleum and asphalt manufacture is, in essence, a matter of distilling everything possible from crude petroleum and modification of the residue (typically by air blowing) until a residue with the desired properties is obtained (Figure 1.1) (see Chapter 2) (Speight, 2014). Petroleum distillation at atmospheric pressure removes the lower boiling fractions and yields reduced crude that may contain higher boiling (lubricating) oils, asphalt, and even wax. Distillation of the reduced crude under vacuum removes the oils (and wax) as overhead products and the asphalt remains as a bottom (or residual) product. The majority of the polar functionality in the original crude oil end is non-volatile and concentrates first in the atmospheric residuum and then in the vacuum residuum (Figure 1.2). It is this concentration effect that confers upon

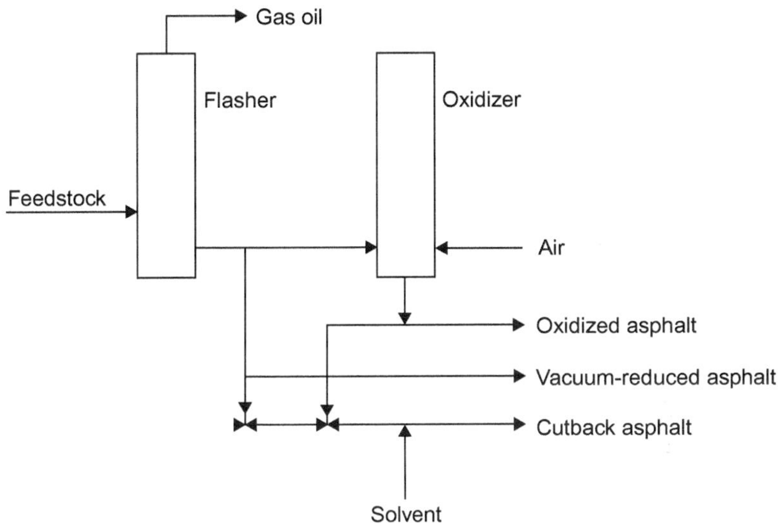

Figure 1.1 General representation of asphalt manufacture: the feedstock can be a residuum from the atmospheric distillation tower or from the vacuum distillation tower.

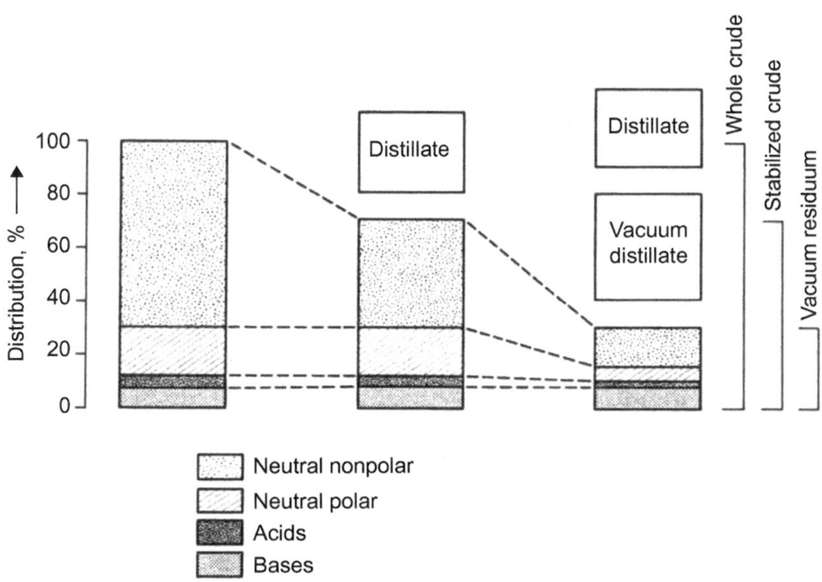

Figure 1.2 Accumulation of polar functions in residua during distillation.

asphalt some of its unique advantageous properties (such as the adhesion of asphalt to aggregate) and some of the disadvantageous properties (such as the type and level of sulfur that is prone to oxidation of sulfur oxides while in service). At this stage the asphalt is frequently and incorrectly referred to as pitch and has a softening point related to the amount of oil removed: the more oil distilled from the residue, the higher the softening point.

However, since there are wide variations in refinery operations and crude petroleum, asphalt with softening points ranging from 25°C to 55°C (80–130°F) may be produced. Blending with higher and lower softening point asphalt may make asphalt of intermediate softening points. If lubricating oils are not required, the reduced crude may be distilled in a flash drum, which is similar to a bubble tower but has few, if any, trays. Asphalt descends to the base of the drum as the oil components pass out of the top of the drum. If the asphalt has a relatively low softening point, it can be hardened by further distillation with steam or by oxidation.

Asphalt is also produced by propane deasphalting (see Chapter 2) and the asphalt so produced may have a softening point of about 95°C (200°F). Softer grades are made by blending *hard asphalt* with the extract obtained in the solvent treatment of lubricating oils.

If the asphalt requires a change in properties to a harder material, such as is required for roofing, pipe coating, or as an under-sealant or waterproofing material, *soft asphalt* can be converted into harder asphalt by oxidation. This process produces a material that softens at a higher temperature than paving asphalts. It may be air blown at the refinery, at an asphalt processing plant, or at a roofing material plant. The asphalt is heated to approximately 260°C (500°F) and air is bubbled through the hot mix for 1–5 h. The product, when cooled, remains in the liquid phase.

The chemistry of the oxidation process, although more complex that a simple representation, promotes the formation of additional resin constituents and additional asphaltene constituents from the polar and/or higher-molecular-weight aromatic non-resin and non-asphaltene constituents:

Oil (aromatic) constituents → resin constituents → asphaltene constituents

The increase in the proportion of semisolid (resin) and solid (asphaltene) constituents as a result of air blowing is accompanied by an increase in softening point with only a small loss in volume. A similar increase in softening point by removing oily constituents would cause a considerable decrease in volume. Oxidation is carried out by blowing air through asphalt heated to about 260°C (500°F) and is usually done in a tower (an oxidizer) equipped with a perforated pipe at the bottom through which the air is blown. The asphalt, in the batch mode or continuous mode, is heated until the oxidation reaction starts, but the reaction is exothermic, and the temperature is controlled by regulating the amount of air and by circulating oil or water through cooling coils within the oxidizer. Asphalt products with softening points as high as 180°C (350°F) may be produced.

Asphalt, normally a liquid when applied at higher temperature, may be referred to as *non-volatile liquid asphalt*, but semisolid or solid asphalt may be made liquid for easier handling by dissolving in a solvent and are referred to as *cutback asphalt*. The asphalt and solvent (naphtha, kerosene, or gas oil) are heated to about 105°C (225°F) and passed together through a mechanical mixer, and the effluent then enters a horizontal tank, which is used for further mixing. Liquid asphalt is pumped from the top of the tank into a perforated pipe lying at the bottom of the tank, and circulation is continued until mixing is complete. The following terminology is that frequently use in the asphalt industry and is presented below (alphabetically rather than by preference). However, the terminology will vary from region to region and/or from country to country. When using such terminology, a word or two of qualification is always helpful.

1.2.1 Types

Asphalt has several uses, although its use as road construction material is the best known, for which it is better known as *asphalt concrete* (the asphalt–aggregate mix). However, there are several different forms of asphalt: semi-solid to solid asphalt, liquid asphalt such as asphalt emulsion, and cutback asphalt, and bitumen, often referred to as natural asphalt. Another term that is often used is *rolled asphalt*, which is the asphalt–aggregate mix that is used to construct roads and other surfaces, such as parking lots, by being applied in layers and compacted.

Thus, it is the purpose of this section to allay some of the confusion that exists with the file of asphalt science and technology by explanations and descriptions that are in keeping with the origin and nature of the various types of asphalt as well as definitions of tar and pitch.

1.2.1.1 Asphalt Blends

Properties of blends of asphalts from different sources can not necessarily be predicted by averaging properties of the components. Asphalts may be considered to be combinations of a dispersed and a solvent moiety, the nature and amounts of which vary among asphalts (Heithaus, 1962). Asphalt properties will be determined by the manner in which the two moieties interact. In this work, dispersed and solvent moieties were generated from eight asphalts by size exclusion chromatography. Cross-blended mixtures of dispersed and solvent moieties from different asphalts were prepared and their rheological properties measured. The influence of the relative amounts and nature of the two moieties on rheological properties were determined. It was found that rheological properties of the cross-blended mixtures were greatly affected by the relative amounts of the two moieties. The nature of the two moieties also was influential in determining rheological properties, as the solvent components of some types of asphalt were found to be compatible with the dispersed moieties of some asphalt constituents but incompatible with others.

The compatibility of asphalt systems has long been of interest to researchers, especially the difference in properties among what are designated as compatible (sol-type) and non-compatible (gel-type) asphalts (Traxler, 1961). Accordingly, well-dispersed, compatible asphalts exhibit high-temperature susceptibility of viscosity, high ductility, low elasticity, and low rates of oxidative age hardening. Non-compatible asphalts, which are poorly dispersed, are characterized by low-temperature susceptibility, low ductility, pronounced elasticity, and high aging rates. These and other physical properties serve to categorize asphalts according to compatibility. The two fundamental types of asphalts will, of course, vary markedly in pavement performance.

Asphalt compatibility is also related to chemical composition. Non-compatible asphalts tend to be high in asphaltene and sulfur content compared with compatible systems. An initial theory was that asphalt was considered to be a dispersion of asphaltenes in a maltene solvent, and asphalt and asphalt blend properties therefore should be determined by the relative amounts of the two materials and the effectiveness by which maltenes disperse asphaltenes (Heithaus, 1962). The properties of asphalt blends thus are not necessarily computed additively from the properties of their parent asphalts. Properties of blends of various fractions of different asphalts have also been investigated, particularly the rheological properties of various asphaltene–maltene combinations, as well as mixtures prepared from solvent-derived fractions (Traxler, 1961).

Rheological properties of asphalts depend on relative amounts of dispersed and solvent components. The solvent component is much less viscous and less elastic than the dispersed component, which contains the viscosity-building constituents. Presumably, the nature of the two components and the manner of their interaction also affect rheological properties. Another physical property, the glass transition temperature, is determined by the relative amount of the two components and the source of the solvent component, but not the source of the dispersed component. Other properties conceivably might depend more on the source of the dispersed component than the solvent and there must be serious attempts to quantify any effects on asphalt rheological properties by examining factors such as: (i) the source of the solvent, (ii) the type of solvent, (iii) the source of dispersed component, and (iv) the relative amounts of the solvent and the dispersed component.

Cross-blended mixtures of asphalts have been prepared from two size exclusion chromatography fractions derived from eight different asphalts. The rheological properties of the mixtures were determined and it was found that, with one exception, the nature of the fraction assumed to correspond to the asphalt dispersed component and its relative abundance strongly influences rheological properties. These materials, which are the initial size exclusion chromatography eluate fractions, are known to contain most of the polar, aromatic viscosity-building components of asphalts. The nature of the fraction corresponding to asphalt solvent components has a lesser influence on measured rheological properties of most cross-blended mixtures, again with the exception of one set of mixtures.

1.2.1.2 Asphalt Cement

Asphalt is produced in a variety of types and grades ranging from hard-brittle solids to readily fluid liquids (see Chapter 3). The semi-solid form of asphalt—usually known as *asphalt cement*—is the basic material used for the construction of asphalt concrete pavements. Liquid asphalt is produced when asphalt cement is blended or *cut back* with one or more petroleum distillates or emulsified with water and an emulsifying agent. Liquid asphalt products may be produced for various uses and applications.

1.2.1.3 Asphalt Concrete

Asphalt concrete is commonly called asphalt, blacktop, or pavement in North America, and tarmac in Great Britain and Ireland. Asphalt concrete is known by many different names, examples are: (i) hot mix asphalt (HMA), (ii) warm mix asphalt (WMA), (iii) cold mix asphalt (CMA), (iv) plant mix, (v) bituminous mix, and (vi) bituminous concrete. When spilled into the environment, other names may be used. Asphalt concrete is a combination of two primary ingredients: (i) aggregate and (ii) asphalt cement. Typically, the aggregate constitutes 90–95% (w/w) of the total mixture and the asphalt cement (asphalt binder) constitutes 5–10% (w/w) of the total mixture to form the asphalt concrete.

In the production process, the aggregate and binder are combined in an efficient manufacturing plant which produces the specified material. Plant equipment includes: (i) cold bins for storage of graded aggregate, (ii) a dryer for drying and heating aggregates to the required mixing temperature, (iii) tanks for storing the liquid asphalt, and (iv) a pug mill for combining the graded, heated aggregate and liquefied asphalt cement according to the specified mix formula.

Once produced, the asphalt concrete is transported by truck to the paving site where it is spread to a uniform thickness with a mechanical paving or finishing machine. Then the material is compacted to the required degree by heavy, self-propelled rollers, producing a smooth, well-compacted pavement course. The paving or finishing machine places the asphalt concrete at temperatures above 107°C (225°F). The material should be compacted before the temperature of the mix falls below 80°C (175°F) to achieve adequate density.

1.2.1.4 Asphalt Emulsion

Asphalt emulsion (emulsified asphalt), like cutback asphalt, is also designated as a liquid asphalt to distinguish this product from the typical semi-solid to solid asphalt binders. It is a mixture of two normally immiscible components (asphalt and water) and an emulsifying agent (usually soap). However, it is important to note that the asphalt binder is the base material that has been liquefied by cutting back with a solvent.

In the emulsification process, hot binder (the discontinuous phase) is mechanically (using a colloid mill) separated into minute globules and dispersed in water (the continuous phase) treated with a small quantity of emulsifying agent. By proper selection of an emulsifying agent and other manufacturing controls, emulsified asphalts are produced in several types and grades. Through the choice of emulsifying agent, the emulsified asphalt may be: (i) anoxic, in which the binder globules are electro-negatively charged in the water (the continuous phase) and (ii) cationic, in which the binder globules are electro-positively charged. Because particles having a like electrostatic charge repel each other, asphalt globules are kept apart until the emulsion is deposited on the surface of the soil or aggregate particles. At this time, the asphalt globules coalesce through neutralization of the electrostatic charges or water evaporation. Coalescence of asphalt globules occurs in rapid and medium setting grades, resulting in a phase separation between asphalt and water. When this coalescence occurs, it is usually referred to as the break or set.

Emulsified asphalt is used to seal coats on asphalt pavements, built-up roofs, and other waterproof coverings (Speight, 1992). Emulsified asphalt is further graded according to its setting rate (i.e., rapid, medium, and slow). Rapid-setting grades are used for surface treatment and seal coating, whereas medium-setting grades are used for patch mixtures, and slow-setting grades are used for mix-in-place road mixtures, patch mixtures, tack coats, fog coats, slurry seals, and soil stabilization (Speight, 1992).

Asphalt emulsion, as used in road construction and maintenance, is a homogeneous mixture of minute asphalt droplets suspended in a continuous water phase (Dybalski, 1976). The preparation of these oil-in-water (o/w) emulsions involves use of a high-speed, high-shear mechanical device, such as a colloid mill which breaks down molten asphalt into minute droplets in the presence of water and a chemical, surface-active emulsifier. The emulsifier imparts its properties to the dispersed asphalt which is influential in maintaining stable asphalt droplet suspension.

For use in construction operations, asphalt emulsions must remain fluid enough for proper application and stable enough to keep the asphalt particles suspended in the water phase until it is in contact with the aggregate or pavement surface. After it is applied, the asphalt droplets coalesce (come together) because of water evaporation and neutralization of the electrostatic charges—the time in which the asphalt droplets coalesce is referred to as the *breaking time* or *setting time*.

Foamed asphalt is formed by combining hot asphalt binder with small amounts of cold water. When the cold water comes in contact with the hot asphalt binder it turns to steam, which becomes trapped in tiny asphalt binder bubbles. The asphalt expands many times its original volume and the viscosity is lowered significantly. In this condition, the asphalt is ideally suited for mixing with cold and damp aggregates and/or recycled asphalt pavement (RAP). However, the high-volume foam state only lasts for a few minutes, after which the asphalt binder resumes its original properties. Foamed asphalt can be used as a binder in soil or base course stabilization, and is often used as the stabilizing agent in cold in-place recycling (CIPR).

Two parameters: (i) expansion ratio and (ii) half-life measure the quality of foamed asphalt. The *expansion ratio* is the ratio of maximum volume of the foamed asphalt to the original volume of the nonfoamed asphalt. Typically asphalt expands 10–20 times with eight times being the minimum acceptable. The *half-life* is the time it takes for the foamed asphalt to be reduced to half its maximum achieved volume. The minimum acceptable half-life is on the order of 12s, which ensures that in the full-depth reclamation and/or CIPR processes, where foamed asphalt is most commonly used, there will be a sufficient volume of asphalt and sufficient time for mixing and coating. The amount of water to be injected in the field foaming operation is the one that gives the maximum expansion ratio with a half-life that exceeds the 12-s minimum.

1.2.1.5 Cold Mix Asphalt Concrete

Cold mix asphalt concrete (CMA, also called *cold placed mixture*) is typically a mixture produced from emulsified asphalt or cutback asphalt. While in its emulsified state the asphalt is less viscous and the mixture is easy to work and compact. The emulsion will break after enough water evaporates and the cold mix will, ideally, take on the properties of cold hot-mix asphalt concrete. Cold mix is commonly used as a patching material and on lesser trafficked service roads. The term *CMA* has also been applied to the production of an asphalt mix at temperatures on the order of less than 60°C (<140°F). Thus, when using the term it is best to clarify the meaning.

Emulsified asphalt may be an anionic or a cationic grade. Aggregate material may be anything (although the choice should be suitable for the task) from a dense-graded crushed aggregate to a granular soil having a relatively high percentage of dust. At the time of mixing, the aggregate may either be damp, air-dried, or artificially heated and dried. The mixing methods may be performed either in the roadway, on the side of the roadway, or in a stationary mixing facility. The resultant mixtures usually are spread and compacted at ambient temperature.

CMA may be used for surface, base, or sub-base courses if the pavement is properly designed. Cold mix surface courses are suitable for light and medium traffic; however, they normally require a seal coat or hot asphalt concrete overlay as surface protection. When used in the base or sub-base, CMA may be suitable for all types of traffic.

Bituminous treated aggregate base is a type of CMA concrete which can consist of processing gravel, crushed stone, or blends of gravel, sand, and crushed stone material in which each mix is stabilized by a specified percentage of asphalt binder. Specific mix formulas are not often employed and these mixtures are typically used as a base course and stabilized-shoulder surfacing, although other uses may be assigned by special design. All designs should provide for a seal coat or surface course to provide protection from traffic abrasion and weathering.

1.2.1.6 Cutback Asphalt

Cutback asphalt is asphalt that is liquefied by the addition of diluents (typically petroleum solvents) and is generally designated as liquid asphalt to distinguish this product from the typical semisolid to solid asphalt binders. However, it is important to note that the asphalt binder is the base material that has been liquefied by cutting back with a solvent. Cutback asphalt is used in both paving and roofing operations, depending on whether it is used as paving asphalt or as roofing asphalt, and is liquefied (Speight, 1992). It is further classified according to the solvent used to liquefy the asphalt cement to produce rapid-, medium-, or slow-curing asphalt.

A petroleum-based solvent (distillate, diluent, or cutter stock) is used for dissolving the binder and if the solvent used in making the cutback asphalt is highly volatile it will quickly escape by evaporation while solvents of lower volatility evaporate more slowly. On the basis of relative speed of evaporation, cutback asphalts are divided into three types: (i) rapid curing, (ii) medium curing, and (iii) slow curing.

Rapid-curing cutback asphalt is asphalt mixed with a volatile solvent or light distillate, generally in the naphtha boiling point range (approximately 30–190°C, 86–375°F). *Medium-curing cutback asphalt* is asphalt mixed and a solvent of intermediate volatility or medium distillate, generally in the kerosene boiling point range (approximately 190–290°C, 375–550°F). *Slow-curing cutback asphalt* is asphalt and a diluent of low volatility (approximate boiling range is in excess of 260°C, >500°F). Slow-curing asphalt is often referred to as road-oil—this term originated in the early days of asphalt technology when asphaltic residual oil was used to give roads a low-cost, all-weather surface. The degree of fluidity of each cutback asphalt depends on (i) the grade of asphalt cement, (ii) the volatility of the solvent, and (iii) the proportion of solvent to binder. The degree of fluidity results in several grades of cutback asphalt—some may be fluid at ordinary atmospheric temperatures while other cutback asphalts can be much more viscous and may require heating to melt them to be sufficiently suitable for construction operations.

Cutback asphalts can be used with cold aggregates, with a minimum of heat. Rapid-curing and medium-curing cutback asphalts are used in a variety of highway construction methods. Among the more important uses are road mixing operations, stockpiling mixes, and spray applications such as prime (Abraham, 1945; Barth, 1962; Hoiberg, 1964; Speight, 1992). In the early days of the highway paving industry, cutback asphalts were used widely and successfully in many applications. Because the distillates are high-energy products that are lost by evaporation into the atmosphere, environmental concerns and regulations (see Chapter 12) have resulted in emulsified asphalts supplanting cutbacks in most applications. The main use of cutback asphalts in the modern asphalt industry is in the manufacture of asphalt cold-patching materials and as a prime coat spray application.

1.2.1.7 Hot Mix Asphalt

Hot mix asphalt (HMA) *concrete* is a combination of approximately 95% (w/w) aggregate (stone, sand, or gravel) bound together by asphalt cement. In the production of the mix, asphalt cement is heated aggregate, combined, and mixed with the aggregate at an HMA facility. The resulting HMA is loaded into trucks for transport to the paving site. The trucks dump the HMA into hoppers located at the front of paving machines. The asphalt is placed on the base of the roadway, and then compacted using a heavy roller, which is driven over the asphalt. Traffic is generally permitted on the pavement as soon as the pavement has cooled.

Hot mix asphalt concrete (commonly abbreviated as HMAC) is produced by heating the asphalt binder to decrease its viscosity, and drying the aggregate to remove moisture from it prior to mixing. Mixing is generally performed with the aggregate at approximately 150°C (300°F) for virgin asphalt and 165°C (330°F) for polymer-modified asphalt, and asphalt cement at 95°C (200°F). Paving and compaction must be performed while the asphalt is sufficiently hot. In many countries paving is restricted to summer months because in winter the compacted base will cool the asphalt too quickly before it is able to be packed to the required density. HMAC is the form of asphalt concrete most commonly used on high-traffic roadways such as those on major highways, racetracks, and airfields. It is also used as an environmental liner for landfills, reservoirs, and fish hatchery ponds.

1.2.1.8 Oxidized Asphalt

Oxidized asphalt (air-blown asphalt, air-refined asphalt) is asphalt that has been treated by blowing air through it at elevated temperatures to produce the physical properties required for industrial use of the final product. Oxidized asphalt is typically used in roofing operations, pipe coating, undersealing for Portland cement concrete pavements, and hydraulic applications. Oxidized asphalt is usually classified by the penetration value and softening point.

Mastic asphalt concrete or sheet asphalt is produced by heating hard-grade blown asphalt in a green cooker (mixer) until it has become a viscous liquid after which the aggregate mix is then added. The asphalt–aggregate mixture is heated (matured) for approximately 6–8 h and once it is ready the mastic asphalt mixer is transported to the work site where experienced layers empty the mixer and either machine or hand lay the mastic asphalt contents on to the road. Mastic asphalt concrete is generally used in footpath and road applications and for flooring or roof applications. In addition to the asphalt and aggregate, additives, such as polymers, and anti-stripping agents may be added to improve the properties of the final product.

Stone mastic asphalt is a special type of asphalt for wearing courses and consists of a mineral mix with a higher asphalt and stone content—it also has a high content of the coarsest particles of each fraction. Typically, the asphalt used is a road construction type or, for roadways that experience heavy traffic loads, polymer-modified asphalt with stabilizing additives is preferred. The mineral composition follows the principle of gap-grading which results in a self-supporting chipping skeleton the cavities of which are, to a large extent, filled with asphalt. Due to its special composition this mix is highly wear-resistant and less prone to deformation and fatigue cracking. However, this type of mix is difficult to compact and should therefore preferably be compacted with heavy rollers at high temperatures.

1.2.1.9 Penetration-Grade Asphalt

Penetration-grade asphalt is asphalt that has been further processed by air-blowing, solvent precipitation, or propane deasphalting. A combination of these processes may be used to produce different grades that are classified according to their penetration value.

Asphalt grading by means of the penetration test method (AASHTO M20; AASHTO T49; ASTM D5; ASTM D243; ASTM D946) is based on the premise that less viscous asphalt will exhibit a deeper needle penetration than more viscous asphalt. The penetration depth is empirically (albeit only an approximation that is not always accurate or meaningful) correlated with the in-service performance of the asphalt binder (Table 1.3) (Roberts et al., 1996). Therefore, an asphalt binder with a high penetration

Table 1.3 Comments on the Validity of the Penetration Test Method

Advantages:

Specifies a temperature of 25°C (77°F)—supposedly representative of an *average* pavement temperature[a]

Believed to provide a better correlation than the viscosity test method for low-temperature asphalt binder

Temperature susceptibility[b] can be determined by conducting the test at temperatures other than 25°C (77°F)

Easy to use in the field

Disadvantages:

Does not measure any fundamental engineering parameter such as viscosity

Shear rate is variable during the test—test results are not always reliable

Temperature susceptibility cannot be determined by a single test at 25°C (77°F)

Does not provide information that can be used to establish mixing temperature and/or compaction temperature

Notes:
[a]Difficult to define in a meaningful manner.
[b]The change in asphalt binder rheology with temperature.

number (soft asphalt) is used for cold climates, while an asphalt binder with a low penetration number (hard asphalt) is used for warm climates.

1.2.1.10 Reclaimed Asphalt Pavement

The use of reclaimed asphalt pavement (also known as *recycled asphalt pavement*, RAP) material is increasing as local, state, and federal transportation agencies make more efficient use of their resources. RAP material is generated when old, damaged pavement materials are milled and crushed for addition as a component to new mixtures placed in the pavement structure. Historically, old pavement material was removed and disposed of in landfills. As landfilling these materials has become less practical and more expensive and the availability of quality virgin materials declines, the addition of RAP to pavement mixtures has become more and more prevalent. Recycling of pavement material can be done as an in-place process or a central plant process. The in-place process combines the reclamation, mixing, laydown, and compaction procedures into a single paving train in the field.

In-place recycled materials are typically used for base or binder courses and are typically overlaid with a surface course. The central plant process involves stockpiling RAP at the asphalt plant, which is then mixed with virgin materials at the plant and trucked to the construction site for laydown and compaction (McDaniel et al., 2000; McDaniel and Anderson, 2001a, b).

The addition of reclaimed asphalt pavement (RAP) to an asphalt mixture changes the mechanistic properties (i.e., strength, durability) of the mixture and affects its performance (i.e., resistance to cracking and deformation) in the field. The mechanistic properties change as a result of the aged binder introduced to the mixture as part of the reclaimed asphalt pavement. The binder in the reclaimed asphalt pavement will have a different chemical composition and different properties than the virgin binder added during the mixing process. These two binders will mix favorably as long as they are compatible and the blend is stable (see Chapter 6), changing the properties of the mixture containing reclaimed asphalt pavement from one that contains only virgin material. In fact, the addition of reclaimed asphalt pavement increased mixture stiffness, measured by the indirect tension and semicircular bending tests (Huang et al., 2004). As the pavement industry moves towards more realistically based pavement design and analysis methods, it is essential to evaluate the effect of reclaimed asphalt pavement on the properties of asphalt mixtures to ensure compatibility and stability of the blend (see Chapter 5).

1.2.1.11 Refinery Asphalt

Asphalt is the distillation residue of mixed-base and asphalt-base crude oils and is designated as a manufactured product (Table 1.4) (Speight and

Table 1.4 Sub-Division of Petroleum and Other Carbonaceous Materials into Various Sub-Groups

Natural Materials	Derived Materials[a]	Manufactured Materials[b]
Natural gas	Saturate constituents	Liquefied petroleum gas
Petroleum	Aromatic constituents	Naphtha
Heavy oil	Resin	Kerosene
Bitumen[a]	Asphaltene constituents	Atmospheric gas oil
Asphaltite	Carbenes	Vacuum gas oil
Asphaltoid	Carboids	Atmospheric residuum
Ozocerite (natural wax)		Vacuum residuum
Kerogen		Coke
Coal		Tar (from cracking distillation)
		Pitch (from cracking distillation)
		Deasphalter residuum[c]

[a]Produced by fractionation procedures.
[b]Produced by distillation or by refining processes.
[c]Also called deasphalter bottoms or deasphalter tar.

Ozum, 2002; Parkash, 2003; Hsu and Robinson, 2006; Gary et al., 2007; Speight, 2014). It cannot be distilled even under the highest vacuum, because the temperatures required to do this promote formation of coke. Asphalt has complex chemical and physical compositions that usually vary with the source of the crude oil and are considered dispersions of particles (asphaltene constituents, in a high-boiling fluid composed of oil and resin constituents, see Chapters 3 and 4) (Traxler, 1961; Barth, 1962; Hoiberg, 1964; Koots and Speight, 1975).

The nature of the asphalt is determined by such factors as the nature of the medium (paraffinic or aromatic), as well as the nature and proportion of the asphaltenes and the resins (Pauli and Branthaver, 1998). The asphaltenes have been suggested to be lyophobic; the resins are lyophilic, and the *interaction* of the resins with the asphaltenes is responsible for asphaltene dispersion, which seems to exercise marked control on the nature of the asphalt. The asphaltenes vary in character but are of sufficiently high molecular weight to require dispersion as micelles, which are peptized by the resins. If the asphaltenes are relatively low in molecular weight, the resins plentiful, and the medium aromatic in nature, the result may be viscous asphalt without anomalous properties. If, however, the medium is paraffinic and the resins are scarce, and the asphaltenes are high in molecular (or micellar) weight (these conditions are encouraged by vacuum, steam reduction, or air blowing), the asphalt is of the gel type and exhibits the properties that accompany such a structure. A high content of resins imparts to a product desirable adhesive character and plasticity; high asphaltene content is usually responsible for the harder, more brittle, asphalt as evidenced from the structure and rheological properties of modified asphalt (Giavarini et al., 2000).

1.2.1.12 Superpave

Superpave, an abbreviation of the term *superior performing asphalt pavement*, is a pavement system designed to provide longer-lasting roadways (Asphalt Institute, 2003). Key components of the system are careful selection of binders and aggregates, volumetric proportioning of ingredients, and evaluation of the finished product.

1.2.1.13 Warm Mix Asphalt

Warm mix asphalt concrete is produced by adding either zeolites, waxes, asphalt emulsions, or sometimes even water to the asphalt binder prior to mixing (Johnston et al., 2006; Button et al., 2007; Austerman et al., 2009;

Bennert et al., 2010). This allows significantly lower mixing and laying temperatures and results in lower consumption of petroleum resources, thus releasing less carbon dioxide, aerosols, and vapor. The lower temperature not only improves working conditions but also leads to more rapid availability of the surface for use, which is important for construction sites with critical time schedules. The use of these additives in hot mixed asphalt (see above) may afford easier compaction and allow cold-weather paving or longer hauls.

Warm mix asphalt (also generally known as WMA) is a name given to a group of technologies that have the common purpose of reducing the viscosity of the asphalt binders (Abbas and Ali, 2011). This reduction in viscosity offers the advantage of producing asphalt–aggregate mixtures at lower mixing and compaction temperatures, and subsequently reducing energy consumption and pollutant emissions during asphalt mix production and placement. WMA technologies reduce the asphalt binders' viscosity through the addition of organic or chemical additives or by introducing cool water into the heated molten asphalt under controlled temperature and pressure conditions, resulting in so-called foamed asphalt binder. The latter has received increased attention in some states and countries because the technology does not require the use of costly additives.

Another aspect of the warm mix technology includes the product of *warm mix foamed asphalt binders*. These binder mixtures are more workable and easily compacted than HMA mixtures even though they are produced at lower mixing and compaction temperatures. However, warm mix foamed asphalt binders may be somewhat more susceptible to moisture damage than HMA mixtures but often the difference is statistically insignificant. Therefore, if designed properly, both mixtures are expected to meet the minimum tensile strength ratio (TSR) requirement for various traffic levels. In fact, using appropriate aggregate and binder types can help in overcoming any adverse effects that warm mix foamed asphalt binders have on the mixture performance.

1.2.1.14 Miscellaneous Binders

There is currently considerable interest in the development of non-petroleum materials to produce other binders. Sources such as biomass and algae are receiving attention (Allen, 2003; Hill and Jennings, 2011; TRB, 2012). Thus, a biobinder (non-petroleum-related binder) is an asphalt binder alternative made from non-petroleum-based renewable resources, which should not rival any food material, and have environmental and economic benefits (Marek et al., 1972; Peralta et al., 2012).

Currently, biobinders are produced by upgrading bio-oils produced from the rapid heating of biomass in a vacuum condition (Speight, 2008, 2011). Bio-oils are produced from plant matter and residues, such as agricultural crops, municipal wastes, and agricultural and forestry byproducts (Allen, 2003; Hill and Jennings, 2011; Demirbaş and Balat, 2006; Mohan et al., 2006; Williams et al., 2009; Fini et al., 2010; Speight, 2011). Currently, the state of the art for the utilization of bio-oils is concentrated on its use as bio-renewable fuels to replace fossil fuels. However, there has been a limited amount of research investigating the applicability of bio-oils as a bitumen modifier or extender. It has been reported that bio-oils can considerably increase the performance grade of polymer-modified asphalt binders by nearly 6°C (11°F) (Williams et al., 2009) and the use of biomaterials can enhance the asphalt low-temperature performance (Fini et al., 2010). Based on the conclusions of these investigations, the utilization of bio-oils as an asphalt modifier is very promising (Peralta et al., 2012). Nevertheless, until recently, there was no major research effort on the production of asphalt or asphalt modifiers from bio-materials or from biomass-produced products and there is a need for serious investigation into the applicability of using bio-oils as an asphalt substitute (up to 100% replacement of petroleum-based asphalt) that could be used in the roadway paving industry.

However, when evaluating a new material, the binder must meet current binder specifications without any leeway given because it is non-petroleum-related and therefore necessary that the new binder be accepted, whatever the properties. Substantial changes in chemical composition—as dictated by the source materials—can lead to unexpected physical property changes. Current binder specifications are performance-related, so significant chemistry changes may give rise to unusual (even non-acceptable) behavior when tested. Any non-petroleum-related binder must (as is the case with petroleum-related binders) be the focus of extensive testing.

1.3 BITUMEN

For the most part, bitumen is a naturally occurring high-viscosity material (Table 1.4) that is typically associated with various sediments, including sandstones (Meyer and De Witt, 1990; Speight, 2013a–c, 2014). To confuse matters even further, *bitumen* is also the name given in Europe and several other countries) to a manufactured product (asphalt in North America and many other countries). Therefore, it is absolutely essential to investigate the country of origin of the reports to determine the means of production of the material when consulting reports of asphalt and bitumen chemistry and behavior.

1.3.1 Naturally Occurring

The bitumen that occurs in the tar sand deposits (*oil sand deposits*) of north-eastern Alberta is often incorrectly referred to as natural asphalt. In fact, the expression *tar sand* is commonly used in the petroleum industry to describe sandstone reservoirs that are impregnated with a heavy, viscous black crude oil that cannot be retrieved through a well by conventional production techniques (FE-76-4). However, the term *tar sand* is actually a misnomer; more correctly, the name *tar* is usually applied to the heavy product remaining after the destructive distillation of coal or other organic matter (Speight, 2013d).

The term *bitumen* (also, on occasion, referred to as *native asphalt*, *rock asphalt*, and *extra heavy oil*, while in Europe and other countries the term bitumen often refers to *road asphalt*) includes a wide variety of reddish-brown to black materials of semisolid, viscous to brittle character that can exist in nature with no mineral impurity or with mineral matter (sandstone, limestone, or argillaceous sediments) contents that exceed 50% by weight (Abraham, 1945; Barth, 1962; Hoiberg, 1964). For the purposes of this text, tar sand bitumen is a naturally occurring material that is found in deposits where the permeability is low and passage of fluids through the deposit can only be achieved by prior application of fracturing techniques (Speight, 2009, 2013b,c, 2014). Furthermore, it is incorrect to refer to native bituminous materials as *tar* or *pitch*—the word *tar* is descriptive of the black, heavy bituminous material produced in the destructive distillation of such organic substances as coal and pitch is the distillation residue of the various types of tar (Speight, 2013d). Alternative names, such as *bituminous sand* or *oil sand*, are gradually finding usage, with the former name (bituminous sands) more technically correct. However, the term *tar sand* is actually a misnomer; more correctly (as described above), the name *tar* is usually applied to the heavy product remaining after the destructive distillation of coal or other organic matter (Speight, 2013d).

Tar sand bitumen is a high-boiling material with little, if any, material boiling below 350°C (660°F) and the boiling range approximately equivalent to the boiling range of an atmospheric residuum. In order to correct deficiencies in nomenclature of tar sand formations and tar sand bitumen, in the United States, *tar sands* have been defined (FE-76-4) as

> *...the several rock types that contain an extremely viscous hydrocarbon which is not recoverable in its natural state by conventional oil well production methods including currently used enhanced recovery techniques. The hydrocarbon-bearing rocks are variously known as bitumen-rocks oil, impregnated rocks, oil sands, and rock asphalt.*

By inference, heavy oil can be recovered from a reservoir using enhanced recovery techniques, while conventional crude oil does not require enhanced recovery techniques for recovery.

On an international note, the bitumen in tar sand deposits represents a potentially large supply of energy, as well as asphalt. However, many of the reserves are available only with some difficulty and optional refinery scenarios will be necessary for conversion of these materials to liquid products because of the substantial differences in character between conventional petroleum and tar sand bitumen and the high potential for coke production and fouling during processing (Speight, 2014).

1.3.2 Other

As already stated, the term bitumen in European countries and several other countries of the world refers to the product of a refinery—asphalt.

By way of further explanation, but without giving tacit approval to the use of the term, bitumen (as used outside of North America) is the residual product (atmospheric residuum) from the distillation of crude oil (sometimes referred to as *straight run bitumen*) in petroleum refining and is used to meet a variety of specifications based upon physical properties. The term is also used to include the vacuum residuum (sometimes also referred to as *straight run bitumen*) as well as the insoluble residuum from a deasphalting process. The residuum can be further processed by blowing air through it at elevated temperatures (air-blown bitumen) to alter the physical properties and meet specifications for commercial applications.

In summary, bitumen is produced to meet specifications either directly by refining or by blending and should not be confused with coal-derived products, such as coal tar or coal-tar pitch, which are manufactured by the high-temperature processing of bituminous coal and differ substantially in composition and physical characteristics from petroleum bitumen (petroleum asphalt) and insoluble residues produced from biomass

(Speight, 2008, 2011, 2013d, 2014). Similarly, bitumen should not be confused with petroleum pitch, which is a highly aromatic residue that is produced by thermal cracking, coking, or oxidation from selected petroleum fractions (Speight, 2014).

And this is where the confusion really starts, refinery-produced bitumen (as defined in Europe) should not be confused with natural or lake asphalt such as Trinidad Lake asphalt, Gilsonite, rock asphalt, and bitumen from various tar sand (oil sand) deposits. These products are unrefined and not produced by refining of crude oil and often contain a high proportion of mineral matter (up to 37%, w/w), although bitumen deposits some such as the tar sand deposits of Canada and the United States contain bitumen in the minority (<25%, w/w) and mineral matter in the majority (>75%, w/w) in contrast to refinery-produced bitumen (which originated as a distillation residue), naturally occurring bitumen may also contain low-boiling components, leading to a low-to-significant loss of mass when heated.

To add further confusion to the system of nomenclature, the term *bituminous materials* is used to describe a class of black or dark-colored (solid, semi-solid, or viscous) cementitious substances, natural or manufactured, composed principally of high-molecular-weight hydrocarbons, of which asphalt, tar, pitch, Gilsonite, asphaltoid, and asphaltite are typically included under the umbrella term.

1.3.2.1 Asphaltite

Asphaltite is a naturally occurring solid hydrocarbonaceous natural, solid hydrocarbon having a melting point higher than that of (refinery) asphalt or natural bitumen (such as tar sand bitumen or Trinidad Lake bitumen). It is a hydrocarbonaceous material that may have originated along similar pathways to petroleum but at some stage of the evolution became detached from the petroleum formation pathway.

The term *asphaltite* is applicable to any one of several naturally occurring solid bituminous materials whose chief constituents are asphaltene constituents. Asphaltites are dark brown to black in color and are insoluble in petroleum naphtha, thus requiring heat (thermal decomposition) to release distillates and gases. Though related to bitumen, asphaltites differ from bitumen chemically and physically in some ways. For example, asphaltites usually contain little or no inorganic minerals, whereas bitumen (such as tar sand bitumen) may have a relatively large percentage of such matter. Also, unlike bitumen, asphaltites do not fuse readily.

Asphaltites are commonly classified into three groups: (i) Gilsonite, also called uintaite, (ii) glance pitch, also called Manjak, and (iii) Grahamite. These substances differ from one another in terms of specific gravity and the softening point (the temperature at which they soften).

Asphaltites are a variety of naturally occurring, dark brown to black, solid, non-volatile bituminous substances that are differentiated from *bitumen* primarily by their high content of material insoluble in the common organic solvents and high yields of thermal coke. The resultant high temperature of fusion (approximate range 115–330°C, 240–625°F) is characteristic. The names applied to the two rather distinct types included in this group that are now accepted and used for the most part without ambiguity.

Gilsonite was originally known as *uintaite* from its discovery in the Uinta Basin of western Colorado and eastern Utah. It is characterized by a bright luster and a carbon residue in the range 10–20% by weight. The mineral occurs in nearly vertical veins varying from about an inch to many feet in width and is relatively free of occluded inorganic matter. Samples taken from different veins and across the larger veins may vary somewhat in softening point, solubility characteristics, sulfur content, and so on, but the variation is not great. It is evident in all instances that it is essentially the same material, and it is therefore appropriate to apply a single name to this mineral. However, caution should be exercised in using the same term without qualification for similar materials until it can be shown that they are equivalent to *Gilsonite*.

The second recognized type in this category is *Grahamite*, which is very much like Gilsonite in external characteristics but is distinguished from the latter by its black streak, relatively high fixed carbon value (35–55%), and high temperature of fusion, which is accompanied by a characteristic intumescence. The undifferentiated term *Grahamite* must be used with caution; similarities in the characteristics of samples from different areas do not necessarily imply any chemical or genetic relationship.

A third but rather broad category of asphaltite includes a group of bituminous materials known as *glance pitch*, which physically resemble Gilsonite but have some of the properties of *Grahamite*. They have been referred to as intermediates between the two, although the possibility exists that they are basically different from *Gilsonite* and may represent a material that has properties which lie between tar sand *bitumen* and *Grahamite*.

1.3.2.2 Asphaltoid

Asphaltoid is the name applied to a group of brown to black, solid bituminous materials of which the members are differentiated from the *asphaltites* by their infusibility and low solubility in carbon disulfide. These substances have also been designated *asphaltic pyrobitumen*, as they decompose on heating into bitumen-like materials. However, the term *pyrobitumen* does not convey the impression intended; thus the members of this class are referred to as *asphaltoids* since they closely resemble the asphaltites.

1.3.2.3 Pyrobitumen

Pyrobitumen (often called *asphaltic pyrobitumen*) is a widespread bituminous (or hydrocarbonaceous) mineral of variable composition and is a type of solid, amorphous organic matter. It is a naturally occurring solid organic substance that is distinguishable from bitumen by being infusible and insoluble. Thus, pyrobitumen is mostly insoluble in carbon disulfide (and other organic solvents), reputedly because of molecular cross-linking which renders previously soluble organic matter (such as tar sand bitumen or Trinidad Lake bitumen) insoluble. However, not all naturally occurring solid bitumen is pyrobitumens, insofar as that some solid bitumen types (such as Gilsonite) are soluble in common organic solvents. Pyrobitumen differs from the typical natural bitumen by being brittle, infusible and relatively insoluble in organic solvents and may be relatively enriched in nitrogen, sulfur, and trace metals such as nickel and vanadium, suggesting that it has a similar origin to petroleum and heavy oil. In addition, the pyrobitumen Wurtzilite occurs in veins in an oil shale zone of the lacustrine facies of the Uinta Formation. The Wurtzilite deposits appear to be restricted to a stratigraphic interval of about 50 feet and are located in the area including Indian, Lake, and Sams Canyons in the western part of the Uinta Basin. The asphaltic pyrobitumens are differentiated from the non-asphalts or coals, in that the former generally contain less than 5% oxygen. Wurtzilite, Ingramite, and Albertite are the principal asphaltic pyrobitumens found in the Uinta Basin. When heated, pyrobitumen generates, or transform into, bitumen-like liquid and gaseous hydrocarbon compounds. Pyrobitumen may be either asphaltic or non-asphaltic. The asphaltic pyrobitumens are derived from petroleum, are relatively hard, and have a specific gravity below 1.25. They do not melt when heated but swell and decompose (*intumesce*).

The terms bitumen and pyrobitumen have related definitions as the definitions pertain to the crust of the Earth and in the laboratory.

In geology, bitumen is the product of deposition and maturation of organic matter, and the extractable organic material by organic solvents in petroleum source rocks and reservoir rocks is often defined as bitumen. Upon exposure to high regional temperatures over geological time, bitumen is converted to pyrobitumen as a result of the thermally activated reactions that drive off lighter oil and gas products and leave an insoluble, carbon-rich residue. Pyrobitumen represents a significant fraction of the ultimate fate of petroleum liquids formed from kerogen during catagenesis. In the laboratory, experiments on organic-rich rocks (oil shale and petroleum source rocks), decomposition of the initially insoluble organic matter (defined as kerogen) produces gaseous and liquid products. The soluble fluid that remains in the heated rock is often defined (incorrectly) as bitumen. Upon further thermal exposure, this type of bitumen continues to evolve and reacts further to produce a harder pyrobitumen along with volatile products (liquid distillate and gas).

1.3.2.4 Wax

Although not related to the bituminous materials described above, a description of naturally occurring wax is warranted. There have been occasions when these natural products have been confused with bituminous natural products.

Ozokerite (*ozocerite*), from the Greek meaning *odoriferous wax*, is a naturally occurring hydrocarbon material composed chiefly of solid paraffins and cycloparaffins (i.e., hydrocarbons). Ozocerite usually occurs as stringers and veins that fill rock fractures in tectonically disturbed areas. It is predominantly paraffinic material (containing up to 90% non-aromatic hydrocarbons) with a high content (40–50%) of normal or slightly branched paraffins as well as cyclic paraffin derivatives. Ozocerite contains approximately 85% carbon, 14% hydrogen, and 0.3% each of sulfur and nitrogen and is, therefore, predominantly a mixture of pure hydrocarbons; any non-hydrocarbon constituents are in the minority. Ozocerite is soluble in solvents that are commonly employed for dissolution of petroleum derivatives, e.g., toluene, benzene, carbon disulfide, chloroform, and ethyl ether.

Montan wax (also known as *lignite wax*) is a hard wax obtained by solvent extraction of certain types of lignite (brown coal). Its color ranges from dark brown to light yellow when crude, or white when refined. Its composition is non-glyceride long-chain (C_{24}–C_{30}) carboxylic acid esters (62–68%, w/w), free long-chain organic acids (22–26%, w/w), long-chain alcohol derivatives, ketone derivatives, and hydrocarbons (7–15%, w/w), and other unknown constituents generally called *resins* as well as ill-defined *bitumen*. It is in effect a fossilized plant wax with a melting range on the order of 82–95°C (180–205°F). Crude Montan wax can be de-resined by cooling the extraction solvent (such as toluene).

1.4 TAR AND PITCH

Tar and pitch are viscous, dark-brown to black substances obtained by the destructive distillation of coal, wood, petroleum, peat, and certain other organic materials. The heating or partial burning of wood to make charcoal yields tar as a byproduct and is an ancient method for the production of both tar and pitch. By the application of heat, tar is separated into several materials: distillable tar and non-distillable pitch.

The terms *tar* and *pitch* are loosely applied to the many varieties of the two substances, sometimes interchangeably. For example, bitumen, which is naturally occurring, is sometimes referred to (incorrectly) as *mineral tar* and *mineral pitch*—the word *mineral* indicating the natural origin of the material. Typically, tar is fluid, depending upon its origin and the temperature to which it is exposed. Pitch tends to be more solid. Tar from pine wood is used in making soap and medicinal preparations. Pitch is used in the manufacture of roofing paper, in varnishes, as a lubricant, and as a binder for coal dust in the making of briquettes used as fuel. Coal-tar derivatives are used in the manufacture of dyes, cosmetics, and synthetic flavoring extracts.

Coal tar is a brown or black liquid of extremely high viscosity and is among the by-products from the carbonization of coal in the production of coke and/or coal gas as well as from various other carbonaceous feedstocks (Speight, 2008, 2011, 2013d). Coal tar is a complex and variable composition mixture of phenol derivatives, polynuclear aromatic derivatives (PNAs), and heterocyclic compounds. Depending upon the thermal conditions, coal tar typically contains 20–35% (w/w) non-distillable coal-tar pitch—the remainder of the distillable tar. Being flammable, coal tar is sometimes used for heating or to fire boilers and, like most heavy fuels, requires heat before it will flow easily. Coal tar was a component of the first sealed roads—in its early use, tarmac was tar covered with granite chips. Coal tar is also used to manufacture paints, synthetic dyes, and photographic materials.

In the past, coal tar has been used as a seal for water-proofing the hulls of ships and is still used to seal traditional wooden boats and the roofs of historical shingle-roofed churches, as well as painting the exterior walls of log buildings. Tar is also a general disinfectant and pine tar oil, or wood tar oil, is a pure natural product used for the surface treatment of wooden shingle roofs, boats, buckets, and tubs, and in the medicine, soap, and rubber industries. Pine tar has good penetration on rough wood. In the current context of environmental legislation, the use of coal tar is not readily advised to be used in the manner in which it was used in the eighteenth, nineteenth, and the early-to-mid twentieth centuries.

Thus, refinery asphalt, tar, and pitch are manufactured products and it is incorrect to refer to native bituminous materials as *tar* or *pitch*—the word *tar* is descriptive of the black, heavy bituminous material produced in the destructive distillation of such organic substances as coal, and pitch is the distillation residue of the various types of tar (Speight, 2013d, 2014). When a residuum is obtained from a crude oil and thermal decomposition has commenced during production of the residuum, it is more usual to refer to this product as *pitch* and the high-boiling constituents of the residue are often referred to collectively as *tar*. The differences between the original parent petroleum and the atmospheric residuum and the vacuum residuum are due to the relative amounts of various constituents present, which are removed or remain by virtue of their relative volatility (Speight and Ozum, 2002; Parkash, 2003; Hsu and Robinson, 2006; Gary et al., 2007; Speight, 2014).

More specifically, *tar* is a product of the destructive distillation of many bituminous or other organic materials and varies in color from brown to black and in physical character from an oily, viscous liquid to a semi-solid material. *Tar* is most commonly produced from *bituminous coal* and is generally understood to refer to the product from coal, although it is advisable to specify *coal tar* if there is the possibility of ambiguity (Speight, 2008, 2013d). The most important factor in determining the yield and character of coal tar is the temperature at which the tar was produced (the carbonizing temperature). Thus, three general temperature ranges are recognized, and the products have acquired the designations: *low-temperature tar* (approximately 450–700°C; 540–1290°F); *mid-temperature tar* (approximately 700–900°C; 1290–1650°F); and *high-temperature tar* (approximately 900–1200°C; 1650–2190°F). Tar released during the early stages of the decomposition of the organic material (*primary tar*) represents a product that has been recovered without the secondary alteration that is a result of prolonged residence of the vapor in the heated zone. Treatment of the distillate (boiling up to 250°C, 480°F) of the tar with caustic soda causes separation of a fraction known as *tar acids*; acid treatment of the distillate produces a variety of organic nitrogen compounds known as *tar bases*. The residue left following removal of the heavy oil, or distillate, is *pitch*, a black, hard, and highly ductile material.

In the chemical-process industries, *pitch* is the black or dark-brown residue obtained by distilling coal tar, wood tar, fats, fatty acids, or fatty oils. *Coal tar pitch* is a soft to hard and brittle substance containing chiefly aromatic resinous compounds along with aromatic and other hydrocarbons and their derivatives; it is used chiefly as road tar, in waterproofing roofs and other structures, and to make electrodes. *Wood tar pitch* is a bright, lustrous substance containing resin acids; it is used chiefly in the manufacture of plastics and insulating materials and in caulking seams. *Pitches* derived from fats, fatty acids, or fatty oils by distillation are usually soft substances containing polymers and decomposition products; they are used chiefly in varnishes and paints and in floor coverings.

Pitch Lake is the name that has been applied to a large surface deposit of bitumen found in the south-west part of the island of Trinidad (Republic of Trinidad and Tobago). Although most pitch lakes are fossils of formerly active seeps, some (the Trinidad Pitch Lake) continue to be supplied with fresh crude oil seeping from a subterranean source. Pitch Lake covers 115 acres, contains an estimated 40,000,000 barrel (bbl) of bitumen, and is used commercially for the manufacture of asphalt binder. Another so-called pitch lake, Guanoco Lake in Venezuela, covers more than 1100 acres (445 hectares) and contains an estimated 35,000,000 bbl of bitumen. It was used as a commercial source of asphalt from 1891 to 1935. Smaller deposits occur commonly where Tertiary marine sediments outcrop on the surface; an example is the tar pits at Rancho La Brea in Los Angeles (*brea* and *tar* have been used synonymously with *bitumen*) (Abraham, 1945; Traxler, 1961; Barth, 1962; Hoiberg, 1964).

The island of Trinidad lies off the north coast of South America, between 10° and 11° latitude and 61° and 62° longitude. It is bounded on the north by the Caribbean Sea, on the east by the Atlantic, on the south by a narrow channel, into which flows the waters of the northern and most westerly mouths of the Orinoco, and on the west by the Gulf of Paria, the two latter bodies of water separating it from the mainland of Venezuela. It is of an irregular rectangular shape, with promontories extending from its southwestern and northwestern corners which are several miles in length, between which and the mainland are the narrow straits known as the Dragon's and Serpent's Mouths. These promontories from a large portion of the northern and southern boundaries of the shallow rectangular Gulf of Paria, whose outlets to the ocean are through the Dragon's and Serpent's Mouths. The island has an average length of 48 miles and breadth of 36, containing about 1750 square miles, and being about one fifth the area of the state of Vermont. It is, as a whole, a flat country, with a high and striking mountain chain descending abruptly into the sea along its northern shore, and with low central and southern ranges of less importance. Its coasts are naturally abrupt on the north, consist of low bluffs on the south and are flat on the east and west. The only harbors are on the western coast.

Guanaco Lake (also called *Bermudez Lake* or Largo La Brea—Lake of Pitch) is a natural Pitch Lake found in Venezuela. It is the largest pitch lake (asphalt lake) in area in the world. It lies between the edge of a swamp and the foothills near Guariquen and has an irregular-shaped surface with a width of about a mile and a half from north to south and about a mile east to west. The area of the lake is in excess of 1000 acres and the depth only varies between 3 and 10 feet. The lake is covered with sparse vegetation, grass, and shrubs, and occasional groves of large palms (*morichales*). The asphalt exudes from springs and remains soft and semiliquid under a hard crust.

Guanaco Lake is found at a similar latitude to the more well-known and documented famous Pitch Lake (La Brea)—thanks to Sir Walter Raleigh—on the island of Trinidad in the West Indies. Both pitch lakes continually produce water and bitumen but never seem to decrease in size—replenishing/refilling themselves quickly from an underground sources (or sources).

A similar asphaltic lake and tar deposit exists in the Cretaceous Oran Basin in northern Argentina. It is named Laguna de la Brea; the name dates back to the seventeenth century. This asphaltic deposit was subject to rudimentary exploitation in 1885.

1.5 AGGREGATES

The aggregate (mineral aggregate) is the hard inert material, such as sand, gravel, crushed stone, slag, or rock dust that is mixed with the asphalt (binder) for the construction of roadways. However, the choice of aggregate is not an easy "pick-any-material" choice and is far from being a simple procedure. The aggregate must be selected according to the properties of the asphalt binder as well as the conditions that will exist when the roadway is completed—this includes environmental (weather) conditions, traffic loading, and the desired in-serve lifetime of the roadway. Once selected, the aggregate is mixed with the asphalt binder and laid down according to established procedures. The aggregate is the principal load-supporting component of an *asphalt concrete* (binder plus aggregate) pavement and is typically on the order of 90–95% (w/w) (and 75–85%, v/v) of the pavement mixture.

Selection of an aggregate material for use in an asphalt concrete pavement depends on several factors, including: (i) availability, (ii) cost, (iii) quality of the material, as well as (iv) the type of construction for which it is intended. To determine whether an aggregate material is suitable for use in asphalt roadway construction, it must be evaluated in terms of the following properties: (i) size and grading, (ii) cleanliness or dirtiness, (iii) toughness, (iv) soundness, (v) particle shape, (vi) surface texture, and (vii) absorption properties.

In terms of *size and grading*, the maximum size of an aggregate is the smallest sieve through which 100% of the material will pass. The use of the asphalt concrete determines not only the maximum aggregate size, but also the desired gradation (distribution of sizes smaller than the maximum size). *Cleanliness* of the aggregate refers to the lack of foreign or deleterious substances—typically, a high degree of cleanliness is required for the aggregate to be suitable for use in a paving mixture. By inference, the *dirtiness* of the aggregate refers to the presence of foreign or deleterious substances that make the potential aggregate unsuitable for paving mixtures. The *toughness* or *hardness* of the aggregate is the ability of the aggregate to resist crushing or disintegration during mixing, placing, and compacting; or under traffic loading. The *soundness* of the aggregate, although similar to toughness of the aggregate, is the ability of the aggregate to resist deterioration caused by natural elements, of which variation in the weather pattern is the prime example.

Particle shape refers to the shapes of the aggregate particles which will influence the overall strength and workability of the asphalt mixture as well as the density achieved during compaction of the mixture. When compacted, irregular particles such as crushed stone tend to *lock* together (fitting together like the pieces of a jigsaw puzzle) and resist displacement. The *surface texture* of the aggregate influences the workability and pavement strength—a rough, sandpaper-type texture is preferred over a smooth surface texture. Although smooth-faced aggregates are easy to coat with an asphalt film, they are generally not as good as rough surfaces—it is more difficult for the asphalt binder to adhere to the smooth surface and produce a strong bond. In addition to surface adhesion (*adsorption*), the *absorption* properties of the aggregate also influence the strength and quality of the product. For example, the porosity of the aggregate permits absorption of the binder into the pore system of the aggregate, which results in the formation of a superior bonding relationship between the aggregate particle and the asphalt binder. While a degree of porosity is desired in the aggregate, an aggregate that is highly absorbent is not always suitable for the task at hand.

Finally, separation of the asphalt binder from the aggregate (*stripping*) typically occurs through the action of water. An aggregate coated with too much dust can result in poor bonding between the binder and the aggregate which results in stripping of the binder from the aggregate. An aggregate that is readily susceptible to stripping action is not suitable for the production of road asphalt.

1.5.1 Types

Moreover, in order to produce quality asphalt roadways, the binder *and* the aggregate should be of high quality. Just as any tar-like residue does not produce quality asphalt concrete, use of any geological mineral does not serve the purpose of producing a high-quality pavement.

The aggregate should have the desired properties which typically include a single aggregate or a mix of aggregates from the following types: the aggregates should be clean, hard, sound, and durable, with a majority of crushed particles. Aggregates fall into several categories. The major categories are granite, gravel, limestone, pit aggregate, sand, slag, as well as secondary aggregates such as mineral filler.

1.5.1.1 Granite

Granite aggregate is crushed hard rock of granular structure and originates from magma that erupted on the ground surface and then hardened. In terms of its technical characteristics granite is solid (grades 800–1200) and highly solid (grades 1400–1600), and frost resistant (grades 300–400). It is the best aggregate for high-grade concrete. Granite is also used as a decorative stone. It can be gray, red, or pink and has many shades. After polishing it acquires a beautiful smooth surface. Granite is mostly composed of feldspar, quartz, and mica crystals, the content of which influences the color and shade of the stone.

The fraction known as granite sand (0–2 mm, 0–5 mm) is used for roadway construction and for reinforced concrete structures. They are used for deicing (decrease the slip coefficient). The *2–5 mm fraction* is used mainly to produce paving slab and other similar construction products where such fraction is crucial. The *3–10 mm fraction*, like the 5–10 mm fraction, is used for road construction and production of concrete and concrete structures, construction of roads, buildings, and bridges as well as for different types of fillings and drainage structure. However, the main application of aggregates of this fraction is road surfacing. The quality of round or cubic aggregates creates the highest compaction, while the presence of needle shapes leads to the creation of voids and increases in binding component volumes and thus reduces the concrete strength. The *5–20 mm fraction* is mainly used in road concrete mixes and this size provides the most adhesion in the mixture, which subsequently makes the most solid type of concrete and also provides surface evenness of concrete blocks. The *10–20 mm fraction* is in demand for road construction and aggregates of this fraction are used for laying both asphalt and concrete roads. It is used as filler for sub-base preparation for the road surface. The *20–40 mm fraction* is a medium fraction that is used for road construction as a lower layer beneath the asphalt surface. Aggregates of this fraction are also used as sub-bases in construction of highways and railways and in production of concrete and massive structures from reinforced concrete.

1.5.1.2 Gravel

Gravel aggregates are aggregates acquired as the result of sifting quarried rock and by crushing natural stone rock. In strength, gravel aggregates are inferior to granite aggregates, but they also have their advantages: their radioactive background is usually very low and they are cheaper than granite aggregates. Gravel aggregates are used for foundations and concretes and in products made of reinforced concrete and in road construction. There are two types of gravel aggregate: (i) scabbled stone—regular natural or crushed which has been worked and dressed roughly and (ii) gravel—rounded pebbles usually of river or sea origin. Fractions of gravel aggregates are in a variety of ranges: 3–10 mm, 5–20 mm, 5–40 mm, and 20–40 mm.

1.5.1.3 Limestone

Limestone aggregates (sometimes called lime or dolomite aggregate) are a product of the crushing of sedimentary rock—limestone, which is composed mainly of calcite (calcium carbonate, $CaCO_3$). Limestone aggregate is one of the main types of aggregate that besides the gravel and the granite types is used in road construction and in the production of reinforced concrete objects.

Typically, limestone is not used in asphalt concrete wearing courses in many states or regions because of potential low skid resistance. It is generally considered to be polish susceptible and hence many available sources of limestone are not utilized on roadway construction.

1.5.1.4 Pit Aggregate

Pit or bank-run aggregate includes both gravel and sand, which are typically pit or bank-run natural aggregates. This type of aggregate is usually screened to the proper size and washed to remove dirt before being used for asphalt concrete paving purposes. On the other hand, a *processed aggregate* is a natural pit or bank-run aggregate that has been crushed and screened to make it suitable for use in asphalt concrete pavements. It is considered a processed aggregate since crushing typically improves the particle shape by making the rounded particles more angular and crushing also improves the size distribution and range. Crushed stone is also a processed aggregate and is produced when the fragments of bedrock and large stones are crushed so that all particle faces are fractured and variation in particle size is achieved by screening. Aggregates that have received little or no screening are known as *crusher run aggregate*. The crusher run aggregate is generally more economical than screened aggregate and can be used for the production of asphalt concrete pavements. In the processing of crushed limestone, the rock dust produced is separated from the other crushed aggregate and may be used as crushed sand or as mineral filler for the pavement.

1.5.1.5 Sand

Sand is the degradation product of the breakdown of rock or sandstone and the result is a fine granular material. Different types of sand are blow sand, lake sand, gravel pit sand, bank run sand, manufactured sand, and river sand. On the other hand, *gravel* generally results from a glacial deposit. Different types of gravel are pea gravel, river gravel, and bank gravel. Before use, the gravel is usually crushed and graded into various sizes. *Crushed stone* is produced by crushing rock or granite to produce crushed aggregate of various sizes. *Slag* is a nonmetallic product containing silicates, derived from the production of steel.

In addition, aggregates used on the production of asphalt concrete paving are classified according to the source or means of preparation. For example, aggregates are typically classified as (i) pit or bank-run aggregate, (ii) processed aggregate, and (iii) synthetic aggregate.

1.5.1.6 Slag

Slag aggregates are obtained by crushing of smelter slag or by special treatment of fire-liquid slag melt (molten slag aggregates). Currently there are many types of concrete that are developed and used in construction that include metallurgical slags both as a binding component and as fillers.

Slag, generally classed as a *synthetic aggregate* (also called *artificial aggregate*), is produced by altering both physical and chemical properties of a parent material, which may be produced and processed specifically for use as synthetic aggregate. A synthetic aggregate may also be the byproduct of manufacturing and a final burning process—blast furnace slag is an example of a synthetic aggregate. Briefly, *blast furnace slag* is formed when iron ore or iron pellets, coke, and a flux (such as limestone, $CaCO_3$, or dolomite, $CaCO_3{\cdot}MgCO_3$) are melted together in a blast furnace. When the metallurgical smelting process is complete, the lime in the flux has been chemically combined with the aluminates and silicates of the ore and coke ash to form the non-metallic blast furnace slag. During the period of cooling and hardening from its molten state, blast furnace slag can be cooled in several ways to form any of several types of ballast furnace products, including aggregate for the manufacture of road asphalt.

In addition, bottom ash is also used. Bottom ash is a waste product with few commercial uses. It is produced in coal-burning power plants (Speight, 2013d) and may prove to be a suitable replacement for a portion of the aggregate commonly used in asphalt pavement. Increasing environmental awareness makes it difficult to find landfill sites that will accept bottom ash, therefore necessitating the discovery of a suitable use for it. Possible transportation-related applications of the ash that have been considered include use as a material for embankment fill, roadway fill, subbase, and base courses (Pandeline et al., 1997).

1.5.1.7 Secondary Aggregate

Secondary aggregate comes from crushing construction waste—concrete, bricks, and asphalt. The main benefit of secondary aggregate is its low cost: on average, it is two times cheaper than the granite type. In terms of strength, frost resistance, and certain other characteristics secondary aggregate is inferior to aggregates made of natural stones; nonetheless it has wide application: (i) as a large-scale filler for concrete, (ii) in road construction as a filler for roads and ramps, use as a lower layer for roads that are not the typical high-traffic highway, (iii) in maintenance works as a filler under non-asphalt areas and sidewalks, and (iv) as a reinforcing agent for reinforcing weak soils that may be a sub-grade for an asphalt roadway.

Mineral filler, rightly or wrongly despite its importance in the roadway, is often classed as a secondary aggregate and is a fine-grained material (that usually passes the #200 sieve) such as granular dust, or powdered rock, such as limestone screenings. Types of mineral filler are granular dust, or powdered rock, such as limestone screenings. In the processing of crushed limestone, the rock dust produced is separated from the other crushed aggregate and may be used as crushed sand or as mineral filler in asphalt concrete pavements.

1.5.2 Aggregate Selection

Choice of an aggregate should not be haphazard. As much thought and consideration should be put into aggregate choice as is put into binder choice. Thus, before the commencement of a project, samples of all aggregates to be used in cold mixes should be analyzed for conformance to the specifications, and a mix design developed by the contractor or asphalt supplier. This includes: construction procedures, and quality control/quality assurance procedures. An inspection should be made of the equipment to be used, and the mixing equipment must be calibrated. Prior to the start of the project, aggregate gradation certificates should be obtained for all aggregates.

During the project each load of asphalt emulsion should be accompanied by a certification of compliance to the specifications. The owner should sample the materials. Random samples should be taken of each aggregate. Each load of emulsion asphalt and each day's mix should be sampled. These samples should be taken according to standard methods, labeled, and retained. Emulsions should be stored in clean plastic containers and kept from freezing. The owner may opt to test all or a portion of the samples to assure quality. As part of the quality control procedure, samples of aggregates, asphalt emulsion, and the cold mix should be taken once each day, at a minimum, and analyzed for compliance.

REFERENCES

AASHTO M20, 2014. Specification for Penetration-Graded Cement for Use in Pavements Construction. American Association of State Highway and Transportation Officials, Washington, DC.

AASHTO T49, 2014. Test Method for Penetration of Bituminous Materials. American Association of State Highway and Transportation Officials, Washington, DC.

Abbas, A.R., Ali, A., 2011. Mechanical properties of warm mix asphalt prepared using foamed asphalt binders Report No. FHWA/OH-2011/6. University of Akron/Ohio Department of Transportation, Akron, OH/Columbus, OH.

Abraham, H., 1945. Asphalts and Allied Substances: Their Occurrence, Modes of Production, Uses in the Arts and Methods of Testing, fifth ed. Van Nostrand, New York, NY.

Allen, B.R., 2003. Pretreatment methods for the degradation of lignin Proceedings. Superpave Performance Graded Asphalt Binder Specification and Testing, Superpave Series No 1(SP-1). The Asphalt Institute, Lexington, KY.

Asphalt Institute, 2003. Superpave Performance Graded Asphalt Binder Specification and Testing (SP-1). Asphalt Institute, Lexington, KY.

ASTM D243, 2014. Standard test method for residue of specified penetration Annual Book of Standards. ASTM International, West Conshohocken, PA.

ASTM D5, 2014. Standard test method for penetration of bituminous materials Annual Book of Standards. ASTM International, West Conshohocken, PA.

ASTM D946, 2014. Standard specification for penetration-graded asphalt cement for use in pavement construction Annual Book of Standards. ASTM International, West Conshohocken, PA.

Austerman, A., Mogawer, W., Bonaquist, R., 2009. Evaluating the effects of warm mix asphalt technology additive dosages on the workability and durability of asphalt mixtures containing recycled asphalt pavement Proceedings. 88th Annual Meeting. Transportation Research Board, Washington, DC.

Barth, E.J., 1962. Asphalt: Science and Technology. Gordon and Breach Publishers, Taylor & Francis Group, Abingdon, Oxfordshire.

Bell, H.S., 1945. American Petroleum Refining. Van Nostrand Scientific Publishers, New York, NY.

Bennert, T., Reinke, G., Mogawer, W., Mooney, K., 2010. Assessment of workability/compactability of warm mix asphalt Proceedings. 89th Annual Meeting. Transportation Research Board, Washington, DC.

Button, J., Estakhri, C., Wimsatt, A., 2007. A synthesis of warm mix asphalt Report No. FHWA/TX-07/0-5597-1. Texas Department of Transportation, Austin, TX.

Corbett, L.W., 1975. Reaction variables in the air blowing of asphalt. Ind. Eng. Chem., Process Des. Dev. 14 (2), 181–187.

Demirbaş, M.F., Balat, M., 2006. Recent advances on the production and utilization trends of biofuels: a global perspective. Energy Conserv. Manage. 47, 2371–2381.

Dybalski, J.N., 1976. The chemistry of asphalt emulsions. The chemistry of asphalt emulsions Proceedings. 55th Annual Meeting. Transportation Research Board, Washington, DC, January.

FE-76-4, 1976. US Congress, 1976. Public Law FEA 76-4. Congress of the United States, Washington, DC.

Fini, E.H., Yang, S.H., Xiu, S., 2010. Characterization and application of manure-based biobinder in asphalt industry Proceedings. 89th Annual Meeting. Transportation Research Board, Washington, DC.

Gary, J.G., Handwerk, G.E., Kaiser, M.J., 2007. Petroleum Refining: Technology and Economics, fifth ed. CRC Press, Taylor & Francis Group, Boca Raton, FL.

Giavarini, C., Mastrofini, D., Scarsella, M., 2000. Macrostructure and rheological properties of chemically modified residues and bitumens. Energy Fuels 14, 495–502.

Heithaus, J.J., 1962. Measurement and significance of asphaltene peptization. J. Inst. Petrol. 48, 45–53.

Hill, D.R., Jennings, A.A., 2011. Bioasphalt from urban yard waste carbonization: a student study Report No. FHWA/)H-2011/13. Ohio Department of Transportation, Office of Research and Development, Columbus, OH.

Hoiberg, A.J., 1964. Bituminous Materials: Asphalts, Tars, and Pitches. John Wiley & Sons Inc., New York, NY.

Hsu, C.S., Robinson, P.R., 2006. Practical Advances in Petroleum Processing, vols. 1 and 2. Springer, New York, NY.

Huang, B., Egan, B.K., Kingery, W.R., Zhang, Z., Zuo, Z., 2004. Laboratory Study of Fatigue Characteristics of HMA Surface Mixtures Containing RAP. Transportation Research Board, National Research Council, Washington, DC.

Johnston, A., Da Silva, M., Soleymani, H., Yeung, C., 2006. An early evaluation of the initial Canadian experience with warm asphalt mix. In: Proceedings. Annual General Conference of the Canadian Society for Civil Engineering, May.

Koots, J.A., Speight, J.G., 1975. The relation of petroleum resins to asphaltenes. Fuel 54, 179.

Marek, C.R., Herrin, H., Kesler, C.L., Barenberg, E.J., 1972. Promising replacements for conventional aggregates for highway use Report No. 135. National Cooperative Highway Research Program. Transportation Research Board, National Research Council, Washington, DC.

McDaniel, R.S., Soleymani, H., Anderson, R.M., Turner, P., Peterson, R., 2000. Recommended Use of Reclaimed Asphalt Pavement in the Superpave Mix Design Method. Contractor's Final Report, Project No. D9-12. National Cooperative Highway Research Program, Transportation Research Board, National Research Council, Washington, DC.

McDaniel, R.S., Anderson, R.M., 2001a. Recommended Use of Reclaimed Asphalt Pavement in the Superpave Mix Design Method: Technician's Manual. Report No. 452. National Cooperative Highway Research Program, Transportation Research Board, National Research Council, Washington, DC.

McDaniel, R.S., Anderson, R.M., 2001b. Recommended Use of Reclaimed Asphalt Pavement in the Superpave Mix Design Method: Guidelines. Research Results Digest No. 253. National Cooperative Highway Research Program, Transportation Research Board, National Research Council, Washington, DC.

Meyer, R.F., De Witt Jr., W., 1990. Definition and world resources of natural bitumen Bulletin No. 1944. United States Geological Survey, Reston, VA.

Mohan, D., Pittman, C.U., Steele, P.H., 2006. Pyrolysis of wood/biomass for bio-oil: a critical review. Energy Fuels 20 (3), 848–889.

Pandeline, D., Cosentino, P., Kalajian, E., Chavez, M., 1997. Shear and deformation characteristics of municipal waste combustor bottom ash for highway applications Transportation Research Record No. 1577. Transportation Research Board, National Research Council, Washington, DC.

Parkash, S., 2003. Refining Processes Handbook. Gulf Professional Publishing, Elsevier, Amsterdam, Netherlands.

Pauli, A.T., Branthaver, J.F., 1998. Relationship between asphaltenes, heithaus compatibility parameters, and asphalt viscosity. Petrol. Sci. Technol. 16 (9&10), 1125–1147.

Peralta, J., Williams, R.C., Rover, M., Silva, H.M.R.D., 2012. Development of rubber-modified fractionated bio-oil for use as non-crude petroleum binder in flexible pavements Proceedings. Alternative Binders for Sustainable Asphalt Pavements. Transportation Research Circular E-C165. E-C147. Transportation Research Board, Washington, DC, December, pp. 23–36.

Puzinauskas, V.P., Corbett, L.W., 1978. Differences between petroleum asphalt, coal-tar pitch and road tar Report No. RR-78-1. Asphalt Institute, Lexington, KY, January.
Roberts, F.L., Kandhal, P.S., Brown, E.R., Lee, D.Y., Kennedy, T.W., 1996. Hot Mix Asphalt Materials, Mixture Design, and Construction. National Asphalt Pavement Association Education Foundation, Lanham, MD.
Speight, J.G., 1992. Asphalt Kirk-Othmer Encyclopedia of Chemical Technology, fourth ed. John Wiley & Sons Inc., New York, NY, pp. 689–724.
Speight, J.G., 2008. Synthetic Fuels Handbook: Properties, Processes, and Performance. McGraw-Hill, New York, NY.
Speight, J.G., 2009. Enhanced Recovery Methods for Heavy Oil and Tar Sands. Gulf Publishing Company, Houston, TX.
Speight, J.G. (Ed.), 2011. The Biofuels Handbook Royal Society of Chemistry, London.
Speight, J.G., 2013a. Heavy and Extra Heavy Oil Upgrading Technologies. Gulf Professional Publishing, Elsevier, Oxford.
Speight, J.G., 2013b. Heavy Oil Production Processes. Gulf Professional Publishing, Elsevier, Oxford.
Speight, J.G., 2013c. Oil Sand Production Processes. Gulf Professional Publishing, Elsevier, Oxford.
Speight, J.G., 2013d. The Chemistry and Technology of Coal, third ed. CRC Press, Taylor & Francis Group, Boca Raton, FL.
Speight, J.G., 2014. The Chemistry and Technology of Petroleum, fifth ed. CRC Press, Taylor & Francis Group, Boca Raton, FL.
Speight, J.G., Ozum, B., 2002. Petroleum Refining Processes. Marcel Dekker Inc., New York, NY.
Traxler, R.N., 1961. Asphalt: Its Composition, Properties, and Uses. Reinhold Publishing Corp., New York, NY.
TRB, 2012. Alternative binders for sustainable asphalt pavements Transportation Research Circular No. E-C165. Transportation Research Board, National Research Council, Washington, DC, January 22.
Williams, R.C., Satrio, J., Rover, M., Brown, R.C., Teng, S., 2009. Utilization of fractionated bio-oil in asphalt. Presented at 88th Annual Meeting of the Transportation Research Board, Washington, DC.
Zakar, P., 1971. Bitumen (American Edition). Chemical Publishing Company Inc., Los Angeles, CA.

CHAPTER 2

History, Use, and Production

2.1 INTRODUCTION

Asphalt—the black cementing agent and the name given to the collective high-boiling constituents of petroleum—is widely used in the modern world and is produced by the refining industry, although there are natural deposits of similar bituminous materials throughout the world. One of the key characteristics and advantages for the use of asphalt as an engineering construction and maintenance material is its versatility, as well as it being highly resistant to the actions of most acids, alkalis, and salts. Asphalt is typically a semi-solid to near solid at ambient temperature and may be changed to the liquid state by the application of heat, dissolution in solvents, or emulsification with water. Asphalt is strong cement that is readily adhesive and highly waterproof and durable, making it particularly useful for modern roadways, and covers more than 90% of the paved highways in the United States. Because of the versatility, durability, and ease of use in highway construction, asphalt is unique and has no equal.

However, the use of asphalt is not a new technology and has been practised for millennia (Abraham, 1945; Barth, 1962). While the fuels that are derived from petroleum supply more than half of the world's total supply of energy, the collective higher boiling constituents—often considered by the modern industry as being of little value—have risen to be of prime importance. In fact it is the documented history of asphalt use that has led to its popularity in the modern word.

The *history* of any subject is the means by which the subject is studied with the justifiable belief that much can be learned from the events of the past. In the current context, the occurrence and use of petroleum, petroleum derivatives (naphtha), heavy oil, and bitumen is not new and was practiced in pre-Christian times and is known largely through historical use in many of the older civilizations (Henry, 1873; Abraham, 1945; Forbes, 1958a,b, 1959, 1964; Barth, 1962; James and Thorpe, 1994). Thus, the use of petroleum and the development of related technology is not such a modern subject as we are inclined to believe. However, the petroleum-derived asphalt industry is essentially a twentieth-century

Asphalt Materials Science and Technology.
DOI: http://dx.doi.org/10.1016/B978-0-12-800273-5.00002-7

industry, and to understand the evolution of the industry it is essential to have a brief understanding of the first uses of petroleum.

Briefly and by way of clarification to avoid confusion, the term *asphalt* as used in this book refers to the residuum from petroleum that has been subsequently processed to meet necessary specifications. It is, in essence, the binder for products such as road asphalt, which is a mixture of aggregates, binder and filler, used for constructing and maintaining roads and parking areas. Aggregates (see Chapter 5) used for asphalt mixtures could be crushed rock, sand, gravel, or slag. A typical asphalt pavement consists of the road structure above the formation level which includes bound and unbound materials. This gives the pavement the ability to distribute the loads of the traffic before it arrives at the formation level.

2.2 SOURCES

The source of asphalt has, for the most part, been petroleum, but coal tar was used as the source in the eighteenth, nineteenth, and early-to-mid twentieth centuries. From the mid-to-late twentieth century as many countries developed an environmental conscience (see Chapter 12), the use of coal tar has diminished and other sources have been sought (see Chapter 1) (Speight 2008, 2011a). However, other sources have been sought and are being developed along with assiduous test protocols to ensure that these new courses do not cause harm to the flora and fauna (including humans) in the environment. For the most part, petroleum remains the major source of asphalt.

2.3 PETROLEUM

Asphalt manufacture—a part of the modern refinery (Figure 2.1)—is a matter of distilling everything possible from crude petroleum until a residue with the desired properties is obtained. This is usually done by stages involving atmospheric distillation (Figure 2.2) and vacuum distillation (Figure 2.3). Once the residua are produced they can be sent individually to the asphalt plant for further individual processing or blended at the asphalt plant.

Petroleum distillation at atmospheric pressure (Figure 2.2) removes the lower boiling fractions and yields reduced crude that may contain higher boiling (lubricating) oils, asphalt, and even wax. Distillation of the reduced crude under vacuum (Figure 2.3) removes the oils (and wax) as

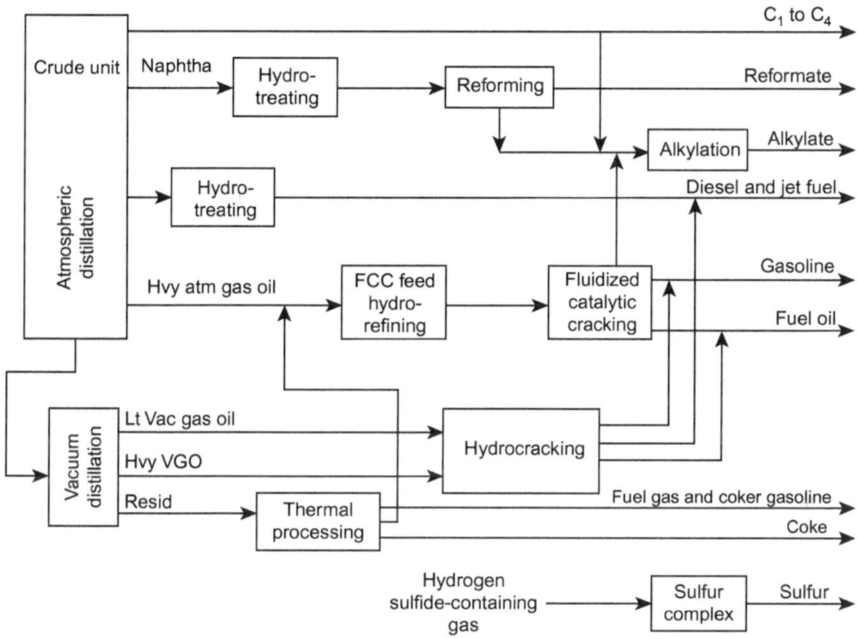

Figure 2.1 Schematic of a modern refinery.

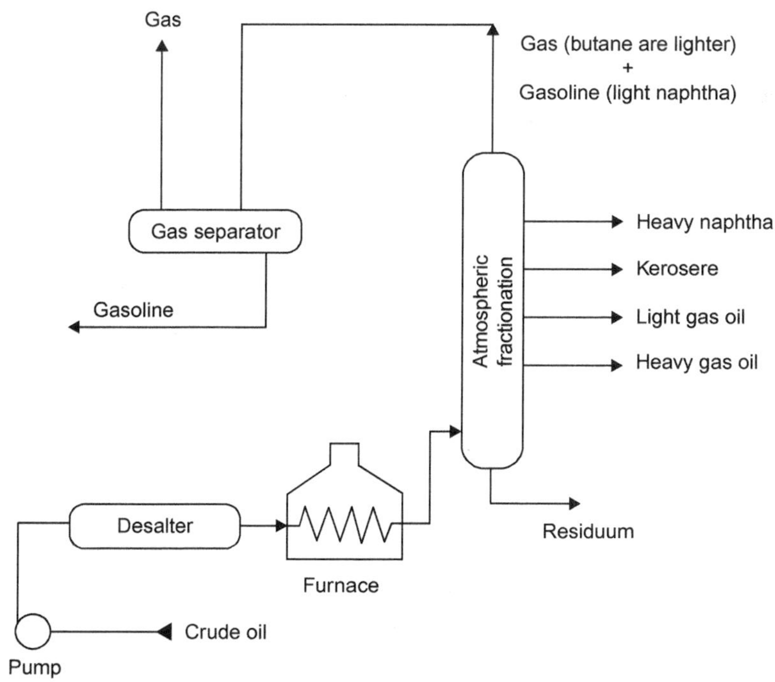

Figure 2.2 An atmospheric distillation unit. *Source: OSHA Technical Manual, Section IV, Chapter 2: Petroleum Refining Processes; http://www.osha.gov/dts/osta/otm/otm_iv/otm_iv_2.html.*

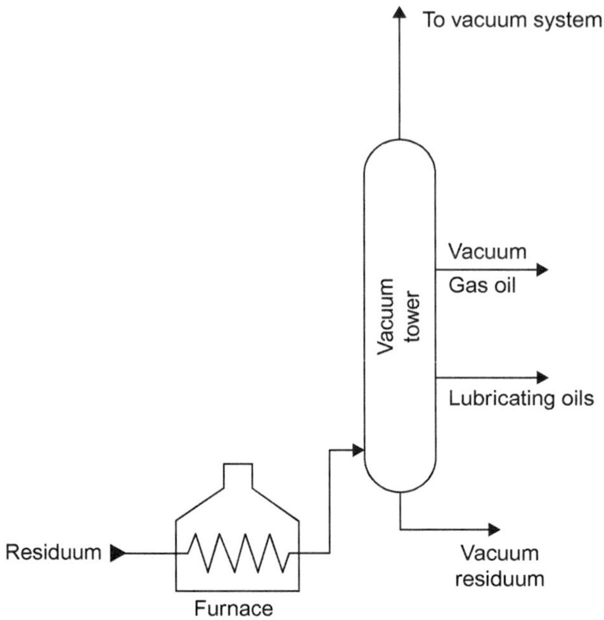

Figure 2.3 A vacuum distillation unit. *Source: OSHA Technical Manual, Section IV, Chapter 2: Petroleum Refining Processes; http://www.osha.gov/dts/osta/otm/otm_iv/otm_iv_2.html.*

overhead products and the asphalt remains as a bottom (or residual) product. The majority of the polar functionality in the original crude oil (typically conferred upon the residuum by the presence of nitrogen, oxygen, and sulfur constituents) tends to concentrate in the atmospheric residuum and thence into the vacuum residuum (Tables 2.1 and 2.2). It is this concentration effect that confers upon asphalt some of its unique properties (Table 2.3). At this stage the asphalt is frequently and incorrectly referred to as pitch and has a softening point related to the amount of oil removed: the more oil distilled from the residue, the higher the softening point.

Several manufacturing methods are available to produce specification asphalt depending on the (i) the source of the petroleum and (ii) the available processing capabilities (Speight and Ozum, 2002; Parkash, 2003; Hsu and Robinson, 2006; Gary et al., 2007; Speight, 2014). There is, in modern refineries, an additional factor that influences asphalt production and properties: the modern refinery (Figure 2.1) accepts several crude oils that are sent into the refining system as a blend as opposed to the refinery of the early-to-mid twentieth century which accepted one crude oil (at the

Table 2.1 Properties of Atmospheric and Vacuum Residua

Feedstock	Gravity API	Sulfur (wt. %)	Nitrogen (wt. %)	Nickel (ppm)	Vanadium (ppm)	Asphaltenes (heptane) (wt. %)	Carbon Residue (Conradson) (wt. %)
Arabian light, > 650°F	17.7	3.0	0.2	10.0	26.0	1.8	7.5
Arabian light, > 1050°F	8.5	4.4	0.5	24.0	66.0	4.3	14.2
Arabian heavy, > 650°F	11.9	4.4	0.3	27.0	103.0	8.0	14.0
Arabian heavy, > 1050°F	7.3	5.1	0.3	40.0	174.0	10.0	19.0
Alaska, north slope, > 650°F	15.2	1.6	0.4	18.0	30.0	2.0	8.5
Alaska, north slope, > 1050°F	8.2	2.2	0.6	47.0	82.0	4.0	18.0
Lloydminster (Canada), > 650°F	10.3	4.1	0.3	65.0	141.0	14.0	12.1
Lloydminster (Canada), > 1050°F	8.5	4.4	0.6	115.0	252.0	18.0	21.4
Kuwait, > 650°F	13.9	4.4	0.3	14.0	50.0	2.4	12.2
Kuwait, > 1050°F	5.5	5.5	0.4	32.0	102.0	7.1	23.1
Tia Juana, > 650°F	17.3	1.8	0.3	25.0	185.0		9.3
Tia Juana, > 1050°F	7.1	2.6	0.6	64.0	450.0		21.6
Taching, > 650°F	27.3	0.2	0.2	5.0	1.0	4.4	3.8
Taching, > 1050°F	21.5	0.3	0.4	9.0	2.0	7.6	7.9
Maya, > 650°F	10.5	4.4	0.5	70.0	370.0	16.0	15.0

Table 2.2 Petroleum Products

Product	Lower Carbon Limit	Upper Carbon Limit	Lower Boiling Point (°C)	Upper Boiling Point (°C)	Lower Boiling Point (°F)	Upper Boiling Point (°F)
Refinery gas	C1	C4	−161	−1	−259	31
Liquefied petroleum gas	C3	C4	−42	−1	−44	31
Naphtha	C5	C17	36	302	97	575
Gasoline	C4	C12	−1	216	31	421
Kerosene/diesel fuel	C8	C18	126	258	302	575
Aviation turbine fuel	C8	C16	126	287	302	548
Fuel oil	C12	>C20	216	421	>343	>649
Lubricating oil	>C20		>343		>649	
Wax	C17	>C20	302	>343	575	>649
Asphalt	>C20[a]		>343		>649	
Coke	>C50[a]		>1000[a]		>1832[a]	

[a]Estimate of the carbon number and boiling point are inserted for illustrative purposes only.

Table 2.3 Test Methods Used to Determine Asphalt Properties

Test	Number	Description
Adsorption	ASTM D4469	Calculation of degree of adsorption of asphalt by an aggregate
Bond and adhesion	ASTM D1191	Used primarily to determine whether an adhesion asphalt has bonding strength at low temperatures See also ASTM D3141 and ASTM D5078
Compatibility	ASTM D1370	Indicates whether asphalts are compatible and dis-bond under stress See also ASTM D3407
Distillation	ASTM D402	Determination of volatiles content; applicable to road oil and cutback asphalt
Ductility	ASTM D113	The distance in cm which a standard briquette can he elongated before breaking; also an indication of cohesion and shear susceptibility
Emulsified asphalt	ASTM D244	Covers the composition and classification, storage, use, and specifications See also ASTM D977 and ASTM D118
Flash point	ASTM D92	Cleveland open cup method Tag open cup (ASTM D3143) applicable to cutback asphalt
Float test	ASTM D139	Normally used for asphalts that are too soft for the penetration test
Penetration	ASTM D5	The extent to which a needle penetrates asphalt under specified conditions of load, time, and temperature; units are mm/10 measured from 0 to 300 See also ASTM D243
Sampling	ASTM D140	Provides guidance for the sampling of asphalts
Softening point	ASTM D36	Ring and ball method; the temperature at which asphalt attains a particular degree of softness under specified conditions: used to classify asphalt grades See also ASTM D2389
Solubility in CS_2	ASTM D4	Determination of the carbon amount of carboids and/or carbenes and mineral matter: trichloroethylene and 1,1,1-trichloroethane are suitable solvents See also ASTM D2042
Specific gravity	ASTM D70	See ASTM D3142

(*Continued*)

Table 2.3 Continued

Test	Number	Description
Stain	ASTM D1328	Measures the amount of stain on paper or other cellulosic materials
Temp./volume	ASTM D-1250	Allows the conversion of volumes of asphalts from one temperature to another See also ASTM D4311
Thin film oven test	ASTM D1754	Determines the hardening effect of heat and air on a film of asphalt See also ASTM D2872
Viscosity	ASTM D2170	A measure of resistance to flow See also ASTM D88 (now discontinued but a useful reference), ASTM D1599, ASTM D2171, ASTM D2493, ASTM D3205, ASTM D3381, ASTM D4402, ASTM D-4957
Water content	ASTM D95	Determines the water content by distillation with a Dean and Stark receiver
Weathering	ASTM D529	Used for determining the relative weather resistance of asphalt See also ASTM D1661, ASTM D1670

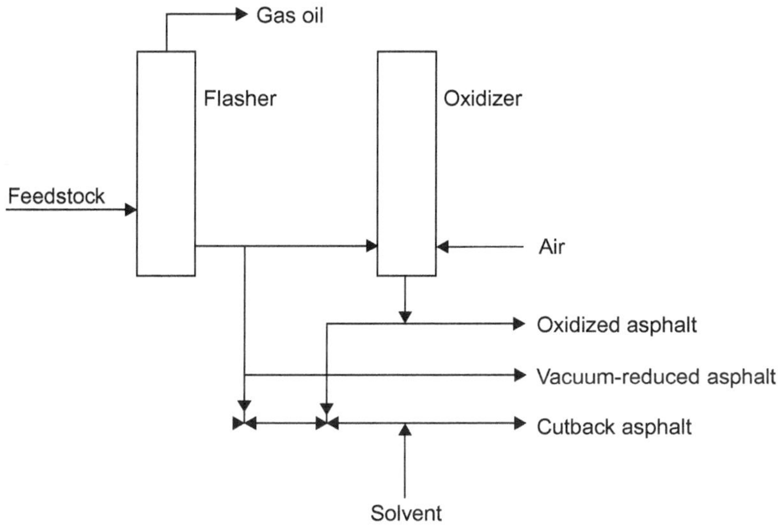

Figure 2.4 General representation of asphalt manufacture: the feedstock can be a residuum from the atmospheric distillation tower or from the vacuum distillation tower.

most, two crude oils) for processing. In any case a combination of processes is used for asphalt production (Figure 2.4).

The most common refining process used for producing asphalt from the refinery feedstock involves atmospheric and vacuum distillation. Atmospheric distillation is used to remove gases and lower boiling liquid fractions from the feedstock leaving atmospheric residuum (atmospheric resid, reduced crude). To remove the last traces of the lower boiling constituents and avoid thermal degradation of the high boiling constituents, the atmospheric residuum is introduced into a vacuum distillation unit (Speight and Ozum, 2002; Parkash, 2003; Hsu and Robinson, 2006; Gary et al., 2007; Speight, 2014). At reduced pressure, the boiling temperatures are lower and unwanted thermal cracking of the high-boiling constituents is avoided. For example, vacuum gas oil is removed at atmospheric equivalent temperatures of 345–40°C (650–750°F) and 370–450°C (700–850°F), leaving a high-boiling-point residuum (vacuum residuum, vacuum resid). The atmospheric equivalent temperature to yield the vacuum residue is up to 535°C (1000°F). This residuum (as opposed to the atmospheric residuum) is often preferred for asphalt manufacture and, depending upon the specification requirements, the vacuum residuum may even be used directly as a component of the final asphalt blend. In addition, the properties of the vacuum residuum can be modified by application of the solvent deasphalting process

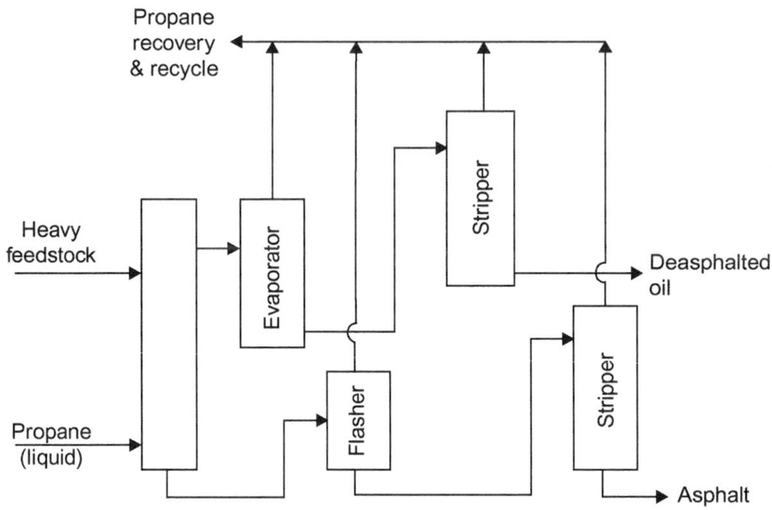

Figure 2.5 Schematic of the propane deasphalting process: the feedstock can be a residuum from the atmospheric distillation tower or from the vacuum distillation tower or a heavy oil or tar sand bitumen.

which employs solvents such as liquid propane, liquid butane, and liquid pentane (Figure 2.5) (Speight, 2014). In the process, asphaltene constituents and resin constituents (along with high-boiling aromatic constituents), are discharged from the feedstock on addition of the hydrocarbon liquid. In addition, the amount of precipitate varies with the amount of added paraffin liquid (Speight and Ozum, 2002; Parkash, 2003; Hsu and Robinson, 2006; Gary et al., 2007; Speight, 2014), which may be more representative of the situation that occurs in many reactors during refinery operations.

However, since there are wide variations in refinery operations and crude petroleum, asphalt with softening points ranging from 25°C to 55°C (80–130°F) may be produced and require further processing. The final asphalt product is processed to meet technical specifications, either directly in the refining process, or by blending of asphalts with different physical properties (Karl, 1986). Blending may take place at the refinery, at terminals, or at a third-party facility where blend components and finished products can be easily transported by truck, rail, or barge to their final locations. Higher viscosity products or asphalt may be blended with lower viscosity products or asphalt in suitable proportions to satisfy final product specification requirements.

Blending of higher and lower softening point asphalt can be used to produce asphalt that has intermediate softening points. If lubricating oil is not a required product, the atmospheric residuum may be distilled in a flash drum, which is similar to a bubble tower but has few, if any, trays (Speight, 2014). Asphalt descends to the base of the drum as the oil components pass out of the top of the drum. If the asphalt has a relatively low softening point, harder asphalt can be created by the use of further distillation with steam or by oxidation.

Oxidized asphalt (also known commonly as blown asphalt) is manufactured by treatment with oxygen (air) in the asphalt blowing unit (air blowing unit, oxidizer). The process involves introducing air under pressure into asphalt feedstock, usually heated to 220–300°C (430–570°F) and sometimes in the presence of catalyst, to produce a harder, less temperature-sensitive product, by oxidation and condensation polymerization. The asphaltene content is increased while the cyclic aromatic content decreases. Moderate blowing is used to obtain hard road asphalt or viscosity-grade asphalts from vacuum residues. Severe treatment produces oxidized asphalts suitable for a wide range of building and industrial applications. Depending on the feedstock viscosity and the processing conditions, the process produces two types of product with distinctly different characteristics: (i) air rectified asphalt, also known as semi-blown asphalt, and (ii) oxidized asphalt, more commonly known as blown asphalt.

The processes involve passing air through asphalt feedstock at *elevated* temperature (typically on the order of 250°C, 480°F) in order to manipulate the physical properties of the product. The main purpose of the process is to (i) harden the asphalt, (ii) increase the softening point, and (iii) change the temperature-viscosity properties of the asphalt. The process *achieves* these goals through varying degrees of chemical reactions which result in (i) an increase in the oxygen functional groups of the product (compared to the starting material) as well as (ii) an increase in the *average* molecular weight of the asphalt (Corbett, 1975; Moschopedis and Speight, 1973, 1975, 1977). These changes can also be achieved in the presence of various metal salts (Moschopedis and Speight, 1978) and all such processes typically lead to higher viscosity asphalt.

A relatively mild degree of air blowing (air rectification) is commonly used to adjust the physical properties of a given feedstock in order to manufacture substances used to make products suitable for use in road asphalt. These products are used at the same application temperatures as conventional vacuum-distilled asphalt as well as in roofing applications and also for a variety of industrial applications. However, an intense degree of air blowing (oxidation) is also commonly used to adjust the physical properties of a feedstock in order to manufacture suitable products for roofing. A flux oil may be used to soften the feedstock to the process and a catalyst (such as ferric chloride, hydrochloric acid, phosphorous pentoxide, or phosphoric acid) may also be used to increase the speed of reaction and improve temperature susceptibility relative to oxidation without a catalyst.

The air blowing unit typically consists of (i) a reactor, (ii) an air blower, (iii) an off-gas treatment facility, and (iv) temperature control equipment (Speight, 2014). The reactor is often an empty *vessel*, but may contain baffles or a mechanical agitation system to ensure turbulent mixing of the asphalt with air. The oxidation reaction is generally exothermic; therefore the reactor may be fitted with a water jacket and/or a water spray facility at the head of the reactor to control the asphalt temperature. Injection of steam/water into the reactor head space may also be used to reduce the oxygen content of the off-gases, in order to manage the risk of fire or explosion.

The chemical changes that take place during oxidation confer changes to the physical characteristics of the feedstock that are dependent upon the severity of the oxidation process. As the reaction proceeds naphthene-aromatic constituents and resin constituents that make up the maltene (non-asphaltene) phase are converted into asphaltenes resulting in hardening of the asphalt (reduced penetration) and an increase in softening point. As the asphaltene concentration increases beyond a certain point (limiting concentration), the ambient temperature flow properties of the asphalt change from viscoelastic to practically pure elastic behavior. The *severity* of oxidation determines the temperature susceptibility of the product—the products of mild oxidation (air rectification) have similar properties to those of straight-run vacuum-distilled (paving) asphalts. For this reason, asphalt produced by mild oxidation and by vacuum distillation residues is used either directly or in blends to produce finished products that have the necessary end use specifications. Oxidized asphalt has significantly different temperature susceptibility and a substantially higher softening point (for a given penetration value) as a consequence of the higher asphaltene content. The combination of these properties cannot be obtained by simple distillation or by mild oxidation.

The rate at which the oxidation reaction occurs is affected by feedstock properties and the conditions in the blowing unit such as: (i) viscosity or penetration of the feedstock, (ii) reactivity of the feedstock, (iii) temperature, (iv) air flow rate, (v) degree of agitation, (vi) pressure, (vii) air to feed ratio, and (viii) whether or not a catalyst was employed. Furthermore, the products from the mold oxidation and severe oxidation processes can be clearly differentiated using characteristics such as: (i) input feed to asphalt blowing unit, (ii) penetration, (iii) change of properties: input feed versus output asphalt blowing unit, (iv) change in softening point, (v) ratio of penetration, (vi) output of asphalt blowing unit, (vii) penetration index, and (viii) softening point.

Asphalt is also produced by propane deasphalting (Figure 2.5) (Speight, 2014) and the asphalt so produced may have a softening point of about 95°C (200°F). Solvent deasphalting (solvent precipitation) is employed to remove asphaltic compounds from certain vacuum residues to leave valuable high-viscosity base oil (bright stock) for lubricants (Speight, 2014; Speight and Exall, 2014). Residuum is dissolved in liquid propane or a propane/butane mixture, and the aliphatic fraction is precipitated and drawn off. Solvent precipitated asphalts have a higher content of asphaltenes than the vacuum residuum from which they are produced but a lower content of saturates than would be obtained by distillation of the vacuum residue.

Softer grades of asphalt are produced by blending *hard asphalt* with the extract obtained in the solvent treatment of lubricating oils. Modification of the properties of the vacuum residue by solvent deasphalting was originally (and still is in many refineries) the use of solvents to remove asphaltene fractions from distillation residues for the production of lubricating oil base stocks. The residue remaining after solvent deasphalting is a type of hard asphalt that can be used in blends to produce specification-grade asphalt. The principal deasphalting processes use propane, butane, isobutene, pentane, or supercritical solvent extraction (the ROSE process).

In practice, asphalt produced from all of the above processes can be blended to produce asphalt having the desired specification. Thus, ancillary processes (which are processes used to produce specification-grade asphalt and additional product flexibility) include (i) blending, (ii) cutback mixing with volatile petroleum diluents, or (iii) fluxing, which is the addition of high boiling (>350°C, >650°F) distillates or industrial process oils.

2.4 NATURALLY OCCURRING MATERIALS

Large amounts of bitumen (sometimes called *natural asphalt*) occur throughout the world. Typically, these naturally occurring bitumen deposits are formed from the remains of ancient, microscopic algae (diatoms) and other once-living things. These remains were deposited in the mud on the bottom of the ocean or lake where the organisms lived. Under the heat and pressure of burial deep in the Earth, the remains were transformed into materials such as petroleum and bitumen. There are also instances where bitumen appears at the surface of the Earth in seepages in various locations, such as: (i) the Trinidad Pitch Lake deposits on the Caribbean island of Trinidad, Republic of Trinidad and Tobago; (ii) the tar sands of western Canada, called oil sands in Canada, (iii) the Gilsonite deposits of Utah; and (iv) the La Brea Tar Pits of Southern California. Although these deposits are referred to as *tar pits* or *pitch lakes* they are, in fact, bitumen deposits which exist in hard or soft solid form. In historical times, these natural deposits of bitumen were the most common source for use as *asphalt mastic*, the more modern source of asphalt is from petroleum refining. In terms of the origin of these deposits, it is believed that this type of bitumen (rather than buried bitumen) arises from petroleum and is the result of the evaporation of the volatile constituents that leaves the bituminous residue (Abraham, 1945; Forbes, 1958a,b, 1959, 1964; Barth, 1962; Hoiberg, 1964).

2.5 OTHER SOURCES

Other sources of asphalt include the production of cracked residua (often called *tar*) in the refinery.

For example, in a visbreaking unit (either a coil visbreaker or a soaker visbreaker) (Figures 2.6 and 2.7) (Speight, 2014), a residue stream (either atmospheric or vacuum residue) is heated to temperatures between 440°C

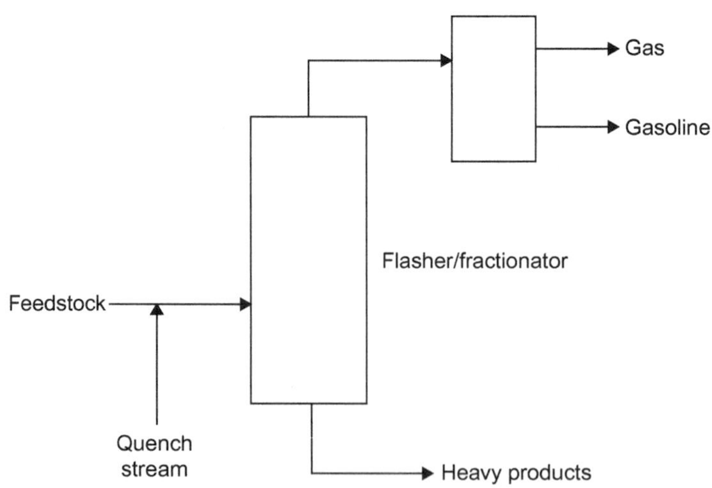

Figure 2.6 Schematic of a coil visbreaking unit.

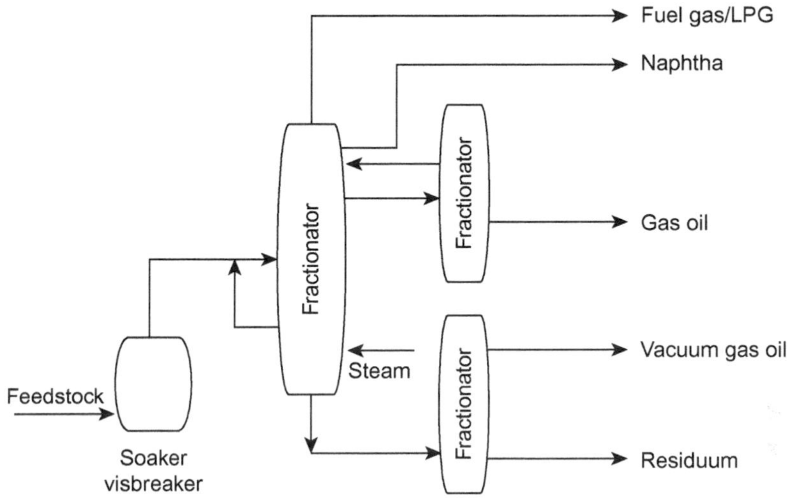

Figure 2.7 Schematic of a soaker visbreaking unit.

and 500°C (825°F and 930°F), although process conditions can vary depending upon the feedstock and the desired properties of the thermally cracked material, to avoid coke formation. Long paraffinic side chains attached to aromatic rings are the primary cause of the high pour point and viscosity seen with residue streams. Visbreaking is carried out under conditions optimized to break off these long side chains and subsequently transform them to form shorter molecules with lower viscosity and pour point. It is believed that the process occurs via a free radical mechanism. The number of hydrogen atoms in the parent molecule is insufficient to provide the full complement for each carbon atom and therefore unsaturated molecules are formed. When used for asphalt production the thermally cracked residue is subjected to vacuum distillation to remove the distillate fractions which are then further treated and used in the production of fuels. The residuum (often called visbreaker tar) is subjected to (for example) vacuum distillation to remove all volatile constituents after which the non-volatile product is typically a hard material which can be used as a blending component for asphalt production (Poirier, 1986; Belokon et al., 2001).

A number of other refinery processes are used to produce small amounts of residual materials that can be used in the production of asphalt. These are primarily further treatment or extraction processes applied to residual materials, to remove or convert constituents that are unsuitable for asphalt product performance and produce feedstocks for other processes. The processes, including solvent extraction, hydrodesulfurization and hydrogenation, are not commonly used and hence represent only a minor part of the overall asphalt production.

Coal tar has also been used as a road paving material (Yan, 1986) but the properties of the tar are not consistent with the desired properties of modern roadways and the longevity of the roadways, especially as the product relates to environmental issues and regulations (see Chapter 12). Coal tar was a component of the first sealed roads and was used with granite chips as the aggregate. A large part of the binders used in the graphite industry for making *green blocks* are coke oven volatiles (coal tar). During the baking process of the green blocks as a part of commercial graphite production, most of the coal tar binders are vaporized and are generally burned in an incinerator to prevent release into the atmosphere—coke oven volatiles and coal tar can be injurious to health.

Coal tar was used in the paving industry to varying degrees in Europe, southern Africa, Australia, and the United States. Because of their similar appearance, little distinction was made between asphalt and tar as a construction material in the past. However, their origins, and consequently the chemical composition, are quite different.

Generally, coal tar and coal tar pitch are obtained as a byproduct of the destructive distillation of bituminous coal to produce coke by thermal cracking at high temperatures (460–1215°C, 850–2200°F). One process, which results in coke oven tar, is the processing of coal by thermal degradation in a coking plant, used in steel manufacture. The second process yields coal tar as a byproduct of making oil from coal. This is sometimes known as the Sasol process, and the product is sometimes called Lurgi tar (Jamieson, 1979). As a result of the destructive distillation of coal, coal tar contains polycyclic aromatic hydrocarbons (PAHs). It is well recognized that the PAH content—such as that of B(a)P (Benzo[a]Pyrene)—in coal tar is far higher than the PAH content in asphalt.

Coal tar contains relatively high levels of condensed-ring aromatic compounds with a greater proportion of unsubstituted polycyclic nuclear aromatic compounds (PNAs) in the toxicologically active three-to-seven ring size range (Blackburn et al., 1999; Speight, 2013). In contrast, asphalt contains much larger proportions of high-molecular-weight paraffinic and naphthenic hydrocarbons and their derivatives that, because of their size, viscosity, and limited solubility are not readily bioavailable and have minimal toxicological activity. Thus, measurements of routinely monitored polycyclic nuclear aromatic hydrocarbons such as benzo(a)pyrene, as is commonly done for coal-derived product, are not useful indicators of potential carcinogenic activity of asphalts because the carcinogenic polycyclic nuclear aromatic hydrocarbons are present in extremely low concentrations and most asphalt polycyclic nuclear aromatic hydrocarbons are alkylated derivatives.

Coal tar has been used in all layers of pavement applications in Europe. It was sometimes used at 100%, sometimes as a mixture of petroleum-derived asphalt along with tar, and sometimes in a blend with polymers. Some of the products had brand names like Carbo-asphalt (a product of asphalt blended with tar) that contributed to confusion with regard to the difference between petroleum-derived asphalt and coal-derived tar.

Only after asphalt had replaced coal tar almost completely in Europe during the 1970s and 1980s—due to increasing oil production and declining coke usage and the related economic factors—was the hazard of coal tar to human health and to the environment realized. In Europe, by the early 1990s, the use of coal tar in road paving had been generally discontinued. Unfortunately, many people are still confused by the terminology relating to the historic use of the term "tar." Coal tar was used in the following European countries: Belgium (until 1992), Czech Republic (until 1999), Germany (until 1995), Denmark (until 1975), Finland (until the 1960s), France (until 1970), the Netherlands (until 1991), Norway (until 1960), Sweden (until 1974), Slovakia (until 1980), Turkey (until 1979), and the United Kingdom (until 1999). Controls on coal tar use in Europe since about 1990 are intended to prevent the significant presence of coal tar in pavements as a result of recycling.

In contrast, coal tar has not been used much in asphalt pavement applications in the the United States since World War II. Throughout this time, the economics of petroleum-derived asphalt have been favorable, while the sourcing of coal tar was on the decline. Following World War II, there was an increase in traffic volume, travel speeds, and axle loads concurrent with an increased demand for asphalt road construction and maintenance. There is evidence of very limited coal tar use as late as 1965 in areas of the country where coal and steel production were prominent. More recent applications for coal tar have been limited to a few non-road applications such as airfields and emulsion application as a pavement sealer for parking lots, driveways, and bridges. State specifications typically prohibit the use of recycled asphalt pavement known to contain coal tar (Mundt et al., 2009).

The primary uses of coal tar in South Africa and Australia have been in primers and chip seals. Some tar mixes have been used in the late 1960s and 1970s in base courses and surface courses including container terminals which are subject to fuel spills, car parks, and bus terminals. In South Africa, the estimated use of coal tar in the 1970s was less than 25% of all road binders (Jamieson, 1979) and it declined significantly after that time. Road agencies and contractors in South Africa have indicated an intention to discontinue its use (SABITA, 2005).

However, coal tar may still be incorporated into some parking-lot seal-coat products, which are used to protect and beautify the underlying pavement—seal-coat products that are coal-tar-based typically contain 20–35% (w/w) coal-tar pitch. Asphalt sealcoats are used to improve the appearance and prolong the life of driveways and parking lots.

Finally, as refineries approach the next century and supplies of petroleum dwindle, the need for asphalt will increase. A future source of asphalt could well be biomass that may increase in its appearance as a refinery feedstock as the current century matures and refineries evolve (Speight, 2011b). The choices of biomass will expand (Table 2.4) and the desirability of non-petroleum sources for asphalt production will become a reality (Mills-Beale et al., 2014).

The asphalt industry is faced with increasing prices and decreasing availability of the petroleum used in making conventional asphalt. Rising asphalt binder prices on the world market, diminishing crude oil reserves from which asphalt is derived, and a quest for a more environmentally friendly and energy-efficient asphalt binder are key challenges the asphalt pavement industry is facing. One of the most promising ways to address this issue is producing binders from biomass resources, such as the production of bio-binders from vegetation and forest waste, yard waste, sugar cane molasses, and other forms of biomass (Fini et al., 2010, 2011a,b; Mills-Beale et al., 2014).

In addition, wood cellulose, wood lignin, bottom ash, fly ash, waste tires, and coal mine waste are used as pavement components (Table 2.4). Of these wastes, *bio-oil* produced from cellulose and waste tires was thought to represent the most environmentally friendly, abundant and cost-effective opportunity for reducing the amount of petroleum-based materials in asphalt (Sundstrom et al., 1983; Williams et al., 2008). There have also been investigations into the use of lignin as a biological polymer in retarding the aging (oxidation) of asphalt pavements (Dizhbite et al., 2004; Ouyang, 2006). This function of bio-oil serves to prolong asphalt

Table 2.4 Choice of Biomass for Asphalt Production

Source/Type	Comment
Lignin asphalt[a]	Similar functional group structures to the resin fractions of asphalt binders and mixtures
	Contains significant unsaturated aromatic rings joined by alkyl chains similar to petroleum-based asphalt
	Similar adhesive properties to petroleum-based asphalt
Lignin asphalt[b]	Good coating, workability, compaction, and fatigue resistance achieved
	Optimum paving mixture needs more asphalt with lignin–asphalt binders than with asphalt alone
Lignin asphalt[c,d,e]	Has antiaging properties in asphalt pavements
Various sources[f]	Needs heat upgrading before use in asphalt binders and pavements
Oakwood	Mixing temperature range may be lower than that of petroleum-based binders by approximately 30–40°C
Corn stover	The rheological properties are improved after addition of modifiers to bio-asphalt
Switchgrass	High-temperature performance grade similar to that of traditional petroleum-based binders
	Low-temperature performance grade varies from typical petroleum-based binder due to high oxygen content

[a]Barth (1962).
[b]Terrel and Rimsritong (1979).
[c]Dizhbite et al. (2004).
[d]Ouyang et al. (2006).
[e]McCready and Williams (2007).
[f]Metwally and Williams (2010).

pavement life by reducing aging-related failures such as thermal and fatigue cracking. However, the properties of bio-oil are influenced by the source of the biomass, and by the conditions under which they are manufactured. This is especially noteworthy given the wide range of variability in the materials that make up feedstocks for bio-oil production.

In addition, lignin and lignin products have potential for use in the asphalt industry (Table 2.4) (Sundstrom et al., 1983; Dizhbite et al., 2004; Ouyang et al., 2006; McCready and Williams, 2007; Williams et al., 2008). This function of bio-oil serves to prolong asphalt pavement life by reducing aging-related failures such as thermal and fatigue cracking. Furthermore, lignin was found to have a profound effect on widening the performance-grade range of asphalt binders. However, the properties of bio-oils produced are influenced by the source of the biomass, and by the conditions under which they are manufactured. This is especially noteworthy given the wide range of variability in the materials that make up yard waste (Allen, 2003; Hill and Jennings, 2011).

2.6 USE IN ANCIENT TIMES

Asphalt is one of oldest engineering materials known to humanity, having been used since the beginning of civilization. The asphalt used by early civilizations occurred naturally and was formed when crude petroleum seeped up through cracks and fissures to the surface of the earth and (after evaporation of the volatile constituents) appeared either as soft workable mortars or as hard brittle black veins of rock formations (the latter was sometime referred to as *asphaltic coal* in more modern times).

However, the use of bitumen extends back into ancient times—the first evidence for the use of bitumen comes from the appearance of bitumen on stone tools from Neanderthal sites in Syria dating from approximately 40,000 years ago. In the period 5000–6000 BC, in the Tigris-Euphrates valley, in what is now Iraq, lived the Sumerians who established one of the first great cultures of the civilized world—they devised the cuneiform script, built the temple-towers known as ziggurats, and created an impressive law, literature, and mythology. The Sumerians had a thriving shipbuilding industry that produced and used asphalt for caulking and waterproofing. By 4000 BC, there is evidence for the widespread use of bitumen in Sumeria and there is evidence for the use of bitumen at the ziggurat near to the old city of Ur (Speight, 1978).

Herodotus (in the *Histories*) claims, with some justification, that bitumen was used in the construction of buildings in Babylon (Herodotus, 1909). Bitumen was mined at Ardericca (an unidentified site, supposedly near to the city of Babylon), and asphalt was also recovered from sites on the Euphrates river. Herodotus also describes the construction of the city wall of Babylon using hot bitumen (as the cement) and interposing a layer of wattled reeds at every 30th course of the bricks (Rawlinson, 1885). This corresponds to the use of bitumen as a road-building material in Babylon around 625 BC, in the reign of King Nabopolassar. A preserved inscription records the paving of Procession Street in Babylon, which led from the palace to the north wall of the city, with *asphalt* (i.e., bitumen) and burned brick (Speight, 1978). Not to be excluded from this natural wonder (bitumen), in 2600 BC, the Egyptians used asphalt as a waterproofing material and also to impregnate the wrappings of mummies as a preservative. Other civilizations widely used asphalt as a mortar for building and paving blocks used in temples, irrigation systems, reservoirs, and highways. For example, in the sixteenth century, Sir Walter Raleigh stepped ashore and observed the huge pitch lake in the southwest peninsular of Trinidad (Trinidad Pitch Lake, Trinidad Asphalt Lake) and used the asphalt to caulk his ocean-weary ship.

Moving further to the north west, during the eighteenth century when Spanish settlers first moved into the area of what is now Los, they came upon an alluvial, grass-covered plain about 45 miles in circumference at the foot of the Santa Monica Mountains.

By way of explanation, an alluvial plain is a largely flat landform created by the deposition of sediment over a long period of time by one or more rivers coming from highland regions, from which alluvial soil forms. A flood plain is part of the process, being the smaller area over which the rivers flood at a particular period of time, whereas the alluvial plain is the larger area representing the region over which the floodplains have shifted over geological time.

In the middle of the Los Angeles alluvial plain was an area of approximately 20 acres (the approximate area was estimated by visual observation) where bitumen flowed out of numerous vents and made black puddles amidst the sandy dirt and tufts of grass. The thick and viscous bitumen formed volcano-shaped mounds that generally measured several inches above the surrounding level of the ground. The bitumen was a useful substance to Indians in the region, who used it to caulk their canoes, waterproof their baskets, and attach wooden handles to stone blades. The area moved further into development when, in 1828, a Portuguese former sailor (Antonio de Rocha) became the owner of the tar springs through a land grant (from the Government of Mexico) of one square league (one square league = 4439 acres; one league is equivalent to approximately three miles or the distance a person could walk in 1 h) and the estate became known as Rancho La Brea (Tar Ranch). One of the conditions of the grant was that Rocha allow local inhabitants to come to the ranch and obtain as much bitumen as they needed—generally, the bitumen was used to waterproof the roofs of adobe houses.

As cultures developed, *bitumen* or *asphalt* was frequently used in construction and in ornamental works. Although it is possible to differentiate between the words *bitumen* and *asphalt* in modern use the occurrence of these words in older texts offers no such possibility (Table 2.5). It is significant that the early use of bitumen was in the nature of cement for securing or joining together various objects, and it thus seems likely

Table 2.5 Petroleum and Derivatives such as Asphalt and Natural Bitumen Have Been Known and Used for Almost 6000 Years

3800 BC	Documented use of asphalt for caulking reed boats
3500 BC	Use as cement for jewelry and for ornamental applications
3000 BC	Use as a construction cement by Sumerians; also believed to be used as a road material; asphalt used to seal bathing pool or water tank at Mohenjo Daro
2500 BC	Use in the embalming process; asphalt believed to be widely used for caulking boats
1500 BC	Use for medicinal purposes and (when mixed with beer) as a sedative for the stomach; continued reference to use of asphalt liquids (oil) as illuminant in lamps
1000 BC	Use as a waterproofing agent by lake dwellers in Switzerland
650 BC	Use of asphalt as a road-building material was in Babylon
500 BC	Use (mixed with sulfur) as an incendiary device in Greek wars; also use of asphalt liquid (oil) in warfare
250 BC	Occurrences in several areas of the Fertile Crescent (Mesopotamia); repeated use of liquid asphalt (oil) as an illuminant in lamps
40 AD	Bitumen floating on the Dead Sea described by Dioscorides as *Judaicum bitumen*
750 AD	Use in Italy as a color in paintings
950 AD	Report of destructive distillation to produce distillate; reference to distillate as nafta (naft, naphtha)
1500 AD	Discovery in the Americas; first attempted documentation of the relationship of asphalt and naphtha (petroleum)
1600 AD	Used for a variety of tasks; relationship to coal tar and wood tar studied; used for paving
1859 AD	Discovery of petroleum in North America; birth of modern-day petroleum science and refining for the production of derivatives such as asphalt

that the name itself was expressive of this application (Speight, 1978). The bitumen was most likely acquired from the city of Hit (Oyun Hit—the Fountains of Hit, also called Tuttul, near Raqqa in northern Syria) which, in Middle and Neo-Assyrian texts (circa 1100–600 BC) is cited as a source of the bitumen used in building as a waterproof lining, and to caulk boats. Herodotus mentions Hit as the source of bitumen used by Nebuchadnezzar for his architectural activities at Babylon (Rawlinson, 1885), and this association is further reinforced by the relationship between its name and the Akkadian word for bitumen *Ittu*. In modern times, bitumen is still produced there, as well as salt and quarried stone.

The word *asphalt* is derived from the Akkadian term *asphaltu* or *sphallo*, meaning *to split*. It was later adopted by the Homeric Greeks in the form of the adjective ασφαλής ες signifying *firm*, *stable*, *secure*, and the corresponding verb ασφαλίζω ίσω meaning *to make firm or stable*, *to secure*. It is a significant fact that the first use of asphalt by the ancients was in the nature of cement for securing or joining together various objects, such as the bricks used for building and it thus seems likely that the name itself was expressive of this application. From the Greek, the word passed into late Latin (*asphaltum, aspaltum*), and thence into French (*asphalte*) and English (*aspaltoun*).

The origin of the word *bitumen* is more difficult to trace and subject to considerable speculation. The word was proposed to have originated in the Sanskrit, where we find the words *jatu*, meaning *pitch*, and *jatukrit*, meaning *pitch creating*. From the Sanskrit, the word *jatu* was incorporated into the Latin language as *gwitu* and is reputed to have eventually become *gwitumen* (pertaining to pitch)—although the construction of these words in not a usual Latin word form. Another word, *pixtumen* (exuding or bubbling pitch) is also reputed to have been in the Latin language, although the construction of this Latin word form from which the word *bitumen* was reputedly derived, is certainly suspect. There is the suggestion that subsequent derivation of the word led to a shortened version (which eventually became the modern version) *bitûmen* thence passing via French into English. From the same root is derived the Anglo Saxon word *cwidu* (*mastic, adhesive*), the German work *kitt* (cement or mastic), and the equivalent word *kvada* which is found in the old Norse language as being descriptive of the material used to waterproof the *long ships* and other seagoing vessels. It is just as (perhaps even more than) likely that the word is derived from the Celtic *bethe* or *beithe* or *bedw* that was the birch tree that was used as a source of resin (tar). The word appears in Middle English as *bithumen*. In summary, a variety of terms exist in ancient language from which, from their described use in texts, they can be proposed as having the meaning bitumen or asphalt (Table 2.5) (Abraham, 1945).

There are several references to bitumen (albeit under different names) in the Bible. Although the exact classification of the hydrocarbon material is not known, there is little doubt that bitumen was the product named. The earliest reference in the Bible to asphalt is its use as a mortar in the construction of the Tower of Babel (Genesis, 11:3). This reference agrees with historical and archeologic evidence on the widespread use of asphalt in ancient Mesopotamia for building purposes (Abraham, 1945; Forbes, 1964). However, by the time the various books of the Bible were written as a collection (in approximately the ninth century BC), the use of petroleum and bitumen was established. Nevertheless, these writings do offer documented examples of the use of petroleum and related materials.

As an example of pre-Biblical use of bitumen, in the Epic of Gilgamesh (as well as in the older Epic of Atrahasis) written more than 4500 years ago, a great Flood causes the hero to build a boat that is caulked with bitumen and pitch (Kovacs, 1990). And, in a related story (it is not the intent here to discuss the similarities of the two stories) of Mesopotamia and just prior to the Flood, Noah is commanded to build an ark that also includes instructions for caulking the vessel with pitch (Genesis, 6:14):

> *Make thee an ark of gopher wood; rooms shalt thou make in the ark, and shalt pitch it within and without with pitch.*

The occurrence of *slime* (bitumen) *pits* in the Valley of Siddim (Genesis, 14:10), a valley at the southern end of the Dead Sea, is reported. There is also reference to the use of tar as a mortar when the Tower of Babel was under construction (Genesis, 11:3):

> *And they said one to another, Go to, let us make brick, and burn them thoroughly. And they had brick for stone, and slime had they for mortar.*

In the Septuagint, or Greek version of the Bible, this work is translated as *asphaltos*, and in the Vulgate or Latin version, as *bitumen*. In the Bishop's Bible of 1568 and in subsequent translations into English, the word is given as *slime*. In the Douay translation of 1600, it is *bitume*, while in Luther's German version, it appears as *thon*, the German word for clay. Another example of the use of pitch (and slime) is given in the story of Moses (Exodus, 2:3):

> *And when she could no longer hide him, she took for him an ark of bulrushes, and daubed it with slime and with pitch, and put the child therein; and she laid it in the flags by the river's brink.*

Perhaps the *slime* was a lower-melting bitumen (bitumen mixed with solvent), whereas the *pitch* was a higher-melting material; the one (*slime*) acting as a flux for the other (*pitch*). The lack of precise use of the words for *bitumen* and *asphalt* as well as for *tar* and *pitch* even now makes it unlikely that the true nature of the biblical *tar*, *pitch*, and *slime* will ever be known, but one can imagine their nature! In fact, even modern Latin dictionaries give the word *bitumen* as the Latin word for *asphalt*!

It is most probable that, in both these cases, the *pitch* and the *slime* were obtained from the seepage of oil to the surface, which was a fairly common occurrence in the area. And during biblical times, bitumen was exported from Canaan to various parts of the countries that surround the Mediterranean (Armstrong, 1997). In terms of liquid products, there is an interesting reference (Deuteronomy, 32:13) to bringing *oil out of flinty rock.* The exact nature of the oil is not described nor is the nature of the rock. The use of *oil* for lamps is also referenced (Matthew, 23:3) but whether it was mineral oil (a petroleum derivative such as naphtha) or whether it was vegetable oil is not known.

Excavations conducted at Mohenjo-Daro, Harappa, and Nal in the Indus Valley indicated that an advanced form of civilization existed there. An asphalt mastic composed of a mixture of asphalt, clay, gypsum, and organic matter was found between two brick walls in a layer about 25 mm thick, probably a waterproofing material. Also unearthed was a bathing pool that contained a layer of mastic on the outside of its walls and beneath its floor.

In the Bronze Age, dwellings were constructed on piles in lakes close to the shore to better protect the inhabitants from the ravages of wild animals and attacks from marauders. Excavations have shown that the wooden piles were preserved from decay by coating with asphalt, and posts preserved in this manner have been found in Switzerland. There are also references to deposits of bitumen at Hit (the ancient town of Tuttul on the Euphrates river in Mesopotamia) and the bitumen from these deposits was transported to Babylon for use in construction (Herodotus, *The Histories*, Book I). There is also reference (Herodotus, *The Histories*, Book IV) to a Carthaginian story in which birds' feathers smeared with *pitch* are used to recover gold dust from the waters of a lake.

One of the earliest recorded uses of asphalt was by the pre-Babylonian inhabitants of the Euphrates valley in southeastern Mesopotamia, present-day Iraq, formerly called Sumer and Akkad and, later, Babylonia. In this region there are various asphalt deposits, and uses of the material have become evident. For example, King Sargon Akkad (Agade) (ca. 2550 BC) was (for reasons that are lost in the annals of time) set adrift by his mother in a basket of bulrushes on the waters of the Euphrates, he was discovered by Akki the husbandman (the irrigator), whom he brought up to serve as gardener in the palace of Kish. Sargon eventually ascended to the throne. On the other hand, the bust of Manishtusu, King of Kish, an early Sumerian ruler (about 2270 BC), was found in the course of excavations at Susa in Persia, and the eyes, composed of white limestone, are held in their sockets with the aid of bitumen. Fragments of a ring composed of asphalt have been unearthed above the flood layer of the Euphrates at the site of the prehistoric city of Ur in southern Babylonia, ascribed to the Sumerians of about 3500 BC.

An ornament excavated from the grave of a Sumerian king at Ur consists of a statue of a ram with the head and legs carved out of wood over which gold foil was cemented by means of asphalt. The back and flanks of the ram are coated with asphalt in which hair was embedded. Another piece of decorative art consisted in beating thin strips of gold or copper, which were then fastened to a core of asphalt mastic. An alternative method was to fill a cast metal object with a core of asphalt mastic, and such specimens have been unearthed at Lagash and Nineveh. Excavations at Tell-Asmar, the ruins of which are located approximately 50 miles northeast of Baghdad (Iraq), revealed the use of asphalt by the Sumerians for building purposes.

Mortar composed of asphalt has also been found in excavations at Ur, Uruk, and Lagash, and excavations at Khafaje have uncovered floors composed of a layer of asphalt that has been identified as asphalt, mineral filler (loam, limestone, and marl), and vegetable fibers (straw). Excavations at the city of Kish (Persia) in the palace of King Ur-Nina showed that the foundations consist of bricks cemented together with an asphalt mortar. Similarly, in the ancient city of Nippur (approximately 60 miles south of Baghdad), excavations show Sumerian structures composed of natural stones joined together with asphalt mortar. Excavation has uncovered an ancient Sumerian temple in which the floors are composed of burnt bricks embedded in an asphalt mastic that still shows impressions of reeds with which it must originally have been mixed.

The Epic of Gilgamesh (written before 2500 BC and transcribed on to clay tablets during the time of Ashurbanipal, king of Assyria (668–626 BC)) make reference to the use of asphalt for building purposes. In the eleventh tablet, Ut-Napishtim relates the well-known story of the Babylonian flood, stating that he smeared

> *...the inside of a boat with six sar of kupru and the outside with three sar...*

The word *kupru* may have meant that the pitch or bitumen was mixed with other materials (perhaps even a solvent such as distillate from petroleum) to give it the appearance of *slime* as mentioned in the Bible. In terms of measurement, *sar* is a word of mixed origin and appears to mean an interwoven or wickerwork basket. Thus, an approximate translation is that:

> *the inside of the boat was smeared (coated, caulked) with six baskets full of pitch and the outside of the boat was smeared (coated, caulked) with three baskets full of pitch.*

There are also indications from these texts that asphalt mastic was sold by volume (by the *gur*). On the other hand, bitumen was sold by weight (by the *mina* or *shekel*).

Use of asphalt by the Babylonians (1500–538 BC) is also documented. The Babylonians were well versed in the art of building, and each monarch commemorated his reign and perpetuated his name by the construction of buildings or other monuments. For example, bitumen was used as mastic and as a sealant for water pipes, water cisterns, and in outflow pipes leading from flush toilets cities such as Babylon, Calah, Nineveh, and Ur.

Bitumen was used as mortar from very early times, and sand, gravel, or clay was employed in preparing these mastics. Asphalt-coated tree trunks were often used to reinforce wall corners and joints, for instance in the temple tower of Ninmach in Babylon. In vaults or arches, a mastic–loam composite was used as mortar for the bricks, and the keystone was usually dipped in asphalt before being set in place. The use of bituminous mortar was introduced into the city of Babylon by King Hammurabi, but the use of bituminous mortar was abandoned toward the end of Nebuchadnezzar's reign in favor of lime mortar to which varying amounts of asphalt were added. The Assyrians recommended the use of asphalt for medicinal purposes, as well as for building purposes, and perhaps there is some merit in the fact that the Assyrian moral code recommended that asphalt, in the molten state, be poured onto the heads of delinquents. Pliny, the Roman author, also notes that bitumen could be used to stop bleeding, heal wounds, drive away snakes, treat cataracts as well as a wide variety of other diseases, and straighten out eyelashes which inconvenience the eyes. One can appreciate the use of bitumen to stop bleeding but its use to cure other ailments is questionable and one has to consider what other agents were being used concurrently with bitumen.

The Egyptians were the first to adopt the practice of embalming their dead rulers and wrapping the bodies in cloth. Before 1000 BC, asphalt was rarely used in mummification, except to coat the cloth wrappings and thereby protect the body from the elements. After the viscera had been removed, the cavities were filled with a mixture of resins and spices, the corpse immersed in a bath of potash or soda, dried, and finally wrapped. From 500 to about 40 BC, asphalt was generally used both to fill the corpse cavities, as well as to coat the cloth wrappings. The word *mûmûia* first made its appearance in Arabian and Byzantine literature about 1000 AD, signifying *bitumen*. In fact, it was through the spread of the Islamic Empire that, it is believed, brought Arabic science and the use of bitumen to western Europe.

In Persian, the term *bitumen* is believed to have acquired a meaning equivalent to *paraffin wax* that might be symptomatic of the nature of some of the crude oils in the area. Alternatively, it is also possible that the destructive distillation of bitumen to produce pitch produced paraffins that crystallized from the mixture over time. In Syriac, the term alluded to *substances used for mummification*. In Egypt, resins were used extensively for the purposes of embalming up to the Ptolemaic period, when asphalts gradually came into use.

The product *mûmûia* was used in prescriptions, as early as the twelfth century, by the Arabian physician Al Magor, for the treatment of contusions and wounds. Its production soon became a specialized industry in Alexandria. The scientist Al-Kazwînî alluded to the healing properties of *mûmûia*, and Ibn Al-Baitâr gives an account of its source and composition. Engelbert Kämpfer (1651–1716), in his treatise *Amoenitates Exoticae*, gives a detailed account of the gathering of *mûmûia*, the different grades and types, and its curative properties in medicine. As the supply of mummies was, of course, limited, other expedients came into vogue. The corpses of slaves or criminals were filled with asphalt, swathed, and artificially aged in the sun. This practice continued until the French physician, Guy de la Fontaine, exposed the deception in 1564 AD.

Many other references to bitumen occur throughout the Greek Empire and the Roman Empire, and from then to the Middle Ages early scientists (alchemists) frequently alluded to the use of bitumen. In later times, both Christopher Columbus and Sir Walter Raleigh (depending upon the country of origin of the biographer) have been credited with the discovery of the asphalt deposit on the island of Trinidad and apparently used the material to caulk their ships.

The use of petroleum has also been documented in China: as early as 600 BC (Owen, 1975), petroleum was encountered when drilling for salt and mention of petroleum as an impurity in the salt is also noted in documents of the third century AD. In a more national context, there is also the suggestion of pre-Columbian oil mining in ancient Pennsylvania. The precise nature of the oil is unknown although the method of recovery indicates light oil. The oil was recovered by oil mining in pits, which were oblong, approximately four by six feet in depth, and which were operated by allowing them to fill with water overnight then skimming the oil on the surface into containers.

There was also an interest in the thermal product of petroleum (nafta; naphtha) when it was discovered that this material could be used as an illuminant and as a supplement to asphalt incendiaries in warfare. For example, there are records of the use of mixtures of pitch and/or naphtha with sulfur as a weapon of war during the Battle of Palatea, Greece, in the year 429 BC (Forbes, 1959). There are references to the use of a liquid material, *naft* (presumably the volatile fraction of petroleum which we now call *naphtha* and which is used as a solvent or as a precursor to gasoline), as an incendiary material during various battles of the pre-Christian era (James and Thorpe, 1994). This is the so-called Greek fire, a precursor and chemical cousin to napalm. Greek fire is also recorded as being used in the period 674–678 when the city of Constantinople was saved by the use of the fire against an Arab fleet (Davies, 1996). In 717–718 AD, Greek fire was again used to save the city of Constantinople from attack by another Arab fleet; again with deadly effect (Dahmus, 1995). After this time, the Byzantine navy of 300 triremes frequently used Greek fire against all-comers (Davies, 1996). This probably represents the first documented use of the derivatives of petroleum that led to a continued interest in petroleum.

Greek fire was a viscous liquid (probably a mix of naphtha and asphalt) that ignited on contact with water and was sprayed from a pump-like device on to the enemy. One can imagine the early users of *the fire* attempting to ignite the liquid before hurling it towards the enemy. However, the hazards that can be imagined from such tactics could become very real, and perhaps often fatal, to the users of the Greek fire if any spillage occurred before ejecting the fire towards the enemy. The later technology for the use of Greek fire probably incorporated a heat-generating chemical such as quicklime (CaO) (Cobb and Goldwhite, 1995), which was suspended in the liquid and which, when coming into contact with water (to produce [$Ca(OH)_2$]), released heat that was sufficient to cause the liquid to ignite. One assumes that the users of the fire were extremely cautious during periods of rain or, if at sea, during periods of turbulent weather.

As an aside, the use of powdered lime in warfare is also documented. The English used it against the French on August 24, 1217, with disastrous effects for the French. As was usual for that time, there was a difference of opinion between the English and the French that resulted in their respective ships meeting at the east end of the English Channel. Before any other form of engagement could occur, the lime was thrown from the English ships and carried by the wind to the French ships where it made contact with the eyes of the French sailors. The burning sensation in the eyes was too much for the French sailors and the English prevailed with the capture of much booty (Powicke, 1962).

The combustion properties of bitumen (and its fractions) were known in Biblical times. There is reference to these properties (Isaiah, 34:9) when it is stated that:

> *And the stream thereof shall be turned into pitch, and the dust thereof into brimstone, and the land thereof shall become burning pitch.*
> *It shall not be quenched night nor day; the smoke thereof shall go up forever: from generation to generation it shall lie waste; none shall pass through it forever and forever.*

One might surmise for this wording that the effects of the burning bitumen and sulfur (brimstone) were long-lasting and quite devastating.

Approximately 2000 years ago, Arabian scientists developed methods for the distillation of petroleum, which were introduced into Europe by way of Spain. This represents another documented use of the volatile derivatives of petroleum which led to a continued interest in petroleum and its derivatives as medicinal materials and materials for warfare, in addition to the usual construction materials.

The Baku region of northern Persia was also reported (by Marco Polo in 1271–1273) as having an established commercial petroleum industry. It is believed that the prime interest was in the kerosene fraction that was then known for its use as an illuminant. By inference, it has to be concluded that the distillation, and perhaps the thermal decomposition, of petroleum were established technologies. If not, Polo's diaries might well have contained a description of the stills or the reactors. In addition, bitumen was investigated in Europe during the Middle Ages (Bauer, 1546, 1556), and the separation and properties of bituminous products were thoroughly described. Other investigations continued, leading to a good understanding of the sources and use of this material even before the birth of the modern petroleum industry (Forbes, 1958a,b).

To continue such references is beyond the scope of this book, although they do give a flavor of the developing interest in petroleum. However, it is sufficient to note that there are many other references to the occurrence and use of bitumen and asphalt as well as other petroleum derivatives up to the beginning of the modern petroleum industry (Mallowan and Rose, 1935; Mallowan, 1954; Marschner et al., 1978).

In summary, the use of bitumen (naturally occurring), asphalt (manufactured), and other petroleum-related materials has been observed for almost 6000 years. During this time, the use of petroleum has progressed from the relatively simple use of asphalt from Mesopotamian seepage sites to the present-day refining operations that yield a wide variety of products and petrochemicals (Speight and Ozum, 2002; Parkash, 2003; Hsu and Robinson, 2006; Gary et al., 2007; Speight, 2014).

However, despite these early uses of asphalt, several hundred years passed before European or American builders tried it as a paving material—a good or satisfactory method of road building had not yet been developed. Then in the mid-eighteenth century, John Metcalf built 180 miles of roads in Yorkshire (England). His insistence on good drainage, requiring a foundation of large stones covered with excavated road material to raise the roadbed, followed by a layer of gravel probably led to the first guidelines or specifications for the construction of asphaltic roadways. Later (in 1803–1821), Thomas Telford built more than 900 miles of roads in Scotland. He perfected the method of building roads with broken stones, laid to a depth according to the weight and volume of traffic that the roads would carry. John Loudon McAdam, a contemporary of Telford, taught himself engineering after being appointed a trustee of a Scottish turnpike. McAdam observed that it was the native soil that supports the weight of traffic, and that while in a dry state could carry any weight without sinking. To construct his roads, McAdam used broken stone to form a hard surface. Later, to reduce dust and maintenance, builders used hot tar to bond the broken stones together, to produce tarmac (tarmac pavements, tarmac roadways—unlike the United States, the term *pavement* in many European countries refers to the sidewalk or footpath and not the roadway).

2.7 USE IN THE NINETEENTH AND TWENTIETH CENTURIES

In the early nineteenth century there was a surge of interest, and asphalt became widely used for pavements, flat roofs, and the lining of cisterns (Forbes, 1958b, 1959). The discovery of petroleum and subsequent commercialization of petroleum in Pennsylvania in 1859 led to further innovations and developments in petroleum refining (Bell, 1945). For example, modern refining commenced in 1862 with the first appearance of petroleum distillation (Table 2.6).

This was due, in no small part, to the efforts of Benjamin Silliman, who developed an interest in geological phenomena (of which, petroleum was such a phenomenon) and passed this interest on to his son (Benjamin Silliman Jr.). When Benjamin Sr. retired in 1853, his son took up where he had left off, as Professor of General and Applied Chemistry at Yale (this time, with a degree in the subject). After writing a number of chemistry books and being elected to the National Academy of Sciences, Benjamin,

Table 2.6 Process Development since the Commencement of the Modern Refining Era

Year	Process Name	Purpose	By-Products
1862	Atmospheric distillation	Produce kerosene	Naphtha, cracked residuum
1870	Vacuum distillation	Lubricants	Asphalt, residua
1913	Thermal cracking	Increase gasoline yield	Residua, fuel oil
1916	Sweetening	Reduce sulfur	Sulfur
1930	Thermal reforming	Improve octane number	Residua
1932	Hydrogenation	Remove sulfur	Sulfur
1932	Coking	Produce gasoline	Coke
1933	Solvent extraction	Improve lubricant viscosity index	Aromatics
1935	Solvent dewaxing	Improve pour point	Wax
1935	Catalytic polymerization	Improve octane number	Petrochemical feedstocks
1937	Catalytic cracking	Higher octane gasoline	Petrochemical feedstocks
1939	Visbreaking	Reduce viscosity	Increased distillate yield
1940	Alkylation	Increase octane number	High-octane aviation fuel
1940	Isomerization	Produce alkylation feedstock	Naphtha
1942	Fluid catalytic cracking	Increase gasoline yield	Petrochemical feedstocks
1950	Deasphalting	Increase cracker feedstock	Asphalt
1952	Catalytic reforming	Convert low-quality naphtha	Aromatics
1954	Hydrodesulfurization	Remove sulfur	Sulfur
1956	Inhibitor sweetening	Remove mercaptans	Disulfides and sulfur
1957	Catalytic isomerization	Convert to high-octane products	Alkylation feedstocks
1960	Hydrocracking	Improve quality and reduce sulfur	Alkylation feedstocks
1974	Catalytic dewaxing	Improve pour point	Wax
1975	Resid hydrocracking	Increase gasoline yield	Cracked residua

Jr., took up lucrative consulting posts, as his father had done, with the Boston City Water Company and various mining enterprises.

In 1855, he was asked to research and report on some mineral samples from the new Pennsylvania Rock Oil Company. After several months work Benjamin, Jr., announced that about 50% of the black tar-like substance could be distilled into first-rate burning oils (which would eventually be called kerosene and paraffin) and that an additional 40% of what was left could be distilled for other purposes, such as lubrication and gaslight. On the basis of this single report a company was launched to finance the drilling of the Drake Well at Oil Creek, PA, and in 1857 it became the first well to produce petroleum. It would be another 50 years before Silliman's reference to *other fractions* available from the oil through extra distillation would provide gasoline for the combustion engine of the first automobile. Silliman's report changed the world because it made possible an entirely new form of transportation and helped turn the United States into an industrial superpower.

After completion of the first well (by Edwin Drake, the self-styled *Colonel Drake*), the surrounding areas were immediately leased and extensive drilling took place. Crude oil output in the United States increased from approximately 2000 barrels (1 barrel, bbl = 42 US gallons = 35 imperial gallons = 5.61 foot^3=158.8 liters) in 1859 to nearly 3,000,000 bbl in 1863 and approximately 10,000,000 bbl in 1874. In 1861 the first cargo of oil, contained in wooden barrels, was sent across the Atlantic to London (UK), and by the 1870s, refineries, tank cars, and pipelines had become characteristic features of the industry, mostly through the leadership of Standard Oil that was founded by John D. Rockefeller (Johnson, 1997). Throughout the remainder of the nineteenth century the United States and Russia were the two areas in which the most striking developments took place (Yergin, 1991).

The first asphalt mixtures produced in the United States were used for sidewalks, crosswalks, and even roads starting in the late 1860s. In 1870, a Belgian chemist (Edmund J. DeSmedt) laid the first true asphalt pavement—a sand mix—in front of the City Hall in Newark, NJ. The design was patterned after a natural asphalt pavement placed on a French highway in 1852. DeSmedt went on to pave Pennsylvania Avenue in Washington, DC, a project that included 54,000 square yards paved with sheet asphalt imported from Asphalt Lake (Pitch Lake) on the island of Trinidad (now the Republic of Trinidad and Tobago). The durability of this pavement proved that the quality of the asphalt found in the Americas was as good as that imported from Europe. In fact, the bitumen from Pitch Lake is exported to many countries of the world as a binder for asphalt roadways.

The work of DeSmedt led to a burst in activity in the use of asphalt for roadway construction. As a result of the competition among asphalt producers, cities started to require more stringent requirements for their asphalt roadways. In 1896, for example, New York City adopted asphalt paving in place of brick, granite, and wood block as well as 15-year warranties on workmanship and materials. However, the long-term warranties, which did not recognize pavement failures caused by factors beyond the asphalt contractor's control, bankrupted many roadway builders and the result was fewer and higher bids for asphalt pavements.

As a result, almost all of the binder for the construction of asphaltic roadways in the United State came from the natural sources of Lake Trinidad and Bermudez Lake in Venezuela. Refined petroleum asphalts, used initially as an additive to soften the natural asphalt for handling and placing, made an appearance in the mid-1870s and slowly gained acceptance. By 1907, production of refined asphalt had outstripped the use of naturally occurring bitumen. As the automobile grew in popularity, local and state governments were besieged by requests for more and better roads. This demand led to innovations in both the production and properties of asphalt.

The first asphalt facility to contain virtually all (it lacked a cold feed component and pollution control equipment) the basic components of a modern asphalt production facility was built in 1901 by Warren Brothers in East Cambridge, MA. The first drum mixers and drum dryer-mixers, which came into use around 1910, were Portland cement concrete mixers that were adapted for use with hot mix asphalt production. Mechanization took another step forward in the 1920s with the improvement of cold feed systems for portable and semi-portable systems followed by the addition of vibrating screens and pressure injection systems in the 1930s. Furthermore, roadway construction methods improved at an even faster pace and the earliest items of equipment used for laying asphalt were hand equipment such as brooms and tampers in what was a highly labor-intensive process. Only after the asphalt was dumped, spread, and smoothed by hand did the horse-drawn roller, and later the steam roller, come into use to complete the roadway construction.

At this time, the construction of asphaltic roadways was improving and demand for such roadways was increasing. This demand was answered by the discovery of new oil fields. For example, at the outbreak of World War

I in 1914, the two major producers were the United States and Russia, but supplies of oil were also being obtained from Indonesia, Romania, and Mexico. During the 1920s and 1930s, attention was also focused on other areas for oil production, such as the United States, the Middle East, and Indonesia. However, asphalt production facilities could not at first keep up with the additional demand for hot mix asphalt that resulted from paving improvements, but the gap was finally closed around 1930 when it was recognized in the United States that asphalt is an essential material in nearly every form of highway construction and maintenance. In the four years from 1934 to 1937, asphalt entered into the construction of more than four-fifths of the mileage of highways completed in those years under state highway direction.

During World War II, asphalt technology improved at a great pace, spurred in part by the need of military aircraft for surfaces that could stand up to heavier loads. After the war ended, and families moved to the suburbs, road building became an extremely large industry. In order to meet the demand, contractors needed bigger and better equipment. In the post-1945 era Middle Eastern countries continued to rise in importance because of new discoveries of vast reserves. The United States, although continuing to be the biggest producer, was also the major consumer and thus was not a major exporter of oil. At this time, oil companies began to roam much farther in the search for oil, and significant discoveries in Europe, Africa, and Canada led to further development of petroleum resources and increased production of asphalt (Yergin, 1991). In order to accommodate the increasing demand for asphaltic roadways, engineers added electronic leveling controls in the 1950s, and automated screed (base) controls in the early 1960s. Extra-wide finishers, capable of paving two lanes at once, made a debut in 1968. Furthermore, the introduction of pickup machines allowed bottom-dump trailers to windrow the material in advance of the paver so that it could operate continuously.

The asphalt plants of the early 1950s typically included a dryer, a tower with a screed, and a mixer. By the mid-1960s, with air pollution a serious concern across the country, many had added wet scrubbers and some plants had added baghouses. The other major change in the mid-to-late 1960s was the addition of surge bins and storage bins—prior to that, everything was loaded right from the plant into the truck. Then bins for storing the mix for short periods of time were added to accommodate increased production capacity. Throughout this time, concern for the environment has encouraged the construction of better plants.

Furthermore, the trend toward continuous improvement has also allowed the industry to embrace recycling. Recycling was common in the early twentieth century, but the practice fell out of use as new asphalt refineries were built, increasing supplies and decreases in the price of petroleum and asphalt. The energy crisis of the 1970s, however, demonstrated the need for conservation of natural resources and, as a result, both base and surface roadway courses have incorporated an increasing amount of recycled asphalt in their mixes.

In addition, the industry led the drive for innovation and quality in design and construction of asphalt roadways. The National Asphalt Pavement Association was established in 1955 and one of the first activities undertaken by the fledgling organization was a Quality Improvement Program, which sponsored asphalt testing at universities and private testing labs, then shared the results with members. Furthermore, since the mid-1980s, the industry has developed advanced pavement materials including *open graded friction course* (OGFC), Superpave, and *stone matrix asphalt* (SMA), also called gap-graded Superpave (see Chapters 1 and 9). Engineering control systems placed on asphalt pavers beginning in 1997 have improved conditions for workers at the paving site.

2.8 USE IN THE TWENTY-FIRST CENTURY

Asphalt technology has evolved over the past century to the point where the asphalt of the nineteenth and twentieth centuries has become more complex and more efficient in terms of roadway longevity in the twenty-first century. Asphalt use is no longer restricted to use for roadways and waterproofing (caulking). In fact, it is the versatility of asphalt that makes it such a widely used material. Among others, it can be used in the following sectors: (i) transportation—in areas such as roads, railway beds, or airport runways, (ii) recreation—in areas such as playgrounds, bicycle paths, running tracks, and tennis courts, (iii) agriculture—in areas such as barn floors and greenhouse floors, (iv) industrial—in areas such as ports, landfill caps, and work sites, and (v) building construction—in areas such as floorings. For these uses, asphalt can be produced in a fixed plant or even in a mobile mixing plant. In addition, asphalt used in roadways is sometimes referred to as a flexible pavement due to its ability to resist the stress imposed by slight settlements of the roadway subgrade without cracking.

To be able to provide the best performance to different sectors, a large variety of asphalt mixes can be offered. Due to the different requirements (e.g., a road needs to fulfill: high traffic, tough weather conditions, etc.) the respective mix used needs to have a sufficient stiffness and resistance to deformation in order to cope with the applied pressure from vehicle wheels on the one hand, yet on the other hand, the need to have an adequate flexural strength to resist cracking caused by the varying pressures exerted on them. Moreover, good workability during application is essential in order to ensure that they can be fully compacted to achieve optimum durability.

Hot mix asphalt was, at one time, produced at a temperature between 150°C and 190°C (300°F and 375°F), whereas a typical warm mix asphalt was produced at a temperature on the order of around 20–40°C (68–104°F), but the two types of mix are often difficult to differentiate because of the lowering of the temperature of the hot mix and increasing the temperature of the warm mix. The principle behind the use of cold mix asphalt (CMA) was (i) lower energy requirement, (ii) lower temperature in the mix, resulting in improved working conditions for the crew, and (iii) an earlier opening of the road. On the other hand, CMA (which is particularly recommended for light-traffic roads) is produced without heating the aggregate. This is only possible due to the use of a specific bitumen emulsion which breaks either during compaction or during mixing. After breaking, the emulsion coats the aggregate and, over time, increases its strength. Depending on the designated use, different asphalt mixtures can be used, thus: porous asphalt, stone mastic asphalt, mastic asphalt, hot rolled asphalt, soft asphalt, asphalt concrete, asphalt concrete for very thin layers, asphalt concrete for ultra-thin layers, and double-layered porous asphalt.

Briefly, asphalt concrete (commonly called asphalt, blacktop, or pavement in North America, and tarmac in the United Kingdom and other European countries) is a composite material commonly used to surface roadways and (although the quality may differ) parking lots, and airport runways. The concrete consists of mineral aggregate bound together with asphalt, laid in layers, and compacted. The terms *asphalt concrete* (*asphaltic concrete*), *bituminous asphalt concrete*, and *bituminous mixture* are also used and further define concrete as any composite material composed of mineral aggregate adhered with a binder.

Thus, over the past 50 years, the versatility of asphalt has led to its increased use in other applications. Asphalt has been approved by the Environmental Protection Agency and used successfully as a primary liner for both sanitary and hazardous-waste landfills. It is also used to line drinking water reservoirs and fish hatcheries in California and Washington. Meanwhile, the industry continues to benefit from improvements in design and production.

2.9 PRODUCTION

Unlike the typical refinery processes and products, asphalt is more difficult to place on an individual refinery evolutionary scale. Processes changed and evolved to accommodate the demand for, say, higher octane fuel or longer-lasting asphalts or lower-sulfur coke. The majority of asphalt that is used commercially is obtained from petroleum. Furthermore, the facile production of asphalt in a refinery makes this product an ideal choice for rehabilitation of busy highways and aviation facilities. Whether on the highway or airport, asphalt pavements add up to the best value for the taxpayer dollar, with the lowest life cycle cost and the highest residual value. However, asphalt is a complex product that is used with aggregate, as one example, to build roads. It is the purpose of this chapter to present a brief description of the alternative sources of asphalt and the processes by which asphalt can be produced from various sources.

The production of asphalt is dictated by performance specifications and not by chemical composition. To meet the necessary performance specifications, the asphalt is produced from a petroleum residuum or mixed residua and then may be air-blown or further processed by solvent precipitation or propane deasphalting to produce the desired product (Corbett, 1976). Additionally, the products of other refining processes may be blended with the asphalt to achieve the desired performance specifications. Therefore, the exact chemical composition of asphalt is dependent on the chemical complexity of the original crude petroleum and the manufacturing process. Crude petroleum consists mainly of aliphatic compounds, cyclic alkanes, aromatic hydrocarbons, PACs (a class of chemicals that includes PAHs and heterocyclic derivatives in which one or more of the carbon atoms in the PAH ring system have been replaced by a heteroatom of nitrogen [N-PAC], oxygen [O-PAC], or sulfur [S-PAC]), and metals (e.g., iron, nickel, and vanadium). The proportions of these chemicals can vary greatly because of significant differences in crude petroleum from oil field to oil field or even at different locations in the same oil field (Speight, 2001, 2014).

The manufacturing process can cause significant changes to the physical properties of asphalt but the chemical nature of the asphalt changes to the extent that additional oxygen functional groups are introduced into the constituents of the asphalt. The so-called polymerization of the constituents is considered unlikely and this theory was based on the increased molecular weight of the oxidized product without the investigators realizing that additional oxygen functions can increase the extent of the intermolecular interactions of the constituents (through hydrogen bonding) and thereby increase the molecular weight (Moschopedis and Speight, 1973, 1975, 1977, 1978; Corbett, 1975; Speight, 1992, 2014). However, air blowing can also, but not always, decrease saturation and increase cross-linking within and between different molecular constituents of asphalt molecules. In this sense, the use of the word *polymerization* to describe the chemistry of the air-blowing process is incorrect.

Raising the temperature of the asphalt feedstock either during distillation or during oxygen-blowing will increase the likelihood of cracking and cause more volatiles and even higher-boiling components to be formed during the process (Speight, 2014). Solvent precipitation (usually using propane or butane) removes high-boiling components from vacuum-processed asphalt, which are then used to make other products. Solvent precipitation results in the production of a hard asphalt that is less resistant to temperature changes and which is often used as blend stock, along with blended straight-run or vacuum-processed asphalts.

The air-blowing process can be a continuous or batch operation and, since the continuous operation is faster and results in softer asphalt, the continuous process is preferred for processing paving asphalts (Speight, 1992, 2014). The process is exothermic (heat-producing) and may cause a series of chemical reactions, such as oxidation, condensation, dehydration, dehydrogenation, and polymerization. These reactions cause the amount of asphaltene constituents (pentane-insoluble or heptane-insoluble constituents) to increase and the amounts of resin constituents (polar aromatic constituents), cycloalkanes, and non-polar aromatic constituents to decrease, while the amount of aliphatic constituents (oils and waxes) changes very little (Speight, 1992, 2014). At the same time, the oxygen content of the asphalt increases (Moschopedis and Speight, 1973, 1975, 1977, 1978; Corbett, 1975; Speight, 1992).

Although no two asphalts are chemically identical and chemical analysis cannot be used to define the exact chemical structure or chemical composition of asphalt, elemental analyses indicate that most asphalts contain 79–88% (w/w) carbon, 7–13% (w/w) hydrogen, trace to 3% (w/w) nitrogen, trace to 8% (w/w) sulfur, trace to 8% (w/w) oxygen, and trace to 2000 ppm (w/w) metals (nickel, vanadium) in some cases the residuum may also contain iron, copper, and calcium, which are often artifacts of the crude oil transportation process. With the exception of oxygen in the asphalt, the other elements are derived from the residuum (or residua) from which the asphalt was produced (Table 2.2) (Speight, 1992, 2014). While heteroatoms (i.e., nitrogen, oxygen, and sulfur) make up only a minor component of asphalt, the heteroatoms profoundly influence the differences in physical properties of asphalts from different crude sources (Speight, 1992).

In terms of the classification of asphalt, there is no formal and genially accepted procedure. Reference to a physical property or two is often used but without any accuracy or satisfaction. More recently, there has been the tendency to classify asphalt according to whether or not the asphalt has been oxidized. On the other hand, asphalt has been pseudo-classified according to the performance specifications for which the asphalt was manufactured (e.g., paving asphalt, and roofing asphalt, cutback asphalt, and asphalt paint). This adds complications to attempts to understand the chemistry of asphalt—most asphalt used in paving is not manufactured from oxidized asphalt, but most asphalt used in roofing is manufactured from oxidized asphalt (Speight, 1992). The situation is further complicated by the addition of additives and modifiers as well as differences in application temperatures (Terrel and Eps, 1988).

In terms of paving asphalt, the category can be generally divided into three sub-categories: (i) asphalt cement, (ii) cutback asphalt, and (iii) emulsified asphalt. The two latter sub-categories (cutback asphalt and emulsified asphalt) are also called liquid asphalt because they are liquid at ambient temperatures. As mentioned previously, most of the asphalt used in paving operations is not oxidized. The asphalt is heated to approximately 150–175°C (300–350°F) and mixed with mineral aggregate that has been heated to approximately 145–165°C (290–330°F). Once transported to the worksite, the hot-mix asphalt is applied to the roadway, usually at a temperature on the order of 110–160°C (230–320°F) (Speight, 1992).

Oxidized asphalts may or may not be used in roofing manufacturing plants to produce shingles, roll goods, built-up roofing felts, and underlayment felts; these asphalts are shipped hot and kept hot until used in the manufacturing process. In addition, some cutback and emulsified asphalts are also used in roofing operations (Speight, 1992). However, most oxidized asphalts are used to produce *mopping-grade* roofing asphalts, which are typically shipped as a solid and heated in a kettle at the worksite until they become a liquid.

Differences in the way in which asphalts are handled during paving and roofing operations probably influence the composition of asphalt fumes and vapors. When hot-mix paving asphalt arrives at the worksite, the asphalt has been cooling since leaving the plant and may not be used immediately upon arrival. Conversely, roofing asphalt is heated continuously and stirred occasionally at the worksite until the asphalt is needed for application.

A caution that needs to be addressed here and elsewhere (see Chapters 6 and 12) is the behavior of asphalt when heated. Upon heating, asphalt releases vapors which cool and condense—the vapors are chemically and potentially toxicologically distinct from the parent material because of the presence of the more volatile components of the asphalt. However, because all of the constituents of the vapor do not condense at the same time, potential exposure to fumes and vapors can occur. The physical nature of the so-called *fumes* and *vapors* has not been well characterized, but the fumes may contain asphalt particles (or droplets) that have collided and coalesced, thereby making it difficult to characterize the fume particle size. Some of the vapors may condense only to the liquid phase, thus forming a viscous liquid with some solids. In either case, the potential danger of exposure to toxic chemicals cannot be over-emphasized.

Thus, since the compositions of asphalts and asphalt fumes and vapors vary depending on (i) the temperature, (ii) the manufacturing process, (iii) the presence of additives and modifiers, and (iv) the work practices, it is questionable whether or not laboratory-generated asphalt fumes that mimic asphalt fumes in the environment are difficult to produce; the answer lies on the negative side since experimental conditions used to generate the fumes can affect the composition. Using a variety of analytical techniques, such as gas chromatography with flame ionization detection (GC/FID), gas chromatography with flame photometric detection, gas chromatography with atomic emission detection, and gas chromatography with mass spectrometry (GC/MS)—laboratory-generated asphalt fumes can be compared to fumes collected from, for example, the headspace in a storage tank at a hot-mix plant (paving asphalt) or from the headspace in roofing kettles. The general conclusion is that the temperature, rate of stirring, and the means of collecting the sample affect the chemical composition of the fumes. If such sample collection is to be employed, strict protocols must be followed for the analysis to have any meaning (Speight, 2001, 2015).

Thus, each step in the refining process, beginning with the residuum from atmospheric distillation, is designed to extract the maximum high-value distillates from the residue until only the high-boiling high-molecular-weight components remain to be thermally cracked in a coking unit, marketed as commercial asphalt, or as blending components of asphalts or fuel oils (Speight and Ozum, 2002; Parkash, 2003; Hsu and

Robinson, 2006; Gary et al., 2007; Speight, 2014). With heavy crude oils, the vacuum residuum can often be commercial asphalt. With lighter crude oils, these residues are feedstock for further processing.

These streams and variations of them are typically used to produce the three main types of commercial asphalts: (i) penetration-grade asphalt, (ii) hard asphalt, and (iii) oxidized asphalt.

Penetration-grade asphalt (asphalt cement, viscosity-grade asphalt) is produced from crude oil atmospheric distillation residues by further processing, such as vacuum distillation (straight run asphalts), thermal conversion, partial oxidation (air rectification/semi-blowing), or solvent precipitation. A combination of these processes can be used to meet application specifications for road surfacing or in roofing applications.

Hard asphalt is manufactured using processes similar to penetration grades but has a lower penetration value and higher softening point. The asphalt is hard and more brittle, and is used primarily in the manufacture of asphalt paint and enamel. *Oxidized asphalt* (air-blown asphalt) is produced by passing air through hot, soft asphalt feedstock under controlled conditions, producing a higher softening point material with reduced susceptibility to changes in temperature and greater resistance to imposed stress. Applications include roofing materials, waterproof papers, electrical components, pipe coating, undersealing of concrete pavements, hydraulic applications, membrane envelopes, and the manufacture of paint.

REFERENCES

Abraham, H., 1945. Asphalts and Allied Substances: Their Occurrence, Modes of Production, Uses in the Arts and Methods of Testing, fifth ed. Van Nostrand, New York, NY.

Allen, B.R., 2003. Pretreatment methods for the degradation of lignin Proceedings Superpave Performance Graded Asphalt Binder Specification and Testing, Superpave Series No 1(SP-1). The Asphalt Institute, Lexington, KY.

Armstrong, K., 1997. Jerusalem: One City, Three Faiths. Ballantine Books, New York, NY, p. 3.

Barth, E.J., 1962. Asphalt: Science and Technology. Gordon and Breach Publishers, Taylor & Francis Group, Abingdon, Oxfordshire.

Bauer, G. (Georgius Agricola), 1546. De Natura Fossilium. Book IV. H. Frobenius, Basel, Switzerland. Published in English by Dover Publications, Mineola, NY.

Bauer, G. (Georgius Agrícola), 1556. De Re Metallica. Book XII. H. Frobenius, Basel, Switzerland. Published in English by Dover Publications, Mineola, NY.

Bell, H.S., 1945. American Petroleum Refining. Van Nostrand Publishers, New York, NY.

Belokon, N.Yu., Kompaneets, V.G., Stepanova, T.M., Shabalina, L.N., 2001. Visbreaking residues as components of feedstock for paving asphalts. Chem. Technol. Fuels Oils 37 (6), 388–392.

Blackburn, G.R., Kriech, A.J., Kurek, J.T., Osborn, L.A., 1999. Detection of coal tar materials in asphalt pavements using chemical and biological methods Transportation

Research Record, No. 1266. Transportation Research Board, National Research Council, Washington, DC.
Cobb, C., Goldwhite, H., 1995. Creations of Fire: Chemistry's Lively History from Alchemy to the Atomic Age. Plenum Press, New York, NY.
Corbett, L.W., 1975. Reaction variables in the air blowing of asphalt. Ind. Eng. Chem. Process Design Dev. 14 (2), 181–187.
Corbett, L.W., 1976. Asphalt Paving Compositions from Positive Spot Asphalt Fractions. US Patent 3,939,004, February 17.
Dahmus, J., 1995. A History of the middle Ages. Barnes and Noble, New York, NY, Pages 135 and 136; previously published with the same title by Doubleday Book Co., 1968.
Davies, N., 1996. Europe: A History. Oxford University Press, Oxford, Pages 245 and 250.
Dizhbite, T.G., Galina, T., Vilhelmina, J., Uldis, V., 2004. Characterization of the radical scavenging activity of lignins-natural antioxidants. Bioresource Technol. 95, 309–317.
Fini, E., Yang, S., Xiu, S., 2010. Characterization and application of manure-based biobinder in the asphalt industry Proceedings 89th Transportation Research Board Annual Meeting. Transportation Research Board, Washington, DC.
Fini, E.H., Kalberer, E.W., Shahbazi, A., 2011a. Biobinder from swine manure: sustainable alternative for asphalt binder. In: Transportation Research Board 90th Annual Meeting, Washington, DC.
Fini, E., Yang, S., Xiu, S.X., 2011b. Chemical characterization of biobinder from swine manure: sustainable modifier for asphalt binder. J. Mater. Civ. Eng. 10, 1506–1513.
Forbes, R.J., 1958a. A History of Technology. Oxford University Press, Oxford.
Forbes, R.J., 1958b. Studies in Early Petroleum Chemistry.. E.J. Brill, Leiden, Netherlands.
Forbes, R.J., 1959. More Studies in Early Petroleum Chemistry. E.J. Brill, Leiden, Netherlands.
Forbes, R.J., 1964. Studies in Ancient Technology. E.J. Brill, Leiden, Netherlands.
Gary, J.G., Handwerk, G.E., Kaiser, M.J., 2007. Petroleum Refining: Technology and Economics, fifth ed. CRC Press, Taylor & Francis Group, Boca Raton, FL.
Henry, J.T., 1873. The Early and Later History of Petroleum. Volumes I and II. APRP Co., Philadelphia, PA.
Herodotus, 1909. The Histories. Translated by G. Rawlinson. The Tandy-Thomas Company, New York.
Hill, D.R., Jennings, A.A., 2011. Bioasphalt from Urban Yard Waste Carbonization: A Student Study. Ohio Department of Transportation, Office of Research and Development, Columbus, OH, Report No. FHWA/)H-2011/13.
Hoiberg, A.J., 1964. Bituminous Materials: Asphalts, Tars, and Pitches. John Wiley and Sons, New York, NY.
Hsu, C.S., Robinson, P.R. (Eds.), 2006. Practical Advances in Petroleum Processing Volume 1 and Volume 2 Springer Science, New York, NY.
James, P., Thorpe, N., 1994. Ancient Inventions. Ballantine Books, New York, NY.
Jamieson, I., 1979. A study of lurgi road tar and its use in double seals. In: Proceedings 3rd Conference on Asphalt Pavement for Southern Africa, Durban, South Africa.
Johnson, P., 1997. A History of the American People. Harper Collins Publishers Inc., New York, NY, p. 602.
Karl, W.J., 1986. Effects of construction practices on asphalt properties in the mix. In: Proceedings Canadian Technical Asphalt Association, XXVII, pp. 321–334.
Kovacs, M.G., 1990. The Epic of Gilgamesh.. Stanford University Press, Stanford, Tablet XI.
Mallowan, M.E.L., 1954. The excavations at Nimrud (Kahlu), 1953. Iraq, 16: 59-114, and 16: 115-163.
Mallowan, M.E.L., Rose, J.C., 1935. Excavations at tell arpachiyah, 1933. Iraq 2 (1), 1–178.
Marschner, R.F., Duffy, L.J., Wright, H., 1978. Asphalts from ancient towns in Southwestern Iran. Paléorient 4, 97–112.

McCready, N.S., Williams, R.C., 2007. The utilization of agriculturally derived lignin as an antioxidant Asphalt Binder. Proceedings 2007 Mid-Continent Transportation Research Symposium. Iowa State University, Ames, IA.
Metwally, M.A.R.M., Williams, C.R., 2010. Development of nonpetroleum binders for use in flexible pavements IHRB Project TR-594, InTrans Project 08-133. Institute for Transportation. Iowa State University, Ames, IA.
Mills-Beale, J., You, Z., Fini, E., Zada, B., Lee, C.H., Yap, Y.K., 2014. Aging Influence on rheology properties of petroleum-based asphalt modified with biobinder. J. Mater. Civ. Eng. 26, 358–366.
Moschopedis, S.E., Speight, J.G., 1973. The oxidation of petroleum fractions. Fuel 52, 83.
Moschopedis, S.E., Speight, J.G., 1975. Oxidation of bitumen. Fuel 54, 210.
Moschopedis, S.E., Speight, J.G., 1977. The effect of air blowing on the properties and constitution of natural bitumen. J. Mater. Sci. 12, 990.
Moschopedis, S.E., Speight, J.G., 1978. The influence of metal salts on bitumen oxidation. Fuel 57, 235.
Mundt, D.J., Marano, K.M., Nunes, A.P., Adams, R.C., 2009. A review of changes in composition of hot mix asphalt in the United States. J. Occup. Environ. Hyg. 6 (11), 714–725.
Ouyang, C., Wang, S., Zhang, Y., Zhang, Y., 2006. Improving the aging resistance of asphalt by addition of zinc dialkyldithiophosphate. Fuel 85, 1060–1066.
Owen, E.W., 1975. In a history of exploration for petroleum Memoir No. 6. American Association of Petroleum Geologists, Tulsa, OK, p. 2.
Parkash, S., 2003. Refining Processes Handbook. Gulf Professional Publishing, Elsevier, Amsterdam, Netherlands.
Poirier, M.A., 1986. Road Asphalt Compositions Containing Visbreaking Residues. US Patent 4,631,088, December 23.
Powicke, M., 1962. The Thirteenth Century. 1216–1307, second ed. Oxford University Press, Oxford, p. 13.
Rawlinson, G. (Ed.), 1885. The History of Herodotus. Book I Appleton and Company, New York, NY. pp. 178–200.
SABITA, 2005. Memorandum on the Discontinuation of the Use of Coal Tar in Road Construction. Southern African Bitumen Association, Pinelands, South Africa.
Speight, J.G., 1978. The Use of Bitumen as an Adhesive and Building Material at the Ancient Cities of Babylon, Calah, Nineveh, and Ur. Report of Investigations. College of Science, University of Mosul, Iraq.
Speight, J.G., 1992. A Chemical and Physical Explanation of Incompatibility during Refining Operations. Proceedings. 4th International Conference on the Stability and Handling of Liquid Fuels. U.S. Department of Energy (DOE/CONF-911102). p. 169.
Speight, J.G., 2001. Handbook of Petroleum Analysis. John Wiley & Sons, Hoboken, NJ.
Speight, J.G., 2008. Synthetic Fuels Handbook: Properties, Processes, and Performance. McGraw-Hill, New York, NY.
Speight, J.G. (Ed.), 2011a. The Biofuels Handbook Royal Society of Chemistry, London.
Speight, J.G., 2011b. The Refinery of the Future. Gulf Professional Publishing, Elsevier, Oxford.
Speight, J.G., 2013. The Chemistry and Technology of Coal, third ed. CRC Press, Taylor & Francis Group, Boca Raton, FL.
Speight, J.G., 2014. The Chemistry and Technology of Petroleum, fifth ed. CRC Press, Taylor & Francis Group, Boca Raton, FL.
Speight, J.G., 2015. Handbook of Petroleum Product Analysis, second ed. John Wiley & Sons, Hoboken, NJ.
Speight, J.G., Ozum, B., 2002. Petroleum Refining Processes. Marcel Dekker Inc., New York, NY.

Speight, J.G., Exall, D.I., 2014. Refining used Lubricating Oils. CRC Press, Taylor and Francis Group, Boca Raton, FL.
Sundstrom, D.W.K., Klei, E., Daubenspeck, T.H., 1983. Use of byproduct lignins as extenders in asphalt. Ind. Eng. Chem. Prod. Res. Dev. 22, 496–500.
Terrel, R.L., Epps, J.A., 1988. Using additives and modifiers in hot mix asphalt Quality Improvement Series No. 114. National Asphalt Pavement Association, Lanham, MD.
Terrel, R.L., Rimsritong, S., 1979. Proc. Assoc. Asph. Paving Technol. 48, 111.
Williams, C.R.S., Rover, M., Brown, R.C., Teng, S., 2008. Utilization of fractionated bio oil in asphalt. J. Trans. Res. Board 9 (3187), 1–19.
Yan, T.-Y., 1986. Manufacture of road-paving asphalt using coal tar. Ind. Eng. Chem. Prod. Res. Dev. 25 (4), 637–640.
Yergin, D., 1991. The Prize: The Epic Quest for Oil, Money, and Power. Simon & Schuster, New York, NY.

CHAPTER 3

Residua and Asphalt

3.1 INTRODUCTION

Previously, the production of asphalt from petroleum has been described. Currently, the major source of asphalt is from the processing of crude oil (see Chapter 2) and the increased demand for liquid fuels is a factor that affects asphalt production. The increased conversion of residua (the precursors to asphalt) is causing competition for the use and ultimate fate of residua: to produce liquid fuels or to produce asphalt? It is, therefore pertinent to enter into a description of the differences between residua (the precursors to asphalt) and asphalt itself and to understand the means of competition between the production of liquid fuels and the production of asphalt. In addition, understanding the differences or similarities between residua and asphalt may assist in the design of alternate production methods of asphalt from other sources (see Chapters 1 and 2).

Approximately 80% of the asphalt that is removed during road construction is recycled to be used as (i) part of a new road, (ii) roadbeds, (iii) shoulders, or (iv) embankments. As the costs for liquid asphalt continue to rise, it is likely that the amount of asphalt reused will increase. In addition to the recycling of the used asphalt, asphalt shingles can also be recycled, through grinding, and mixed with the aggregate. This use of recycled material is expected to increase as agencies are required to find ways to repair and maintain the current infrastructure in spite of increased costs. In addition, the growing concern about the environmental impact of using fossil fuels and solid-fuel-derived products such as asphalt (see Chapter 12), has meant that environmentally friendly, non-toxic alternatives to asphalt are being introduced (see Chapters 1 and 2). Asphalt can now be manufactured from alternative sources, such as used lubricating oil and non-petroleum-based sources (bioasphalt), such as sugar, molasses, rice, corn, potatoes, and vegetable oil (Hesp and Shurvell, 2010; Hill and Jennings, 2011; TRB, 2012). In fact, renewable energy sources are developed worldwide because of the variability of petroleum prices and to limit greenhouse gas emissions. In this context, some groups have focused their work on non-petroleum sources of asphalt. Furthermore, the comparison

Asphalt Materials Science and Technology.
DOI: http://dx.doi.org/10.1016/B978-0-12-800273-5.00003-9

of rheological and chemical analyses shows that the various products exhibit behavior (properties) comparable to petroleum-based asphalt (TRB, 2012).

Asphalt is, with the exception of coke, carbon, and graphite, the highest boiling product of petroleum. Assuming that thermal decompositon—cracking—has not occurred during the distillation, asphalt derives its characteristics from the nature of the residuum from which it was produced, which in turn derives its characteristics from the original crude oil. Although there are a number of refineries or refinery units whose prime function is to produce asphalt, petroleum asphalt is primarily a byproduct of integrated refinery operations (Figure 3.1). Crude oil may be selected for these refineries for a variety of other product requirements and the asphalt produced may vary somewhat in characteristics from one refinery-crude system to another.

Briefly, and by way of explanation, a *resid* (*residuum*, pl. *residua*) is the residue obtained from petroleum after non-destructive distillation has removed all the volatile materials (Speight and Ozum, 2002; Parkash, 2003; Hsu and Robinson, 2006; Gary et al., 2007; Speight, 2014).

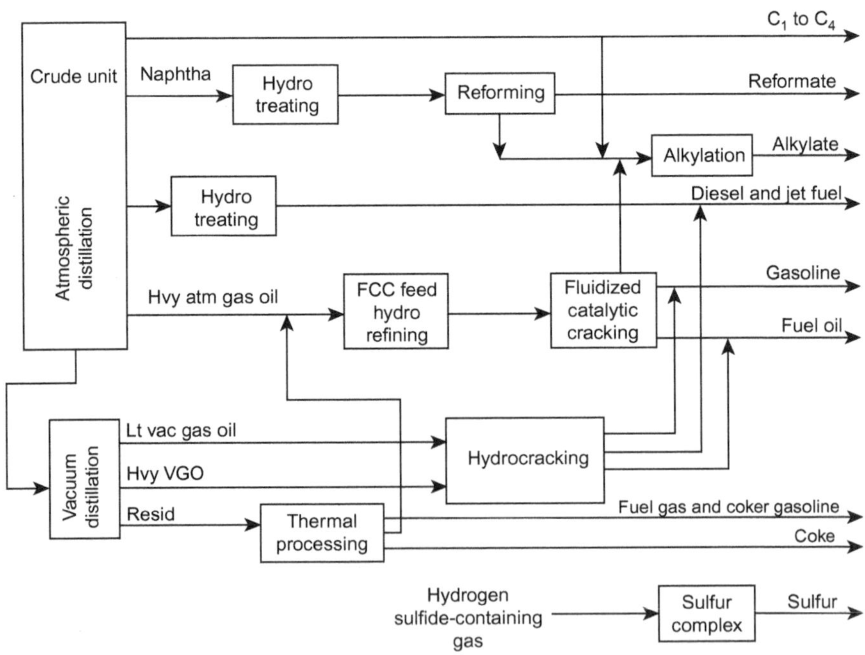

Figure 3.1 Schematic of a petroleum refinery.

The temperature of the distillation is usually maintained below 350°C (660°F) since the rate of thermal decomposition of petroleum constituents is minimal below this temperature but the rate of thermal decomposition of petroleum constituents is substantial above 350°C (660°F). A residuum that has held above this temperature so that thermal decompositon has occurred is known as a *cracked resid*. *Resids* are black, viscous materials and are obtained by distillation of a crude oil under atmospheric pressure (atmospheric residuum) or under reduced pressure (vacuum residuum). They may be liquid at room temperature (generally atmospheric residua) or almost solid (generally vacuum residua) depending upon the nature of the crude oil (Speight and Ozum, 2002; Parkash, 2003; Hsu and Robinson, 2006; Gary et al., 2007; Speight, 2014). When a residuum is obtained from a crude oil and thermal decomposition has commenced, it is more usual to refer to this product as *pitch* (Speight, 2014). The differences between the *parent* petroleum and the residua are due to the relative amounts of various constituents present, which are removed or remain by virtue of their relative volatility.

The chemical composition of asphalt produced from the residuum is extremely complex. Physical methods of fractionation usually indicate high proportions of asphaltene constituents and resin constituents, even in amounts up to 50% (or higher) of the asphalt (Speight, 1992, 2014, 2015). In addition, the presence of ash-forming metallic constituents, including such organometallic compounds as those of vanadium and nickel, is also a distinguishing feature of asphalt produced from residua. Furthermore, the deeper the *cut* into the crude oil, the greater is the concentration of heteroatoms and metals in the residuum (Table 3.1) and the greater the heteroatoms and metals in the asphalt produced therefrom (Reynolds, 1998; Speight and Ozum, 2002; Parkash, 2003; Hsu and Robinson, 2006; Gary et al., 2007; Speight, 2014).

Asphalt is a black or brown material that has a consistency varying from a viscous liquid to a glassy solid. To a point, asphalt can resemble tar sand bitumen, hence the tendency to refer to bitumen (incorrectly) as *native asphalt*. It is recommended that there be differentiation between asphalt (manufactured) and bitumen (naturally occurring) other than by use of the qualifying terms *petroleum* and *native* since the origins of the materials may be reflected in the resulting physicochemical properties of the two types of materials. It is also necessary to distinguish between the asphalt which originates from petroleum by refining and the product in which the source of the asphalt is a material other than petroleum,

Table 3.1 Heteroatom Content of Different Fractions of Kern River Crude Oil; Distillation by ASTM D-86 and D-1160

Fraction	Crude	IBP to 400°F	400–500°F	500–650°F	650–800°F	800–1000°F	1000°F+
%	100	1.7	6.0	15.3	15.6	23.9	37.5
S (wt %)	1.1	0.07	0.17	0.63	1.0	1.1	1.3
N (wt %)	0.7	0.0	0.0	0.06	0.38	0.56	1.4
Fe (ppm)	28	0	0	0	0	0	90
Ni (ppm)	63	0	0	0	0	0	110
V (ppm)	35	0	0	0	0	0	95

e.g., *Wurtzilite asphalt* (Speight, 2014). In the absence of a qualifying word, it is assumed that the term *asphalt* refers to the product manufactured from petroleum.

When the asphalt is produced simply by distillation and without further processing of the residuum, the product can be referred to as *residual asphalt* or *straight run asphalt*. However, if the asphalt is produced from the residuum by solvent extraction or by precipitation using a low-boiling hydrocarbon such as propane or if the residuum is oxidized, the nomenclature should be modified accordingly to qualify the product (e.g., *propane asphalt, blown asphalt*) (Speight and Ozum, 2002; Parkash, 2003; Hsu and Robinson, 2006; Gary et al., 2007; Speight, 2014).

Asphalt softens when heated and is elastic under certain conditions. The mechanical properties of asphalt are of particular significance when it is used as a binder or adhesive. The principal application of asphalt is in road surfacing, which may be done in a variety of ways. Light oil *dust layer* treatments may be built up by repetition to form a hard surface, or a granular aggregate may be added to an asphalt coat, or earth materials from the road surface itself may be mixed with the asphalt.

Other important applications of asphalt include canal and reservoir linings, dam facings, and sea works. The asphalt so used may be a thin, sprayed membrane, covered with earth for protection against weathering and mechanical damage, or thicker surfaces, often including riprap (crushed rock). Asphalt is also used for roofs, coatings, floor tiles, soundproofing, waterproofing, and other building-construction elements and in a number of industrial products, such as batteries. For certain applications an asphaltic emulsion is prepared, in which fine globules of asphalt are suspended in water.

3.2 PROPERTIES

Contrary to the use of confusing terminology, residua and asphalts are different materials. Residua are the non-volatile materials from the distillation proceeds while asphalts are manufactured products—they are generally residua that have had the properties midwifed to meet the necessary specification for service. Thus, the term *natural asphalt* which is often used to describe (tar sand) bitumen—a natural occurring non-volatile material (see Chapter 1)—is incorrect.

3.2.1 Residua

A residuum (pl. *residua*) is the non-volatile residue from either atmospheric distillation or vacuum distillation. Either the atmospheric residuum or the vacuum residuum, depending upon the properties, can be the starting material for asphalt production and, therefore, the properties of the asphalt are dependent on the properties of the residuum from which the asphalt is manufactured. And residua properties can vary, depending upon the cut point of the residuum (Table 3.2).

Not only do residua constituents have high molecular weight (>500) but they also contain a polynuclear aromatic system (PNA)—the systems with three to four aromatic rings (or more) provide the greatest hindrance to the conversion of residua because of high thermal stability. In addition, the high concentrations of heteroatoms (nitrogen, oxygen, sulfur, and metals such as vanadium and nickel) in residua will (along with PNA) poison catalysts. However, no matter which type of thermal or catalytic process is used for resid processing, a substantial portion of the residuum feedstock (at least 30–50%, w/w, or more, depending on the process) can be converted to liquid products that are suitable for production of liquid fuels. Gases are also produced during resid processing along with non-volatile coke.

The residua from which asphalt is produced were once considered the unusable remnants of petroleum refining and had little value and little use, other than as a (passable in some cases but not in every case) road oil. In fact, the delayed coking process was developed with the purpose of converting residua to liquids (valuable products) and coke (fuel). In addition, recognition that road asphalt—a once maligned product of comparatively little value—is now a valuable refinery product, has led to a renewed importance for residua. However, detailed specifications (based on selected analytical test methods) are necessary for asphalt to be used for highway paving as well as for other uses.

Table 3.2 Properties of Different Petroleum Residua

	API Gravity	Sulfur (% w/w)	Nitrogen (% w/w)	Nickel (% w/w)	Vanadium (% w/w)	Asphaltenes (C_7, % w/w)	Carbon Residue (% w/w)
Atmospheric Residua							
Alaska, North Slope crude oil, >650°F	15.2	1.6	0.4	18.0	30.0	2.0	8.5
Arabian heavy crude oil, >650°F	11.9	4.4	0.3	27.0	103.0	8.0	14
Arabian light crude oil, >650°F	17.7	3	0.2	10.0	26.0	1.8	7.5
Kuwait, crude oil, >650°F	13.9	4.4	0.3	14.0	50.0	2.4	12.2
Lloydminster heavy oil (Canada), >650°F	10.3	4.1	0.3	65.0	141.0	14.0	12.1
Taching crude oil, >650°F	27.3	0.2	0.2	5.0	1.0	4.4	3.8
Vacuum Residua							
Alaska, North Slope crude oil, >1050°F	8.2	2.2	0.6	47.0	82.0	4.0	18
Arabian heavy crude oil, >1050°F	7.3	5.1	0.3	40.0	174.0	10.0	19
Arabian light crude oil, >1050°F	8.5	4.4	0.5	24.0	66.0	4.3	14.2
Kuwait crude oil, >1050°F	5.5	5.5	0.4	32.0	102.0	7.1	23.1
Lloydminster heavy oil (Canada) >1050°F	8.5	4.4	0.6	115.0	252.0	18.0	21.4
Taching crude oil, >1050°F	21.5	0.3	0.4	9.0	2.0	7.6	7.9

However, the production of atmospheric and vacuum residua by distillation is not the final step. In addition to asphalt production, there are many options for upgrading residua to lower-boiling products (Speight, 2011, 2014). This is typically referred to as *bottom of the barrel processing*. The price differentials between the various crude oil feedstocks and their respective products creates a strong incentive for refiners to lower their feed cost by using residua–gas oil blends as feedstocks to thermal cracking processes, catalytic cracking processes, and hydrocracking–hydrotreating processes (Speight, 2011, 2014). This creates a need for additional *bottom of the barrel processing*, both for expansion and for yield improvement. Traditionally, this would automatically call for the addition of crude and/or vacuum units as a starting point.

3.2.1.1 Thermal Cracking

Distillation has remained a major refinery process and a process to which just about every crude oil that enters the refinery is subjected (Speight and Ozum 2002; Parkash, 2003; Hsu and Robinson, 2006; Gary et al., 2007; Speight, 2014). However, not all crude oils yield the same distillation products. In fact, the nature of the crude oil dictates the processes that may be required for refining, and balancing product yield with product demand is a necessary part of refinery operations.

After 1910, the demand for automotive fuel began to outstrip the market requirements for kerosene, and refiners were pressed to develop new technologies to increase gasoline yields. The earliest process, called thermal cracking, consisted of heating residua oils (for which there was a low market requirement) in pressurized reactors and thereby cracking, or splitting, their large molecules into the smaller ones that form the lighter, more valuable fractions such as gasoline, kerosene, and light industrial fuels.

During the 1930s and World War II, improved refining processes involving the use of catalysts led to further improvements in the quality of transportation fuels and further increased their supply. These improved processes, including catalytic cracking of residua, enabled the petroleum industry to meet the demands for increasing quantities of transportation fuels for public and military use. The 1950s and 1960s brought large-scale demand for jet fuel and high-quality lubricating oils. The continuing increase in demand for petroleum products also heightened the need to process a wider variety of crude oils into high-quality products. Catalytic reforming of naphtha replaced the earlier thermal reforming process and became the leading process for upgrading fuel qualities to meet the needs of higher compression engines (Speight and Ozum, 2002; Parkash, 2003; Hsu and Robinson, 2006; Gary et al., 2007; Speight, 2014). Hydrocracking, a catalytic cracking process conducted in the presence of hydrogen, was developed to be a versatile manufacturing process for increasing the yields of either gasoline or jet fuels (Speight and Ozum, 2002; Parkash, 2003; Hsu and Robinson, 2006; Gary et al., 2007; Speight, 2014).

Balancing product yield and market demand, without the manufacture of large quantities of fractions having low commercial value, has long required processes for the conversion of hydrocarbons of one molecular weight range and/or structure into some other molecular weight range and/or structure. Basic processes for this are still the so-called cracking processes in which relatively high-boiling constituents are cracked (thermally decomposed) into lower-molecular-weight, lower-boiling molecules, although reforming, alkylation, polymerization, and hydrogen-refining processes have wide applications in producing premium-quality products (Speight and Ozum, 2002; Parkash, 2003; Hsu and Robinson, 2006; Gary et al., 2007).

It is generally recognized that the most important part of any refinery is its gasoline (and liquid fuels) manufacturing facilities; other facilities are added to manufacture additional products as indicated by technical feasibility and economic gain. More equipment is used in the manufacture of gasoline, the equipment is more elaborate, and the processes more complex than for any other product. Among the processes that have been used for liquid fuel production are thermal cracking, catalytic cracking, thermal reforming, catalytic reforming, polymerization, alkylation, coking, and distillation of fractions directly from petroleum. Each of these processes may be carried out in a number of ways, which differ in details of operation, or essential equipment, or both.

When kerosene was the major product, gasoline was the portion of crude petroleum too volatile to be included in kerosene. The first refiners had no use for it and often dumped an accumulation of gasoline into the stream or river that was always nearby. As the demand for gasoline increased, more and more of the lighter kerosene components were included in gasoline, but the maximum suitable portion depended on the kind of crude oil and rarely exceeded 20% of the crude oil. The problem of how to get more gasoline from less crude oil was solved in 1913 by the use of cracking in which fractions higher boiling than gasoline were converted into gasoline.

Thermal processes are essentially processes that decompose, rearrange, or combine hydrocarbon molecules by the application of heat. The major variables involved are feedstock type, time, temperature, and pressure and, as such, are usually considered in promoting cracking (thermal decomposition) of the higher-molecular-weight constituents to lower-boiling products and in minimizing coke formation.

When residua are heated to temperatures in excess of 350°C (660°F), the rates of thermal decomposition proceed at significantly higher rates (Speight and Ozum, 2002; Parkash, 2003; Hsu and Robinson, 2006; Gary et al., 2007; Speight, 2014). Thermal decomposition does not require the addition of a catalyst. Therefore, this approach is the oldest technology available for residue conversion. The severity of thermal processing determines the conversion and the product characteristics. Thermal treatment of residues ranges from mild treatment for reduction of viscosity to *ultrapyrolysis* (high-temperature cracking at very short residence time) for better conversion to overhead products. A higher temperature requires a shorter time to achieve a given conversion but, in many cases, there is *a change in the chemistry of the reaction.* The severity of the process conditions is the combination of reaction time and temperature to achieve a given conversion.

Sufficiently high temperatures convert oils entirely to gases and coke; cracking conditions are controlled to produce as much as possible of the desired product, which is usually gasoline but may be cracked gases for petrochemicals or lower-viscosity oil for use as a fuel oil. The feedstock, or cracking stock, may be almost any fraction obtained from crude petroleum, but the greatest amount of cracking is carried out on gas oils, a term that refers to the portion of crude petroleum boiling between the fuel oils (kerosene and/or stove oil) and the residuum. Residua are also cracked, but the processes are somewhat different from those used for gas oils.

Thus, thermal conversion processes are designed to increase the yield of lower-boiling products obtainable from petroleum either directly (by means of the production of gasoline components from higher-boiling feedstocks, such as residua) or indirectly (by production of olefins and the like, which are precursors of the gasoline components). These processes may also be characterized by the physical state (liquid and/or vapor phase) in which the decomposition occurs. The state depends on the nature of the feedstock as well as conditions of pressure and temperature (Speight and Ozum, 2002; Parkash, 2003; Hsu and Robinson, 2006; Gary et al., 2007; Speight, 2014).

From the chemical viewpoint, the products of cracking are very different from those obtained directly from crude petroleum. When a 12-carbon atom hydrocarbon typical of straight-run gas oil is cracked, there are several potential reactions that can occur that lead to a variety of products, for example:

$$CH_3(CH_2)_{10}CH_3 \rightarrow CH_3(CH_2)_8CH_3 + CH_2{=}CH_2$$
$$CH_3(CH_2)_{10}CH_3 \rightarrow CH_3(CH_2)_7CH_3 + CH_2{=}CHCH_3$$
$$CH_3(CH_2)_{10}CH_3 \rightarrow CH_3(CH_2)_6CH_3 + CH_2{=}CHCH_2CH_3$$
$$CH_3(CH_2)_{10}CH_3 \rightarrow CH_3(CH_2)_5CH_3 + CH_2{=}CH(CH_2)_2CH_3$$
$$CH_3(CH_2)_{10}CH_3 \rightarrow CH_3(CH_2)_4CH_3 + CH_2{=}CH(CH_2)_3CH_3$$
$$CH_3(CH_2)_{10}CH_3 \rightarrow CH_3(CH_2)_3CH_3 + CH_2{=}CH(CH_2)_4CH_3$$
$$CH_3(CH_2)_{10}CH_3 \rightarrow CH_3(CH_2)_2CH_3 + CH_2{=}CH(CH_2)_5CH_3$$
$$CH_3(CH_2)_{10}CH_3 \rightarrow CH_3CH_2CH_3 + CH_2{=}CH(CH_2)_6CH_3$$
$$CH_3(CH_2)_{10}CH_3 \rightarrow CH_3CH_3 + CH_2{=}CH(CH_2)_7CH_3$$
$$CH_3(CH_2)_{10}CH_3 \rightarrow CH_4 + CH_2{=}CH(CH_2)_8CH_3$$

The products are dependent on temperature and residence time and the simple reactions shown above do not take into account the potential for isomerization of the products or secondary and even tertiary reactions that can (and do) occur.

The hydrocarbons with the least thermal stability are the paraffins, and the olefins produced by the cracking of paraffins are also reactive. Cycloparaffins (naphthenes) are less easily cracked, their stability depending mainly on any side chains present, but ring splitting may occur, and dehydrogenation can lead to the formation of unsaturated naphthenes and aromatics. Aromatics are the most stable (*refractory*) hydrocarbons, the stability depending on the length and stability of side chains. Very severe thermal cracking of high-boiling feedstocks can result in condensation reactions of ring compounds, yielding a high proportion of coke.

The gas oil produced by cracking is, in fact, a further important source of gasoline. In a once-through cracking operation, all the cracked material is separated into products and may be used as such. However, cracked gas oil is more resistant to cracking (more refractory) than straight-run gas oil but can still be cracked to produce gasoline. This is done in a recycling operation in which the cracked gas oil is combined with fresh feed for another trip through the cracking unit. The operation may be repeated until the cracked gas oil is almost completely decomposed (*cracking to extinction*) by recycling (*recycling to extinction*) the higher-boiling product, but it is more usual to withdraw part of the cracked gas oil from the system according to the need for fuel oils. The extent to which recycling is carried out affects the amount or yield of cracked gasoline resulting from the process.

The gases formed by cracking are particularly important because of their chemical properties and their quantity. Only relatively small amounts of paraffinic gases are obtained from crude oil, and these are chemically inactive. Cracking produces both paraffinic gases (e.g., propane, C_3H_8) and olefin gases (e.g., propene, C_3H_6); the latter are used in the refinery as the feed for polymerization plants where high-octane polymer gasoline is made. In some refineries, the gases are used to make *alkylate*, a high-octane component for aviation gasoline and for motor gasoline. In particular, the cracked gases are the starting points for many petrochemicals (Speight and Ozum, 2002; Parkash, 2003; Hsu and Robinson, 2006; Gary et al., 2007; Speight, 2014).

3.2.1.2 Catalytic Cracking

Catalytic cracking is a conversion process that can be applied to a variety of feedstocks ranging from gas oil to residua and in a variety of configurations (Speight and Ozum, 2002; Speight, 2014). Fluid catalytic cracking units are currently in place in approximately 400 refineries around the world and the units are considered to be one of the most important achievements of the twentieth century.

Catalytic cracking is basically the same as thermal cracking, but it differs by the use of a catalyst that is not (in theory) consumed in the process and it is one of several practical applications used in a refinery that employ a catalyst to improve process efficiency. The original incentive to develop cracking processes arose from the need to increase gasoline supplies and to increase the octane number of gasoline, while maintaining the yields from high-boiling stocks using catalysts (Speight and Ozum, 2002; Parkash, 2003; Hsu and Robinson, 2006; Gary et al., 2007; Speight, 2014). Since cracking could virtually double the volume of gasoline from a barrel of crude oil, the purpose of cracking was wholly justified.

This history of catalytic cracking started in the early part of the twentieth century. In the 1930s, thermal cracking units produced about half the total gasoline manufactured, the octane number of which was about 70 compared to 60 for straight-run gasoline. These were usually blended together with light ends and sometimes with *polymer gasoline* and *reformate* to produce gasoline base stock with an octane number of about 65. The addition of *tetraethyl lead* (*ethyl fluid*) increased the octane number to about 70 for the *regular grade* gasoline and 80 for *premium-grade* gasoline. The thermal reforming and polymerization processes that were developed during the 1930s could be expected to further increase the octane number of gasoline to some extent, but something new was needed to break the octane barrier that threatened to stop the development of more powerful automobile engines. In 1936, a new cracking process opened the way to higher-octane gasoline; this process was *catalytic cracking*. Since that time, the use of catalytic materials in the petroleum industry has spread, and will continue to spread, to other processes (Speight and Ozum, 2002; Parkash, 2003; Hsu and Robinson, 2006; Gary et al., 2007; Speight, 2011, 2014).

The last 70 years have seen substantial advances in the development of catalytic processes. This has involved not only rapid advances in the chemistry and physics of the catalysts themselves but also major engineering advances in reactor design, for example, the evolution of the design of the catalyst beds from *fixed beds* to *moving beds* to *fluidized beds* as well as the different designs of the fluidized bed reactors currently in use. Catalyst chemistry and physics and bed design have allowed major improvements in process efficiency and product yields.

An important purpose evolved from the ability of cracked gasoline to resist detonation that is the cause of knocking in automobile engines. In the early days of automobile engine development, *straight-run gasoline* had a lower end point (165°C, 330°F) than that of *cracked gasoline* (205°C, 400°F) and, furthermore, did not color when exposed to sunlight. Similar exposure caused the cracked gasoline to turn brown, and hence straight-run gasoline was regarded as the premium material.

Gasoline is now produced, for the most part, by *catalytic cracking* processes or by *reforming* processes. In catalytic cracking, the petroleum (or petroleum-derived feedstock) is fed into a reaction vessel containing a catalyst. In reforming processes, naphtha (refined or unrefined) that may have been produced by catalytic cracking of higher-molecular-weight feedstocks, is heated with hydrogen in the presence of a catalyst (Speight and Ozum, 2002; Parkash, 2003; Hsu and Robinson, 2006; Gary et al., 2007; Speight, 2014). Reforming causes a rearrangement of the structures of the molecular constituents and creates a gasoline product.

Catalytic cracking has a number of advantages over thermal cracking (Speight and Ozum, 2002; Parkash, 2003; Hsu and Robinson, 2006; Gary et al., 2007; Speight, 2014). Gasoline produced by catalytic cracking has a higher octane number and consists largely of *iso*-paraffins and aromatics. The *iso*-paraffins and aromatic hydrocarbons have high-octane numbers and greater chemical stability than mono-olefins (R–CH=CH2 or R–CH=CH–R′) and diolefins (R–CH=CH–CH=CH_2 or R–CH=CH(CH_2)$_n$ CH=CH=R′). The olefins and diolefins are present in much greater quantities in gasoline produced by thermal cracking processes. Furthermore, substantial quantities of olefin gases and smaller quantities of methane (CH_4), ethane (CH_3CH_3), and ethylene (CH_2=CH_2) are produced by catalytic cracking. The olefin gases are suitable for polymer gasoline manufacture (Speight and Ozum, 2002; Parkash, 2003; Hsu and Robinson, 2006; Gary et al., 2007; Speight, 2014).

Sulfur compounds are changed in such a way that the sulfur content of gasoline produced by catalytic cracking gasoline is lower than the sulfur content of gasoline produced by thermal cracking. Catalytic cracking produces less residuum and more of the useful gas oil constituents than thermal cracking. Finally, the process has considerable flexibility, permitting the manufacture of both motor gasoline and aviation gasoline and a variation in the gas oil production to meet changes in the fuel oil market (Speight, 2014).

The feedstocks for catalytic cracking can be any one (or blends) of the following: (i) straight-run gas oil, (ii) vacuum gas oil, (iii) atmospheric residuum, and (iv) vacuum residuum. If blends of the above feedstocks are employed, compatibility of the constituents of the blends must be assured under reactor conditions or excessive coke will be laid down on to the catalyst. In addition, there are several pretreatment options for the feedstocks for catalytic cracking units and these are: (i) deasphalting to prevent excessive coking on catalyst surfaces, (ii) demetallation, i.e., removal of nickel, vanadium, and iron to prevent catalyst deactivation, (iii) use of a short residence time as a means of preparing the feedstock, and (iv) hydrotreating or mild hydrocracking to prevent excessive coking in the fluid catalytic cracking unit.

In addition, the goal is to reduce the sulfur content of the naphtha and there is a number of process options for reducing the sulfur level in fluid catalytic cracker naphtha, the two options that are generating the most interest are: (i) hydrotreat the feedstock to the catalytic cracking unit; and (ii) hydrotreat the naphtha from the catalytic cracker.

Hydrotreating the fluid catalytic cracker feed improves naphtha yield as well as quality and reduces the sulfur oxide (SO_x) emissions from the catalytic cracker unit, but it is typically a high-pressure process and, furthermore, manipulation of feedstock sulfur alone may not be sufficient to meet future gasoline performance standards (Speight and Ozum, 2002; Parkash, 2003; Hsu and Robinson, 2006; Gary et al., 2007; Speight, 2011, 2014). Refineries wishing to process residua may only have the option to desulfurize the resulting high-sulfur naphtha. Hydrodesulfurization of catalytic cracker naphtha is a low-pressure process. Obviously, the selection of an optimum hydrotreating process option for reducing sulfur in catalytic cracker naphtha is determined by economic factors specific to a refinery and to the feedstock. Hydrotreating catalytic cracker feedstock can be very profitable for a refiner despite its large capital investment. By taking advantage of feedstock hydrotreating, margins can be optimized by considering the feedstock hydrotreater unit, the fluid catalytic cracker, and any post-catalytic cracker hydrotreaters as one integrated upgrading step.

Catalytic cracking in the usual commercial process involves contacting a feedstock (usually a gas oil fraction) with a catalyst under suitable conditions of temperature, pressure, and residence time. By this means, a substantial part (>50%) of the feedstock is converted into gasoline and lower-boiling products, usually in a single-pass operation. However, during the cracking reaction, carbonaceous material is deposited on the catalyst, which markedly reduces its activity, and removal of the deposit is very necessary. The carbonaceous deposit arises from the thermal decomposition of high–molecular-weight polar species in the feedstock (Speight and Ozum, 2002; Parkash, 2003; Hsu and Robinson, 2006; Gary et al., 2007; Speight, 2014). Removal of the deposit from the catalyst is usually accomplished by burning in the presence of air until catalyst activity is reestablished.

3.2.1.3 Hydroprocesses

The trend of processing more residua, the shift in demand away from gasoline toward more distillate, and more stringent fuel qualities change the fuel and hydrogen balances in most refineries (Speight and Ozum, 2002; Ancheyta et al., 2005; Parkash, 2003; Hsu and Robinson, 2006; Gary et al., 2007; Speight, 2011, 2014). In addition, the hydrogen requirement for product improvement, which is the hydrotreatment of petroleum products, to insure that they meet utility and performance specifications is also increasing. Product improvement can not only involve hydrotreatment but also changes in molecular shape (*reforming* and *isomerization*) or change in molecular size (*alkylation* and *polymerization*) and hydrotreating can play a major role in product improvement (Speight and Ozum, 2002; Parkash, 2003; Hsu and Robinson, 2006; Gary et al., 2007; Speight, 2014).

Hydrogenation processes for the conversion of petroleum fractions, including residua, may be classified as destructive and non-destructive. Destructive hydrogenation (*hydrogenolysis* or *hydrocracking*) is characterized by the cleavage of carbon–carbon linkages accompanied by hydrogen saturation of the fragments to produce lower-boiling products. Such treatment requires severe processing conditions and the use of high hydrogen pressures to minimize polymerization and condensation that lead to coke formation. Many other reactions, such as isomerization, dehydrogenation, and cyclization, occur under the drastic conditions employed.

Furthermore, a growing dependence on residua has emerged as a result of continuing decreasing availability of conventional crude oil through the depletion of reserves in the various parts of the world. Thus, the ever-growing tendency to convert as much as possible of lower-grade feedstocks to liquid products is causing an increase in the total sulfur content in refined products. Refiners must, therefore, continue to remove substantial portions of sulfur from the lighter products, but residua pose a particularly difficult problem. Indeed, it is now clear that there are other problems involved in the processing of residua, which are emerging as the liquid fuel supply of the future and need special attention.

Catalytic hydrotreating is a hydrogenation process used to remove about 90% of contaminants such as nitrogen, sulfur, oxygen, and metals from liquid petroleum fractions. These contaminants, if not removed from the petroleum fractions as they travel through the refinery processing units, can have detrimental effects on the equipment, the catalysts, and the quality of the finished product. Typically, hydrotreating is done prior to processes such as catalytic reforming so that the catalyst is not contaminated by untreated feedstock. Hydrotreating is also used prior to catalytic cracking to reduce sulfur and improve product yields, and to upgrade middle-distillate petroleum fractions into finished kerosene, diesel fuel, and heating fuel oils. In addition, hydrotreating converts olefins and aromatics to saturated compounds.

Hydrotreating (*non-destructive hydrogenation*) is generally used for the purpose of improving product quality without appreciable alteration of the boiling range. Mild processing conditions are employed so that only the more unstable materials are attacked. Thus nitrogen, sulfur, and oxygen compounds undergo hydrogenolysis to split out ammonia, hydrogen sulfide, and water, respectively. Olefins are saturated, and unstable compounds, such as diolefins, which might lead to the formation of gums or insoluble materials, are converted to more stable compounds.

Hydrotreating catalysts are usually cobalt plus molybdenum or nickel plus molybdenum (in the sulfide) forms, impregnated on an alumina base. The hydrotreating operating conditions (1000–2000 psi hydrogen and about 370°C, 700°F) are such that appreciable hydrogenation of aromatics does not occur (Speight, 2000). The desulfurization reactions are invariably accompanied by small amounts of hydrogenation and hydrocracking, the extent of which depends on the nature of the feedstock and the severity of desulfurization.

One of the problems in the processing of high-sulfur and high-nitrogen feeds, such as residua, is the large quantity of hydrogen sulfide (H_2S) and ammonia (NH_3) that are produced. Substantial removal of both compounds from the recycle gas can be achieved by the injection of water in which, under the high-pressure conditions employed, both hydrogen sulfide and ammonia are very soluble compared with hydrogen and hydrocarbon gases. The solution is processed in a separate unit for the recovery of anhydrous ammonia and hydrogen sulfide.

Hydrotreating is carried out by charging the feed to the reactor, together with a portion of all the hydrogen produced in the catalytic reformer. Suitable catalysts are tungsten–nickel sulfide, cobalt–molybdenum–alumina, nickel oxide–silica–alumina, and platinum–alumina. Most processes employ cobalt–molybdenum catalysts, which generally contain about 10% by weight molybdenum oxide and <1% by weight cobalt oxide supported on alumina. The temperatures employed are in the range of 300–345°C (570–850°F), and the hydrogen pressures are about 500–1000 psi.

Furthermore, it has been found to be more economical to hydrotreat high-sulfur feedstocks such as residua before catalytic cracking than to hydrotreat the products from catalytic cracking. The advantages are: (i) the products require less finishing, (ii) sulfur is removed from the catalytic cracking feedstock, and corrosion is reduced in the cracking unit, (iii) coke formation during cracking is reduced and higher conversions result, and the catalytic cracking quality of the gas oil fraction is improved.

Although hydrocracking will occur during hydrotreating, attempts are made to minimize such effects, but the degree of cracking is dependent on the nature of the feedstock. For example, decalin (decahydronaphthalene) cracks more readily than the corresponding paraffin analog, *n*-decane, [$CH_3(CH_2)_8CH_3$], to give higher *iso*-paraffin to *n*-paraffin product ratios than those obtained from the paraffin. A large yield of single-ring naphthenes is also produced, and these are resistant to further hydrocracking and contain a higher than equilibrium ratio of methylcyclopentane to cyclohexane.

When applied to residua the hydrotreating processes can be used for processes such as: (i) fuel oil desulfurization and (ii) residuum hydrogenation, which is accompanied by hydrodesulfurization, hydrodenitrogenation, as well as partial conversion to produce products suitable as feedstocks for other processes, such as catalytic cracking. One of the chief problems with hydroprocessing residua is the deposition of metals, in particular vanadium, on the catalyst. It is not possible to remove vanadium from the catalyst, which must therefore be replaced when deactivated, and the time taken for catalyst replacement can significantly reduce the unit time efficiency. Fixed-bed catalysts tend to plug owing to solids in the feed or carbon deposits when processing residual feeds. As mentioned previously, the highly exothermic reaction at high conversion gives difficult reactor design problems in heat removal and temperature control.

The choice of processing schemes for a given hydrotreating application depends upon the nature of the feedstock as well as the product requirements. For residua, the process is usually hydrocracking and can be simply illustrated as a single-stage or as a two-stage operation (Speight and Ozum, 2002; Parkash, 2003; Hsu and Robinson, 2006; Gary et al., 2007; Speight, 2014). Variations to the process are feedstock-dependent. For example, the single-stage process can be used to produce gasoline but is more often used to produce middle distillate from residua. The two-stage process was developed primarily to produce high yields of gasoline from straight-run gas oil, and the first stage may actually be a purification step to remove sulfur-containing (as well as nitrogen-containing) organic materials. Both processes use an extinction-recycling technique to maximize the yields of the desired product. Significant conversion of residua can be accomplished by hydrocracking at high severity. For some applications, the products boiling up to 340°C (650°F) can be blended to give the desired final product.

Excessive contact time and/or temperature will create coking. Precautions need to be taken when unloading coked catalyst from the unit to prevent iron sulfide fires. The coked catalyst should be cooled to below 49°C (<120°F) before removal, or dumped into nitrogen-blanketed bins where it can be cooled before further handling. Antifoam additives may be used to prevent catalyst poisoning from silicone carryover in the coker feedstock. There is a potential for exposure to hydrogen sulfide or hydrogen gas in the event of a release, or to ammonia should a sour-water leak or spill occur. Phenol also may be present if high-boiling-point feedstocks are processed.

Attention must also be given to the coke mitigation aspects of hydrotreating as a preliminary treatment option of feedstocks for other processes, especially residua. Although the visbreaking process reduces the viscosity of residua and partially converts the residue to lighter hydrocarbons and coke (Speight and Ozum, 2002; Parkash, 2003; Hsu and Robinson, 2006; Gary et al., 2007; Speight, 2014), the process can also be used to remove the undesirable higher-molecular-weight polar constituents before sending the visbroken feedstock to a catalytic cracking unit. The solvent deasphalting (SDA) process separates the higher-value liquid product (deasphalted oil, DAO) using a light paraffinic solvent from the low-value asphaltene-rich pitch stream. Various residuum hydrotreating and/or hydrocracking processes in which the feedstock is processed under high temperature and pressure using a robust catalyst to remove sulfur, metals, condensed aromatic, or nitrogen, and increase the hydrogen content of the residuum to a desired degree are also available (Speight and Ozum, 2002; Parkash, 2003; Hsu and Robinson, 2006; Gary et al., 2007; Speight, 2014).

However, an increased number of options are becoming available in which the residuum is first hydrotreated (under milder conditions to remove heteroatoms and mitigate the effects of the asphaltene constituents and resin constituents) before sending the hydrotreated product to, for example, a fluid catalytic cracking unit. However, there is also the potential for hydrotreating to change the stability of the asphaltene constituents in the feedstock.

Hydrocracking is a refining technology that, like hydrotreating falls under the general umbrella of *hydroprocessing*. The outcome is the conversion of a variety of feedstocks to a range of products, and units to accomplish this goal can be found at various points in a refinery (Speight and Ozum, 2002; Parkash, 2003; Hsu and Robinson, 2006; Gary et al., 2007; Speight, 2014).

Hydrocracking is a more recent process development compared to the older thermal cracking, visbreaking, coking, and catalytic cracking processes. In fact, the use of hydrogen in thermal processes is perhaps the single most significant advance in refining technology during the twentieth century and the ability of refiners to cope with the renewed trend toward distillate production from residua has created a renewed interest in hydrocracking. To maintain current gasoline and middle distillate production levels, additional conversion capacity is required because of the differential in the amount of distillates produced from light crude oil and the distillate products produced from residua.

The concept of hydrocracking allows the refiner to produce products having a lower molecular weight with higher hydrogen content and a lower yield of coke. In summary, hydrocracking facilities add flexibility to refinery processing and to the product slate. Hydrocracking is more chemically severe than hydrotreating there being the intent, in hydrocracking processes, to convert the feedstock to lower-boiling products rather than to treat the feedstock for heteroatom and metals removal only (Speight and Ozum, 2002; Parkash, 2003; Hsu and Robinson, 2006; Gary et al., 2007; Speight, 2014). Process parameters (Speight and Ozum, 2002; Parkash, 2003; Hsu and Robinson, 2006; Gary et al., 2007; Speight, 2014) emphasize the relatively severe nature of the hydrocracking process.

Hydrocracking is an extremely versatile process that can be utilized in many different ways, and one of the advantages of hydrocracking is its ability to break down high-boiling aromatic stocks produced by catalytic cracking or coking. To take full advantage of hydrocracking, the process must be integrated in the refinery with other process units (Speight and Ozum, 2002; Parkash, 2003; Hsu and Robinson, 2006; Gary et al., 2007; Speight, 2014). In gasoline production, for example, the hydrocracker product must be further processed in a catalytic reformer as it has a high naphthene content and relatively low octane number. The high naphthene content makes the hydrocracker gasoline an excellent feed for catalytic reforming, and good yields of high-octane-number gasoline can be obtained.

The mechanism of hydrocracking is basically similar to that of catalytic cracking, but with concurrent hydrogenation. The catalyst assists in the production of carbonium ions via olefin intermediates and these intermediates are quickly hydrogenated under the high-hydrogen partial pressures employed in hydrocracking. The rapid hydrogenation prevents adsorption of olefins on the catalyst and, hence, prevents their subsequent dehydrogenation, which ultimately leads to coke formation so that long on-stream times can be obtained without the necessity of catalyst regeneration.

When applied to residua the hydrocracking process can be used for processes such as: (i) fuel oil desulfurization and (ii) conversion to lower-boiling distillates. The products from hydrocracking are composed of either saturated or aromatic compounds; no olefins are found. In making gasoline, the lower paraffins formed have high octane numbers; for example, the five- and six-carbon number fractions have leaded research octane numbers of 99–100. The remaining gasoline has excellent properties as a feed to catalytic reforming, producing a highly aromatic gasoline that is capable of a high octane number. Both types of gasoline are suitable for premium-grade motor gasoline. Another attractive feature of hydrocracking is the low yield of gaseous components, such as methane, ethane, and propane, which are less desirable than gasoline. When making jet fuel, more hydrogenation activity of the catalysts is used, since jet fuel contains more saturates than gasoline.

Like many refinery processes, the problems encountered in hydrocracking residua can be directly equated to the amount of complex, higher-boiling constituents that may require pretreatment (Speight and Ozum, 2002; Parkash, 2003; Hsu and Robinson, 2006; Gary et al., 2007; Speight, 2014). Processing these feedstocks is not merely a matter of applying know-how derived from refining *conventional* crude oils but requires knowledge of composition and properties (Speight and Ozum, 2002; Parkash, 2003; Hsu and Robinson, 2006; Gary et al., 2007; Speight, 2014). The materials are not only complex in terms of the carbon number and boiling point ranges but also because a large part of this *envelope* falls into a range of model compounds where very little is known about the properties. It is also established that the majority of the higher-molecular-weight materials produce coke (with some liquids) but the majority of the lower-molecular-weight constituents produce liquids (with some coke). It is to both of these trends that hydrocracking is aimed.

It is the physical and chemical composition of the residuum feedstock that plays a large part not only in determining the nature of the products that arise from refining operations but also in determining the precise manner by which a particular feedstock should be processed. Furthermore, it is apparent that the conversion of residua requires new lines of thought to develop suitable processing scenarios. Indeed, the use of thermal (*carbon rejection*) processes and of hydrothermal (*hydrogen addition*) processes, which were inherent in the refineries designed to process lighter feedstocks, has been a particular cause for concern. This has brought about the evolution of processing schemes that accommodate the residua (Speight and Ozum, 2002; Parkash, 2003; Hsu and Robinson, 2006; Gary et al., 2007; Speight, 2011, 2014).

The hydrogen addition concept is illustrated by the hydrocracking process in which hydrogen is used in an attempt to *stabilize* the reactive fragments produced during the cracking, thereby decreasing their potential for coke formation. The choice of processing schemes for a given hydrocracking application depends upon the nature of the feedstock as well as the product requirements (Speight and Ozum, 2002; Parkash, 2003; Hsu and Robinson, 2006; Gary et al., 2007; Speight, 2014).

The hydrogenation of sulfur and nitrogen compounds in the first-stage reactor requires the consumption of hydrogen. Likewise, the saturation of olefins and aromatics, in both the first-stage and second-stage reactors, to form paraffinic hydrocracked products consumes hydrogen. To a large extent, the amount of hydrogen consumption depends on the feedstock content of sulfur, nitrogen, olefins, and aromatics. The consumption of hydrogen in a hydrocracker may range from 1000 to 3000 standard cubic feet per barrel of feedstock.

In terms of sulfur removal, it appears that non-asphaltene sulfur in non-asphaltene constituents may be removed before the more refractory sulfur in asphaltene constituents (Speight, 2014), thereby requiring thorough desulfurization. This is a good reason for processes to use an extinction-recycling technique to maximize desulfurization and the yields of the desired product. Significant conversion of residua can be accomplished by hydrocracking at high severity and, for some applications, the products boiling up to 340°C (650°F) can be blended to give the desired final product (Speight and Ozum, 2002; Parkash, 2003; Hsu and Robinson, 2006; Gary et al., 2007; Speight, 2011, 2014).

Fundamentally, the trend toward lower API gravity feedstocks such as residua is related to an increase in the hydrogen/carbon atomic ratio of crude oils because of higher content of residuum. This can be overcome by upgrading methods that lower this ratio by adding hydrogen, rejecting carbon, or using a combination of both methods. Though several technologies exist to upgrade residua (Speight and Ozum, 2002; Parkash, 2003; Hsu and Robinson, 2006; Gary et al., 2007; Speight, 2011, 2014), selection of the optimum process units is very much dependent on each refiner's needs and goals, with the market pull being the prime motivator. Furthermore, processing option systems to *dig deeper into the barrel* by converting more of the higher-boiling materials to distillable products should not only be cost-effective and reliable, but also flexible.

Hydrocracking adds that flexibility and offers the refiner a process that can handle varying feeds and operate under diverse process conditions. Utilizing different types of catalyst can modify the product slate produced. the reactor design and number of processing stages play a role in this flexibility.

Finally, a word about conversion measures for upgrading processes. Such measures are necessary for *any conversion process* but more particularly for hydrocracking processes where hydrogen management is an integral, and essential, part of process design.

3.2.1.4 Deasphalting

SDA processes are a major part of refinery operations and are not often appreciated for the tasks for which they are used (Speight and Ozum, 2002; Parkash, 2003; Hsu and Robinson, 2006; Gary et al., 2007; Speight, 2014). In the SDA processes, an alkane is injected into the feedstock to disrupt the dispersion of components and causes the polar constituents to precipitate. Propane (or sometimes propane/butane mixtures) is extensively used for deasphalting and produces a DAO and propane deasphalter asphalt (PDA or PD tar). Propane has unique solvent properties; at lower temperatures (38–60°C; 100–140°C), paraffins are very soluble in propane and at higher temperatures (about 93°C; 200°F) all hydrocarbons are almost insoluble in propane.

An *SDA* unit processes the residuum from the vacuum distillation unit and produces DAO, used as feedstock for a fluid catalytic cracking unit, and the asphaltic residue (deasphalter tar, deasphalter bottoms) which, as a residual fraction, can only be used to produce asphalt or as a blend stock or visbreaker feedstock for low-grade fuel oil. SDA processes have not realized their maximum potential. With on-going improvements in energy efficiency, such processes would display their effects in a combination with other processes. SDA allows removal of sulfur and nitrogen compounds as well as metallic constituents by balancing yield with the desired feedstock properties (Speight and Ozum, 2002; Parkash, 2003; Hsu and Robinson, 2006; Gary et al., 2007; Speight, 2014).

Most of the metals present in refinery feedstocks are concentrated in the asphaltene fraction and, by using SDA it is possible to recover significant amounts of partially demetallized and deasphalted oil (DAO) from the residues which can be processed in fluid catalytic cracking units or in hydrocracking units (Speight and Ozum, 2002; Parkash, 2003; Hsu and Robinson, 2006; Gary et al., 2007; Speight, 2014).

The main advantages of SDA are relatively low investment and operational costs, while its principal limitation is that to obtain DAO with a low content of pollutants (in particular, sulfur, nitrogen, metals, and coke-forming constituents), the yield must be limited. As a result, significant quantities of by-products are generated (asphaltene constituents plus resin constituents) that can be used as components of low-quality fuels or as sources of carbon for the production of synthesis gas and, therefore, hydrogen in suitable gasification units (Speight and Ozum, 2002; Parkash, 2003; Hsu and Robinson, 2006; Gary et al., 2007; Speight, 2014). Furthermore, the application of the SDA process to residua upgrading has led to the development of a variety of technologies in which the typical SDA unit is combined with other processes (Speight and Ozum, 2002; Parkash, 2003; Hsu and Robinson, 2006; Gary et al., 2007; Speight, 2014).

3.2.2 Asphalt

Asphalt is a manufactured product insofar as the constituents of the asphalt may have been changed somewhat from the original constituents of the original petroleum feedstock. Furthermore, in order to assure in-service performance, there are specifications for the use of asphalt in highway pavement construction and the asphalt is selected on the basis of the results of standard test methods. However, many of the tests used to determine the suitability of asphalt use are empirical in nature (TRB, 2010). Thus, there is the need to develop new performance-based test procedures which can provide the necessary information leading to selection of the most appropriate material to maintain asphalt-based highway.

The manufacturing process involves distilling everything possible from crude petroleum until a residuum with the desired properties is obtained. This is usually achieved in stages in which distillation at atmospheric pressure removes the lower-boiling fractions and yields an atmospheric residuum (*reduced crude*) that may contain higher-boiling (lubricating) oils, wax, and asphalt (Figure 3.2). Distillation of the reduced crude under vacuum

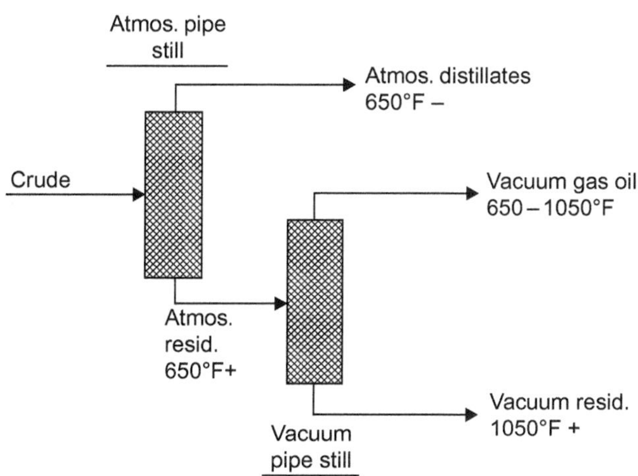

Figure 3.2 The distillation section of a refinery.

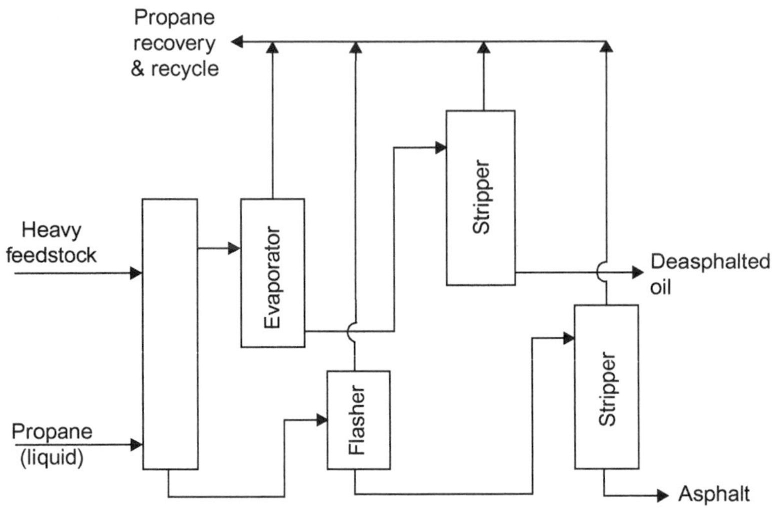

Figure 3.3 Propane deasphalting.

removes the oils (and wax) as overhead products and the asphalt remains as a bottom (or residual) product. The majority of the polar functionalities and high-molecular-weight species in the original crude oil, which tend to be non-volatile, concentrate in the vacuum residuum, thereby conferring desirable or undesirable properties on the asphalt.

At this stage the asphalt is frequently—for unknown reasons other than the similar appearance of the asphalt to coal tar pitch (the non-volatile residue when coal tar is distilled) (Speight, 2013)—and incorrectly referred to as *pitch* and has a softening point (ASTM D36; ASTM D61; ASTM D2319; ASTM D3104; ASTM D3461) related to the amount of oil removed and increases with increasing overhead removal. In character with the elevation of the softening point, the pour point is also elevated (Table 3.2): the more oil that is distilled from the residue, the higher the softening point.

Asphalt is also produced by propane deasphalting (Figure 3.3) and there are differences in the properties of asphalts prepared by propane deasphalting and those prepared by vacuum distillation from the same feedstock. Propane deasphalting also has the ability to reduce a residuum even further and produce an asphalt product having a lower viscosity, higher ductility, and higher temperature susceptibility than other asphalts—such properties are correctly anticipated to be very much crude-oil-dependent. Propane deasphalting is conventionally applied to low-asphalt-content crude oils, which are generally different in type and

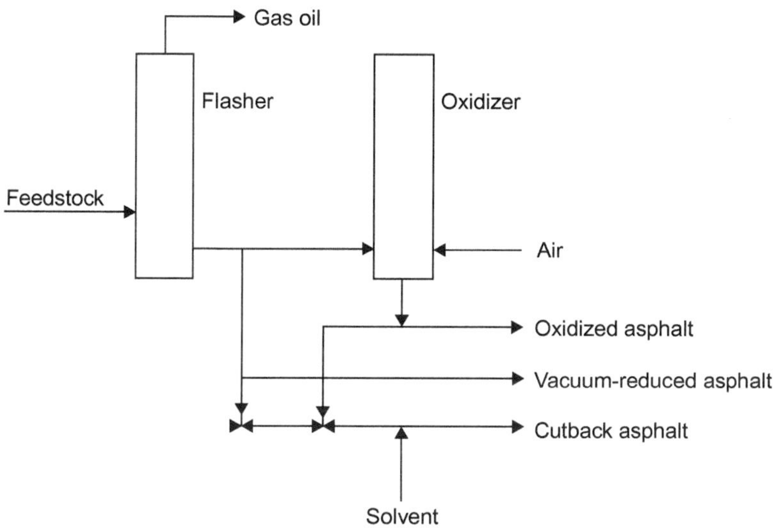

Figure 3.4 Asphalt manufacture including air blowing: the feedstock can be a residuum from the atmospheric distillation tower or from the vacuum distillation tower or the deasphalter.

source from those processed by distillation of higher-yield crude oils. In addition, the properties of asphalt can be modified by air blowing in batch and continuous processes (Figure 3.4) (Speight, 1992, 2014). On the other hand, the preparation of asphalts in liquid form by blending (cutting back) asphalt with a petroleum distillate fraction is customary and is generally accomplished in tanks equipped with coils for air agitation or with a mechanical stirrer or a vortex mixer.

Propane has been used extensively in deasphalting residua, especially in the preparation of high-quality lubricating oils and feedstocks for catalytic cracking units (Speight, 2011, 2014). The use of propane has necessitated elaborate solvent cooling systems utilizing cold water, which is a relatively expensive cooling agent. In order to circumvent such technology, future units will use solvent systems that will allow operation at elevated temperatures relative to conventional propane deasphalting temperatures, thereby permitting easy heat exchange. In addition, it may be found desirable to integrate dewaxing operations with deasphalting operations by having a common solvent recovery system. This will require changes to the solvent composition and the inclusion of solvents not usually considered to be deasphalting solvents.

Furthermore, as a means of energy reduction for the process, in future deasphalting units the conventional solvent recovery scheme will be retrofitted with a supercritical solvent recovery scheme to reap the benefits of higher energy efficiency. Other improvements will include variations in the extraction column internals. For example, the three major properties, which influence the design of extraction column, are interfacial tension, viscosity, and density of phases. SDA extraction column is characterized by low interfacial tension, high viscosities of asphaltene phase, and a density difference between the phases. Extension of these property correlations for SDA applications will be suitably validated and corrections made where necessary to improve extraction performance and yields of the products.

An *asphalt emulsion* is a mixture of asphalt and an anionic agent such as the sodium or potassium salt of a fatty acid. The fatty acid is usually a mixture and may contain palmitic, stearic, linoleic, and abietic acids and/or high-molecular-weight phenols. Sodium lignate is often added to alkaline emulsions to affect better emulsion stability. Non-ionic cellulose derivatives are also used to increase the viscosity of the emulsion if needed. The acid number of asphalt is an indicator of its emulsifiability and reflects the presence of high-molecular-weight naphthenic acids. Diamines, frequently used as cationic agents, are made from the reaction of tallow acid amines with acrylonitrile, followed by hydrogenation. The properties of asphalt emulsions (ASTM D977; ASTM D2397) allow a variety of uses. As with other petroleum products, sampling is an important precursor to residuum and asphalt analysis and a standard method (ASTM D140) is available that provides guidance for the sampling of asphalts, liquid and semisolid, at the point of manufacture, storage, or delivery.

3.3 TEST METHODS

Residua are not typically sold as a refinery product and are, for the most part, used within the refinery to produce (other than asphalt) valuable distillate products (Speight and Ozum, 2002; Parkash, 2003; Hsu and Robinson, 2006; Gary et al., 2007; Speight, 2014). As such, that is not a call for detailed analysis of residua in terms of specification and the required analyses are usually limited to such properties as: (i) API gravity and density, (ii) asphaltene content, (iii) carbon residue, (iv) elemental analysis, with special focus on the sulfur content, nitrogen content, and metals content, (v) molecular weight, as a means of determining the moles of hydrogen required for hydroconversion processes, and (vi) viscosity. As brief as this may seem, these data (usually referred to as a *preliminary assay*) are sufficiently adequate to allow the process chemist and process engineer a measure of predictability as to how the residuum will behave in the next process down the line (after distillation) (Speight, 2014, 2015).

The significance of a particular test is not always apparent by reading the procedure, and sometimes can only be gained through working familiarity with the test. The following tests are commonly used to characterize asphalts but these are not the only tests used for determining the property and behavior of an asphaltic binder. As in the petroleum industry, a variety of tests are employed having evolved through local, or company, use.

The following test methods (listed alphabetically and not in order of preference) are a sample of the test that might be required to determine the suitability of properties of residua and their suitability as feedstocks for conversion processes.

3.3.1 API Gravity and Density

For clarification, it is necessary to understand the basic definitions that are used: (i) *density* is the mass of liquid per unit volume at 15.6°C (60°F), (ii) *relative density* is the ratio of the mass of a given volume of liquid at 15.6°C (60°F) to the mass of an equal volume of pure water at the same temperature, (iii) *specific gravity* is the same as the relative density and the terms are used interchangeably.

Density (ASTM D1298) is an important property of petroleum products since petroleum and especially petroleum products are usually bought and sold on that basis or if on a volume basis then converted to mass basis via density measurements. This property is almost synonymously termed density, relative density, gravity, and specific gravity, all terms related to each other. Usually a hydrometer, pycnometer, or more modern digital density meter is used for the determination of density or specific gravity.

In the most commonly used method (ASTM D1298), the sample is brought to the prescribed temperature and transferred to a cylinder at approximately the same temperature. The appropriate hydrometer is lowered into the sample and allowed to settle and, after temperature equilibrium has been reached, the hydrometer scale is read and the temperature of the sample is noted.

Although there are many methods for the determination of density due to the different nature of petroleum itself and the different products, one test method (ASTM D5002) is used for the determination of the density or relative density of petroleum that can be handled in a normal fashion as liquids at test temperatures between 15°C and 35°C (59°F and 95°F). This test method applies to petroleum oils with high vapor pressures provided appropriate precautions are taken to prevent vapor loss during transfer of the sample to the density analyzer. In the method, approximately 0.7 ml of crude oil sample is introduced into an oscillating sample tube and the change in oscillating frequency caused by the change in mass of the tube is used in conjunction with calibration data to determine the density of the sample.

Another test determines density and specific gravity by means of a digital densimeter (ASTM D4052). In the test, a small volume (approximately 0.7 ml) of liquid sample is introduced into an oscillating sample tube and the change in oscillating frequency caused by the change in the mass of the tube is used in conjunction with calibration data to determine the density of the sample. The test is usually applied to petroleum, petroleum distillates, and petroleum products that are liquids at temperatures between 15°C and 35°C (59°F and 95°F) that have vapor pressures below 600 mm Hg and viscosities below about 15,000 cSt at the temperature of test. However, the method should not be applied to samples so dark in color that the absence of air bubbles in the sample cell cannot be established with certainty.

Accurate determination of the density or specific gravity of crude oil is necessary for the conversion of measured volumes to volumes at the standard temperature of 15.56°C (60°F) (ASTM D1250). The specific gravity is also a factor reflecting the quality of crude oils.

The accurate determination of the API gravity of petroleum and its products (ASTM D287) is necessary for the conversion of measured volumes to volumes at the standard temperature of 60°F (15.56°C). Gravity is a factor governing the quality of crude oils. However, the gravity of a petroleum product is an uncertain indication of its quality. Correlated with other properties, gravity can be used to give approximate hydrocarbon composition and heat of combustion. This is usually accomplished through use of the API gravity that is derived from the specific gravity:

$$\text{API gravity (degrees)} = (141.5/\text{sp gr } 60/60°\text{F}) - 131.5$$

The API gravity is also a critical measure for reflecting the quality of petroleum (Speight, 2001, 2014, 2015).

API gravity or density or relative density can be determined using one of two hydrometer methods (ASTM D287; ASTM D1298). The use of a digital analyzer (ASTM D5002) is finding increasing popularity for the measurement of density and specific gravity.

In this method (ASTM D287), the API gravity is determined using a glass hydrometer for petroleum and petroleum products that are normally handled as liquids and that have a Reid vapor pressure of 26 psi (180 kPa) or less. The API gravity is determined at 15.6°C (60°F), or converted to values at 60°F, by means of standard tables. These tables are not applicable to non-hydrocarbons or essentially pure hydrocarbons such as the aromatics.

This test method is based on the principle that the gravity of a liquid varies directly with the depth of immersion of a body floating in it. The API gravity is determined using a hydrometer by observing the freely floating API hydrometer and noting the graduation nearest to the apparent intersection of the horizontal plane surface of the liquid with the vertical scale of the hydrometer, after temperature equilibrium has been reached. The temperature of the sample is determined using a standard test thermometer that is immersed in the sample or from the thermometer that is an integral part of the hydrometer (thermohydrometer).

For solid and semisolid asphalt a pycnometer is generally used (ASTM D70) and a hydrometer is applicable to liquid asphalt (ASTM D3142).

3.3.2 Asphaltene Content

The asphaltene fraction (ASTM D2007; ASTM D3279; ASTM D4124; ASTM D6560) is the highest molecular weight and most complex fraction in petroleum (Figure 3.5). In the simplest sense, the amount of asphaltene

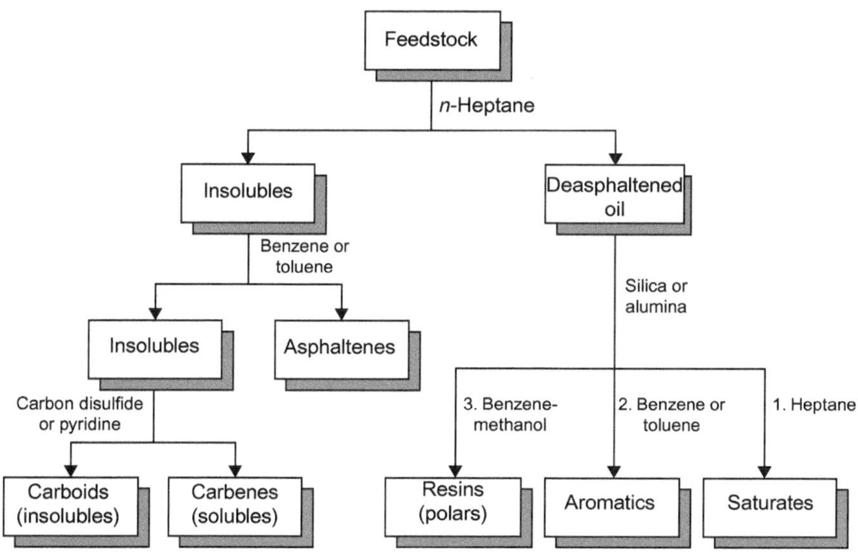

Figure 3.5 Feedstock fractionation.

constituents (the asphaltene fraction) gives an indication of the amount of coke that can be expected during processing (Speight, 2001; Speight and Ozum, 2002). In the present context, the presence of asphaltene constituents in asphalt can be beneficial insofar as the asphalt *may* form stronger bonds with the mineral aggregate because of the presence of the asphaltene constituents. On the other hand, the asphaltene constituents are often subject to rapid oxidation, as occurs during weathering which may lead to incompatibility of the oxidized asphaltene constituents with the remainder of the asphalt and ensuing failure of the asphalt–aggregate bond.

One of the standard test methods (ASTM D6560) has some versatility insofar as it is applicable to the determination of the heptane insoluble asphaltene content of feedstocks such as gas oil, diesel fuel, residual fuel oils, lubricating oil, bitumen, and crude petroleum that has been topped to an oil temperature of 260°C (500°F).

In any of the methods for the determination of the asphaltene content, the crude oil or product (such as asphalt) is mixed with a large excess (usually >30 volumes hydrocarbon per volume of sample) low-boiling hydrocarbon such as *n*-pentane or *n*-heptane. For an extremely viscous sample, a solvent such as toluene may be used prior to the addition of the low-boiling hydrocarbon but an additional amount of the hydrocarbon (usually >30 volumes hydrocarbon per volume of solvent) must be added to compensate for the presence of the solvent. After a specified time, the insoluble material (the asphaltene fraction) is separated (by filtration) and dried. The yield is reported as percentage (% w/w) of the original sample.

It must be recognized that, in any of these tests, different hydrocarbons (such as *n*-pentane or *n*-heptane) will give different yields of the asphaltene fraction and if the presence of the solvent is not compensated for by the use of additional hydrocarbon the yield will be erroneous. In addition, if the hydrocarbon is not present in a large excess, the yields of the asphaltene fraction will vary and will be erroneous (Speight, 2014).

The *precipitation number* is often equated to the asphaltene content but there are several issues that remain obvious in its rejection for this purpose. For example, the method to determine the precipitation number (ASTM D91) advocates the use of naphtha for use with black oil or lubricating oil and the amount of insoluble material (as a % v/v of the sample) is the precipitating number. In the test, 10 ml of sample is mixed with 90 ml of ASTM precipitation naphtha (that may or may not have a constant chemical composition) in a graduated centrifuge cone and centrifuged for 10 min at 600–700 rpm. The volume of material on the bottom of the centrifuge cone is noted until repeat centrifugation gives a value within 0.1 ml (the precipitation number). Obviously, this can be substantially different to the asphaltene content.

In another test method (ASTM D4055), pentane insoluble materials above 0.8 μm in size can be determined. In the test method, a sample of oil is mixed with pentane in a volumetric flask, and the oil solution is filtered through a 0.8-μm membrane filter. The flask, funnel, and the filter are washed with pentane to completely transfer the particulates onto the filter that is then dried and weighed to give the yield of pentane-insoluble materials.

Another test method (ASTM D893) that was originally designed for the determination of pentane- and toluene-insoluble materials in used lubricating oils can also be applied to asphalt. However, the method may need modification by first adding a solvent (such as toluene) to the asphalt before adding pentane. The pentane-insoluble constituents can include oil-insoluble materials. The toluene-insoluble materials can come from external contamination and products from degradation of asphalt. A significant change in the pentane-insoluble constituents or toluene-insoluble constituents indicates a change in asphalt that could lead to performance problems.

There are two test methods used. Procedure A covers the determination of insoluble constituents without the use of coagulant in the pentane and provides an indication of the materials that can be readily separated from the asphalt–solvent mixture by centrifugation. Procedure B covers the determination of insoluble constituents in asphalt containing additives and employs a coagulant. In addition to the materials separated by using Procedure A, this coagulation procedure separates some finely divided materials that may be suspended in the asphalt. The results obtained by Procedures A and B should not be compared since they usually give different values. The same procedure should be applied when comparing results obtained periodically on oil in use, or when comparing results determined in different laboratories.

In Procedure A, a sample is mixed with pentane and centrifuged. The asphalt solution is decanted, and the precipitate washed twice with pentane, dried, and weighed. For toluene-insoluble constituents, a separate sample of the asphalt is mixed with pentane and centrifuged. The precipitate is washed twice with pentane, once with toluene–alcohol solution, and once with toluene. The insoluble material is then dried and weighed. In Procedure B, Procedure A is followed except that instead of pentane, a pentane-coagulant solution is used.

In addition to determining the asphaltene content of residua, a test method is available to determine the maximum flocculation ratio and peptizing power in residua using an optical device (ASTM D7060). The method is applicable to atmospheric or vacuum distillation residues, thermally cracked residue, intermediate and finished residual fuel oils, containing at least 1% (w/w) asphaltene constituents. The flocculation ratio is an important aspect of the behavior of petroleum products that contain asphaltene constituents—it is necessary to know the point at which asphaltene constituents will start to deposit as a solid phase due to changes (thermal or physical) in the product when in use.

In this method, portions of the sample are diluted with various ratios with 1-methylnaphthalene and each solution is inserted into the automatic apparatus, and titrated with cetane until flocculation of the asphaltene constituents is detected by the optical probe. The four flocculation ratios at critical dilution, measured during the fine determinations, are used to calculate the maximum flocculation ratio of the asphaltene constituents in the sample and the peptizing power of the oil medium (non-asphaltene constituents) of the sample.

3.3.3 Carbon Residue

The *carbon residue* of asphalt serves as an indication of the propensity of the sample to form carbonaceous deposits (thermal coke) under the influence of heat. The residue produced is also often used to provide thermal data that give an indication of the composition of the asphalt (Speight, 2001, 2014, 2015).

Tests for Conradson carbon residue (ASTM D189), the Ramsbottom carbon residue (ASTM D524), the microcarbon carbon residue (ASTM D4530), and asphaltene content (ASTM D2007; ASTM D3279; ASTM D4124; ASTM D6560) are sometimes included in inspection data on petroleum. The data give an indication of the amount of coke that will be formed during thermal processes as well as an indication of the amount of high-boiling constituents in petroleum.

The determination of the *carbon residue* of petroleum or a petroleum product is applicable to relatively non-volatile samples that decompose on distillation at atmospheric pressure. Samples that contain ash-forming constituents will have an erroneously high carbon residue, depending upon the amount of ash formed. All three methods are applicable to relatively non-volatile petroleum products that partially decompose on distillation at atmospheric pressure. Crude oils having a low carbon residue may be distilled to a specified residue and the carbon residue test of choice then applied to the residue.

In the Conradson carbon residue test (ASTM D189), a weighed quantity of sample is placed in a crucible and subjected to destructive distillation for a fixed period of severe heating. At the end of the specified heating period, the test crucible containing the carbonaceous residue is cooled in a desiccator and weighed and the residue is reported as a percentage (% w/w) of the original sample (Conradson carbon residue).

In the Ramsbottom carbon residue test (ASTM D524), the sample is weighed into a glass bulb that has a capillary opening and is placed into a furnace (at 550°C, 1020°F). The volatile matter is distilled from the bulb and the non-volatile matter that remains in the bulb cracks to form thermal coke. After a specified heating period, the bulb is removed from the bath, cooled in a desiccator, and weighed to report the residue (Ramsbottom carbon residue) as a percentage (% w/w) of the original sample.

In the microcarbon residue test (ASTM D4530), a weighed quantity of the sample placed in a glass vial is heated to 500°C (930°F) under an inert (nitrogen) atmosphere in a controlled manner for a specific time and the carbonaceous residue (*carbon residue (micro)*) is reported as a percent (% w/w) of the original sample.

The data produced by the microcarbon test (ASTM D4530) are equivalent to those by the Conradson carbon method (ASTM D189). However, this microcarbon test method offers better control of test conditions and requires a smaller sample. Up to 12 samples can be run simultaneously. This test method is applicable to petroleum and to petroleum products that partially decompose on distillation at atmospheric pressure and is applicable to a variety of samples that generate a range of yields (0.01–30% w/w) of thermal coke.

Other test methods that are used for determining the coking value of tar and pitch (ASTM D2416; ASTM D4715) that indicates the relative coke-forming properties of tars and pitches might also be applied to asphalt. Both test methods are applicable to tar and pitch having an ash content ≤0.5% (ASTM D2415). The former test method (ASTM D2416) gives results close to those obtained by the Conradson carbon residue test (ASTM D189). However, in the latter test method (ASTM

D4715), a sample is heated for a specified time at 550 ± 10°C (1022 ± 18°F) in an electric furnace. The percentage of residue is reported as the coking value.

Finally, a method that is used to determine pitch volatility (ASTM D4893) might also be used, on occasion, to determine the non-volatility of asphalt. In the method, an aluminum dish containing about 15 g of accurately weighed sample is introduced into the cavity of a metal block heated and maintained at 350°C (660°F). After 30 min, during which any volatiles are swept away from the surface of the sample by preheated nitrogen, the residual sample is taken out and allowed to cool down in the desiccator. The non-volatility is determined by the sample weight remaining and reported as % w/w residue.

In any of the above tests, ash-forming constituents (ASTM D482) or non-volatile additives present in the sample will be included in the total carbon residue reported, leading to higher carbon residue values and erroneous conclusions about the coke-forming propensity of the sample.

3.3.4 Elemental Analysis

Residua are not composed of a single chemical species, but are a complex mixture of organic molecules that vary widely in composition and are composed of carbon, hydrogen, nitrogen, oxygen, and sulfur as well as trace amounts of metals, principally vanadium and nickel. The heteroatoms, although a minor component compared to the hydrocarbon moiety, can vary in concentration over a wide range depending on the source of the asphalt and be a major influence on asphalt properties. Generally, most residua contain 79–88% (w/w) carbon, 7–13% (w/w) hydrogen, trace–8% (w/w) sulfur, 2–8% (w/w) oxygen, and trace–3% (w/w) nitrogen. Trace metals such as iron, nickel, vanadium, calcium, titanium, magnesium, sodium, cobalt, copper, tin, and zinc occur in crude oils.

The sulfur and nitrogen contents of resids are important insofar as they also allow prediction of the amounts of sulfur oxides (SO_x) and nitrogen oxides (NO_x) that will be released during combustion. The metals content can provide valuable information about the origin of the resids, potentially aiding in assessing catalyst tolerance to the feedstock during processing as well as identifying the source of spills or the potential for metals leaching from asphalt products.

Thus, elemental analysis is still of considerable value to determine the amounts of elements in asphalt and the method chosen for the analysis may be subject to the peculiarities or character of the feedstock under investigation and should be assessed in terms of accuracy and reproducibility. The methods that are designated for elemental analysis are:

1. *Carbon* and *hydrogen content* (ASTM D1018; ASTM D3343; ASTM D3701; ASTM D5291; ASTM E777);
2. *Nitrogen content* (ASTM D3228; ASTM E258; ASTM D5291; ASTM E778);
3. *Oxygen content* (ASTM E385);
4. *Sulfur content* (ASTM D129; ASTM D1266; ASTM D1552; ASTM D2622; ASTM D3120; ASTM D4045; ASTM D4294);
5. *Metals content* (ASTM D5185).

The determination of *nitrogen* has been performed regularly by the Kjeldahl method (ASTM D3228) and the Dumas method. The chemiluminescence method is the most recent technique applied to nitrogen analysis for petroleum. The chemiluminescence method determines the amount of chemically bound nitrogen in liquid hydrocarbon samples. In this method, the samples are introduced to the oxygen-rich atmosphere of a pyrolysis tube maintained at 975°C (1785°F). Nitrogen in the sample is converted to nitric oxide during combustion and the combustion products are dried by passage through magnesium perchlorate [$Mg(ClO_4)_2$] before entering the reaction chamber of a chemiluminescence detector.

Oxygen is one of the five (C, H, N, O, and S) major elements in asphalt and although the level rarely exceeds 1.5% by weight, it may still be critical to performance. Many petroleum products do not specify the oxygen content but if the oxygen compounds are present as acidic compounds such as phenols (ArOH) and naphthenic acids (cycloalkyl-COOH), they are controlled in different specifications by a variety of tests.

The metals assay (ASTM D5185) typically includes nickel and vanadium contents due to the detrimental effects of these metals on catalysts used in cracking and desulfurization processes. In lubricating oils, metal contents can provide information on both the types of additives used in the oil and on the wear history of the equipment being lubricated. ASTM D5185—Standard Test Method for Determination of Additive Elements, Wear Metals, and Contaminants in Used Lubricating Oils and Determination of Selected Elements in Base Oils by Inductively Coupled Plasma Atomic Emission Spectrometry (ICP-AES)—can be used to determine over 20 different metals in a variety of petroleum products.

Vanadium and nickel are bound in organic complexes and, by virtue of the concentration (distillation) process by which asphalt is manufactured, are also found in asphalt. The catalytic behavior of vanadium has prompted investigations into the relation between vanadium content and the sensitivity of asphalt to oxidation but, although advances have been made through the various investigations of metals in petroleum, (Reynolds, 1998), the true effects and significance of metallic constituents in the chemical and physical behavior of asphalt are not yet well understood and need further clarification.

3.3.5 Flash Point

The *flash point* is the lowest temperature at which application of a test flame causes the vapor of a sample to ignite under specified test conditions. The flash point measures the tendency of a sample to form a flammable mixture with air under controlled laboratory conditions. Flash point data are used in shipping and safety regulations to define *flammable* and *combustible* materials as well as to indicate the possible presence of highly volatile and flammable material in a relatively non-volatile or non-flammable material. The flash point should not be confused with auto-ignition temperature (ASTM E659) that measures spontaneous combustion with no external source of ignition.

The Pensky-Martens Closed Tester (ASTM D93) and the Tag Closed Tester (ASTM D56) are normally employed for determining the flash point of fuel oil and similar products. The Cleveland open cup method (ASTM D92) is most commonly used, although the Tag open cup method (ASTM D3143) is applicable to cutback asphalt. As noted above, the flash point of asphalt is an indication of fire hazard and is frequently used to indicate whether asphalt has been contaminated with materials of a lower flash point.

3.3.6 Molecular Weight

The molecular weight of asphalt is not always (in fact, rarely) used in specifications. Nevertheless, there may be occasions when the molecular weight of asphalt is desired (Speight, 2014).

Currently, of the methods available, several standard methods are recognized as being useful for determining the molecular weight of petroleum fractions (ASTM D2502; ASTM D2503; ASTM D2878). Each method has proponents and opponents because of assumptions made in the use of the method or because of the mere complexity of the sample and the nature of the inter- and intramolecular interactions. Before application of any one or more of these methods, consideration must be given to the mechanics of the method and the desired end result.

Methods for molecular weight measurement are also included in other more comprehensive standards (ASTM D128; ASTM D3712) and several indirect methods have been proposed for the estimation of molecular weight by correlation with other, more readily measured physical properties (Speight, 2001, 2014, 2015). They are satisfactory when dealing with the conventional type of crude oils or their fractions and products and when approximate values are desired.

The molecular weights of the individual fractions of asphalt have received more attention and have been considered to be of greater importance, than the molecular weight of the asphalt itself (Speight, 2001, 2014, 2015). The components that make up the asphalt influence the properties of the material to an extent that is dependent upon the relative amount of the component, the molecular structure of the component, and the physical structure of the component that includes the molecular weight.

Asphaltene constituents have a wide range of molecular weights, from 500 to at least 2500, depending upon the method (Speight, 1994, 2014). Asphaltene constituents associate in dilute solution in non-polar solvents giving higher molecular weights than is actually the case on an individual molecule basis. The molecular weights of the resin constituents are somewhat lower than those of the asphaltene constituents and usually fall within the range of 500–1000. This is due not only to the absence of association but also to a lower absolute molecular size. The molecular weight of the oil fraction (i.e., the asphalt minus the asphaltene constituents and minus the resin constituents) is usually <500, but often in the range of 300–400.

3.3.7 Viscosity

The viscosity of a residuum is a measure of the flow characteristics. It is generally the most important controlling property for manufacture and for selection to meet a particular application.

A number of instruments are in common use with resids for this purpose. The vacuum capillary (ASTM D2171) is commonly used to classify paving asphalt at 60°C (140°F). Kinematic capillary instruments (ASTM D2170; ASTM D4402) are commonly used in the 60–135°C (140–275°F) temperature range for both liquid and semisolid asphalts in the range of 30–100,000 cSt. Saybolt tests (ASTM D88) are also used in this temperature range and at higher temperature (ASTM E102). At lower temperatures, the cone and plate instrument (ASTM D4287; ASTM D7395) has been used extensively in the viscosity range 1000 poises to 1,000,000 poises. Other instrumentation techniques are based on (i) the sliding plate microviscometer—the use of machined aggregate plates is to investigate the effects of aggregate-surface-induced structure on the rheological properties of asphalt binders, and (ii) the rheogoniometer—a type of rheometer that can be used to measure the viscous and elastic flow properties of a fluid.

3.4 USE OF THE DATA

Residua are complex mixtures of hydrocarbon compounds and non-hydrocarbon compounds—further refining is required to produce saleable products (such as liquid fuels and asphalt) that meet the specifications required for the designated use. However, analytical methods are also used to describe and then monitor the progress of the refining processes.

However, not all residua are the same or even equivalent. Thus, the procedure and the resulting data will (i) reflect the background of the investigator, (ii) reflect the ability of the investigator to interpret the results in relation to the application, (iii) be dependent upon the analytical facilities available, and (iv) be shown at the required levels of accuracy and precision. Indeed, whatever the source, petroleum and petroleum products are complex mixtures, ranging from low-molecular-weight, low-boiling volatile organic compounds to high-molecular-weight, high-boiling constituents (even non-volatile under achievable temperatures). Furthermore, the composition of residua varies depending upon (i) the source of the crude oil—crude oil is derived from a variety of source materials (Speight, 2014) which vary greatly in chemical composition, and (ii) the refining processes used to produce the residuum.

When the data from the various analytical test and investigations have been collected, the obvious question that comes to mind relates to use of the data. For the most part, the data are used to determine whether or not a crude oil may produce a certain product and whether or not the product meets specifications. However, since the early days of analysis, there has been a growing tendency to use the analytical data as a means of more detailed and accurate projections of (i) the refinery processes, (ii) product yields, (iii) product properties, and (iv) predictability. And nowhere has this been more important than when heavy crude oil and bitumen entered the refinery scene. In fact, predictability in many forms is spread throughout the field of petroleum refining and product properties. Thus, the data derived from any one, or more, of the analytical techniques described in this chapter not only give an indication of the product characteristics but also whether or not the residuum is suitable for the proposed use.

Finally, each process chemist, process engineer, and refiner has his/her own preferences and such efforts are feedstock-dependent as well as being refinery-dependent. It is, in fact, the purpose of this chapter to introduce the reader to the potential that exists in this area of science and engineering. There is also the need to recognize that what is adequate for one refinery and one feedstock (or feedstock blend provided that the blend composition does not change significantly) will not be suitable for a different refinery with a different feedstock (or feedstock blend).

3.4.1 Residuum Evaluation

The use of physical properties for residuum evaluation and product yields and, in some cases, product properties has continued in refineries and in process research laboratories to the present time and will continue for some time. However, the emphasis for residuum evaluation and product yields (as well as product properties) has taken a turn in the direction of residuum mapping. In such procedures, properties of residuum are mapped to show characteristics that are in visual form rather than in tabular form, in order to facilitate evaluation and prediction of the behavior of the feedstock in various refining scenarios (Figure 3.6) (Speight, 2001; Speight and Ozum, 2002; Hsu and Robinson, 2006; Gary et al., 2007; Speight, 2014).

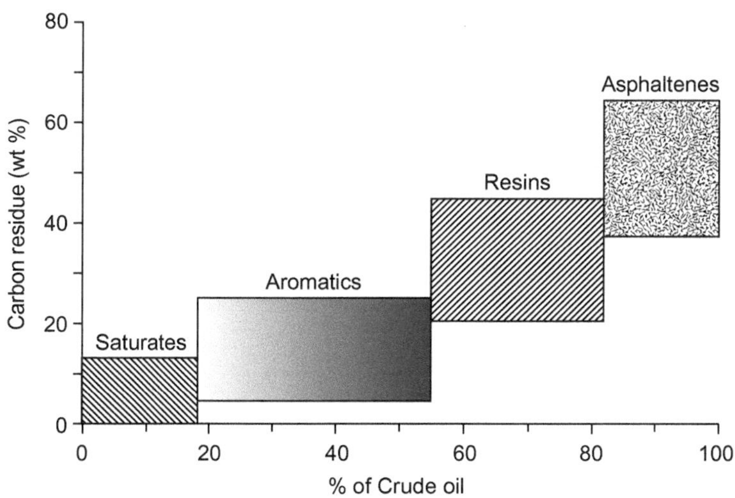

Figure 3.6 Carbon residue data showing overlap of the fractions (for clarity any horizontal overlap is not shown).

Whether or not such methods will supersede the simpler form of property correlations remains to be determined, although it is likely that both will continue to be used in a complimentary fashion for some time to come.

Thus, using the data derived from the test assay, it is possible to assess product quality and acquire a degree of predictability of performance during use. However, knowledge of the basic concepts of refining will help the analyst understand the production and, to a large extent, the anticipated properties of the product, which in turn are related to storage, sampling, and handling of the products. In addition, there are many instances in which inter-relationships of the analytical data enable properties to be predicted from the measured properties with as good precision as can be obtained by a single test. It would be possible to examine in this way the relationships between all the specified properties of a product and to establish certain key properties from which the remainder could be predicted, but this would be a tedious task.

In summary, residuum analysis is a complex discipline involving a variety of standard test methods, some of which have been mentioned above, and which needs a multi-dimensional approach. No single technique should supersede the other without adequate testing, along with an explanation of the data that are obtained along with adequate interpretation.

3.4.2 Residuum Mapping

Beyond using the analytical data to meet specification, the data can also be used to investigate the chemical and physical structure of residua. However, it is now recognized that investigating resid structure in the interactions of the various bulk fractions and their respective constituents can play an important role in refinability and product production. For example, the changes caused by thermal treatment can result in instability of the liquid and incompatibility of a separate phase and can greatly affect the product yield as well as the product properties (Speight, 2014).

Thus, in order to discuss the chemical and physical structure of petroleum it is necessary to give consideration to the chemical and physical nature of the constituents of petroleum. Furthermore, it is essential that the concept of the resid being a *continuum* is accepted not only among the lower-molecular-weight ranges but that the continuum is complete and continues into the higher-molecular-weight ranges. Recent investigations have focused not only on the composition of petroleum and its major operational fractions but also on further fractionation (Speight, 2001, 2014), which allows different feedstocks to be compared on a relative basis to provide a very simple but convenient feedstock *map* that gives an indication of the chemistry that is instrumental in leading to the various products.

Thus, it is not surprising that one of the most effective means of feedstock mapping has arisen through the use of fractionation methods in which the feedstock is sub-divided into several fractions by standard procedures (Speight, 2001, 2014, 2015). The simplest map that can be derived from such methods shows the feedstock as a three-phase system based on asphaltene separation. Indeed, the constituents of the asphaltene fraction can be differentiated on the basis of the solubility parameter of the solvent or solvent used for the separation as well as on the basis of separation of the constituents by adsorption (Mitchell and Speight, 1973; Speight, 2001; Speight, 2014).

However, it must be recognized that such *maps* do not give any indication of the complex intermolecular and intramolecular interactions that occur between, for example, such fractions as the resin constituents and asphaltene constituents (Koots and Speight, 1975; Speight, 1994, 2014), but it does allow predictions of resid behavior leading to identification of product yields and properties. By taking the approach one step further and by assiduous collection of various subfractions it becomes possible to develop a resid map and add an extra dimension to investigation of petroleum composition and also adds the possibility of using a three-dimensional map, thereby enhancing knowledge and predictability.

Thus, advanced resid fractionation and evaluation has played a significant role,along with the physical testing methods (Speight, 2001, 2014; Speight and Ozum, 2002; Speight, 2004a,b; Hsu and Robinson, 2006; Gary et al., 2007) and will no doubt continue to pay a role in the refinery of the future (Speight, 2011). For example, fractionation has been, has continues to be, used as a means of describing a feedstock in the form of a molecular profile. This was a very simple, but adequate, *map* from which the yields of the primary products (i.e. distillation fractions) were presented and allowed refiners to be prepared for a variety of straight-run hydrocarbon products than had previously been thought. At the same time, and as feedstock quality changed, the hydrogen content and molecular weight were considered to be important parameters and worth *mapping* as a match to other properties, such as nitrogen content and viscosity (Speight, 2001, 2014; Speight and Ozum, 2002; Hsu and Robinson, 2006; Gary et al., 2007). In many cases the properties were inserted on the correct numbered scale as one of the *map* coordinates.

The reason for the importance of the hydrogen content and the molecular weight is that both are used to give an evaluation of the residuum and product yields in terms of the hydrogen required to upgrade the feedstock to specific products. In addition, nitrogen content indicated not only the hydrogen required to remove it as ammonia but also the type of catalyst necessary for processing. Viscosity is an indicator of the ability of the feedstock to flow, or more correctly to resist flow. Other properties such as pour point were also used to predict feedstock behavior in a refinery. A combination of viscosity, API gravity, and sulfur—i.e., the preliminary assay (Speight, 2001, 2014, 2015)—also presented indications of residuum behavior in a refinery. Indeed, consideration of pour point and reservoir temperature together—which gives a real picture of whether or not the oil is in the liquid state—is the best gauge of fluidity and flow in a reservoir as well as in the refinery (Speight, 2009, 2014, 2015).

Thus, the data derived from any one, or more, of the evaluation techniques described in this text give an indication of the characteristics of the feedstock and products as well as options for feedstock processing and for the prediction of product properties. Other properties may also be required for more detailed evaluation of the feedstock and for comparison between feedstocks and product yields/properties even though they may not play any role in dictating which refinery operations are necessary for processing.

However, proceeding from the raw evaluation data to full-scale production is not the preferred step. Further evaluation of feedstock processability is usually through the use of a pilot-scale operation followed by scale-up to a demonstration-size plant. It will then be possible to develop accurate and realistic relationships using the data obtained from the actual plant operations. After that, feedstock mapping can play an important role in assisting in the various *tweaks* that are needed to maintain a healthy process to produce saleable products with the necessary properties.

Indeed, the use of physical properties for feedstock evaluation and product slate has continued in refineries and in process research laboratories to the present and will continue for some time. It is, of course, a matter of choosing the relevant and meaningful properties to meet the nature of the task.

REFERENCES

ASTM D36. Standard Test Method for Softening Point of Bitumen (Ring-and-Ball Apparatus). Annual Book of Standards. ASTM International, West Conshohocken, PA.
ASTM D56. Standard Test Method for Flash Point by Tag Closed Cup Tester. Annual Book of Standards. ASTM International, West Conshohocken, PA.

ASTM D61. Standard Test Method for Softening Point of Pitches (Cube-in-Water Method). Annual Book of Standards. ASTM International, West Conshohocken, PA.
ASTM D70. Standard Test Method for Density of Semi-Solid Bituminous Materials (Pycnometer Method). Annual Book of Standards. ASTM International, West Conshohocken, PA.
ASTM D88. Standard Test Method for Saybolt Viscosity. Annual Book of Standards. ASTM International, West Conshohocken, PA.
ASTM D91. Standard Test Method for Precipitation Number of Lubricating Oils. Annual Book of Standards. ASTM International, West Conshohocken, PA.
ASTM D92. Standard Test Method for Flash and Fire Points by Cleveland Open Cup Tester. Annual Book of Standards. ASTM International, West Conshohocken, PA.
ASTM D93. Standard Test Methods for Flash Point by Pensky-Martens Closed Cup Tester. Annual Book of Standards. ASTM International, West Conshohocken, PA.
ASTM D128. Standard Test Methods for Analysis of Lubricating Grease. Annual Book of Standards. ASTM International, West Conshohocken, PA.
ASTM D129. Standard Test Method for Sulfur in Petroleum Products (General High Pressure Decomposition Device Method). Annual Book of Standards. ASTM International, West Conshohocken, PA.
ASTM D140. Standard Practice for Sampling Bituminous Materials. Annual Book of Standards. ASTM International, West Conshohocken, PA.
ASTM D189. Standard Test Method for Conradson Carbon Residue of Petroleum Products. Annual Book of Standards. ASTM International, West Conshohocken, PA.
ASTM D287. Standard Test Method for API Gravity of Crude Petroleum and Petroleum Products (Hydrometer Method). Annual Book of Standards. ASTM International, West Conshohocken, PA.
ASTM D482. Standard Test Method for Ash from Petroleum Products. Annual Book of Standards. ASTM International, West Conshohocken, PA.
ASTM D524. Standard Test Method for Ramsbottom Carbon Residue of Petroleum Products. Annual Book of Standards. ASTM International, West Conshohocken, PA.
ASTM D893. Standard Test Method for Insolubles in Used Lubricating Oils. Annual Book of Standards. ASTM International, West Conshohocken, PA.
ASTM D977. Standard Specification for Emulsified Asphalt. Annual Book of Standards. ASTM International, West Conshohocken, PA.
ASTM D1018. Standard Test Method for Hydrogen in Petroleum Fractions. Annual Book of Standards. ASTM International, West Conshohocken, PA.
ASTM D1250. Standard Guide for Use of the Petroleum Measurement Tables. Annual Book of Standards. ASTM International, West Conshohocken, PA.
ASTM D1266. Standard Test Method for Sulfur in Petroleum Products (Lamp Method). Annual Book of Standards. ASTM International, West Conshohocken, PA.
ASTM D1298. Standard Test Method for Density, Relative Density, or API Gravity of Crude Petroleum and Liquid Petroleum Products by Hydrometer Method. Annual Book of Standards. ASTM International, West Conshohocken, PA.
ASTM D1552. Standard Test Method for Sulfur in Petroleum Products (High-Temperature Method). Annual Book of Standards. ASTM International, West Conshohocken, PA.
ASTM D2007. Standard Test Method for Characteristic Groups in Rubber Extender and Processing Oils and Other Petroleum-Derived Oils by the Clay-Gel Absorption Chromatographic Method. Annual Book of Standards. ASTM International, West Conshohocken, PA.
ASTM D2170. Standard Test Method for Kinematic Viscosity of Asphalts (Bitumens). Annual Book of Standards. ASTM International, West Conshohocken, PA.
ASTM D2171. Standard Test Method for Viscosity of Asphalts by Vacuum Capillary Viscometer. Annual Book of Standards. ASTM International, West Conshohocken, PA.

ASTM D2319. Standard Test Method for Softening Point of Pitch (Cube-in-Air Method). Annual Book of Standards. ASTM International, West Conshohocken, PA.
ASTM D2397. Standard Specification for Cationic Emulsified Asphalt. Annual Book of Standards. ASTM International, West Conshohocken, PA.
ASTM D2415. Standard Test Method for Ash in Coal Tar and Pitch. Annual Book of Standards. ASTM International, West Conshohocken, PA.
ASTM D2416. Standard Test Method for Coking Value of Tar and Pitch (Modified Conradson). Annual Book of Standards. ASTM International, West Conshohocken, PA.
ASTM D2502. Standard Test Method for Estimation of Mean Relative Molecular Mass of Petroleum Oils from Viscosity Measurements. Annual Book of Standards. ASTM International, West Conshohocken, PA.
ASTM D2503. Standard Test Method for Relative Molecular Mass (Molecular Weight) of Hydrocarbons by Thermoelectric Measurement of Vapor Pressure. Annual Book of Standards. ASTM International, West Conshohocken, PA.
ASTM D2622. Standard Test Method for Sulfur in Petroleum Products by Wavelength Dispersive X-ray Fluorescence Spectrometry. Annual Book of Standards. ASTM International, West Conshohocken, PA.
ASTM D2878. Standard Test Method for Estimating Apparent Vapor Pressures and Molecular Weights of Lubricating Oils. Annual Book of Standards. ASTM International, West Conshohocken, PA.
ASTM D3104. Standard Test Method for Softening Point of Pitches (Mettler Softening Point Method). Annual Book of Standards. ASTM International, West Conshohocken, PA.
ASTM D3120. Standard Test Method for Trace Quantities of Sulfur in Light Liquid Petroleum Hydrocarbons by Oxidative Microcoulometry. Annual Book of Standards. ASTM International, West Conshohocken, PA.
ASTM D3142. Standard Test Method for Specific Gravity, API Gravity, or Density of Cutback Asphalts by Hydrometer Method. Annual Book of Standards. ASTM International, West Conshohocken, PA.
ASTM D3143. Standard Test Method for Flash Point of Cutback Asphalt with Tag Open-Cup Apparatus. Annual Book of Standards. ASTM International, West Conshohocken, PA.
ASTM D3228. Standard Test Method for Total Nitrogen in Lubricating Oils and Fuel Oils by Modified Kjeldahl Method. Annual Book of Standards. ASTM International, West Conshohocken, PA.
ASTM D3279. Standard Test Method for n-Heptane Insolubles. Annual Book of Standards. ASTM International, West Conshohocken, PA.
ASTM D3343. Standard Test Method for Estimation of Hydrogen Content of Aviation Fuels. Annual Book of Standards. ASTM International, West Conshohocken, PA.
ASTM D3461. Standard Test Method for Softening Point of Asphalt and Pitch (Mettler Cup-and-Ball Method). Annual Book of Standards. ASTM International, West Conshohocken, PA.
ASTM D3701. Standard Test Method for Hydrogen Content of Aviation Turbine Fuels by Low Resolution Nuclear Magnetic Resonance Spectrometry. Annual Book of Standards. ASTM International, West Conshohocken, PA.
ASTM D3712. Standard Test Method of Analysis of Oil-Soluble Petroleum Sulfonates by Liquid Chromatography. Annual Book of Standards. ASTM International, West Conshohocken, PA.
ASTM D4045. Standard Test Method for Sulfur in Petroleum Products by Hydrogenolysis and Rateometric Colorimetry. Annual Book of Standards. ASTM International, West Conshohocken, PA.
ASTM D4052. Standard Test Method for Density, Relative Density, and API Gravity of Liquids by Digital Density Meter. Annual Book of Standards. ASTM International, West Conshohocken, PA.

ASTM D4055. Standard Test Method for Pentane Insolubles by Membrane Filtration. Annual Book of Standards. ASTM International, West Conshohocken, PA.
ASTM D4124. Standard Test Method for Separation of Asphalt into Four Fractions. Annual Book of Standards. ASTM International, West Conshohocken, PA.
ASTM D4287. Standard Test Method for High-Shear Viscosity Using a Cone/Plate Viscometer. Annual Book of Standards. ASTM International, West Conshohocken, Pennsylvania.
ASTM D4294. Standard Test Method for Sulfur in Petroleum and Petroleum Products by Energy Dispersive X-ray Fluorescence Spectrometry. Annual Book of Standards. ASTM International, West Conshohocken, PA.
ASTM D4402. Standard Test Method for Viscosity Determination of Asphalt at Elevated Temperatures Using a Rotational Viscometer. Annual Book of Standards. ASTM International, West Conshohocken, PA.
ASTM D4530. Standard Test Method for Determination of Carbon Residue (Micro Method). Annual Book of Standards. ASTM International, West Conshohocken, PA.
ASTM D4715. Standard Test Method for Coking Value of Tar and Pitch (Alcan). Annual Book of Standards. ASTM International, West Conshohocken, PA.
ASTM D4893. Standard Test Method for Determination of Pitch Volatility. Annual Book of Standards. ASTM International, West Conshohocken, PA.
ASTM D5002. Standard Test Method for Density and Relative Density of Crude Oils by Digital Density Analyzer. Annual Book of Standards. ASTM International, West Conshohocken, PA.
ASTM D5185. Standard Test Method for Multielement Determination of Used and Unused Lubricating Oils and Base Oils by Inductively Coupled Plasma Atomic Emission Spectrometry (ICP-AES). Annual Book of Standards. ASTM International, West Conshohocken, PA.
ASTM D5291. Standard Test Methods for Instrumental Determination of Carbon, Hydrogen, and Nitrogen in Petroleum Products and Lubricants. Annual Book of Standards. ASTM International, West Conshohocken, PA.
ASTM D6560. Standard Test Method for Determination of Asphaltenes (Heptane Insolubles) in Crude Petroleum and Petroleum Products. Annual Book of Standards. ASTM International, West Conshohocken, PA.
ASTM D7060, 2015. Standard Test Method for Determination of the Maximum Flocculation Ratio and Peptizing Power in Residual and Heavy Fuel Oils (Optical Detection Method). Annual Book of Standards. ASTM International, West Conshohocken. Pennsylvania.
ASTM D7395. Standard Test Method for Cone/Plate Viscosity at a 500 s-1 Shear Rate. Annual Book of Standards. ASTM International, West Conshohocken, PA.
ASTM E102. Standard Test Method for Saybolt Furol Viscosity of Bituminous Materials at High Temperatures. Annual Book of Standards. ASTM International, West Conshohocken, PA.
ASTM E258. Standard Test Method for Total Nitrogen in Organic Materials by Modified Kjeldahl Method. Annual Book of Standards. ASTM International, West Conshohocken, PA.
ASTM E385. Standard Test Method for Oxygen Content Using a 14-MeV Neutron Activation and Direct-Counting Technique. Annual Book of Standards. ASTM International, West Conshohocken, PA.
ASTM E777. Standard Test Method for Carbon and Hydrogen in the Analysis Sample of Refuse-Derived Fuel. Annual Book of Standards. ASTM International, West Conshohocken, PA.
ASTM E778. Standard Test Methods for Nitrogen in the Analysis Sample of Refuse-Derived Fuel. Annual Book of Standards. ASTM International, West Conshohocken, PA.

Gary, J.G., Handwerk, G.E., Kaiser, M.J., 2007. Petroleum Refining: Technology and Economics, fifth ed. CRC Press, Taylor & Francis Group, Boca Raton, FL.

Hesp, S.A.M., Shurvell, H.F., 2010. X-ray fluorescence detection of waste engine oil residue in asphalt and its effect on cracking in service. Int. J. Pavement Eng. 11 (6), 541–553.

Hill, D.R., Jennings, A.A., 2011. Bioasphalt from Urban Yard Waste Carbonization A Student Study. Report FHWA/OH-2011/13. Case Western Reserve University, Cleveland, Ohio and Office of Research and Development, Ohio Department of Transportation, Columbus, OH.

Hsu, C.S., Robinson, P.R. (Eds.), 2006. Practical Advances in Petroleum Processing vols 1 and 2 Springer Science, New York.

IP 80, 2014. Breaking Point of Bitumen—Fraass Method. IP Standard Methods. The Energy Institute, London, United Kingdom.

Koots, J.A., Speight, J.G., 1975. The relation of petroleum resins to asphaltenes. Fuel 54, 179.

Mitchell, D.L., Speight, J.G., 1973. The solubility of asphaltenes in hydrocarbon solvents. Fuel 52, 149.

Parkash, S., 2003. Refining Processes Handbook. Gulf Professional Publishing, Elsevier, Amsterdam, Netherlands.

Reynolds, J.G., 1998. Metals and heteroatoms in heavy crude oils. In: Speight, J.G. (Ed.), Petroleum Chemistry and Refining Taylor & Francis, Philadelphia, PA, pp. 63–120. (Chapter 3).

Speight, J.G., 1992. Asphalt. Kirk-Othmer Encyclopedia of Chemical Technology, fourth ed., vol. 3, pp. 689–724.

Speight, J.G., 1994. Chemical and physical studies of petroleum asphaltenes In: Yen, T.F. Chilingarian, G.V. (Eds.), Asphaltenes and Asphalts, I. Developments in Petroleum Science, 40 Elsevier, Amsterdam, Netherlands. (Chapter 2).

Speight, J.G., 2000. The Desulfurization of Heavy Oils and Residua, second ed. Marcel Dekker Inc., New York.

Speight, J.G., 2001. Handbook of Petroleum Analysis. John Wiley & Sons Inc., New York, NY.

Speight, J.G., 2004a. Petroleum asphaltenes. Part 1: asphaltenes, resins and the structure of petroleum. Oil Gas Sci. Technol. Rev. IFP 59 (5), 467–477.

Speight, J.G., 2004b. Petroleum asphaltenes. Part 2: the effect of asphaltenes and resin constituents on recovery and refining processes. Oil Gas Sci. Technol. Rev. IFP 59 (5), 479–488.

Speight, J.G., 2009. Enhanced Recovery Methods for Heavy Oil and Tar Sands. Gulf Publishing Company, Houston, TX.

Speight, J.G., 2011. The Refinery of the Future. Gulf Professional Publishing, Elsevier, Oxford, United Kingdom.

Speight, J.G., 2013. The Chemistry and Technology of Coal, third ed. CRC Press, Taylor & Francis Group, Boca Raton, FL.

Speight, J.G., 2014. The Chemistry and Technology of Petroleum, fifth ed. CRC Press, Taylor & Francis Group, Boca Raton, FL.

Speight, J.G., 2015. Handbook of Petroleum Product Analysis, second ed. John Wiley & Sons Inc., Hoboken, NJ.

Speight, J.G., Ozum, B., 2002. Petroleum Refining Processes. Marcel Dekker Inc., New York, NY.

TRB, 2010. Development in Asphalt Binder Specifications. Transportation Research Circular E-C147. Transportation Research Board, Washington, DC.

TRB, 2012. Alternative Binders for Sustainable Asphalt Pavements. Transportation Research Circular E-C165. E-C147. Transportation Research Board, Washington, DC.

CHAPTER 4

Test Methods for Asphalt Binders

4.1 INTRODUCTION

The chemical composition of asphalt and related products (see Chapters 2 and 3) is complex, which makes it essential that the most appropriate analytical methods are selected from a comprehensive list of methods and techniques that are used for the analysis of samples (Goodrich et al., 1986; Speight, 2001, 2014, 2015a). Furthermore, samples may be disturbed during sampling, storage, and pre-treatment (Speight, 2015a,b). Also, most laboratory experiments impose steady environmental conditions while in an outdoor climate these conditions show dynamic behavior. However, the manner in which the corrections are applied must involve quality data (ASTM D6299; ASTM D6792).

In the early days of asphalt use, there was not the need to understand the character and behavior of asphalt and the asphalt–aggregate mix in the detail that is currently required (Abraham, 1945; Barth, 1962; Hoiberg, 1964). Asphalt product and asphalt–aggregate mixing were relatively simple but as the twentieth century matured and with the increased popularity of the automobile and the need for well-constructed roadway systems, the desired product was an asphalt and an asphalt–aggregate mix that could withstand not only increased traffic loads but also variations in climatic conditions.

To this end, there are several national organizations that develop specifications. Countries, states, counties, and cities usually adopt these specifications, but may adapt them with some changes based on the particular local conditions of terrain and environment. The two most common groups who write standard specifications for asphalt are: (i) the American Society for Testing and Materials—ASTM International—which is made up of users such as government engineers and producers–suppliers, and (ii) the American Association of State Highway and Transportation Officials—AASHTO—made up of representatives from 50 states (of the United States) as well as the Federal Highway Administration. The test methods designated as standard test methods by both organizations are tried and true and are also used by many international groups (Table 4.1). In

Asphalt Materials Science and Technology.
DOI: http://dx.doi.org/10.1016/B978-0-12-800273-5.00004-0

Table 4.1 Commonly used standard testing methods as designed by AASHTO and ASTM for asphalt[a]

AASHTO Test Methods
AASHTO CT366 Method of Test for Stabilometer Value
AASHTO CT371 Method of Test for Resistance of Compacted Bituminous Mixture to Moisture Induced Damage
AASHTO M81: Cutback Asphalt (Rapid Curing)
AASHTO M82: Cutback Asphalt (Medium Curing)
AASHTO M140: Emulsified Asphalt
AASHTO T166 Bulk Specific Gravity of Compacted Asphalt Mixtures
AASHTO M208: Cationic Emulsified Asphalt
AASHTO M316: Polymer-Modified Cationic Emulsified Asphalt
AASHTO M320: Standard Specification for Performance Graded Asphalt Binder
AASHTO R18: Establishing and Implementing a Quality Management System for Construction Materials Testing Laboratories
AASHTO T27: Sieve Analysis of Coarse and Fine Aggregate
AASHTO T30: Mechanical Analysis of Extracted Aggregate
AASHTO T44: Solubility of Bituminous Materials
AASHTO T48: Flash and Fire Points by Cleveland Open Cup Tester
AASHTO T49: Test Method Penetration of Bituminous Materials
AASHTO T51: Test Method for Ductility of Bituminous Materials
AASHTO T53: Softening Point of Bitumen (Ring and Ball Apparatus)
AASHTO T55: Test Method for Water in petroleum Products and Bituminous Materials by Distillation
AASHTO T84: Specific Gravity and Absorption of Fine Aggregate
AASHTO T85: Specific Gravity and Absorption of Coarse Aggregate
AASHTO T96: Los Angeles Abrasion
AASHTO T102: Spot Test of Asphaltic Materials
AASHTO T104: Sulfate Soundness
AASHTO T110: Test for Moisture or Volatile Distillates in Hot-Mix Asphalt
AASHTO T112: Clay Lumps and Friable Particles
AASHTO T113: Lightweight Particles
AASHTO T164: Quantitative Extraction of Asphalt Binder from Hot-Mix Asphalt
AASHTO T166: Bulk Specific Gravity of Compacted Hot-Mix Asphalt Using Saturated Surface-Dry Specimens
AASHTO T170: Recovery of Asphalt from Solution By Abson Method
AASHTO T176: Sand equivalent
AASHTO T179: Test Method for Effects of Heat and Air on Asphaltic Materials (Thin Film Oven Method)
AASHTO T182: Coating and Stripping of Bitumen-Aggregate Mixtures
AASHTO T201: Test Method for Kinematic Viscosity of Asphalts
AASHTO T202: Test Method for Viscosity of Asphalts by Vacuum Capillary Viscometer

(Continued)

Table 4.1 Commonly used standard testing methods as designed by AASHTO and ASTM for asphalt[a] *(Continued)*

AASHTO T209: Theoretical Maximum Specific Gravity and Density of Hot-Mix Asphalt
AASHTO T228: Density of Semi-Solid Bituminous Materials (Pycnometer Method)
AASHTO T240: Test Method for Effect of Heat and Air on a Moving Film of Asphalt (RTFOT Method)
AASHTO T245: Resistance to Plastic Flow of Bituminous Mixtures Using Marshall Apparatus
AASHTO T269: Percent Air Voids in Compacted Dense and Open Asphalt Mixtures
AASHTO T275: Bulk Specific Gravity of Compacted Bituminous Mixtures Using Paraffin-Coated Specimens
AASHTO T283: Resistance of Compacted Asphalt Mixtures to Moisture-Induced Damage
AASHTO T304: Uncompacted Void Content of Fine Aggregate
AASHTO T308: Determining the Asphalt Binder Content of Hot-Mix Asphalt (HMA) by the Ignition Method
AASHTO T310: Standard Test Method for In-Place Density and Water Content of Soil and Soil-Aggregate by Nuclear Methods (Shallow Depth)
AASHTO T312: Preparing and Determining the Density of Hot-Mix Asphalt (HMA) Specimens by Means of the Superpave Gyratory Compactor
AASHTO T316: Viscosity Determination of Asphalt Binders using Rotational Viscometer
AASHTO T320: Standard Method of Test for Determining the Permanent Shear Strain and Stiffness of Asphalt Mixtures using the Superpave Shear Tester
AASHTO T321: Standard Method of Test for Determining the Fatigue Life of Compacted Hot-Mix Asphalt subjected to Repeated Flexural Bending
AASHTO T324: Standard Method of Test for Hamburg Wheel-Track Testing of Compacted Hot-Mix Asphalt (HMA)
AASHTO T329: Moisture Content of Hot-Mix Asphalt (HMA) by Oven Method
ASTM Test Methods
ASTM C29: Bulk Density ("Unit Weight") and Voids in Aggregate
ASTM C88: Soundness of Aggregates by Use of Sodium Sulfate or Magnesium Sulfate
ASTM C123: Lightweight Particles in Aggregate
ASTM C127: Density, Relative Density (Specific Gravity), and Absorption of Coarse Aggregate
ASTM C128: Density, Relative Density (Specific Gravity), and Absorption of Fine Aggregate
ASTM C131: Resistance to Degradation of Small-Size Coarse Aggregate by Abrasion and Impact in the Los Angeles Machine

(Continued)

Table 4.1 Commonly used standard testing methods as designed by AASHTO and ASTM for asphalt[a] *(Continued)*

ASTM C136: Sieve Analysis of Fine and Coarse Aggregates
ASTM C142: Clay Lumps and Friable Particles in Aggregates
ASTM C1252: Uncompacted Void Content of Fine Aggregate (as Influenced by Particle Shape, Surface Texture, and Grading)
ASTM D5: Test Method for Penetration of Bituminous Materials
ASTM D6: Loss on Heating of Oil and Asphaltic Compounds
ASTM D36: Softening Point of Bitumen (Ring & Ball Apparatus
ASTM D70: Density of Semi-Solid Bituminous Materials (Pycnometer Method)
ASTM D75: Sampling Aggregates
ASTM D92: Flash and Fire Points by Cleveland Open Cup
ASTM D113: Ductility of Bituminous Materials
ASTM D977: Emulsified Asphalt
ASTM D1074: Compressive Strength of Bituminous Mixtures
ASTM D1188: Bulk Specific Gravity and Density of Compacted Bituminous Mixtures Using Coated Samples
ASTM D1461: Moisture or Volatile Distillates in Hot-Mix Asphalt (HMA)
ASTM D1754: Test Method for Effects of Heat and Air on Asphaltic Materials (Thin Film Oven Method)
ASTM D1856: Test for Recovery of Asphalt from Solution By Abson Method
ASTM D2026: Cutback Asphalt (Slow-Curing)
ASTM D2027: Cutback Asphalt (Medium-Curing)
ASTM D2028: Cutback Asphalt (Rapid-Curing)
ASTM D2041: Theoretical Maximum Specific Gravity and Density of Bituminous Paving Mixtures
ASTM D2042: Solubility of Asphalt Material in Trichloroethylene (also other solvents
ASTM D2170: Test Method for Kinematic Viscosity of Asphalts
ASTM D2171: Viscosity of Asphalts by Vacuum Capillary Viscometer
ASTM D2172: Quantitative Extraction of Bitumen From Bituminous Paving Mixtures
ASTM D2397: Cationic Emulsified Asphalt
ASTM D2419: Sand Equivalent Value of Soils and Fine Aggregate
ASTM D2726: Bulk Specific Gravity and Density of Non-Absorptive Compacted Bituminous Mixtures
ASTM D2746: Staining Tendency of Asphalt
ASTM D2872: Test Method for Effect of Heat and Air on a Moving Film of Asphalt (RTFOT Method)
ASTM D3203: Percent Air Voids in Compacted Dense and Open Bituminous Paving Mixtures
ASTM D3320: Emulsified Coal Tar Pitch
ASTM D3666: Minimum Requirements for Agencies Testing and Inspecting Road and Paving Materials

(Continued)

Table 4.1 Commonly used standard testing methods as designed by AASHTO and ASTM for asphalt[a] *(Continued)*

ASTM D4402: Viscosity Determinations of Unfilled Asphalts Using the Brookfield Thermosel Apparatus
ASTM D4791: Flat Particles, Elongated Particles, or Flat and Elongated Particles in Coarse Aggregate
ASTM D4867: Effect of Moisture on Asphalt Concrete Paving Mixtures
ASTM D5444: Mechanical Size Analysis of Extracted Aggregate
ASTM D5727: Emulsified Refined Coal Tar
ASTM D6307: Asphalt Content of Hot-Mix Asphalt by Ignition Method
ASTM D6925: Preparation and Determination of the Relative Density of Hot Mix Asphalt (HMA) Specimens by Means of the Superpave Gyratory Compactor
ASTM D6926: Preparation of Bituminous Specimens Using Marshall Apparatus
ASTM D6927: Marshall Stability and Flow of Bituminous Mixtures
ASTM D6938: Standard Test Method for In-Place Density and Water Content of Soil and Soil-Aggregate by Nuclear Methods (Shallow Depth)
ASTM E329: Agencies Engaged in Construction Inspection and/or Testing
ASTM D7064 Standard Practice for Open-Graded Friction Course (OGFC) Mix Design

[a]Listed alphanumerically and not by preference.

addition, once a test method has been accepted it is not included in the list of test methods there for eternity. Suggested changes in the method of analysis will be tried and, if verified, the test method will be modified or a new test method may be put in place to supersede the older test method.

In addition, with the increased demands on the petroleum industry during, and after, World War II, and the emergence of the age of petrochemicals and plastics, the petroleum industry needed to produce materials not even considered as products in the decade before the war. Thus, petroleum refining took on the role of technological innovator as new and better processes were invented and advances in the use of materials for reactors were developed. In addition, there became a necessity to find out more about petroleum-based asphalt so that refiners might be able to enjoy the luxury of predictability and plan a product slate that was based on market demand. This was a difficult task when the character of the crude oil was unknown and the ability to produce quality asphalt became a hit-and-miss affair. The idea that petroleum refining—leading to the production of quality asphalt—should be a *hit and miss* affair was not acceptable.

The production of asphalt requires not only knowledge of its chemical and physical properties but also knowledge of its chemical and physical reactivity. Furthermore, because petroleum varies markedly in its properties and composition according to the source, it also varies in its chemical and physical reactivity. Thus, knowledge of petroleum reactivity is required for optimization of already-existing asphalt production processes as well as for the development and design of new asphalt production processes.

For example, valuable information can be obtained from the true boiling point (TBP) curve, which is a function of percent weight distilled and temperature, that is, a boiling point distribution (Speight, 2001, 2014, 2015a). However, the boiling range does not convey much detail about the chemical reactivity of crude oil or the properties of any particular product, especially asphalt. In addition to the boiling point distribution, it is possible to measure bulk physical properties, such as specific gravity and viscosity that have assisted in the establishment of certain empirical relationships for petroleum processing from the *TBP* curve. Many of these relationships include assumptions that are based on experience with a range of feedstocks. However, the chemical aspects of refining feedstocks that contain different proportions of chemical species emphasize the need for more definitive data that would enable more realistic predictions to be made of crude oil behavior in refinery operations.

Although asphalts have many physical properties, it is the chemical forces at the molecular level that are responsible for the fact that these properties differ significantly. Asphalt is a complex mixture of molecules, ranging from rather non-polar hydrocarbons similar in composition to waxes to highly polar or polarizable hydrocarbon molecules containing condensed aromatic ring systems that incorporate heteroatoms such as nitrogen, oxygen, sulfur, and metals (particularly nickel and vanadium). The properties of the higher-molecular-weight constituents (resin constituents and asphaltene constituents) are particularly important and it is the molecular association of such constituents that is responsible for the formation of molecular agglomerates at the nanoscale level.

These agglomerates are bonded together, not by primary chemical bonds but by polar association forces such as hydrogen bonding and dipole interactions and the strength of these bonding forces is highly dependent on temperature. Thus, as the temperature increases, these reversible bonds are broken and the sizes of the agglomerates are reduced. It is the reversible nature of these interactions that gives asphalt its unique high-temperature viscosity susceptibility. The use of various standard test methods, as applied to asphalt, also gives an indication of the molecular interactions at the nanoscale level through the physical and chemical properties measured at the macro level (physicochemical characterization) (Planche et al., 1991; Speight, 1992). Chemical and physical property data and methods for their determination are cited in this chapter to illustrate the fundamental effects of oxidation on asphalt properties.

The consistency or viscosity of asphalt cement varies with temperature, and asphalt is graded based on ranges of consistency at a standard temperature. Careless temperature and mixing control can cause more hardening damage to asphalt cement than many years of service on a roadway. A standardized viscosity or penetration test is commonly specified to measure paving asphalt consistency. Air-blown asphalts typically use a softening point test. Purity of asphalt cement can be easily tested since it is composed almost entirely of organic-based asphalt, which is soluble in carbon disulfide. Refined asphalts are usually more than 99.5% soluble in carbon disulfide and any impurities that remain are inert. Because of the hazardous flammable nature of carbon disulfide, trichloroethylene, which is also an excellent solvent for asphalt cement, is used in the solubility purity tests.

Asphalt cement must be free of water or moisture as it leaves the refinery. However, transports loading the asphalt may have moisture present in their tanks which can cause the asphalt to foam when it is heated above 100°C (212°F), which is a safety hazard. Specifications usually require that asphalts do not foam at temperatures up to 175°C (345°F). Asphalt cement, if heated to a high enough temperature, will release fumes which will flash in the presence of a spark or open flame. The temperature at which this occurs is called the flashpoint, and is well above temperatures normally used in paving operations. Because of the possibility of asphalt foaming and to ensure an adequate margin of safety, the flashpoint of the asphalt is measured and controlled.

Another important engineering property of asphalt cement is its ductility, which is a measure of the ability of the asphalt to be pulled, drawn, or deformed. In asphalt cement, the presence or absence of ductility is usually more important than the actual degree of ductility because some types of asphalt cement with a high degree of ductility are also more temperature-sensitive. Ductility is measured by an extension test (AASHTO T51; ASTM D113), whereby a standard asphalt cement briquette molded under standard conditions and dimensions is pulled at a standard temperature (normally 25°C (77°F)) until it breaks under tension. The elongation at which the asphalt cement sample breaks is a measure of the ductility of the sample.

In summary, asphalt testing is a quality control measure (Speight, 2001, 2014, 2015a). The quality of asphalt cement is affected by the inherent properties of the petroleum crude oil and the residuum from which it was produced (see Chapters 2 and 3) (Speight, 2014). Different oil fields and areas produce crude oils with very different characteristics. The refining method also affects the quality of the asphalt cement. For engineering and construction purposes, there are three important factors to consider: consistency, also called the viscosity or the degree of fluidity of asphalt at a particular temperature, purity, and safety. Furthermore, any of the methods described herein might also be applied to the analysis of sample for environmental purposes. However, methods for analytical purpose are not intended to be the focus of this text and are described elsewhere (Speight, 2005, 2014; Speight and Arjoon, 2012).

This chapter presents some of the methods that are generally applied to investigations of the composition of the feedstock in terms of chemical structures as well as methods that might be preferred for refining. There are, of course, many analytical methods that can be applied to the analysis of petroleum and petroleum products but they vary with sample condition and composition. More specifically, this chapter deals with the more common methods used to define chemical and physical properties of the sample.

4.2 PROPERTIES AND PERFORMANCE

Asphalt is used for paving, roofing, industrial, and special purposes. Oxidized asphalt is used in roofing operations, pipe coating, undersealing for concrete pavements, hydraulic applications, membrane envelopes, some paving-grade mixes, and the manufacture of paints (Speight, 1992). First a reminder that asphalt is known by many different names and names such as asphalt concrete, asphalt cement, asphalt binder, hot-mix asphalt, plant mix, bituminous mix, and bituminous concrete are in common use. To be more specific to this text, road asphalt is a combination of two primary ingredients—asphalt and aggregate—both of which require assiduous testing before use.

4.2.1 General Properties

Asphalt binders are typically characterized by their properties at different temperatures and stages of in-service properties (Speight, 1992; Button, 1994). The *purity* of an asphalt binder is defined by the solubility in carbon disulfide and should be more than 99.5% (w/w) soluble in this solvent—impurities, if they are present, are inert. Typically, the binder is free of water or moisture as it leaves the refinery. However, tank transports loading binder may have some moisture present in their tanks. If any water is inadvertently present in the binder, it may cause the binder to foam when it is heated above 100°C (212°F).

Binder foaming can be a safety hazard. Specifications usually require that asphalt does not foam at temperatures up to 177°C (350°F). Binders, if heated to a high enough temperature, will flash in the presence of a spark or open flame. The temperature at which this occurs is well above the temperatures normally used in paving operations. However, to be sure there is an adequate margin of safety, the flash point of the binder should be known.

Durability is the measure of how well a binder retains its original characteristics when exposed to normal weathering and aging processes. The performance grading of current binders includes laboratory tests that simulate the weathering and aging processes and establishes pass/fail limits on the test results. The pavement performance is still greatly affected by several variables such as mix design and aggregate characteristics (see Chapter 5). Related to performance are the terms *adhesion* and *cohesion*—adhesion is the binder's ability to stick to the aggregate in the paving mixture whereas cohesion is the ability of the binder to hold the aggregate particles in place in the finished pavement and conversely the ability of the aggregate to hold the binder.

The temperature susceptibility of the binder (which varies among binders from different petroleum sources, even if the binders are of identical grade) is related to the thermoplastic nature—that is, a binder becomes harder (more viscous) as the temperature is decreased and softer (less viscous) as their temperature is increased. Knowledge of the temperature susceptibility of the binder being used in a paving mixture is important because it indicates the temperature at which to mix the binder with aggregate, and the temperature at which to compact the mixture on the roadbed. The temperature susceptibility is an important property of a binder, which must be sufficiently fluid at elevated temperatures to permit coating of the aggregate particles to occur during mixing and to allow these particles to move past each other during compaction. The binder must then be sufficiently viscous at normal air temperatures to hold the aggregate particles in place in the pavement.

Other common characteristics of asphalt binders are: (i) high solubility in aromatic solvents and temperature-sensitive and (ii) prone to aging through oxidation. At high temperatures, asphalt is highly fluid whereas at low temperatures asphalt is stiff and brittle. This change of consistency with temperature is generally called temperature sensitivity or temperature susceptibility. Over time, asphalt undergoes oxidation and as the oxidation process continues, the asphalt tends to become increasingly brittle and thus more prone to cracking (fracturing). The rate of oxidation increases with the temperature of the bituminous material and there is usually a color change of asphalt (from black to gray) as the oxidation process proceeds.

Asphalt, normally a liquid when applied at higher temperature, may be referred to as *non-volatile liquid asphalt*, but semisolid or solid asphalt may be made liquid for easier handling by dissolving them in a solvent and are referred to as *cutback asphalt*. The asphalt and solvent (naphtha, kerosene, or gas oil) are heated to about 105°C (225°F) and passed together through a mechanical mixer, and the effluent then enters a horizontal tank, which is used for further mixing. Liquid asphalt is pumped from the top of the tank into a perforated pipe lying at the bottom of the tank, and circulation is continued until mixing is complete.

If the asphalt requires a change in properties to a harder material, such as is required for roofing, pipe coating, or as an under-sealant or waterproofing material, *soft asphalt* can be converted into harder asphalt by oxidation. This process produces a material that softens at a higher temperature than paving asphalts. It may be air-blown at the refinery, at an asphalt processing plant, or at a roofing material plant. The asphalt is heated to approximately 260°C (500°F) and air is bubbled through the hot-mix for 1–5 h. The product, when cooled, remains in the liquid phase.

Thus, an asphalt binder will harden in the paving mixture during construction and in the pavement itself. The hardening is caused primarily by oxidation (binder combining with oxygen), a process that occurs most readily at higher temperatures (such as construction temperature) and in thin binder films (such as the film coating aggregate particles). During mixing, the binder is at a high temperature and also in a thin film on the aggregate particles, which subject the asphalt to potential (often a reality) oxidation and causing hardening to occur. However, not all binders harden at the same rate when heated in thin films and, therefore, each binder used should be tested to determine the aging characteristics so that construction techniques can be adjusted to minimize hardening. Such adjustments usually involve mixing the binder with the aggregate at the lowest possible temperature for the shortest practical time. In fact, hardening of the binder can (and often does) commence in the pavement after construction but hardening may even commence during transportation as the mix is subjected to aerial oxidation as the main cause because of the subsequent chemical reactions that lead to the introduction of oxygen functions into the asphalt (Moschopedis and Speight, 1973, 1975, 1977, 1978). These processes can be retarded by keeping the number of connected voids (air spaces) in the final pavement low, and a thick binder coating on the aggregate particles.

The chemistry of the oxidation process, although more complex than a simple representation, promotes the formation of additional resin constituents and additional asphaltene constituents from the polar and/or higher-molecular-weight oil non-resin non-asphaltene constituents:

Oil (aromatic) constituents → resin constituents → asphaltene constituents

The increase in the proportion of semisolid (resin) and solid (asphaltene) constituents as a result of air blowing is accompanied by an increase in softening point with only a small loss in volume. A similar increase in softening point by removing oily constituents would cause a considerable decrease in volume. Oxidation is carried out by blowing air through asphalt heated to about 260°C (500°F) and is usually done in a tower (an oxidizer) equipped with a perforated pipe at the bottom through which the air is blown. The asphalt, in the batch mode or continuous mode, is heated until the oxidation reaction starts, but the reaction is exothermic, and the temperature is controlled by regulating the amount of air and by circulating oil or water through cooling coils within the oxidizer. Asphalt products with softening points as high as 180°C (350°F) may be produced.

Despite the complex reaction mechanisms and kinetics in the blowing unit, the general reaction pathway can be summarized as transformation of cyclic structures into resin constituents (polar aromatics) which are, in turn, reacted to form asphaltene constituents. It is generally believed that the primary oxidative process is carbon–carbon bond formation via oxidative condensation and to a lesser extent the formation of ester (and other oxygen-containing) groups, both of which link smaller molecules into larger ones. However, the chemistry is not always as straightforward as the condensation-to-polynuclear-aromatics proponents would suggest. The introduction of additional oxygen functions into already-polar feedstock constituents also occurs, and may even be the predominant mode of the reaction which can also lead to increased molecular weight through the occurrence of intermolecular bonding (Moschopedis and Speight, 1973, 1975, 1977, 1978).

The use of asphalt—in many cases this was naturally occurring *bitumen* or a residuum (see Chapters 1–3) rather than a processed material going back into antiquity—was in fact the first petroleum derivative that was used extensively. Nowadays, a good portion of the asphalt produced from petroleum is consumed in paving roads; the remainder is employed for roofing, paints, varnishes, insulating, rust-protective compositions, battery boxes, and compounding materials that go into rubber products, brake linings, and fuel briquettes. However, asphalt uses can be more popularly divided into uses as road oils, cutback asphalt, asphalt emulsion, and solid asphalt. The properties of asphalt are defined by a variety of standard tests that can be used to define quality and viscosity specifications (Tables 4.2 and 4.3) (Abraham, 1945; Barth, 1962; Hoiberg, 1964; Speight, 1992).

Asphalt is characterized by the properties at different temperatures and stages of life simulated by laboratory aging. *Consistency* is the term used to describe the degree of fluidity or plasticity of binders at any particular

Table 4.2 ASTM standard methods for sampling asphalt and similar materials[a]

ASTM C702	Practice for Reducing Samples of Aggregate to Testing Size
ASTM D2041	Test Method for Theoretical Maximum Specific Gravity and Density of Bituminous Paving Mixtures
ASTM D2234	Practice for Collection of a Gross Sample of Coal
ASTM D2726	Test Method for Bulk Specific Gravity and Density of Non-Absorptive Compacted Bituminous Mixtures
ASTM D3665	Practice for Random Sampling of Construction Materials
ASTM D5361	Practice for Sampling Compacted Bituminous Mixtures for Laboratory Testing
ASTM D5444	Test Method for Mechanical Size Analysis of Extracted Aggregate
ASTM D6307	Test Method for Asphalt Content of Hot-Mix Asphalt by Ignition Method
ASTM D6925	Test Method for Preparation and Determination of the Relative Density of Hot Mix Asphalt (HMA) Specimens by Means of the Superpave Gyratory Compactor
ASTM D6926	Practice for Preparation of Bituminous Specimens Using Marshall Apparatus
ASTM D6927	Test Method for Marshall Stability and Flow of Bituminous Mixtures
ASTM E105	Practice for Probability Sampling of Materials
ASTM E122	Practice for Calculating Sample Size to Estimate, with Specified Precision, the Average for a Characteristic of a Lot or Process
ASTM E141	Practice for Acceptance of Evidence Based on the Results of Probability Sampling

[a]Listed alphanumerically and not by preference.

Table 4.3 ASTM standard test methods used to determine asphalt properties (Listed Numerically and not Preferentially)

Test	Number	Description
Solubility in CS_2	ASTM D4	Determination of the carbon amount of carboids and/or carbenes in mineral matter; trichloroethylene and 1,1,1-trichloroethane are suitable solvents. See also ASTM D2042
Penetration	ASTM D5	Needle penetration asphalt under specified conditions of load, time, and temperature. See also ASTM D243
Loss on heating	ASTM D6	Loss on heating of oil and asphaltic compounds See also ASTM D1754, ASTM D2872
Softening point	ASTM D36	Ring and ball method; used to classify asphalt grades
Specific gravity	ASTM D70	See ASTM D3142
Flash point	ASTM D92	Cleveland open cup method. See also ASTM D93, ASTM D3143
Flash point	ASTM D93	Pensky–Martens closed cup tester. See also ASTM D92, ASTM D3143
Water content	ASTM D95	Determines the water content by distillation with a Dean and Stark receiver
Ductility	ASTM D113	The distance in cm which a standard briquette can he elongated before breaking; also an indication of cohesion and shear susceptibility
Float test	ASTM D139	Normally used for asphalts that are too soft for the penetration test
Sampling	ASTM D140	Provides guidance for the sampling of asphalts
Penetration	ASTM D243	Primarily used with cutback asphalt. See also ASTM D5 and ASTM D2026
Emulsified asphalt	ASTM D244	Covers the composition and classification, storage, use, and specifications. See also ASTM D977
Distillation	ASTM D402	Determination of volatiles content; applicable to road oil and cutback asphalt. See also ASTM D244
Weathering	ASTM D529	Used for determining the relative weather resistance of asphalt. See also ASTM D1670
Specification	ASTM D977	Emulsified asphalt

(*Continued*)

Table 4.3 ASTM standard test methods used to determine asphalt properties (Listed Numerically and not Preferentially) *(Continued)*

Test	Number	Description
Bond and adhesion	ASTM D1191	Determination of bonding strength at low temperatures. See also ASTM D3141 and ASTM D5078
Temp/volume	ASTM D1250	Allows the conversion of volumes of asphalts from one temperature to another. See also ASTM D4311
Stain	ASTM D18	Measures the amount of stain on paper or other cellulosic materials
Compatibility	ASTM D1370	Indicates whether asphalts are compatible and dis-bond under stress. See also ASTM D3407
Weathering	ASTM D1669	Accelerated and Outdoor Weathering of Bituminous Coatings. See also ASTM D1670, ASTM D529
Weathering	ASTM D1670	Accelerated and outdoor weathering of bituminous materials. See also ASTM D1669, ASTM D529
Thin film oven test	ASTM D1754	Determines the hardening effect of heat and air on a film of asphalt. See also ASTM D2872
Specification	ASTM D2026	Cutback asphalt. See also ASTM D5 and ASTM D243
Solubility	ASTM D2042	Solvent: trichloroethylene; toluene or xylene can also be used. See also ASTM D4
Viscosity	ASTM D2170	A measure of resistance to flow. See also ASTM D2171, ASTM D2493, ASTM D3205, ASTM D3381, ASTM D4402, ASTM D4957, ASTM D4989
Viscosity	ASTM D2171	Viscosity of asphalts by vacuum capillary viscometer. See also ASTM D2170, ASTM D2493, ASTM D3205, ASTM D3381, ASTM D4402, ASTM D4957 ASTM D4989
Viscosity	ASTM D2493	Viscosity-temperature chart. See also ASTM D2170, ASTM D2171, ASTM D3205, ASTM D3381, ASTM D4402, ASTM D4957, ASTM D4989
Effect of heat	ASTM D2872	Loss on heating; rolling thin-film oven test. See also ASTM D6, ASTM D1754

(Continued)

Table 4.3 ASTM standard test methods used to determine asphalt properties (Listed Numerically and not Preferentially) *(Continued)*

Test	Number	Description
Specification	ASTM D3141	Asphalt underseal for concrete pavements. See also ASTM D1191, ASTM D5078
Viscosity	ASTM D3205	Viscosity (Flow) of Roofing Bitumens using cone and plate viscometer. See also ASTM D2170, ASTM D2171, ASTM D2493, ASTM D3381, ASTM D4402, ASTM D4957, ASTM D4989
Density	ASTM D3142	Specific gravity, API gravity, or density of cutback asphalt; hydrometer method. See also ASTM D70
Flash point	ASTM D3143	Tag open cup; applicable to cutback asphalt. See also ASTM D92, ASTM D93
Specification	ASTM D3381	Viscosity-graded asphalt cement for use in pavement construction. See also ASTM D2170, ASTM D2171, ASTM D2493, ASTM D3205, ASTM D4402, ASTM D4957, ASTM D4989
Specification	ASTM D3407	Joint sealants. See also ASTM D1370
Temp/volume	ASTM D4311	Volume correction to base temperature. See also ASTM D1250
Viscosity	ASTM D4402	Determination of viscosity at elevated temperatures using a rotational viscometer. See also ASTM D2170, ASTM D2171, ASTM D2493, ASTM D3205, ASTM D3381, ASTM D4957, ASTM D4989
Adsorption	ASTM D4469	Calculation of degree of adsorption of asphalt by an aggregate
Viscosity	ASTM D4957	Viscosity of asphalt emulsion residue and non-Newtonian asphalt; vacuum capillary viscometer. See also ASTM D2170, ASTM D2171, ASTM D2493, ASTM D3205, ASTM D3381, ASTM D4402, ASTM D4989
Viscosity	ASTM D4989	Viscosity of roofing bitumen using the parallel plate plastometer. See also ASTM D2170, ASTM D2171, ASTM D2493, ASTM D3205, ASTM D3381, ASTM D4402, ASTM D4957
Specification	ASTM D5078	Crack filler for asphalt concrete and Portland cement concrete pavements. See also ASTM D1191, ASTM D3141

temperature. The consistency of binder varies with temperature. Binders are graded based on ranges of consistency at a standard temperature. When the asphalt binder is exposed to air in thin films and is subjected to prolonged heating, i.e., during mixing with aggregate, the binder tends to harden. This means that the consistency (viscosity) of the binder has increased for any given temperature. A limited increase is allowable. However, careless temperature and mixing control can cause more damage to the binder, through hardening, than many years of service on the finished roadway.

Asphalt is composed almost entirely of resid, which by definition is non-volatile and soluble in carbon disulfide. Refined binders are almost pure resid and are usually more than 99.5% (w/w) soluble in carbon disulfide—any impurities, if they are present, are inert. Typically, the asphalt is free of water or moisture as it leaves the refinery. However, tank transports loading binder may have some moisture present in their tanks. If any water is inadvertently present in the binder, it may cause the binder to foam when it is heated above 100°C (212°F).

Asphalt foaming can be a safety hazard. Specifications usually require that asphalt does not foam at temperatures up to 177°C (350°F). Asphalt, if heated to a sufficiently high temperature, will flash in the presence of a spark or open flame. The temperature at which this occurs is well above the temperatures normally used in paving operations. However, to be sure there is an adequate margin of safety, the flash point of the binder should be known.

Durability is the measure of how well a binder retains its original characteristics when exposed to normal weathering and aging processes. The performance grading of current binders include laboratory tests that simulate the weathering and aging processes and establishes pass/fail limits on the test results.

All asphalt binders are thermoplastic; that is, they become harder (more viscous) as their temperature decreases and softer (less viscous) as their temperature increases. This characteristic is known as temperature susceptibility, and is one of a binder's most valuable assets. Knowing the temperature susceptibility of the binder being used in a paving mixture is important because it indicates the proper temperature at which to mix the binder with aggregate, and the proper temperature at which to compact the mixture on the roadbed. It is important for asphalt to be temperature-susceptible—the asphalt must be fluid enough at elevated temperatures to permit coating of the aggregate particles during mixing and to allow these particles a degree of mobility during compaction. The asphalt must then become viscous enough at normal air temperatures to hold the aggregate particles in place in the pavement.

Asphalt hardens in the paving mixture during construction and in the pavement itself. The hardening is caused primarily by oxidation (binder combining with oxygen), a process that occurs most readily at higher temperatures (such as construction temperature) and in thin binder films (such as the film coating aggregate particles). During mixing, binder is both at a high temperature and in thin films as it coats the aggregate particles. This makes mixing the stage at which the most severe oxidation and hardening usually occurs.

Not all asphaltic binders harden at the same rate when heated (as required by the thin-film test method) in thin films. Therefore, asphalt used should be tested to determine its aging characteristics so that construction techniques can be adjusted to minimize hardening. Such adjustments usually involve mixing the binder with the aggregate at the lowest possible temperature for the shortest practical time. Asphalt hardening continues in the pavement after construction. Again, oxidation and polymerization are the main causes. These processes can be retarded by keeping the number of connected voids (air spaces) in the final pavement low, and the asphalt binder coating on the aggregate particles thick.

Asphalt can be temporarily liquefied for construction operations in three ways: (i) by melting with heat, (ii) by dissolving the binder in selected solvents—this process is called cutting back (hence, *cutback asphalt*), and (iii) by emulsifying the binder with water (hence, *emulsified asphalt*). Cutback asphalt and emulsified asphalt are usually referred to as *liquid asphalt* to distinguish them as a group from the more typical asphalt binder.

Finally, asphalt is soluble in more volatile phases of petroleum, especially certain types of naphtha (the precursor to gasoline). Consequently, when petroleum-based oil or gasoline is spilled onto asphalt pavements, the pavement will be damaged due to the solution of asphalt in the solvent. One possible reason to use tar as opposed to asphalt as a binder in pavements is that tar is not soluble in petroleum products. Indeed, tar-based pavements might be a better choice than asphalt-based pavements where fuel or oil spills are expected but there are disadvantages to the presence of tar in the asphalt mix (Kriech et al., 1997, 2002).

Other properties including the thermal transmissivity of heat through the material (ASTM C177; ASTM C518; ASTM C1045) are necessary knowledge when asphalt is used as a heat transfer medium or as an insulating material. The test methods provide procedures for calculating the thermal transmission properties of asphalt from standard test methods used to determine heat flux and surface temperatures. For example, the data provide a means of developing the thermal conductivity as a function of temperature for an asphalt specimen from data taken at small or large temperature differences and can be used to characterize material for comparison to material specifications.

4.2.2 Specifications

Specifications are a statement of the necessary properties of asphalt before it is sold to the consumer (Newcomb, 2001a,b,c)—the necessary properties being derived from the various test methods outlined above. These relatively simple descriptions contain substantial technological implications. In order to derive specifications, the binder properties linked to asphalt pavement performance must be identified and standardized test methods must be selected to measure these properties. Data are collected from the test methods and field validation is necessary after which the grading system may need to be reviewed and modified in accordance with the new specifications. Above all, the specifications must be strongly indicative of in-service performance. In short, the specification must address the following items: (i) asphalt properties, (ii) mix characteristics, (iii) pavement performance, (iv) surface characteristics, and (v) user requirements.

Performance specifications, not chemical composition, direct asphalt production. To meet these performance specifications, the asphalt may be air-blown or further processed by solvent precipitation or propane deasphalting. Additionally, the products of other refining processes may be blended with the asphalt to achieve the desired performance specifications. Therefore, the exact chemical composition of asphalt is dependent on the chemical complexity of the original crude petroleum and the manufacturing process, no matter what methods are used for investigations of the composition of the asphalt (Jennings, 1980; Jennings and Prabanic, 1985; Kim and Burati, 1993; Speight, 1992, 2014; Kim et al., 1995; Speight, 2013). In addition, the amounts of the various constituents can vary greatly because of significant differences in crude petroleum from oil field to oil field or even at different locations in the same oil field (Speight, 2014).

While the manufacturing process may change the physical properties of asphalt dramatically, the chemical nature of the asphalt does not change unless thermal cracking occurs. Raising the temperature will increase the likelihood of cracking and cause more volatiles and even higher-boiling components to be released from the residuum. Solvent precipitation (usually using propane or butane) removes high-boiling components from vacuum-processed asphalt, which are then used to make other products. Solvent precipitation results in a harder asphalt that is less resistant to temperature changes and often blended with straight-reduced or vacuum-processed asphalts. The air-blowing process can be a continuous or batch operation. Since the continuous operation is faster and results in softer asphalt, a continuous operation is preferred for processing paving asphalts (Speight, 1992, 2014). Air blowing combines oxygen with hydrogen in the asphalt, producing water vapor. This decreases saturation and increases cross-linking within and between different asphalt molecules. The process is exothermic (heat producing) and may cause a series of chemical reactions, such as oxidation, condensation, dehydration, dehydrogenation, and polymerization. These reactions cause the amount of asphaltenes (hexane-insoluble materials) to increase and the amounts of polar aromatics (hard resins), cycloalkanes, and non-polar aromatics (soft resins) to decrease, while the amounts of aliphatic compounds (oils and waxes) remain about the same. At the same time, the oxygen content of the asphalt increases (Moschopedis and Speight, 1973; Speight, 1992, 2014).

Although no two asphalts are chemically identical and chemical analysis cannot be used to define the exact chemical structure or chemical composition of asphalt, elemental analyses indicate that most asphalts contain 79–88% (w/w) carbon, 7–13% (w/w) hydrogen, traces to 3% (w/w) nitrogen, traces to 8% (w/w) oxygen, traces to 8% (w/w) sulfur (Speight, 1992, 2014). While heteroatoms (i.e., nitrogen, oxygen, and sulfur) make up only a minor component of most asphalts, the heteroatoms profoundly influence the differences in physical properties of asphalts from different crude sources (Speight, 1992, 2014).

Asphalt is used for paving, roofing, industrial, and special purposes. Oxidized asphalt is used in roofing operations, pipe coating, undersealing for concrete pavements, hydraulic applications, membrane envelopes, some paving-grade mixes, and the manufacture of paints (Speight, 1992, 2014). For the most part, asphalts are classified according to the performance specifications for which they were manufactured (e.g., paving asphalts and roofing asphalts).

Differences in the way in which asphalts are handled during paving and roofing operations probably influence the composition of asphalt fumes and vapors. When a hot-mix paving asphalt arrives at the worksite, the asphalt has been cooling since leaving the plant and may not be used immediately when it arrives at the worksite. Conversely, roofing asphalts are heated continuously and stirred occasionally at the worksite until the asphalt is needed.

When asphalts are heated, vapors are released which condense on cooling. These vapors are enriched in the more volatile components present in the asphalt and would be expected to be chemically and potentially toxicologically distinct from the parent material. Asphalt fumes are the cloud of small particles created by condensation from the gaseous state after volatilization of asphalt. However, because the components in the vapor do not condense all at once, workers are exposed not only to asphalt fumes but also to vapors. The physical nature of the fumes and vapors has not been well characterized, but the fume should be fairly viscous. The asphalt fume particles may collide and stick together, making it difficult to characterize the fume particle size. Some of the vapors may condense only to the liquid phase, thus forming a viscous liquid with some solids.

While the concept of basing payment upon expected performance is an admirable goal, it is essential in this type of specification that equations exist between pavement material properties and pavement performance. The accuracy of these models must be carefully evaluated if performance-related specifications (PRS) are to be used in determining responses that could then be used in previously developed performance equations to estimate pavement life. To investigate the accuracy of these models, a plan was developed to construct an accelerated pavement test track.

It is appropriate at this point to expand this discussion to illustrate the use of specification in the realm of modern asphalt technology, which is also a part of binder characterization (SHRP, 1993, 1994a,b). Specifications are used to convey information concerning desired products from a buyer to a seller or potential seller, they are used as a basis for competitive bidding for the delivery of products and they are used to measure compliance to contracts. There are four types of specification generally recognized in the construction industry: (i) proprietary product, (ii) method, (iii) end-result, and (iv) performance.

A proprietary product specification is used when a generic description of a desired product or process cannot be easily formulated. It usually contains an "or equivalent" clause to allow for some measure of competition in providing the product. It is generally acknowledged that such a specification severely limits competition which increases cost. It provides very little latitude for innovation, and it puts substantial risk on the owner for product performance. Most agencies avoid this type of specification whenever possible. A method specification outlines a specific materials selection and construction operation process to be followed in providing a product. In the past, many construction specifications were written in this manner. A contractor would be told what type of material to produce, what equipment to use, and in what manner it was to be used in building a structure. In its strictest sense, only the final form of the structure can be stipulated (for instance, the thickness of the pavement layers). This type of specification allows for a greater degree of competition than the proprietary product specification, but as long as the structure is built according to the materials and methods stipulated, the agency bears the responsibility for the performance. Contractor innovation is severely restricted in this environment.

An end-result specification is one in which the final characteristics of the product are stipulated, and the contractor is given considerable freedom in achieving those characteristics. In their roughest form, they specify minimum, maximum, or a range of values for any given characteristic. For instance, they may state a minimum layer thickness or a range of in-place air voids. Normally, agencies try to use statistical techniques in employing end-result specifications. In well-written specifications, the statistics account for sources of variability (sampling and testing) when placing limits on the material variability (ASTM D140). Despite the term *end-result*, some elements of method specifications are usually incorporated to guard against early failure of the product (Speight, 2015a). This is the current general state of highway specifications. In the future, efforts will be made to tie the end-result quality characteristics for the product to its performance; these will be referred to as PRS. Performance-based specifications will rely on characteristics or properties directly related to the expected performance.

Performance specifications are those in which the product payment is directly dependent upon its actual performance. Typical of these specifications are warranty, limited warranty, and design–build–operate contracts. Contractors are held responsible for the product performance within the context of what they have control over. The contractor is given a great deal of leeway in providing the product, as long as it performs according to established guidelines. In this case, the contractor assumes considerable risk for the level of service the product provides by paying for or providing any necessary maintenance or repair within the warranty period.

The use of statistics in specifications is appropriate when considering the variability inherent in any product and the variability associated with sampling and testing activities. When properly formulated, the statistical description of desired characteristics provides the best insurance that the buyer is getting what was paid for and that the seller will be paid a fair price for delivering the product. Depending upon the importance of a characteristic to define the performance of the product, a trade-off will need to be made between the quality of the product (limiting variability) and its associated cost.

As responsibility for the quality of asphaltic pavements continues to shift from the specifying agency to the contractor, it will be critical to better understand the relationships between characteristics of the mixture and the expected pavement performance. However, it must be understood that the performance also depends upon the conditions imposed by the traffic, climate, and underlying materials. PRS is a very desirable goal, but there are important technical issues to address with the current statistical specifications first. These issues need to be resolved as progress is made.

4.3 THE NEED FOR TESTING

The need for testing any product is required to ensure that the product is suitable for the claimed uses of the product and will perform when in use according to the manufacturer's claims. To do this, the product must have specified properties that are suitable to in-service use and can vary from simple properties to a more complex suite of properties that are used as a collection that is typically referred to as the product *specifications*. Asphalt, like other petroleum products, is also required to meet specific properties (specifications) before use. In this case, the specifications will be determined by the user insofar as the specifications will vary from one region to another depending upon climatic and environmental conditions.

The in-service use of an asphalt binder depends upon the chemistry of the crude oil source and the method used to refine it, which dictate the chemical and physical properties of the binder, as any additives used. The additives, in turn, influence the properties of the binder and can also contribute to the in-service performance of the binder as well as enabling the binder to meet the required specifications. Specifications dictate that inferior-quality materials (i.e., materials that do not meet the specifications) are not used and the physical properties (such as strength, stickiness, or elasticity) must match specifications or the result will be poor in-service performance (such as rutting, cracking, and the formation of potholes). This also includes structural design—a pavement structure design and/or poor-quality construction that, for example, does not allow water to drain from the roadway or a roadway that is not strong enough to support the traffic results in a poor-quality roadway and rapid deterioration during service.

In order to assure high-quality roadways, a series of physical properties have been identified (usually by the highway agency) that will provide the best-quality roadways. Thus, a series of standard laboratory and field test methods and sampling procedures (ASTM D140; ASTM D979) have been developed to ensure that the asphalt– binder, the aggregate, and the asphalt–aggregate mix will give the expected performance. Typically, each of the necessary properties will have upper and/or lower limits for those test results and these limits are the specifications that the binder, the aggregate, and the asphalt–aggregate mix must meet to be suitable for roadway (or other) use. Thus, for any roadway paving project to be successful and give the necessary in-service performance, the materials, structural design, and construction should meet standard specifications. Similar conditions also apply to other uses of asphalt.

Water sensitivity of a mix is determined by submerging the samples in water for a specified time at a specified temperature, and then the samples are retested for retained strength. The data from this test method are used to compare the strength of unconditioned mixtures with mixtures that have been conditioned under water. The asphalt binder should be sufficiently adhesive enough not to strip from the aggregate and to give a good retained strength to the mix. If the mix design testing indicates the mix may have a susceptibility to moisture, other tests can be conducted to determine the appropriate additives—such as liquid anti-strip additives for the asphalt binder or hydrated lime for the aggregate—that might be employed to enhance the performance of the asphalt–aggregate mix.

Thus, the importance of chemical and physical properties is dependent upon the purity of the petroleum or petroleum product. In the strictest sense, petroleum and petroleum products are complex chemical mixtures. These mixtures contain hydrocarbons and non-hydrocarbon compounds that confer properties on the mixture that may not be reflected in the composition. Therefore, it is necessary to apply various text methods to petroleum and petroleum products to determine whether the material is suitable for processing and (in the case of products) for sale with a designated use in mind. The more common tests are introduced in the following sub-sections—these tests are presented in alphabetical order with no preference given to any particular test method.

4.4 BINDER SUITABILITY

The purpose of asphalt binder testing is to ensure that the binder conforms to specifications for that binder type throughout the life of the binder. The life of a binder can be categorized into three stages as follows: (i) transport, storage, and handling in which testing is performed on original binder material, (ii) mix production and construction, in which testing simulates the binder as it passes through a plant, and (iii) after a long period in a pavement, in which testing simulates the binder after an extended period of time in the pavement and the data are used to determine whether or not the binder is still suitable for the intended use.

Current binder specifications are, at best, performance-related, so significant chemistry changes may give spurious test behavior. From a scientific perspective, it is always hoped that the results of a specified test method indicate the necessary performance characteristics. As a result, there are many issues to consider, such as: (i) predictable aging characteristics, (ii) predictable rheological properties, (iii) predictable adhesion properties to aggregates, (iv) predictable coating behavior in a mix plant, (v) predictable flow characteristics during construction, (vi) predictable mixing with reclaimed asphalt pavement, and (vii) predictable interaction—compatibility or incompatibility—with other binders. Other characteristics that are not only related to binder as well as binder–aggregate performance but also to environmental issues (see Chapter 12) are: (i) predictable leaching characteristics, (ii) predictable water solubility, (iii) predictable interactions with solvents such as spilled fuels and oils that occur on roadways, and (iv) predictable odor.

At ambient temperature *in situ*, asphaltic binder is solid and virtually non-volatile and requires heating to temperatures in excess of 140°C (>285°F) to become liquid to facilitate transportation and handling (AASHTO R28). Some of the typical physical properties (Table 4.3) are regulated by national or international specifications, while other properties, such as specific gravity or vapor pressure, result from the manufacturing processes used to meet the performance specification. Typical elemental analysis varies over a range (Table 4.3) (Goodrich et al., 1986; Speight, 2014).

As a result of the need for adjustment of the physical properties, asphalt must be available in a variety of grades and the specifications of each type of asphalt are tailored to meet the needs of the asphalt in service. The asphalt grades are based on a series of physical tests that define the safety, solubility, viscoelasticity, and durability of the asphalt in order to optimize performance under high- and low-temperature conditions (TRB, 2004).

4.4.1 Test Methods

Unlike the test methods for residua, in order to assure the performance of refinery-produced asphalt when used as roadway pavements, the material (asphalt) is selected on the results of laboratory tests and performance-related tests (Speight, 2015a). Many of the tests are empirical in nature and have been used over the decades giving rise to the development of relationships between the laboratory test data and asphalt performance in service. Although this system has worked relatively well, the increasing volume of traffic and the increasing traffic loads require new approaches that will assure good performance of the highway system. It is no longer adequate to merely relate asphalt performance in service to, say, the composition of the asphalt in terms of bulk fractions—the end-result may merely be a failed paper exercise. For example, one important characteristic that is often ignored, for no justifiable reason, is the asphalt–aggregate interactions and the occurrence of such interactions during the in-service period.

Indeed, the use of old empirical tests cannot effectively evaluate the new materials, such as polymer-modified asphalt, that are increasingly being used to manufacture asphalt destined for use on highways (Airey, 2002). There is a tremendous need to develop new performance-based test procedures. These new tests are obviously the result of the need to provide the necessary information to select the most performance-effective materials to maintain the highway system. However, it must be recognized that the properties of asphalt are defined by the properties of residua from which the asphalt was produced and that the properties of the residua vary with the cut-point, that is, the % v/v of the crude oil that remains as non-volatile residuum (see Chapter 3), and which may dictate variations in the methods of production of the asphalt.

Specifications for paving asphalt cements usually include five grades differing in either viscosity or penetration level at 60°C (140°F). Susceptibility of viscosity to temperature is usually controlled in asphalt cement by viscosity limits at a higher temperature, such as 135°C (275°F), and a penetration or viscosity limit at a lower temperature, such as 25°C (77°F). Paving cutbacks are also graded at 60°C (140°F) by viscosity, with usually four to five grades of each type. For asphalt cements, the newer viscosity grade designation is the midpoint of the viscosity range. The cutback's grade designation is the minimum kinematic viscosity for the grade, with a maximum grade viscosity of twice the minimum.

Roofing and industrial asphalts are also generally specified in various grades of hardness, usually with a combination of softening point (ASTM D61; ASTM D2319; ASTM D3104; ASTM D3461) and penetration to distinguish grades (ASTM D312; ASTM D449). Temperature susceptibility is usually controlled in these requirements by specifying penetration limits or ranges at 25°C (77°F) and other temperatures as well as softening point ranges at higher temperatures. Asphalt for built-up roof constructions is differentiated according to application, depending primarily on the pitch of the roof and to some extent on whether or not mineral surfacing aggregates are specified (ASTM D312). The damp-proofing grades reflect above- or below-grade construction, primarily, and whether or not a self-healing property is incorporated.

The following standard test methods (Speight, 2001, 2014, 2015a) are a sample of the standard test methods (listed alphabetically and not in order of preference) that might be applied (or required) to determine the suitability of asphalt for the designated use.

4.4.1.1 Acid Number

The acid number is a measure of the acidity of a product and is used as a guide in the quality control of asphalt properties. Since a variety of oxidation products contribute to the acid number and the organic acids vary widely in service properties so the test is not sufficiently accurate to predict the precise behavior of asphalt in service.

Asphalt contains a small amount of organic acids and saponifiable material that is largely determined by the percentage of naphthenic (cycloparaffin) acids of higher molecular weight that were originally present in the crude oil. With increased hardness, asphalt from a particular crude oil normally decreases in acid number as more of the naphthenic acids are removed during the distillation process. Acidic constituents may also be present as additives or as degradation products formed during service, such as oxidation products (ASTM D5770). The relative amount of these materials can be determined by titrating with bases. The acid number is used as a guide in the quality control of lubricating oil formulations. It is also sometimes used as a measure of lubricant degradation in service. Any condemning limits must be empirically established.

In a manner akin to the *acid number*, the *base number* (often referred to as the *neutralization number*) is a measure of the basic constituents in the oil under the conditions of the test. The base number is used as a guide in the quality control of oil formulation and is also used as a measure of oil degradation in service. The *neutralization number* is expressed as the *base number* and is a measure of the amount of basic substance in the oil always under the conditions of the test. The neutralization number is used as a guide in the quality control of lubricating oil formulations. It is also sometimes used as a measure of lubricant degradation in service; however, any condemning limits must be empirically established.

The *saponification number* expresses the amount of base that will react with 1 g of the sample when heated in a specific manner. Since compounds of certain elements are sometimes added to asphalt and also consume alkali and acids, the results obtained indicate the effect of these extraneous materials in addition to the saponifiable material present. In the test method (ASTM D94), a known weight of the sample is dissolved in methyl ethyl ketone or a mixture of suitable solvents and the mixture is heated with a known amount of standard alcoholic potassium hydroxide for between 30 and 90 min at 80°C (176°F). The excess alkali is titrated with standard hydrochloric acid and the saponification number is calculated.

4.4.1.2 Boiling Point

Asphalt is obtained from the residue(s) from the vacuum distillation of crude oil (Speight, 1992, 2014). Thus, a typical boiling range in excess of 450°C (840°F) is to be expected.

In the petroleum refining industry, boiling range distribution data are used (i) to assess petroleum crude quality before purchase, (ii) to monitor petroleum quality during transportation, (iii) to evaluate petroleum for refining, and (iv) to provide information for the optimization of refinery processes. Traditionally, boiling range distributions of the various fractions have been determined by distillation. Yield-on-crude data are still widely reported in the petroleum assay literature, providing information on the yield of specific fractions obtained by distillation.

To some extent in the laboratory, atmospheric and vacuum distillation techniques have largely been replaced by *simulated* distillation methods, which use low-resolution gas chromatography and correlate retention times to hydrocarbon boiling points. Two test methods (ASTM D2887; ASTM D3710) use external standards composed of *n*-alkanes while a similar third test method (ASTM D5307) requires two determinations to be made with each sample, one of which uses an internal standard. The amount of material boiling above 538°C (1000°F) (reported as residue) is calculated from the differences between the two determinations.

From the point of view of petroleum and petroleum product analysis for environmental purposes, boiling range distributions provide an indication of volatility and component distribution. In addition, boiling range distribution data are also useful for the development of equations for predicting evaporative loss (Speight, 2005; Speight and Arjoon, 2012).

4.4.1.3 Bond and Adhesion

The adhesion of asphalt to the mineral aggregate is a fundamental property of road asphalt. Once the adhesion deteriorates, the surface becomes unstable and unusable. There is a test method (ASTM D3409) which covers the determination of the adhesion of asphalt roofing cements to damp, wet, or underwater surfaces. The data are primarily used to determine whether a jointing material possesses an arbitrary amount of bonding strength at low temperatures where Portland cement concrete is being used.

Another test method (AASHTO T283; ASTM D4867) is used to examine the effect of moisture on asphalt concrete paving mixtures. The test method can be used to test asphalt concrete mixtures in conjunction with mixture design testing to determine the potential for moisture damage, to determine whether or not an anti-stripping additive is effective, and to determine what dosage of an additive is needed to maximize the effectiveness. This test method can also be used to test mixtures produced in plants to determine the effectiveness of additives under the conditions imposed in the field.

4.4.1.4 Breaking Point

Brittle asphalt causes pavement instability and the appearance of cracks due to (i) excessive loading, (ii) weak surface, base, or subgrade, (iii) thin surface or base, or (iv) poor drainage. Once cracks appear the asphalt will deteriorate quickly. One particular test method (IP 80) is an approximate indication of the temperature at which asphalt possesses no ductility and would reflect brittle fracture conditions. The Fraass breaking point test (IP 80) is used for determining the breaking point of solid and semi-solid asphalt. The breaking point is the temperature at which asphalt first becomes brittle, as indicated by the appearance of cracks when a thin film of the asphalt on a metal plaque is cooled and flexed in accordance with specified conditions.

4.4.1.5 Carbon Disulfide Insoluble Constituents

The component of highest carbon content is the fraction termed *carboids* and consists of species that are insoluble in carbon disulfide or in pyridine. The fraction that has been called *carbenes* contains molecular species that are soluble in carbon disulfide and soluble in pyridine but which are insoluble in toluene (Figure 4.1). The carbenes and carboids have been represented as having higher polarity and lower molecular weight than the asphaltene constituents (Figure 4.2) because they were first identified and named as thermal products of asphaltene decomposition. The same rationale applies to carbenes and carboids that are oxidation products of the asphaltene constituents. The disruption of the molecular associations by the addition of oxygen functions could well result in lower-molecular-weight products than depicted by standard molecular weight methods that are subject to associative effects.

Asphalt is a hydrocarbonaceous material that is made of constituents (containing carbon, hydrogen, nitrogen, oxygen, and sulfur) that are completely soluble in carbon disulfide (ASTM D4). Trichloroethylene or 1,1,1-trichloroethane has been used in recent years as a solvent for the determination of asphalt solubility (ASTM D2042). The carbene and carboid fractions are generated by thermal degradation or by oxidative degradation and are not considered to be naturally occurring constituents

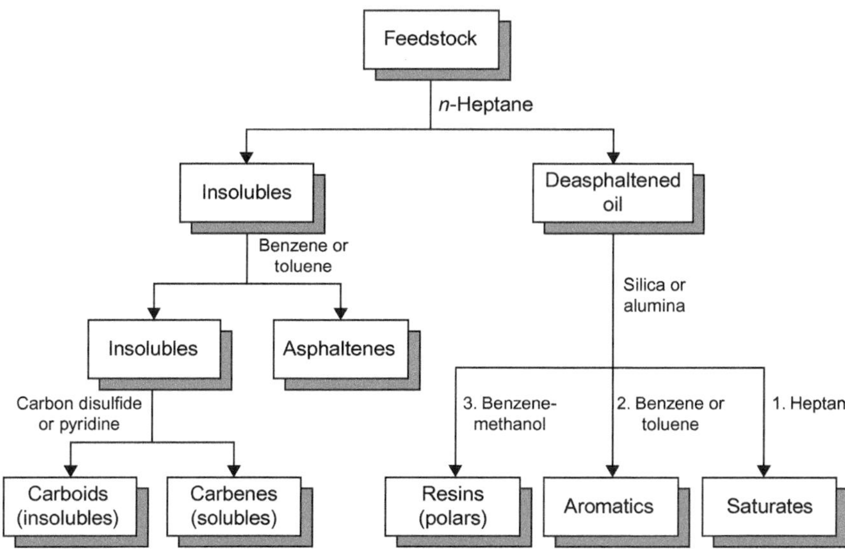

Figure 4.1 Separation of asphalt into bulk fractions.

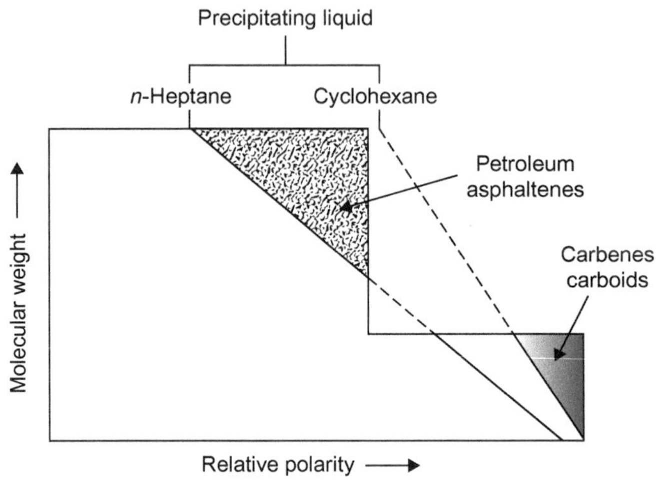

Figure 4.2 Representation of carbene and carboids compared to asphaltene constituents.

of asphalt. The test method for determining the toluene-insoluble constituents of tar and pitch (ASTM D4072; ASTM D4312) can be used to determine the amount of carbenes and carboids (both are ill-defined constituents) in asphalt (Parkash, 2003; Hsu and Robinson, 2006; Gary et al., 2007; Speight, 2014).

Pure asphalt cement is completely soluble in carbon disulfide and, except for carbenes, completely soluble in organic solvents such as carbon disulfide, carbon tetrachloride, or pyridine (Figure 4.1). As mineral matter and other foreign products are insoluble in these solvents, they can, therefore, be used to detect the presence of such matter. For reasons of safety and convenience, carbon tetrachloride is the preferred solvent (ASTM D4). However, if asphalt cements do not meet the specification requirements for solubility when using carbon tetrachloride, the test should also be run with carbon disulfide as the solvent. When differences in solubility are encountered by use of the two solvents, the presence of carbenes may be indicated. If this difference is greater than about 0.5%, it may mean that the asphalt cement has been overheated or cracked in the process of manufacture. On the other hand, cracked products do not always show the presence of carbenes.

At room temperature, the constituents of asphalt exist as semisolid to solid materials and as such they are expected to have extremely low water solubility. However, since substances in this category are often employed in waterproofing applications, there is a potential to leach components from the asphalt into the water (Brandt and De Groot, 2001; Kriech et al., 2002). While the known water solubility value of polynuclear aromatic hydrocarbons can be used to estimate the water solubility of the constituents of asphalt, the data for pure compounds do not always reflect the solubility of the constituents of a complex mixture such as asphalt.

4.4.1.6 Compatibility

When coating asphalt and saturating-grade asphalt are used together, as in prepared roofing, one test method (ASTM D1370) indicates whether they are likely to bleed or dis-bond under stress at the coating–felt interface.

The test method (ASTM D1370) is used to assess the degree to which asphalt interacts with other asphalts and the data can be used to indicate possible future problems, especially blistering, in a roofing product if incompatible asphalts are in contact in the product. The method provides a means for evaluating contact compatibility between asphaltic materials.

4.4.1.7 Composition

Determination of the composition of asphalt has always presented a challenge because of the complexity and high molecular weights of the molecular constituents. The principle behind investigation of asphalt composition is to evaluate asphalt in terms of composition and performance. Thus, the methods employed can be conveniently arranged into a number of categories: (i) fractionation by precipitation; (ii) fractionation by distillation; (iii) separation by chromatographic techniques; (iv) chemical analysis using spectrophotometric techniques (infrared, ultraviolet, nuclear magnetic resource, X-ray fluorescence, emission, neutron activation), titrimetric and gravimetric techniques, elemental analysis; and (v) molecular weight analysis by mass spectrometry, vapor pressure osmometry, and size exclusion chromatography.

However, fractional separation has been the basis for most asphalt composition analysis (Figure 4.1). The separation methods that have been used divide asphalt into operationally defined fractions. Three types of asphalt separation procedures are now in use: (i) chemical precipitation in which *n*-pentane separation of asphaltene constituents is followed by chemical precipitation of other fractions with sulfuric acid of increasing concentration, (ii) adsorption chromatography using a clay-gel procedure where, after removal of the asphaltene constituents, the remaining constituents are separated by selective adsorption/desorption on an adsorbent (ASTM

D2007; ASTM D3279; ASTM D4055; ASTM D4124; ASTM D6560), (iii) size exclusion chromatography in which gel permeation chromatographic separation of asphalt constituents occurs based on their associated sizes in dilute solutions.

The fractions obtained in these schemes are defined operationally or procedurally. The amount and type of the asphaltene fraction in asphalt are, for instance, defined by the solvent used for precipitating them. Fractional separation of asphalt does not provide well-defined chemical components. The materials separated should only be defined in terms of the particular test procedure (Figure 4.1). However, these fractions are generated by thermal degradation or by oxidative degradation and are not considered to be naturally occurring constituents of asphalt. The test method for determining the toluene-insoluble constituents of tar and pitch (ASTM D4072; ASTM D4312) can be used to determine the amount of carbenes and carboids in asphalt.

In these methods, a sample is digested at 95°C (203°F) for 25 min and then extracted with hot toluene in an alundum thimble. The extraction time is 18 h (ASTM D4072) or 3 h (ASTM D4312). The insoluble matter is dried and weighed. Combustion will then show whether the material is truly carbonaceous or if it is inorganic ash from the metallic constituents (ASTM D482; ASTM D2415; ASTM D4628; ASTM D4927; ASTM D5185; ASTM D6443).

Another method (ASTM D893) covers the determination of pentane- and toluene-insoluble constituents in used lubricating oils and can be applied to asphalt. Pentane-insoluble constituents include oil-insoluble materials and toluene-insoluble constituents can come from external contamination and highly carbonized materials from degradation. A significant change in pentane- or toluene-insoluble constituents indicates a change in asphalt properties that could lead to problems in service. The insoluble constituents measured can also assist in evaluating the performance characteristics of asphalt.

There are two test methods used: Procedure A covers the determination of insoluble constituents without the use of coagulant in the pentane and provides an indication of the materials that can be readily separated from the diluted asphalt by centrifugation. Procedure B covers the determination of insoluble constituents in asphalt that contains additives and employs a coagulant. In addition to the materials separated by using Procedure A, this coagulation procedure separates some finely divided materials that may be suspended in the asphalt. The results obtained by

Procedures A and B should not be compared since they usually give different values. The same procedure should be applied when comparing results obtained periodically on asphalt in use, or when comparing results determined in different laboratories.

In Procedure A, a sample is mixed with pentane and centrifuged after which the asphalt solution is decanted, and the precipitate washed twice with pentane, dried and weighed. For toluene-insoluble constituents, a separate sample of the asphalt is mixed with pentane and centrifuged. The precipitate is washed twice with pentane, once with toluene–alcohol solution, and once with toluene. The insoluble material is then dried and weighed. In Procedure B, Procedure A is followed except that instead of pentane, a pentane-coagulant solution is used.

Many investigations into relationships between composition and properties take into account only the concentration of the asphaltene constituents, independently of any quality criterion. However, a distinction should be made between the asphaltene constituents that occur in straight-run asphalts and those which occur in blown asphalts. It cannot be overemphasized that asphaltene fraction is a solubility class rather than a distinct chemical class, meaning that vast differences occur in the make-up of this fraction when it is produced by different procedures.

Finally, composition data should always be applied to in-service performance in order to properly evaluate the behavior of the asphaltic binder under true working conditions.

4.4.1.8 Distillation

Asphalt is prepared from a distillation residuum and therefore the need for distillation data is limited. Vacuum distillation data (ASTM D1160) will be valuable for composition purposes if the asphalt is prepared from an atmospheric residuum. Approximate amounts of volatile constituents can also be determined by test methods developed for other products (ASTM D20; ASTM D402; ASTM D3607; ASTM D4893) but that are particularly applicable to cutback asphalt.

Asphalt can also be examined for evaporative losses using a test method designed for grease (ASTM D2595) or for engine oil (ASTM D5800; ASTM D6375). In another test method (ASTM D972), the evaporative losses at any temperature in the range of 100°C to 150°C (210–300°F) can be determined. A weighed sample is placed in an evaporation cell in an oil bath at the desired test temperature. Heated air at a specified flow rate is passed over the sample surface for 22 h, after which, the loss in sample mass is determined. In yet another method (ASTM D2595) that is used to supplement the original method (ASTM D972), the loss of volatile materials from a grease over a temperature range of 93°C to 316°C (200–600°F) can be determined. This test uses an aluminum block heater, instead of an oil bath (ASTM D972), to achieve higher temperatures.

Another test (ASTM D6) allows the determination of the percent loss of mass when a weighed quantity of water-free material is heated in moving air for 5 h at 163°C (325°F). In the test method, a gravity convection oven with a rotating shelf is used and the method provides only a relative measurement of the volatility of a material under test conditions. It may be required for bituminous coating to be applied to galvanized, corrugated, culvert tube. A test for the loss of heating from a thin film is also available (ASTM D1754).

4.4.1.9 Ductility

The ductility of a solid material is the ability of the material to deform under tensile stress. Malleability, a similar property, is the ability of a material to deform under compressive stress. Thus, the *ductility* of asphalt is a measure of the flexibility of the asphalt and is expressed as the distance in cm that a standard briquette can be elongated before breaking (ASTM D113). Ductility is a combination of flow properties and reflects homogeneity, cohesion, and shear susceptibility; it is an indication of fatigue life and cracking (Sousa, 1994; Castelo Branco et al., 2008; Masad et al., 2008).

In many applications, ductility (AASHTO T51; ASTM D113) is considered an important characteristic of binders. The presence or absence of ductility, however, is usually considered more significant than the actual degree of ductility. Binders possessing ductility are normally more adhesive than binders lacking this characteristic. However, some binders with an exceedingly high degree of ductility are also more temperature-susceptible. That is, the change in consistency is apt to be greater for a change in temperature. In some applications, such as paving mixes, ductility and adhesion are more important. In other situations, such as slab undersealing and crack filling, the more essential property is low-temperature susceptibility.

In the test method (ASTM D113), a briquette of asphalt cement is cast under standard conditions and to specified dimensions and is then brought to a standard test temperature and pulled or extended at a specified rate of speed until the thread of asphalt connecting the two ends breaks. The length of pull (in centimeters) at which the thread of material breaks is designated as its ductility. Unless otherwise specified, the test shall be made at a temperature of 25 ± 0.5°C (77 ± 0.9°F) and with a speed of 5 cm/min ± 5.0%; at other temperatures, the speed should be specified.

4.4.1.10 Durability

The *durability* of asphalt is an indication of the presence of the necessary chemical and physical properties required for the specified pavement performance. The property indicates the resistance of the asphalt to change during the in-service conditions that are prevalent during the life of the pavement. The durability is determined in terms of resistance to oxidation (resistance to weathering) and water resistance (ASTM D1669; ASTM D1670).

4.4.1.11 Emulsified Asphalt

The tendency for asphalt to form an emulsion with water is often used as a means of sales of asphalt emulsion. The premise of using an asphalt emulsion is that the emulsion is stable during storage, mixing, and pumping, but dissociates quickly upon contact with the aggregate to form an asphalt mixture similar to asphalt–aggregate mix (Clyne et al., 2003). However the formation of an emulsion during periods of heavy rain, especially if a roadway is water-logged, can be detrimental to roadway performance and longevity.

Emulsified asphalt is a suspension of small asphalt globules in water, which is assisted by an emulsifying agent (such as soap). The emulsifying agent assists by imparting an electrical charge to the surface of the asphalt globules so that they do not coalesce. Emulsions are used because they effectively reduce asphalt viscosity for lower temperature. Generally, asphalt emulsions appear as a thick brown liquid when initially applied but when the asphalt starts to adhere to the surrounding material (e.g., aggregate) the color changes from brown to black. As water begins to evaporate, the emulsion begins to behave more and more like pure asphalt cement. The time required to break and set depends upon the type of emulsion, the application rate, the temperature of the surface onto which it is applied and environmental conditions.

Unlike conventional asphalt binders, for which performance-based specifications are available (see Chapter 9), there is a need to apply emulsion properties to performance in order to characterize asphalt emulsions and identify potential test methods for characterizing asphalt emulsions as a preliminary action to understand performance. Currently, there are a variety of laboratory test methods for testing asphalt emulsions and the residue. In order to obtain properties that can be related to field performance, it is of critical importance to obtain an emulsion residue that is representative of the emulsion used in the field and which plays an important role in any attempt to relate the residue properties to performance. In terms of recovery of a residue from an asphalt emulsion, the distillation procedures (ASTM D86; ASTM D1160; ASTM D5236) require temperatures that can change the structure of the emulsion. The residues recovered by these processes may not represent the field conditions where construction is done in ambient temperatures and other test methods may be required (Salomon, 2000; Takamura, 2000). Other conventional tests (AASHTO T59; ASTM D244) for the characterization of asphalt emulsions include tests such as the test method for viscosity using the Saybolt–Furol apparatus and specific emulsion properties such as the *particle charge test* to identify cationic emulsions, the *demulsibility* test to check the rate of breaking of an emulsion upon spreading on soil or aggregates, the *sieve test* to quantify the amount of large and unmanageable globules and the test for *miscibility with water.*

4.4.1.12 Fire Point and Flash Point

The *fire point* is the lowest temperature, corrected to one atmosphere pressure (14.7 psi), at which the application of a test flame to the petroleum or petroleum product sample surface causes the vapor of the oil to ignite and burn for at least 5 s.

At any time after a spill of petroleum or a petroleum product, fire should always be considered an imminent hazard. Related to the fire point, the flash point is a measure of the tendency of the petroleum or a petroleum product to form a flammable mixture with air under controlled laboratory conditions. It is only one of a number of properties that should be considered in assessing the overall flammability hazard of a spilled material (ASTM D92). The ignition temperature (sometimes called the *autoignition temperature*) is the minimum temperature at which the material will ignite without a spark or flame being present (ASTM E659). Also related to the fire point, the flammability limits of vapor in air are an expression of the percent concentration in air (by volume) and is given for the lower and upper limits. These values give an indication of relative flammability. The limits are sometimes referred to as the *lower explosive limit* and *upper explosive limit.*

The *flash point* of petroleum or a petroleum product is the temperature to which the sample must be heated to produce a vapor/air mixture above the liquid fuel that is ignitable when exposed to an open flame under specified test conditions. In North America, the flash point is used as an index of fire hazard. The flash point is an extremely important factor in relation to the safety of spill cleanup operations. Gasoline and other low-boiling liquid fuels can be ignited under most ambient conditions and therefore pose a serious hazard when spilled. Many freshly spilled crude oils also have low flash points until the lighter components have evaporated or dispersed.

The *flash point* of asphalt cement is of interest primarily to the user, since the material must be heated for application or mixing and the test method gives an indication of the temperature to which the material can be safely heated without danger of instantaneous flash in the presence of an open flame. This temperature, however, is usually well below that at which the material will burn (the *fire point*). The fire point is seldom included in specifications on asphalt cement.

The flash point test is performed on the original binder and is determined to ensure the binder is safe to work with at production temperatures. Upto the flash point, the binder may be safely heated without instantaneous flash in the presence of an open flame. There are several ASTM methods for measuring flash points (AASHTO T48; ASTM D56; ASTM D93 are among the most commonly used). The minimum flash point that can be determined (ASTM D93, IP 34) is 10°C (50°F). One method (ASTM D56) is intended for liquids with a viscosity <9.5 cSt at 25°C (77°F). The flash point and fire point of lubricating oil are determined by a separate method (ASTM D92, IP 36).

The flash point of asphalt cement is measured by the Cleveland open-cup method (ASTM D92). In this method, a brass cup is partially filled with asphalt and heated at a prescribed rate. At intervals, a small flame is played over the surface of the sample and the temperature at which sufficient vapors are released to produce an instantaneous flash is designated as the flash point of the sample under test.

4.4.1.13 Float Test

The float test is used to determine the consistency of asphalt at a specified temperature. One test method (ASTM D139) is normally used for asphalt that is too soft for the penetration test (ASTM D5; ASTM D217; ASTM D937; ASTM D1403).

The float test characterizes the flow behavior or consistency of certain bituminous materials and is useful for determining the consistency of asphalt as one element in establishing the uniformity of certain shipments or sources of supply.

4.4.1.14 Fractionation

Rather than quantifying a complex petroleum-based mixture (such as asphalt) as a single number, petroleum fractionation methods break the mixture into discrete fractions, thus providing data that can be used in a risk assessment and in characterizing product type and compositional changes such as may occur during refining and during weathering (oxidation) after a spill (Speight and Arjoon, 2012). The fractionation methods can be used to measure both the volatile constituents and the extractable constituents.

In contrast to traditional methods for the analysis of petroleum products that report a single concentration number for complex mixtures, the fractionation methods report separate concentrations for discrete aliphatic and aromatic fractions of the product (Speight and Arjoon, 2012; Speight, 2014). The available petroleum fractionation methods are chromatography-based and are thus sensitive to a broad range of hydrocarbon constituents as well as to constituents containing one or more polar (non-hydrocarbon) functions (Speight, 2014). Identification and quantification of aliphatic and aromatic fractions allow identification of petroleum products and evaluate the extent of contamination by non-hydrocarbon constituents or the extent of participation in product petroleum behavior.

4.4.1.15 Loss on Heating

Since asphalt must be heated for application or mixing, its properties should not change appreciably when it is heated to specified application or mixing temperatures. Such temperatures may vary, but they usually do not exceed 160°C (325°F). The loss-on-heating test was devised to prevent the inclusion of excessive amounts of volatile materials (such as water, gas, and hydraulic fluid), the loss of which might cause appreciable changes in some of the properties of the asphalt cement (AASHTO T240; ASTM D6). The results of this test should be considered only as a general indication of what can be expected under service conditions.

In the test method (ASTM D6), a 50-g sample of material is placed in a metal cup, which in turn is placed on a slowly revolving shelf in a well-ventilated oven maintained at 160°C (325°F) for 5 h. The weight loss of the sample during this period is expressed as a percentage of the original weight. In addition to determination of the weight loss during this test, specifications usually require that the asphalt cement shall retain a given percentage of its original penetration after the loss-on-heating test. A second penetration test is thus conducted on the material subjected to the loss-on-heating test and the final penetration is expressed as a percentage of the original penetration of the sample.

4.4.1.16 Melting Point

Asphalt is a viscoelastic substance and does not have a well-defined melting point and gradually softens and becomes less viscous as the temperature rises. For this reason, the softening point is commonly used as a means of standardizing the classification of the flow characteristics of asphalts. In thia method, asphalt is placed into a vessel of specified dimensions and suspended in a water bath. A standardized steel ball is placed at the center of the sample and the water bath is heated at a controlled rate. The temperature at the instant the steel ball reaches the bottom of the vessel is recorded as the softening point.

4.4.1.17 Metals Content

The metals content of petroleum and petroleum-based product such as asphalt can provide valuable information about the origin of those oils, potentially aiding in identifying the source of spills or the potential for metals leaching from asphalt products. The metals assay typically includes nickel and vanadium due to the detrimental effects of these metals on catalysts used in cracking and desulfurization processes. In lubricating oils, metal contents can provide information on both the types of additives used in the oil and on the wear history of the equipment being lubricated. ASTM D5185—Standard Test Method for Determination of Additive Elements, Wear Metals, and Contaminants in Used Lubricating Oils and Determination of Selected Elements in Base Oils by Inductively Coupled Plasma Atomic Emission Spectrometry (ICP-AES)—can be used to determine over 20 different metals in a variety of petroleum products.

4.4.1.18 Partition Coefficient

Substances in the asphalt category are semisolid to solid at ambient temperatures and have negligible vapor pressures and water solubility. The constituents of asphalt are not expected to exhibit measurable partition coefficients when assessed by standard testing methodology. Some constituents with carbon numbers that fall below the typical asphalt range may have measurable partition coefficients in their pure state, but as constituents in the asphalt matrix their contribution to this property will not be measurable.

4.4.1.19 Penetration

The penetration test is a measure of the consistency or relative hardness of the asphaltic material (ASTM D5) and is an empirical measure of asphalt consistency.

Several test method are available for products such as grease (ASTM D217; ASTM D1403) and petrolatum (ASTM D937) that might be modified for asphalt. The more usual test for asphalt (ASTM D5) is a commonly used consistency test. It involves the determination of the extent to which a standard needle penetrates a properly prepared sample of asphalt under definitely specified conditions of temperature, load, and time (100 g load, 5 s). The distance that the needle penetrates in units of mm/10 measured from 0 to 300, is the penetration value. Soft asphalt has a high penetration value and the converse is true for hard asphalt.

In the test method (ASTM D5), a sample of the asphalt cement is placed in a suitable container and brought to a standard test temperature (normally 25°C, 77°F) in a temperature-controlled water bath. A weighted (100 g) needle is then brought to bear on the surface of the asphalt cement for a given period of time (5 s). For special purposes, however, other temperatures, loadings, and time intervals may be employed. The penetration of the needle into the asphalt cement, in units of 0.1 mm, is termed the penetration of the asphalt and the distance which the needle penetrates into the asphalt cement is the penetration factor or measurement. Occasionally, the penetration test is made at a different temperature. When this occurs, needle load, penetration time, or both may be varied.

For a given set of conditions, the needle will penetrate farther into soft asphalt than a harder grade. Thus soft asphalt cements are indicated by the higher penetration numbers, and the harder grades by the lower penetration numbers. Standard grades are normally identified on the basis of a penetration range. For example, one of the standard grades is the 85–100 penetration grade. In addition to the classification of asphalt cements on the basis of penetration ranges, specifications for these products also include several other test requirements to prescribe more completely the characteristics desired by users of the material.

4.4.1.20 Pour Point

The pour point of petroleum or a petroleum product is the lowest temperature at which the oil will just flow, under standard test conditions (ASTM D97). The failure to flow at the pour point is usually attributed to the separation of waxes from the oil, but can also be due to the effect of viscosity in the case of very viscous oils. Also, particularly in the case of residual fuel oils, pour points may be influenced by the thermal history of the sample, that is, the degree and duration of heating and cooling to which the sample has been exposed. From the asphalt perspective, the pour point should be sufficiently high but not so high as to prevent flow when heated.

4.4.1.21 Rheology

Asphalt is a viscoelastic material whose rheological properties reflect crude type and, to a lesser extent, processing. The ability of asphalt to perform under many conditions depends on flow behavior (Arega and Bhasin, 2012).

One of the simplest tests for characterizing the stress–strain response of a material is the creep test (Anderson et al., 1994). An elastic material, when loaded in creep, will immediately deform to a constant strain. When the load is removed, the material will immediately return to its initial shape. A viscous material, on the other hand, will deform at a constant rate when the load is applied, and will continue to deform at that rate until the load is removed, at which point there is no further deflection or recovery. A viscoelastic material has both elastic and viscous components of response. When loaded in creep, there is an immediate deformation, corresponding to the elastic response, followed by a gradual time-dependent deformation. This time-dependent deformation may further be divided into a purely viscous component and a delayed elastic component. Upon removing the load, none of this deformation is recovered. The delayed elastic deformation is, however, recovered, but not immediately as with purely elastic deformation. Instead, once the load is removed, the delayed elastic deformation is slowly recovered. Asphalt films or coatings showing no appreciable change from original conditions are usually desired, i.e., they should allow some structural movement without permanent deformation.

The viscosity of hydrocarbons and temperature are related by the Walther equation, which gives the limiting viscosity at low shear rates (ASTM D5018):

$$\log_{10}[\log_{10}(\nu + \lambda)] = A - B\log_{10}(T)$$

In the equation, ν is viscosity, λ is a shift constant, T is the absolute temperature, and A and B are empirical parameters that reflect the intercept level and slope (or measure of susceptibility of viscosity to temperature, respectively).

At lower temperatures (60°C, 140°F, and lower) and/or higher shear rates—which are typical of asphalt service conditions after incorporation in a roof or pavement—semisolid and solid asphalts display an increasing elastic component which relates viscosity with shear rate. The constant high viscosity level at lower shear rates is the limiting viscosity. Viscosities in the area where viscosity changes with shear rate are generally termed apparent viscosities.

A number of viscometers have been developed for securing viscosity data at temperatures as low as 0°C (32°F). The most popular instruments in current use are the cone plate (ASTM D4287), parallel plate, and capillary instruments (ASTM D2170; ASTM D2171). The cone plate can be used for the determination of viscosities in the range of 10 to over 10^9 Pa at temperatures of 0–70°C (32–158°F) and at shear rates from 10^{-3} to $10^2\,s^{-1}$. Capillary viscometers are commonly used for the determination of viscosities at 60–135°C (140–275°F).

Tests recently developed for measurement of viscoelastic properties are directly usable in engineering relations. Properties can be related to the inherent structure of bituminous materials. The fraction of highest molecular weight, the asphaltene fraction, is dispersed within the asphalt and is dependent upon the content and nature of the resin and oil fractions. Higher aromaticity of the oil fractions or higher temperatures lead to viscous (sol) conditions. A more elastic (gel) condition results from a more paraffinic nature and is indicated by large elastic moduli or, empirically, by a relatively high penetration at a given softening point. Empirically, the penetration index and penetration temperature susceptibility have been used to measure the degree of dispersion.

Asphalt develops an internal structure with age, steric hardening, in which viscosity can increase upon aging without any loss of volatile material. Those with a particularly high degree of gel structure exhibit thixotropy.

4.4.1.22 Softening Point

The softening point of asphalt (not quite analogous to the melting point) is the temperature at which asphalt attains a particular degree of softness under specified conditions of test.

Asphalt does not go through a solid–liquid phase change when heated, and therefore does not have a true melting point. As the temperature is raised, asphalt gradually softens or becomes less viscous. For this reason, the determination of the softening point must be made by an arbitrary but closely defined method, if the test values are to be reproducible. Softening point determination is useful in determining the consistency as one element in establishing the uniformity of shipments or sources of supply.

There are several tests available to determine the softening point of asphalt (ASTM D36; ASTM D61; ASTM D2319; ASTM D3104; ASTM D3461). In the test method (ASTM D36), a steel ball of specified weight is laid on a layer of sample contained in a ring of specified dimensions. The softening point is the temperature, during heating under specified conditions, at which the asphalt surrounding the ball deforms and contacts a base plate.

4.4.1.23 Stain

The stain index is a measure of the sweating tendency of asphalt and homogeneity. The test is used for oxidized asphalt. The test method (ASTM D2746) is used to measure the amount of stain on paper or other cellulosic materials by asphalt. Variations of the cigarette paper stain procedure include the Barber stain; talc stain tests are also used. The test method may be modified for use with other bituminous materials with softening points lower than 85°C (185°F) by using a different temperature than specified for the test and any such modifications must be clearly described.

The test method measures the tendency for oil components to separate spontaneously from asphalt—any oil present in the asphalt constituents can cause staining in asphalt roofing products and adjacent materials in storage and use. In addition, the stain index is related to the thermal stability of the asphalt—high stain index values indicate lower stability and greater tendency for staining.

4.4.1.24 Temperature–Volume Correction

Tables are provided (ASTM D1250) to allow the conversion of volumes of asphaltic materials from one temperature to another or, as generally used, to adjust volumes to a temperature of 15.6°C (60°F). The value commonly taken for mean coefficient of expansion is 0.00036 in the range 15.6–121.1°C (60–250°F).

4.4.1.25 Sulfur Content

The sulfur content of a crude oil is important for a number of reasons. Downstream processes such as catalytic cracking and refining will be adversely affected by high sulfur contents. If high-sulfur asphalt is oxidized, there is the potential for production of high levels of sulfur dioxide which can lead to acid deposition (acid rain) (Speight, 2005; Speight and Arjoon, 2012).

The total sulfur content of oil can be determined by numerous standard techniques, one of which (ASTM D129) is applicable to petroleum products of low volatility and containing at least 0.1 mass percent sulfur.

Sulfur contents are also determined in accordance with ASTM D4294 (Standard Test Method for Sulfur in Petroleum Products by Energy-Dispersive X-Ray Fluorescence Spectroscopy). This method is applicable to both volatile and non-volatile petroleum products with sulfur concentrations ranging from 0.05 to 5 mass percent.

4.4.1.26 Surface Tension and Interfacial Tension

Interfacial tension is the force of attraction between the molecules at the interface of two fluids. At the air–liquid interface, this force is often referred to as surface tension. The SI units for interfacial tension are milli-Newtons per meter (mN/m). These are equivalent to the former units of dynes per centimeter (dyne/cm).

The surface tension of petroleum (or a petroleum product), together with its viscosity, affects the rate at which an oil spill spreads. Air/oil and oil/water interfacial tensions can be used to calculate a spreading coefficient which gives an indication of the tendency for the oil to spread. It is defined as:

$$\text{Spreading coefficient} = S_{\text{WA}} : S_{\text{OA}} : S_{\text{WO}}$$

S_{WA} is water/air interfacial tension, S_{OA} is oil/air interfacial tension, and S_{WO} is water/oil interfacial tension. Unlike density and viscosity, which show systematic variations with temperature and degree of evaporation, interfacial tensions of crude oils and oil products show no such correlations.

A single test method (ASTM D971: Standard Test Method for Interfacial Tension of Oil against Water by the Ring Method) is applicable to the measurement of oil/water interfacial tensions. Unlike a manually operated ring tensiometer, the maximum deformation of the lamella is detected electronically, and occurs before the ring pulls completely through the interface. This results in interfacial tension that is somewhat lower than that measured manually.

4.4.1.27 Thin-Film Oven Test

Many different factors contribute to asphalt aging and one of the key components in the process is the loss of volatiles. Asphalt binders typically lose volatiles during the manufacturing and placement processes. The elevated temperature of these processes ages the asphalt binder by driving off a substantial amount of volatiles.

The thin-film test method (ASTM D2872) is used to simulate the short-term aging of the binders during the hot-mixing process and has the purpose of determining the hardening effect of heat and air on a static film of asphalt when exposed in a thin film. The data indicate the approximate change in properties of asphalt during conventional hot-mixing at about 302°F (150°C) as indicated by viscosity and other rheological measurements. It yields a residue which approximates the asphalt condition as incorporated in the pavement. If the mixing temperature differs appreciably from the 302°F (150°C) level, more or less effect on properties will occur. This test method also can be used to determine mass change, which is a measure of asphalt volatility.

4.4.1.28 Vapor Pressure

Substances in the asphalt category are semisolid to solid, boil at temperatures above 450°C (840°F), and have negligible vapor pressures at ambient temperatures. The vapor pressure of representative hydrocarbon structures covering the range of carbon numbers typically found in substances in the asphalt category (e.g., 25–50 C atoms) represent generic hydrocarbon structures of the types: (i) paraffin structures, (ii) naphthene structures, (iii) aromatic structures, and (iv) heteroatomic structures.

Although the production of asphalt removes the majority of petroleum constituents that boil typically below 450°C (840°F), some constituents having fewer carbon atoms than those typically found in asphalt may exist in the solid matrix, but at low concentrations. Those constituents may have measurable vapor pressures when they exist in their pure form, but they are not volatile when they are entrained in the asphalt matrix. However, some of these more volatile constituents may have the opportunity to escape the asphalt matrix during the heating process used during roofing and paving applications.

4.4.1.29 Viscosity

Viscosity is a measure of a fluid's resistance to flow; the lower the viscosity of a fluid, the more easily it flows. Like density (ASTM D70; ASTM E102), viscosity is affected by temperature. As temperature decreases, viscosity increases. Viscosity is a very important property of asphalt because it affects the rate at which the asphalt oil will spread, the degree to which it will penetrate the aggregate, and the selection of mechanical equipment to lay the asphalt roadway.

Viscosity measurements may be absolute or relative (sometimes called "apparent"). Absolute viscosities are those measured by a standard method, with the results traceable to fundamental units. Absolute viscosities are distinguished from relative measurements made with instruments that measure viscous drag in a fluid, without known and/or uniform applied shear rates. An important benefit of absolute viscometry is that the test results are independent of the particular type or make of viscometer used. Absolute viscosity data can be compared easily between laboratories worldwide. Modern rotational viscometers (AASHTO T316) are capable of making absolute viscosity measurements for both Newtonian and non-Newtonian fluids at a variety of well-controlled, known, and/or uniform shear rates.

There are standard test methods for measuring the viscosity of oils (such as ASTM D445 and ASTM D4486) that make use of glass capillary kinematic viscometers and will produce absolute measurements in units of centistokes (cSt) only for oils that exhibit Newtonian flow behavior (viscosity independent of the rate of shear). Although now obsolete, at one time the petroleum industry relied on measuring kinematic viscosity with the Saybolt viscometer and expressing kinematic viscosity in Saybolt Universal Seconds (SUS) or Saybolt Furol Seconds (SFS). Occasionally, Saybolt viscosities are still reported in the literature using the official equations relating SUS and SFS to kinematic viscosity (ASTM D2161).

The rotational viscosity test (AASHTO T316; ASTM D4402) performed on the original binder is used to evaluate the high-temperature workability of binders. This ensures that the binder is sufficiently fluid when pumping and mixing. In addition, the viscosity measured is used to establish the temperature–viscosity plot for a binder type.

4.4.1.30 Water Content

The presence of water in asphalt can seriously affect performance insofar as it can affect asphalt–aggregate interactions and asphalt adsorption (ASTM D3625; ASTM D4469). The water content of asphalt can be determined by a test method (ASTM D95) that uses distillation equipment fitted with a Dean and Stark receiver.

In the test, the sample is heated under reflux with a water-immiscible solvent, which co-distills with the water in the sample. Condensed solvent and water are continuously separated in a trap, the water settling in the graduated section of the trap, and the solvent returning to the still.

4.4.1.31 Weathering

Weathering is the change in asphalt properties due to exposure to the ambient atmosphere. Two important classifications of weathering processes exist: (i) physical weathering and (ii) chemical weathering and each may involve a biological component.

Mechanical or physical weathering involves the breakdown of asphalt through direct contact with atmospheric conditions, such as heat, water, ice, and pressure. Chemical weathering involves the direct effect of atmospheric chemicals or biologically produced chemicals also known as biological weathering in the deterioration of asphalt properties. While physical weathering is accentuated in very cold or very dry environments, chemical reactions are most intense where the climate is wet and hot.

The extent of cracking or pitting of asphalt films is a measure of the extent of deterioration due to weathering. Failure due to cracking is more accurately determined electrically than visually. This test method (ASTM D1670) evaluates the relative weather resistance of asphalts used for protective-coating applications, especially for roofing. No direct measure of outdoor life or service can be obtained from this test. Methods for preparing test panels (ASTM D1669) and failure end-point testing (ASTM D1670) are available.

Failure of the asphalt specimen, determined by this test method, depends not only on the characteristics of the bituminous material and the extent of weathering but also on the film thickness, and the amount and type of mineral filler present (Kim and Little, 2005). Tests on a similar material of known weathering characteristics (a control) exposed at the same time as the test material are strongly recommended as a check on the validity of the test results.

4.4.1.32 Other Test Methods

The majority of the asphalt roadways in the United States and in many other countries are constructed using a composite mixture of asphalt binder and the appropriate aggregate to produce an asphalt mix that will meet preset design requirements. This allows the design and construction of better roadways with high performance and better longevity. Thermal (or low-temperature) cracking of the road surface leads to (i) poor performance, (ii) loss of structural integrity and, ultimately, (iii) premature deterioration of the entire pavement section. Low-temperature cracking, while found in most of the states, is more ubiquitous in the northern latitudes, such as the colder climates of North America and Canada. Resistance to thermal cracking is primarily provided by the asphalt binder and cracked pavements require frequent and expensive maintenance.

Thus, in terms of asphalt testing, there is the need to determine the behavior of asphalt binders at low temperature (ASTM D6648; ASTM

D6723; ASTM D6816). The data produced from such test methods allow design and construction of well-performing roadway pavement performance (with only minimal thermal cracking) that is based on specifications derived from the low-temperature properties of the asphalt binder and the aggregate. From these fundamental relationships between the binder and the hot-mix asphalt are obtained and they provide valuable input to pavement design, pavement construction, and pavement performance.

The dynamic shear test method (AASHTO T315) is performed on original samples and rolling thin-film oven (RTFO) samples, as well as on aged binders to determine the stiffness or resistance of the binder to deform under loading. In addition, the creep stiffness test method is used to determine the degree of brittleness in aged binders after an extended period in the pavement. However, there is a basic assumption built into the RTFO and pressure aging vessel (PAV) procedures that asphalt will, within a narrow range, exhibit consistent aging behavior and that aging will have a predictable effect on the performance in mixtures and pavements. An alternative binder may have significantly different aging characteristics. Consequently, the RTFO and PAV procedures may not adequately represent plant and field aging. It is suggested that experimental considerations start with determining an aging index to compare against conventional asphalt. The RTFO procedure should be run over a range of times and temperatures to see if normal time–temperature correlations still hold. The same applies for the PAV procedure which should be run at 60°C (140°F) for extended times and the resulting behavior compared to the results at standard conditions (Kluttz, 2012).

The Abson recovery test (AASHTO T170), in which the completed mix has the asphalt cement extracted by a chemical procedure, does not change the asphalt cement properties. The recovered asphalt cement must meet certain properties, such as penetration and ductility, which indicate damage or aging that may have taken place during the storage and mixing of the asphalt. When recycled asphalt pavement (RAP) is a component of the mix, the aged asphalt in the RAP can become a significant contributor to the aged properties of the mix. If a penetration test fails in accordance with Section 211.06, then the binder will be PG graded to determine acceptability of the binder.

Testing liquid asphalt may require modification of a standard test method or application of a test method specifically designed for such a task. While ASTM standard test methods are available, AASHTO has several such methods. For example, in the viscosity test (AASHTO T72; AASHTO T201) the Asphalt Institute specifications for binders recommend viscosity tests be run at 135°C (275°F). The viscosity test purpose is to provide control of asphalt consistency in the range of temperatures normally associated with construction operations. Measurement of viscosity may be made by the Saybolt furol viscosity test (ASTM D88) or the kinematic viscosity test (ASTM D445).

The flash point test (AASHTO T79) for liquid asphalt is performed on the original liquid binder (cutback), the flash point is measured to ensure the liquid asphalt (cutback) is safe to work with at production temperatures. Again, the flash point is the temperature to which the liquid binder (cutback) may be safely heated without instantaneous flash in the presence of an open flame.

4.4.2 Use of the Data

The changing nature of refinery feedstocks has required that the analytical methods and interpretation of the data evolve to accommodate the new refining processes, process chemistry, and products. In this time period, the *average* quality of petroleum delivered to refineries has diminished by approximately two points on the API gravity scale. This has been accompanied by, among other things, an increase in the sulfur content. However, as a word of caution, use of the word *average* is open to definition, because some refineries accept medium- to high-quality crude oil while others accept low- to medium-quality crude oil. Nevertheless, the general quality of the crude oil accepted by refineries is lower than it was two decades ago. As a result, refineries have had to adapt processing strategies to deliver liquid fuel products from feedstocks that are deteriorating in quality. The same might be stated for products such as asphalt. The complexity of such a product renders a term such as average quality open to question and whether or not such a term is suitable for making a decision for use of the asphalt. Therefore, caution is advised when considering the result(s) of analytical test method(s) and whether or not the sample represents a material suitable for the designated use (Table 4.4).

Thus, recognition that refinery behavior is related to the composition of the feedstock has led to a multiplicity of attempts to establish petroleum and its fractions as compositions of matter. As a result, various analytical techniques have been developed for the identification and quantification of every molecule in the lower boiling fractions of petroleum. Unfortunately for these past efforts, however, it is now generally

Table 4.4 Ranges of elemental composition for asphalt

Carbon (% w/w)	80.0–84.5
Hydrogen (% w/w)	9.8–10.8
Nitrogen (% w/w)	0.2–1.2
Oxygen (% w/w)	0.4–1.0
Sulfur (% w/w)	0.5–8.0
Nickel (ppm)	10–140
Vanadium (ppm)	7–1600
Iron (ppm)	5–150
Sodium (ppm)	5–160
Copper (ppm)	5–150
Manganese (ppm)	0–4.0
Calcium (ppm)	1–350
Magnesium (ppm)	1–150

Also includes ranges for bitumen samples from various locations.

recognized that the name *petroleum* does not describe a composition of matter but rather a mixture of various organic compounds that includes a wide range of molecular weights and molecular types. Because of the complex nature of petroleum, there has been a need to define petroleum in a more meaningful and understandable manner. The general terminology that has been applied to petroleum science and technology has left much to be desired. The terminology has been inherited from the early days of petroleum use and is often misleading. Although there is still much to be accomplished, some success has been achieved. For example, it is now rare that asphaltic deposits are referred to (environmental issues notwithstanding) as *slime pits*!

Asphalt binders are produced mainly by petroleum refiners and, to a lesser extent, by formulators who purchase blending stock from refiners. The production of asphalt binder consists of two phases: refining the asphalt base stock and formulating this stock into a finished asphalt binder. As a result of binder specifications, both phases of production required logistical control and greater attention to quality as well as focus on environmental characteristics of the asphalt which created a demand for asphalt binders with improved, lower volatility, a characteristic that can be affected by both the refining and the formulation phases.

The refining phase uses relatively mature technologies, and significant changes in these technologies are not expected. However, as the specifications evolved there came a need for more judicious crude oil selection and more careful process control. Because most asphalt crude oils are imported and the residua from which asphalt is produced are produced from blends of crude oils (see Chapter 3) (Speight, 2014), any unforeseen governmental interference or governmental instability can affect the future availability of these crude oils thereby introducing unexpected challenges to asphalt production.

As a result, there arose the need to define asphalt on a more logical in-service basis that would also accommodate any changes in the formulation of the asphalt. These changes were driven, to a large extent, by the varying quality of asphalt binders and the (often futile) attempts to assign in-service behavior of the binder to one or more physical properties, often without consideration of the environmental (climatic) conditions that existed at the in-service site. There was a need to develop a range of binder grades by which binder performance could be be closely matched to climatic conditions. Furthermore, binders necessary for severe climate regions cannot be made from crude oil alone and they must be modified to improve the properties of the binder.

Until recently, binder specification systems were relatively lenient, and gave refiners a high level of production flexibility. Therefore, refiners tended to view asphalt as a simple, convenient way to use the residual material from the refinery operation. However, as a result of the new *Superpave* specification system, asphalt refiners increasingly perceive asphalt as a value-added product. It also has caused many refiners to reevaluate their commitment to asphalt production; some have made a strategic decision to de-emphasize or cease asphalt production, though others have renewed their efforts to produce high-quality binders.

Asphalt binders have been, and many still are, most commonly graded by their physical properties, which are significantly influenced by temperature. The physical properties of the binder are an indication of how the binder might perform as a constituent in asphalt pavement. Although asphalt binder penetration and viscosity grading are still commonly used worldwide, new binder tests and specifications have been developed to more accurately characterize asphalt binders for use in roadways. The most advanced asphalt binder grading system is the *performance graded system*, which has been adopted throughout the United States. The performance grading system is based on the idea that asphalt binder properties should be related to the conditions under which it is used. Thus, the asphalt binder is selected to meet expected climatic conditions as well as aging (time) considerations. Therefore, the system uses a common set of tests to measure physical properties of the binder that can be directly related to field performance of the pavement at extreme temperatures.

The system has replaced the penetration and viscosity grading systems used in the past and the climatic conditions (primarily temperature) to which roadways are exposed are a major component of the *performance graded* system. The standard *performance graded* specifications (ASTM D6373; TRB, 2010, 2012) also include assessment of durability of asphalt as well as safety aspects (such as flash point and loss on heating). However, several of the states have modified these requirements based on their past experience with *performance graded* asphalt and/or asphalt supply considerations specific to their area. The modifications to the *performance graded* system incorporated tests are also designed to accommodate the inclusion of temperature (ASTM D6818).

Thus performance grading, as the name implies, is based on the requirement idea that binder properties should be related to the conditions under which it is to be used. This involves expected climatic conditions as well as aging considerations and, therefore, the system uses data from a number of tests (as the older penetration and viscosity grading systems do) but specifies that a particular asphalt binder must pass these tests at specific temperatures that are dependent upon the specific climatic conditions in the area of use.

As an example, since the system is based on climate, the grade notation consists of two portions: (i) high in-service temperature and (ii) low in-service temperature. The major concern for high-temperature performance is rutting, which typically takes time to cumulate, therefore an average of 7-day maximum pavement temperature is used for describing the high-temperature climate. On the low-temperature side, thermal cracking can happen during a cold period and, therefore, the minimum pavement temperature is used for describing the low-temperature climate. The average 7-day maximum pavement temperature typically ranges from 46°C to 82°C (114–180°F), and minimum pavement temperature typically ranges from −46°C to −10°C (−51°F to 14°F). In terms of the grading nomenclature, a binder identified as PG 70-28 must meet performance criteria at a 7-day maximum pavement temperature of 70°C (158°F) and also at a minimum pavement temperature of −28°C (−18°F).

The common minimum reliability used is 98%, so that means when the PG 70-20 binder is selected, the asphalt binder in the pavement should perform satisfactorily under typical traffic condition at the location where the extreme pavement temperatures are within the range of −28°C (−18°F) and 70°C (158°F) throughout service life with a minimum 98% confidence level. Where the traffic condition is not typical, such as the really heavy traffic on an interstate highway, or slow traffic such as a bus stop or intersection area, one or two grades stiffer asphalt binder may be used to help prevent (in this example) the occurrence of rutting. Furthermore, the workability of asphalt varies with the design and type of mix. In fact, there is often an inverse linkage between workability and performance, with a very workable mix being prone to shoving or rutting under traffic, and possibly bleeding. As an extra note, the use of real-life conditions to critically estimate the performance of an asphalt binder service is a major innovation. This is analogous to the realization that several properties must be used to determine the behavior of petroleum during refining and recovery (Speight, 2014)—without applying real-life conditions to the material, any assessment of the behavior of the binder will be locking in reality.

Implementation of the Superpave system has increased the use of unconventional binder additives, air blowing, blending, and chemical modification with the inclusion of a number of test methods that are appropriate to the task (Stuart et al., 2000). As the relationship between binder properties and mixture performance is better understood, the specification will continue to evolve through the development of new modification technologies and the use of custom blending and modification will increase. In combination, these trends will cause additional supply changes, as binder producers seek to match their strategic resources and interests with the most appropriate niches in the asphalt market.

Thus, test methods and specifications for asphalt binders remained relatively unchanged for several decades until the introduction of viscosity-based specifications in the late 1960s, followed by the widespread use of fundamental rheological test methods that became possible in the 1990s (Speight, 1992, 2014). Although rheological properties are important for predicting the performance of asphalt binders, the characterization of failure properties is also necessary. New test methods will be required to define the fatigue behavior as well as the cracking and healing characteristics of asphalt binders. Three modes of distress are recognized in the current binder specifications (i) rutting, (ii) fatigue, and (iii) single-event low-temperature thermal cracking.

Advances in instrumentation will allow the more rapid characterization of the chemical and physical properties of asphalt binders. As a result, binders will be tailored to specific applications. The trend toward performance-based specifications will accelerate, as will the need for increased involvement in understanding of binder properties and pavement performance (Tables 4.5 and 4.6).

Table 4.5 Asphalt standards as performance-based standards

Steps
1. Identify binder properties linked to asphalt pavement performance
2. Select and standardize appropriate test methods to measure these properties
3. Collect data and ensure field validation
4. Review the grading system according to the new specifications
Performance-Based Specifications Are:
1. Asphalt properties
2. Asphalt mix characteristics
3. Identifying binder properties linked to asphalt pavement performance
4. User requirements
5. Surface characteristics
6. Pavement performance
Asphalt Pavement Performance Characteristics (1)
Resistance to:
1. Surface cracking (aging)
2. Wear
3. Stripping and raveling
4. Low temperature cracking
5. Fatigue cracking
6. Reflective cracking
7. Shear-induced cracking
8. Fuel/de-icing fluids/salt
Asphalt Pavement Performance Characteristics (2)
1. Healing properties
2. Structural strength
3. Stiffness
During Construction Phase
1. Compactability
2. Binder drainage
3. Segregation sensitivity

One issue that must be given consideration is the relationship of laboratory data and data from testing of roadway samples. Samples of the asphalt mix prepared in the laboratory may have considerably different properties from mixtures produced in the asphalt mix plant for a number of reasons; for example, as the asphalt mix ages as the mixture goes through the plant, and during storage and transportation, until the mixture cools down. The binder reacts with oxygen in the air and becomes harder and more brittle. Some volatile fractions of the binder may also be driven off at the high temperatures encountered during construction. Absorption of some of the binder into the aggregate may also occur

Table 4.6 AASHTO test methods for use for performance graded modified (PG-M) asphalt binders

Test method	Test name
AASHTO T48	Flash point
AASHTO T44	Solubility of bituminous material
AASHTO T316	Rotational viscosity
AASHTO T315	G*/SinDelta on original binder
AASHTO T240	Relative (plus or minus) mass change after rolling thin film oven
AASHTO T301	Elastic recovery
AASHTO T315	G*/SinDelta on rolling thin film oven residue
AASHTO T315	Delta @ 2.2 kpa on rolling thin film oven residue + 6°
AASHTO T315	G*SinDelta on residue from pressure aging vessel[a]
AASHTO T313	Creep stiffness on residue from pressure aging vessel[a]
AASHTO T313	M-value[b] on residue from pressure aging vessel[a]
AASHTO T228	Specific gravity

[a]Pressure aging vessel (PAV) is designed to simulate in-service oxidative aging of asphalt binder by exposure to elevated temperatures in a pressurized environment.
[b]See ASTM D6648. Standard Test Method for Determining the Flexural Creep Stiffness of Asphalt Binder Using the Bending Beam Rheometer (BBR). ASTM International, West Conshohocken, PA.

during construction while the binder is still fluid enough to migrate into the pores of the aggregate.

Aging continues at a slower rate throughout the service life of the pavement. The aging (or oxidation) reaction proceeds at a higher rate in hot climates or during the summer months when the temperatures are higher. Accounting for the changes in the mixture properties when preparing mixtures in the lab is essential for the production of reliable test data (AASHTO R30). One way to account for these changes is to condition the laboratory mixtures in such a way as to simulate the aging that happens during construction and service. The short-term mixture conditioning procedure is used to simulate the aging that occurs during construction (up to the point of compaction) and is used during the mix design procedure. Long-term aging is used to simulate the aging that occurs over the many years that the pavement is in service. Consequently, long-term aging is used when performing tests to simulate mixture properties late in the life of the pavement, such as when analyzing the resistance of a mixture to low-temperature cracking using Superpave mixture analysis and the Indirect Tension tester. Long-term mixture conditioning follows short-term conditions for laboratory-prepared mixtures.

4.5 VALIDITY OF TEST PROCEDURES

The major issues that arise when considering the properties of the asphalt binder, the aggregate, and the asphalt–aggregate mix are (i) the validity of the testing procedure and (ii) the relationship of properties (as measured in the laboratory) to performance in terms of roadway behavior and longevity. However, there are many issues that are associated with the nonstandard way in which many materials are tested and characterized and for which relevant data or information was available. These problems make it difficult to establish relationships between properties and performance, especially identification of the trends associated with the use of modified materials (Coplantz et al., 1993). For example, while it can be reasonably concluded that modifiers have an influence on the performance-related properties of asphalt cement and asphalt concrete, as measured in the laboratory, they also have an influence on the performance of asphalt pavements as determined from evaluations of in-service roadways. However, whether or not the two sets of data match and any conclusions drawn can be used as a consensus, is not always the case.

The performance of unmodified or additive-modified asphalt binders from in-service roadways is typically measured through the use of special test sections placed in a roadway project where the remainder is a control section using an unmodified asphalt binder. Comparisons are then made between the performance of the test section and the control section. Interpretation of test data relies on site-specific factors including pavement structure characteristics, traffic, and environmental conditions. In fact, the influence of site-specific environmental factors (including traffic loads) makes it difficult to extrapolate the test data to so that the data are more generally applicable to other roadways. Thus, the interpretation of test data should be considered to be valid predominantly (if not only) on a site-specific basis.

It is rare for test data from one investigation to agree with the data from another investigation, and whether or not the test methods are truly standard and can be extrapolated from the laboratory to the field is another issue that is not always easy to answer. The confusing effects of roadway structure and the roadway environment are factors that can account for some of the differences in the test data and in the behavior of the roadway in service. In addition, the base asphalt appears to have a pronounced influence on modifier effectiveness. An important finding from the review is that modifiers may enhance certain properties of a given binder to produce more favorable performance characteristics but not to a level that could be obtained simply by changing asphalt source or grade.

Unless standardized test methods are used, there will be many problems that are associated with the use of nonstandard test methods. In addition, the repeatability and reproducibility (as well as sampling) play an important role in the believability of the data from test methods (ASTM D140; Speight, 2015a).

For example, when sampling asphalt mixtures (ASTM D979; ASTM D140) the sampling is equally as important as the testing, and the sampler should take every precaution to obtain samples that will yield an acceptable estimate of the nature and conditions of the materials which they represent (Table 4.2). Samples for the development of preliminary data are obtained by the party responsible for the development of the data. Samples for control of the product at the source of manufacture or storage, or at the site of use, are obtained by the manufacturer, contractor, or other parties responsible for accomplishing the work. Samples for tests to be used in acceptance or rejection decisions by the purchaser are obtained by the purchaser or his authorized representative.

In short, if nonstandard test methods are uses and the three criteria mentioned above (sampling, repeatability, and reproducibility) are not in order, the material that was tested and characterized and for which data or information was produced, it will be difficult to compare the data from one laboratory to another and, thus, establish relationships and trends associated with the use of the material. Indeed, there is some controversy regarding whether improvement in the properties of asphalt binders, as measured by various particular non-traditional and non-standard test methods, are any type of indicator of actual performance in the field.

Field performance of asphalt binders is generally measured through the use of special test sections placed in a project where a control section using unmodified asphalt is compared to a *test section*. However, the test section must match the control section—for example, comparing the effect of a modified roadway section in a driving lane with the unmodified asphalt in the passing lane (and vice versa, as is also the case when the test sections of the roadway were laid in different years) is a fruitless and incorrect procedure. Comparisons must be made between similar roadway sections and interpretation of the test results based on site-specific factors, including pavement structure characteristics and traffic and environmental conditions, is relevant and meaningful. If the test sections are not chosen wisely, the compounding effects of the site-specific factors make extrapolation of the test data to more generalized conditions very difficult, if not impossible. The interpretation of test results will, more than likely, be valid for only the site-specific conditions.

If the test sections are chosen in a logical manner and if the standard test methods are selected with care and forethought, the data can provide meaningful correlations with the mechanical properties of the asphalt–aggregate system. This will allow the establishment of the links between laboratory data and field performance and may even point to mix properties that are likely to be required for future asphalt roadways. Thus, each test method should be standardized and evaluated on its own merits and findings from one investigation (if the method has been standardized) should be in agreement with the data from the same method applied to the same sample at a different laboratory locale.

Recognizing that problems do exist in attempting to interpret information from application of the various test methods, serious efforts should be made to extract both qualitative and quantitative information from laboratory data and field performance. In fact, in order to predict the behavior of an asphalt pavement at low temperature, more than one test will be required in order to generate meaningful (and believable) data (Schmidt, 1996; Schmidt and Santucci, 1996). It is necessary to take into account all possible factors which will be affected by the composition and structure of the asphalt roadway (Barth, 1962; Halstead et al., 1966; Speight, 2014, 2015a). If the test methods do not give consideration to such phenomena, the data may be of limited value. Asphalt source is probably another variable that affects the glass transition temperature just as the characteristics of the monomer unit in polymers affect their glass transition temperatures (Breen and Stephens, 1967). The effects of aging on the glass transition temperature and strain energy release rate are a problem that is in need of further research. Hopefully, by knowing these quantities, it will be possible to select asphalts that will not be crack-susceptible at low temperatures.

The need for the continuing evolution of current standard test methods and the development of new standard test methods is a necessary part of the asphalt industry, as it is for the petroleum industry and other industries (Dongre and D'Angelo, 1998; Speight, 2015a). As an example, a timely suite of PRS have been developed for asphalt binders as well as mixtures (collectively known as the Superpave pavement design system) (see Chapters 9 and 10).

These new specifications, which are purchase specifications, were adopted by the American Association of State Highway and Transportation Officials (designated as AASHTO M320—formerly designated as MP1—and AASHTO MP1a). In the former test method (AASHTO M320), empirical parameters are related to performance while in the latter test method (AASHTO MP1a) a mechanistically based criterion to define low-temperature performance of asphalt binders is incorporated into the method. In addition, property data (ASTM D6648; ASTM D6723; ASTM D6816) are used to obtain a critical cracking temperature (AASHTO MP1a), whereas other property data (ASTM D6648) are used in the Superpave pavement design system under the purchase specification (AASHTO M320; ASTM D6373).

Thus, the Superpave specification is based on test methods that were adapted and/or specifically designed to measure fundamental material properties as opposed to the empirical test methods employed by the earlier viscosity and penetration-based specification. Two test methods and a calculation practice have been developed for use in the Superpave specification to address low-temperature cracking performance of pavements: (i) a flexural creep test called the bending beam rheometer test method to measure creep properties at low temperatures (AASHTO T313, see also ASTM D6648), (ii) a uniaxial tension test at constant elongation rate—the direct tension test method—to measure the failure properties at low temperature. The constant elongation rate simulates the loading (in uniaxial tension) during thermal cooling of the pavement (see also ASTM D6723), and (iii) the critical cracking temperature which is calculated using an algorithm (AASHTO MP1a; ASTM D6816) (Anderson et al., 1994; Bouldin et al., 2000).

REFERENCES

AASHTO M320, 2014. Standard Specification for Performance-Graded Asphalt Binder. American Association of State Highway and Transportation Officials, Washington, DC.

AASHTO MP1a, 2014. Performance Graded Binder Specification. American Association of State Highway and Transportation Officials, Washington, DC.

AASHTO R28, 2014. Accelerated Aging of Asphalt Binder Using a Pressurized Aging Vessel (PAV). Standard Specifications for Transportation Materials and Methods of Sampling and Testing. American Association of State Highway and Transportation Officials, Washington, DC.

AASHTO R30, 2014. Standard Practice for Mixture Conditioning of Hot Mix Asphalt. American Association of State Highway and Transportation Officials, Washington, DC.

AASHTO T48, 2014. Standard Method of Test for Flash and Fire Points by Cleveland Open Cup (ASTM D92). American Association of State Highway and Transportation Officials, Washington, DC.

AASHTO T51, 2014. Standard Method of Test for Ductility of Asphalt Materials (ASTM D113). American Association of State Highway and Transportation Officials, Washington, DC.
AASHTO T59, 2014. Emulsified Asphalt. American Association of State Highway and Transportation Officials, Washington, DC.
AASHTO T72, 2014. Standard Method of Test for Saybolt Viscosity. American Association of State Highway and Transportation Officials, Washington, DC.
AASHTO T79, 2014. Standard Method of Test for Flash Point with Tag Open-Cup Apparatus for Use with Material Having a Flash Point Less Than 93 degrees Celsius (200 degrees Fahrenheit). American Association of State Highway and Transportation Officials, Washington, DC.
AASHTO T170, 2014. Standard Specification for Recovery of Asphalt Binder from Solution by Abson Method (ASTM D1856). American Association of State Highway and Transportation Officials, Washington, DC.
AASHTO T201, 2014. Standard Method of Test for Kinematic Viscosity of Asphalts (Bitumens) (ASTM D2170). American Association of State Highway and Transportation Officials, Washington, DC.
AASHTO T240, 2014. Standard Method of Test for Effect of Heat and Air on a Moving Film of Asphalt (Rolling Thin-Film Oven Test) (ASTM D 2872). American Association of State Highway and Transportation Officials, Washington, DC.
AASHTO T283, 2014. Standard Method of Test for Resistance of Compacted Asphalt Mixtures to Moisture-Induced Damage. American Association of State Highway and Transportation Officials, Washington, DC.
AASHTO T313, 2014. Determining the Flexural Creep Stiffness of Asphalt Binder Using the Bending Beam Rheometer (BBR). Standard Specifications for Transportation Materials and Methods of Sampling and Testing. American Association of State Highway and Transportation Officials, Washington, DC.
AASHTO T315, 2014. Determining the Rheological Properties of Asphalt Binder Using a Dynamic Shear Rheometer (DSR). American Association of State Highway and Transportation Officials, Washington, DC.
AASHTO T316, 2014. Standard Method of Test for Viscosity Determination of Asphalt Binder Using Rotational Viscometer. Standard Specifications for Transportation Materials and Methods of Sampling and Testing. American Association of State Highway and Transportation Officials, Washington, DC.
Abraham, H., 1945. Asphalts and Allied Substances: Their Occurrence, Modes of Production, Uses in the Arts and Methods of Testing, fifth ed. Van Nostrand, New York, NY.
Airey, G.D., 2002. Rheological evaluation of ethylene vinyl acetate polymer modified bitumens. Constr. Build. Mater. 16, 473–487.
Anderson, D.A., Christensen, D.W., Bahia, H.U., Dongre, R., Sharma, M.G., Antle, C.E., et al., 1994. Binder Characterization and Evaluation, Volume 3: Physical Characterization Report No. RP-A-369.. Strategic Highway Research Program, National Research Council, Washington, DC.
Arega, Z., Bhasin, A., 2012. Binder Rheology and Performance in Warm Mix Asphalt Report No. FHWA/TX-12/0-6591-1. Center for Transportation Research, University of Texas at Austin, Austin, Texas and Research and Technology Implementation Office, Texas Department of Transportation, Austin, Texas.
ASTM C177, 2014. Standard Test Method for Steady-State Heat Flux Measurements and Thermal Transmission Properties by Means of the Guarded-Hot-Plate Apparatus Annual Book of Standards. ASTM International, West Conshohocken, PA.
ASTM C518, 2014. Standard Test Method for Steady-State Thermal Transmission Properties by Means of the Heat Flow Meter Apparatus Annual Book of Standards. ASTM International, West Conshohocken, PA.

ASTM C1045, 2014. Standard Practice for Calculating Thermal Transmission Properties Under Steady-State Conditions Annual Book of Standards. ASTM International, West Conshohocken, PA.

ASTM D4, 2014. Standard Test Method for Bitumen Content Annual Book of Standards. ASTM International, West Conshohocken, PA.

ASTM D5, 2014. Standard Test Method for Penetration of Bituminous Materials Annual Book of Standards. ASTM International, West Conshohocken, PA.

ASTM D6, 2014. Standard Test Method for Loss on Heating of Oil and Asphaltic Compounds Annual Book of Standards. ASTM International, West Conshohocken, PA.

ASTM D20, 2014. Standard Test Method for Distillation of Road Tars Annual Book of Standards. ASTM International, West Conshohocken, PA.

ASTM D36, 2014. Standard Test Method for Softening Point of Bitumen (Ring-and-Ball Apparatus) Annual Book of Standards. ASTM International, West Conshohocken, PA.

ASTM D56, 2014. Standard Test Method for Flash Point by Tag Closed Tester Annual Book of Standards. ASTM International, West Conshohocken, PA.

ASTM D61, 2014. Standard Test Method for Softening Point of Pitches (Cube-in-Water Method) Annual Book of Standards. ASTM International, West Conshohocken, PA.

ASTM D70, 2014. Standard Test Method for Density of Semi-Solid Bituminous Materials (Pycnometer Method) Annual Book of Standards. ASTM International, West Conshohocken, PA.

ASTM D86, 2014. Standard Test Method for Distillation of Petroleum Products at Atmospheric Pressure Annual Book of Standards. ASTM International, West Conshohocken, PA.

ASTM D88, 2014. Standard Test Method for Saybolt Viscosity Annual Book of Standards. ASTM International, West Conshohocken, PA.

ASTM D92, 2014. Standard Test Method for Flash and Fire Points by Cleveland Open Cup Annual Book of Standards. ASTM International, West Conshohocken, Pennsylvania.

ASTM D93, 2014. Standard Test Methods for Flash Point by Pensky-Martens Closed Tester Annual Book of Standards. ASTM International, West Conshohocken, PA.

ASTM D94, 2014. Standard Test Methods for Saponification Number of Petroleum Products Annual Book of Standards. ASTM International, West Conshohocken, PA.

ASTM D95, 2014. Standard Test Method for Water in Petroleum Products and Bituminous Materials by Distillation Annual Book of Standards. ASTM International, West Conshohocken, PA.

ASTM D97, 2014. Standard Test Method for Pour Point of Petroleum Products Annual Book of Standards. ASTM International, West Conshohocken, PA.

ASTM D113, 2014. Standard Test Method for Ductility of Bituminous Materials Annual Book of Standards. ASTM International, West Conshohocken, PA.

ASTM D129, 2014. Standard Test Method for Sulfur in Petroleum Products Annual Book of Standards. ASTM International, West Conshohocken, PA.

ASTM D139, 2014. Standard Test Method for Float Test for Bituminous Materials Annual Book of Standards. ASTM International, West Conshohocken, PA.

ASTM D140, 2014. Standard Practice for Sampling Bituminous Materials Annual Book of Standards. ASTM International, West Conshohocken, PA.

ASTM D217, 2014. Standard Test Methods for Cone Penetration of Lubricating Grease Annual Book of Standards. ASTM International, West Conshohocken, PA.

ASTM D244, 2014. Standard Test Methods and Practices for Emulsified Asphalts Annual Book of Standards. ASTM International, West Conshohocken, PA.

ASTM D312, 2014. Standard Specification for Asphalt Used in Roofing Annual Book of Standards. ASTM International, West Conshohocken, PA.

ASTM D402, 2014. Standard Test Method for Distillation of Cutback Asphaltic (Bituminous) Products Annual Book of Standards. ASTM International, West Conshohocken, PA.

ASTM D445, 2014. Standard Test Method for Kinematic Viscosity of Transparent and Opaque Liquids Annual Book of Standards. ASTM International, West Conshohocken, PA.
ASTM D449, 2014. Standard Specification for Asphalt Used in Damp-proofing and Waterproofing Annual Book of Standards. ASTM International, West Conshohocken, PA.
ASTM D482, 2014. Standard Test Method for Ash from Petroleum Products Annual Book of Standards. ASTM International, West Conshohocken, PA.
ASTM D893, 2014. Standard Test Method for Insolubles in Used Lubricating Oils Annual Book of Standards. ASTM International, West Conshohocken, PA.
ASTM D937, 2014. Standard Test Method for Cone Penetration of Petrolatum Annual Book of Standards. ASTM International, West Conshohocken, PA.
ASTM D971, 2014. Standard Test Method for Interfacial Tension of Oil Against Water by the Ring Method Annual Book of Standards. ASTM International, West Conshohocken, PA.
ASTM D972, 2014. Standard Test Method for Evaporation Loss of Lubricating Greases and Oils Annual Book of Standards. ASTM International, West Conshohocken, PA.
ASTM D979, 2014. Standard Practice for Sampling Bituminous Paving Mixtures Annual Book of Standards. ASTM International, West Conshohocken, PA.
ASTM D1160, 2014. Standard Test Method for Distillation of Petroleum Products at Reduced Pressure Annual Book of Standards. ASTM International, West Conshohocken, PA.
ASTM D1250, 2014. Standard Guide for Use of the Petroleum Measurement Tables Annual Book of Standards. ASTM International, West Conshohocken, PA.
ASTM D1370, 2014. Standard Test Method for Contact Compatibility Between Asphaltic Materials (Oliensis Test) Annual Book of Standards. ASTM International, West Conshohocken, PA.
ASTM D1403, 2014. Standard Test Methods for Cone Penetration of Lubricating Grease Using One-Quarter and One-Half Scale Cone Equipment Annual Book of Standards. ASTM International, West Conshohocken, PA.
ASTM D1669, 2014. Standard Practice for Preparation of Test Panels for Accelerated and Outdoor Weathering of Bituminous Coatings Annual Book of Standards. ASTM International, West Conshohocken, PA.
ASTM D1670, 2014. Standard Test Method for Failure End Point in Accelerated and Outdoor Weathering of Bituminous Materials Annual Book of Standards. ASTM International, West Conshohocken, PA.
ASTM D1754, 2014. Standard Test Method for Effects of Heat and Air on Asphaltic Materials (Thin-Film Oven Test) Annual Book of Standards. ASTM International, West Conshohocken, PA.
ASTM D2007, 2014. Standard Test Method for Characteristic Groups in Rubber Extender and Processing Oils and Other Petroleum-Derived Oils by the Clay-Gel Absorption Chromatographic Method Annual Book of Standards. ASTM International, West Conshohocken, PA.
ASTM D2042, 2014. Standard Test Method for Solubility of Asphalt Materials in Trichloroethylene Annual Book of Standards. ASTM International, West Conshohocken, PA.
ASTM D2161, 2014. Standard Practice for Conversion of Kinematic Viscosity to Saybolt Universal Viscosity or to Saybolt Furol Viscosity Annual Book of Standards. ASTM International, West Conshohocken, PA.
ASTM D2170, 2014. Standard Test Method for Kinematic Viscosity of Asphalts (Bitumens) Annual Book of Standards. ASTM International, West Conshohocken, PA.
ASTM D2171, 2014. Standard Test Method for Viscosity of Asphalts by Vacuum Capillary Viscometer Annual Book of Standards. ASTM International, West Conshohocken, PA.
ASTM D2319, 2014. Standard Test Method for Softening Point of Pitch (Cube-in-Air Method) Annual Book of Standards. ASTM International, West Conshohocken, PA.
ASTM D2415, 2014. Standard Test Method for Ash in Coal Tar and Pitch Annual Book of Standards. ASTM International, West Conshohocken, PA.

ASTM D2595, 2014. Standard Test Method for Evaporation Loss of Lubricating Greases Over Wide-Temperature Range Annual Book of Standards. ASTM International, West Conshohocken, PA.
ASTM D2746, 2014. Standard Test Method for Staining Tendency of Asphalt (Stain Index) Annual Book of Standards. ASTM International, West Conshohocken, PA.
ASTM D2872, 2014. Standard Test Method for Effect of Heat and Air on a Moving Film of Asphalt (Rolling Thin-Film Oven Test) Annual Book of Standards. ASTM International, West Conshohocken, PA.
ASTM D2887, 2014. Standard Test Method for Boiling Range Distribution of Petroleum Fractions by Gas Chromatography Annual Book of Standards. ASTM International, West Conshohocken, PA.
ASTM D3104, 2014. Standard Test Method for Softening Point of Pitches (Mettler Softening Point Method) Annual Book of Standards. ASTM International, West Conshohocken, PA.
ASTM D3279, 2014. Standard Test Method for n-Heptane Insolubles Annual Book of Standards. ASTM International, West Conshohocken, PA.
ASTM D3409, 2014. Standard Test Method for Adhesion of Asphalt-Roof Cement to Damp, Wet, or Underwater Surfaces Annual Book of Standards. ASTM International, West Conshohocken, PA.
ASTM D3461, 2014. Standard Test Method for Softening Point of Asphalt and Pitch (Mettler Cup-and-Ball Method) Annual Book of Standards. ASTM International, West Conshohocken, PA.
ASTM D3607, 2014. Standard Test Method for Removing Volatile Contaminants from Used Engine Oils by Stripping Annual Book of Standards. ASTM International, West Conshohocken, PA.
ASTM D3625, 2014. Standard Practice for Effect of Water on Bituminous-Coated Aggregate Using Boiling Water Annual Book of Standards. ASTM International, West Conshohocken, PA.
ASTM D3710, 2014. Standard Test Method for Boiling Range Distribution of Gasoline and Gasoline Fractions by Gas Chromatography Annual Book of Standards. ASTM International, West Conshohocken, PA.
ASTM D4055, 2014. Standard Test Method for Pentane Insolubles by Membrane Filtration Annual Book of Standards. ASTM International, West Conshohocken, PA.
ASTM D4072, 2014. Standard Test Method for Toluene-Insoluble (TI) Content of Tar and Pitch Annual Book of Standards. ASTM International, West Conshohocken, PA.
ASTM D4124, 2014. Standard Test Method for Separation of Asphalt into Four Fractions Annual Book of Standards. ASTM International, West Conshohocken, PA.
ASTM D4287, 2014. Standard Test Method for High-Shear Viscosity Using a Cone/Plate Viscometer Annual Book of Standards. ASTM International, West Conshohocken, PA.
ASTM D4294, 2014. Standard Test Method for Sulfur in Petroleum and Petroleum Products by Energy Dispersive X-ray Fluorescence Spectrometry Annual Book of Standards. ASTM International, West Conshohocken, PA.
ASTM D4312, 2014. Standard Test Method for Toluene-Insoluble (TI) Content of Tar and Pitch (Short Method) Annual Book of Standards. ASTM International, West Conshohocken, PA.
ASTM D4402, 2014. Standard Test Method for Viscosity Determination of Asphalt at Elevated Temperatures Using a Rotational Viscometer Annual Book of Standards. ASTM International, West Conshohocken, PA.
ASTM D4469, 2014. Standard Practice for Calculating Percent Asphalt Absorption by the Aggregate in an Asphalt Pavement Mixture Annual Book of Standards. ASTM International, West Conshohocken, PA.
ASTM D4486, 2014. Standard Test Method for Kinematic Viscosity of Volatile and Reactive Liquids Annual Book of Standards. ASTM International, West Conshohocken, PA.

ASTM D4628, 2014. Standard Test Method for Analysis of Barium, Calcium, Magnesium, and Zinc in Unused Lubricating Oils by Atomic Absorption Spectrometry Annual Book of Standards. ASTM International, West Conshohocken, PA.

ASTM D4867, 2014. Standard Test Method for Effect of Moisture on Asphalt Concrete Paving Mixtures Annual Book of Standards. ASTM International, West Conshohocken, PA.

ASTM D4893, 2014. Standard Test Method for Determination of Pitch Volatility Annual Book of Standards. ASTM International, West Conshohocken, PA.

ASTM D4927, 2014. Standard Test Methods for Elemental Analysis of Lubricant and Additive Components—Barium, Calcium, Phosphorus, Sulfur, and Zinc by Wavelength-Dispersive X-ray Fluorescence Spectroscopy Annual Book of Standards. ASTM International, West Conshohocken, PA.

ASTM D5018, 2014. Standard Test Method for Shear Viscosity of Coal-Tar and Petroleum Pitches Annual Book of Standards. ASTM International, West Conshohocken, PA.

ASTM D5185, 2014. Standard Test Method for Multielement Determination of Used and Unused Lubricating Oils and Base Oils by Inductively Coupled Plasma Atomic Emission Spectrometry (ICP-AES) Annual Book of Standards. ASTM International, West Conshohocken, PA.

ASTM D5236, 2014. Standard Test Method for Distillation of Heavy Hydrocarbon Mixtures (Vacuum Potstill Method) Annual Book of Standards. ASTM International, West Conshohocken, PA.

ASTM D5307, 2014. Standard Test Method for Determination of Boiling Range Distribution of Crude Petroleum by Gas Chromatography Annual Book of Standards. ASTM International, West Conshohocken, PA.

ASTM D5770, 2014. Standard Test Method for Semiquantitative Micro Determination of Acid Number of Lubricating Oils During Oxidation Testing Annual Book of Standards. ASTM International, West Conshohocken, PA.

ASTM D5800, 2014. Standard Test Method for Evaporation Loss of Lubricating Oils by the Noack Method Annual Book of Standards. ASTM International, West Conshohocken, PA.

ASTM D6299, 2014. Standard Practice for Applying Statistical Quality Assurance and Control Charting Techniques to Evaluate Analytical Measurement System Performance Annual Book of Standards. ASTM International, West Conshohocken, PA.

ASTM D6373, 2014. Standard Specification for Performance Graded Asphalt Binder Annual Book of Standards. ASTM International, West Conshohocken, PA.

ASTM D6375, 2014. Standard Test Method for Evaporation Loss of Lubricating Oils by Thermogravimetric Analyzer (TGA) Noack Method Annual Book of Standards. ASTM International, West Conshohocken, PA.

ASTM D6443, 2014. Standard Test Method for Determination of Calcium, Chlorine, Copper, Magnesium, Phosphorus, Sulfur, and Zinc in Unused Lubricating Oils and Additives by Wavelength Dispersive X-ray Fluorescence Spectrometry (Mathematical Correction Procedure) Annual Book of Standards. ASTM International, West Conshohocken, PA.

ASTM D6560, 2014. Standard Test Method for Determination of Asphaltenes (Heptane Insolubles) in Crude Petroleum and Petroleum Products Annual Book of Standards. ASTM International, West Conshohocken, PA.

ASTM D6648, 2014. Test Method for Determining the Flexural Creep Stiffness of Asphalt Binder Using the Bending Beam Rheometer Annual Book of Standards. ASTM International, West Conshohocken, PA.

ASTM D6723, 2014. Test Method for Determining the Fracture Properties of Asphalt Binder in Direct Tension Annual Book of Standards. ASTM International, West Conshohocken, PA.

ASTM D6792, 2014. Standard Practice for Quality System in Petroleum Products and Lubricants Testing Laboratories Annual Book of Standards. ASTM International, West Conshohocken, PA.

ASTM D6816, 2014. Standard Practice for Determining Low-Temperature Performance Grade (PG) of Asphalt Binders Annual Book of Standards. ASTM International, West Conshohocken, PA.

ASTM D6818, 2014. Practice for Determining Low-Temperature Performance Grade of Asphalt Binders Annual Book of Standards. ASTM International, West Conshohocken, PA.

ASTM E102, 2014. Standard Test Method for Saybolt Furol Viscosity of Bituminous Materials at High Temperatures Annual Book of Standards. ASTM International, West Conshohocken, PA.

ASTM E659, 2014. Standard Test Method for Autoignition Temperature of Liquid Chemicals Annual Book of Standards. ASTM International, West Conshohocken, PA.

Barth, E.J., 1962. Asphalt: Science and Technology. Gordon and Breach Publishers, Taylor & Francis Group, Abingdon, Oxfordshire, UK.

Bouldin, M.G., Dongre, R., Rowe, G.M., Sharrock, M.J., Anderson, D.A., 2000. Predicting thermal cracking of pavements from binder properties: theoretical basis and field validation. J. Assoc. Asphalt Paving Technol. 69, 455–496.

Brandt, H.C.A., De Groot, P.C., 2001. Aqueous leaching of polycyclic aromatic hydrocarbons from bitumen and asphalt. Water Res. 35 (17), 4200–4207.

Breen, J.J., Stephens, J.E., 1967. Report No. JHR-67-16. In: Proceedings. Annual Conference, Canadian Technical Asphalt Association, Halifax Nova Scotia, Canada. October 30–November 1.

Button, J., 1994. Binder Characterization and Evaluation, Volume 4: Binder Test Methods Report No. SHRP A-369.. Strategic Highway Research Program, National Research Council, Washington, DC.

Castelo Branco, V., Masad, E., Bhasin, A., Little, D.N., 2008. Fatigue Analysis of Asphalt Mixtures Independent of Mode of Loading Transportation Research Record No. 2057. Transportation Research Board, National Research Council., pp. 149–156.

Clyne, T.R., Marasteanu, M.O., Basu, A., 2003. Evaluation of Asphalt Binders Used for Emulsions Report No. MN/RC-2003-24. Office of Research Services, Minnesota Department of Transportation, St. Paul, MN.

Coplantz, J.S., Yapp, M.T., Finn, F.N., 1993. Review of Relationships Between Modified Asphalt Properties and Pavement Performance Report No. SHRP-A-631. Strategic Highway Research Program, National Research Council, Washington, DC.

Dongre, R., D'Angelo, J., 1998. Effect of new direct tension test protocol on the superpave low temperature specification for bitumen binders Proceedings. Fifth International Conference on Bearing Capacity of Roads and Airfields, BCRA, vol. 2. Norwegian University of Science and Technology (NTNU), Trondheim, Norway, pp. 1035–1048.

Gary, J.G., Handwerk, G.E., Kaiser, M.J., 2007. Petroleum Refining: Technology and Economics, fifth ed. CRC Press, Taylor & Francis Group, Boca Raton, FL.

Goodrich, J.L., Goodrich, J.E., Kari, W.J., 1986. Asphalt Composition Tests: Their Application and Relation to Field Performance Record No. 1096. Transportation Research Board, Washington, DC, pp. 146–147.

Halstead, W.J., Rostler, F.S., White, R., 1966. Properties of highway asphalts—part III, influence of chemical composition. In: Proceedings, Association of Asphalt Paving Technologists, vol. 35, pp. 91–138.

Hoiberg, A.J., 1964. Bituminous Materials: Asphalts, Tars, and Pitches. John Wiley and Sons, New York, NY.

Hsu, C.S. Robinson, P.R. (Eds.), 2006. Practical Advances in Petroleum Processing, vols 1 and 2 Springer Science, New York, NY.

IP 80, 2014. Breaking Point of Bitumen—Fraass Method. IP Standard Methods. The Energy Institute, London, UK.
Jennings, P.W., 1980. High Pressure Liquid Chromatography as a Method of Measuring Asphalt Composition Publication No. FHWA-MT-7930. Federal Highway Administration, Department of Transportation, Washington, DC.
Jennings, P.W., Prabanic, J.A.S., 1985. The Expanded Montana Asphalt Quality Study Using High Pressure Liquid Chromatography Publication No. FHWA-MT-85-001. Federal Highway Administration, Department of Transportation, Washington, DC.
Kim, K.W., Burati, J.L., 1993. Use of GPC chromatogram to characterize aged asphalt cement. J. Mater. Civil Eng., ASCE 5 (1), 41–52.
Kim, K.W., Burati, J.L., Park, J.S., 1995. Methodology for defining LMS portion in asphalt chromatogram. J. Mater. Civil Eng., ASCE 7 (1), 31–40.
Kim, Y., Little, D.N., 2005. Development of Specification Type Tests to Assess the Impact of Fine Aggregate and Mineral Filler on Fatigue Damage Report No. FHWA/TX-05/0-1707-10. Research and Technology Implementation Office, Texas Department of Transportation, Austin, TX.
Kluttz, R., 2012. Considerations for use of alternative binders in asphalt pavements: material characteristics Proceedings. Alternative Binders for Sustainable Asphalt Pavements. Transportation Research Circular E-C165. E-C147. Transportation Research Board, Washington, DC, December. pp. 2–6.
Kriech, A.J., Kurek, J.T., Osborn, L.V., Blackburn, G.R., 1997. Extraction and analysis of asphalt pavement core samples: detection of coal tar derived species using chemical and biological methods Proceedings. 16th International Symposium on Polycyclic Aromatic Compounds. Asphalt Institute, Lexington, KY, November 4–8. pp. 31–40.
Kriech, A.J., Kurek, J.T., Osborn, L.V., Wissel, H.L., Sweeney, B.J., 2002. Determination of polycyclic aromatic compounds in asphalt and in corresponding leachate water. Polycyclic Aromat. Compd. 22 (3&4), 517–535.
Masad, E., Castelo Branco, V., Little, D.N., Lytton, R., 2008. A unified method for the analysis of controlled-strain and controlled-stress fatigue testing. Int. J. Pavement Eng. 9 (4), 233–246.
Moschopedis, S.E., Speight, J.G., 1973. The oxidation of petroleum fractions. Fuel 52, 83.
Moschopedis, S.E., Speight, J.G., 1975. Oxidation of bitumen. Fuel 54, 210.
Moschopedis, S.E., Speight, J.G., 1977. The effect of air blowing on the properties and constitution of natural bitumen. J. Mater. Sci. 12, 990.
Moschopedis, S.E., Speight, J.G., 1978. The influence of metal salts on bitumen oxidation. Fuel 57, 235.
Newcomb, D.E., 2001a. Statistical Specifications for Hot Mix Asphalt: What Do We Need to Know? Hot Mix Asphalt Technology. National Asphalt Pavement Association, Lanham, MD, January/February: pp. 54–60.
Newcomb, D.E., 2001b. Statistical Specifications for Hot Mix Asphalt: What Do We Need to Know? Hot Mix Asphalt Technology. National Asphalt Pavement Association, Lanham, MD, March/April: pp. 37–40.
Newcomb, D.E., 2001c. Performance Related Specifications Developments. Hot Mix Asphalt Technology. National Asphalt Pavement Association, Lanham, MD, May/June: pp. 5–37.
Parkash, S., 2003. Refining Processes Handbook. Gulf Professional Publishing, Elsevier, Amsterdam, Netherlands.
Planche, J.-P., King, H.W., King, G.N., 1991. A refiner's approach to asphalt characterization. Nature, 24–25. April 18.
Salomon, D., 2000. Analysis of asphalt emulsions by rotational viscometry. In: Proceedings. International Slurry Surfacing Association/Asphalt Emulsion Manufacturers Association (ISSA/AEMA), pp. 71–80. March 12–15.

Schmidt, R.J., 1996. The relationship of the low temperature properties of asphalt to the cracking of pavements. In: Proceedings, Association of Asphalt Paving Technologists, vol. 35, pp. 263–269.

Schmidt, R.J., Santucci, L.E., 1996. A practical method for determining the glass transition temperature of asphalts and calculation of their low temperature viscosities. In: Proceedings, Association of Asphalt Paving Technologists, vol. 35, pp. 61–90.

SHRP, 1993. Binder Characterization and Evaluation, Volume 2: Chemistry Report No. SHRP-A-368. Strategic Highway Research Program, Transportation Research Board, National Research Council, Washington, DC.

SHRP, 1994a. Binder Characterization and Evaluation, Volume I Report No. SHRP-A-367. Strategic Highway Research Program, Transportation Research Board, National Research Council, Washington, DC.

SHRP, 1994b. Binder Characterization and Evaluation, Volume 4: Test Methods Report No. SHRP-370. Strategic Highway Research Program, Transportation Research Board, National Research Council, Washington, DC.

Sousa, J., 1994. Asphalt-aggregate mix design using the repetitive simple shear test (constant height). J. Assoc. Asphalt Paving Technol. 63, 298–333.

Speight, J.G., 1992. Asphalt, fourth ed. Kirk-Othmer's Encyclopedia of Chemical Technology, vol. 3, pp. 689–724.

Speight, J.G., 2001. Handbook of Petroleum Analysis. John Wiley & Sons Inc., Hoboken, NJ.

Speight, J.G., 2005. Environmental Analysis and Technology for the Refining Industry. John Wiley & Sons Inc., Hoboken, NJ.

Speight, J.G., 2013. The Chemistry and Technology of Coal, third ed. CRC Press, Taylor & Francis Group, Boca Raton, FL.

Speight, J.G., 2014. The Chemistry and Technology of Petroleum, fifth ed. CRC Press, Taylor & Francis Group, Boca Raton., FL.

Speight, J.G., 2015a. Handbook of Petroleum Product Analysis, second ed. John Wiley & Sons Inc., Hoboken, NJ.

Speight, J.G., 2015b. Handbook of Coal Analysis, second ed. John Wiley & Sons Inc., Hoboken, NJ.

Speight, J.G., Arjoon, K.K., 2012. Bioremediation of Petroleum and Petroleum Products. Scrivener Publishing, Salem, MA.

Stuart, K.D., Mogawer, W.S., Romero, P., 2000. Evaluation of the superpave asphalt binder specification for high-temperature pavement performance. J. Assoc. Asphalt Paving Technol. 69, 148–173.

Takamura, K., 2000. Comparison of emulsion residues recovered by forced airflow and RTFO drying. In: Proceedings. International Slurry Surfacing Association/Asphalt Emulsion Manufacturers Association (ISSA/AEMA), pp. 1–17. March 12–15.

TRB, 2004. New Simple Performance Tests for Asphalt Mixes Circular No. E-C068. Transportation Research Board, Washington, DC.

TRB, 2010. Development in Asphalt Binder Specifications Transportation Research Circular E-C147. Transportation Research Board, Washington, DC, December.

TRB, 2012. Alternative Binders for Sustainable Asphalt Pavements Transportation Research Circular E-C165. E-C147. Transportation Research Board, Washington, DC, December.

CHAPTER 5

Test Methods for Aggregate and Asphalt Concrete

5.1 INTRODUCTION

The asphalt–aggregate mix used for construction of an asphaltic roadway is typically composed of approximately 85% (v/v) aggregate, and roadway (pavement) performance is greatly influenced by the characteristics of the aggregate. However, it is only recently that the properties of the aggregate in an asphalt–aggregate mix have been identified as having a major effect of on roadway performance (Kandhal and Parker, 1998; White et al., 2006).

Aggregate has been mixed with various types of asphalt (in many cases with naturally occurring bitumen) to form paving materials for at least 4000 years (Speight, 1978). The result has been a useful material that allowed construction (laydown) of paved roadways that were less prone to being water-logged and were able to be counted on for longer in-service use than the previous naturally worn tracks formed by feet and wagons. However, it is only relatively recently that the importance of the aggregate used in asphaltic roadways was realized and that the properties of the aggregate (in addition to the properties of the asphalt) must be carefully controlled in order to ensure good performance. One of the most important characteristics of asphalt–aggregate interaction was that the precise properties of the asphalt–aggregate mix must be sufficient to withstand weather fluctuations (rain, frost, snow) as well as seasonal temperature fluctuations in the surrounding environment. Such extreme changes in properties can cause performance problems in pavements.

By way of review, the aggregates in an asphalt–aggregate mix can be divided into three types according to particle size: (i) coarse aggregate, generally defined as containing particles retained on the 2.36-mm sieve, (ii) fine aggregates, generally defined as containing particles that pass through the 2.36-mm sieve and are retained on the 0.075-mm sieve, and (iii) mineral filler, generally defined as that portion of the aggregate passing the 0.075-mm sieve—mineral filler is a very fine material and is also referred to as mineral dust or rock dust.

Asphalt Materials Science and Technology.
DOI: http://dx.doi.org/10.1016/B978-0-12-800273-5.00005-2

The aggregate may also be often referred to colloquially as gravel, which is actually a coarse aggregate made up mostly of rounded particles. Gravel is often dredged from rivers and is sometimes mined from deposits. Because of the rounded particle size, gravel is not suitable for use in an asphalt–aggregate mix because the particles will not interlock, which imparts poor strength and rut resistance. However, if the gravel is thoroughly crushed and the rounded integrity of the particle is destroyed, it may be suitable for use.

On the other hand, crushed stone is coarse aggregate that is mined and processed by mechanical crushing. The crushed product tends to contain angular material and, depending on its other properties, can be well suited for use in asphalt–aggregate pavements. However, if the particles are flat or elongated or both, the aggregate will be unsuitable for use in an asphalt–aggregate mix—ideally, the particles in the crushed stone aggregate should be cubic and highly angular. Thus, the shape of the aggregate particles influences major factors such as: (i) the asphalt requirement of the asphalt–aggregate mix, and (ii) the workability and the strength of both Portland cement concrete and the asphalt pavement.

The fine aggregate, or sand, used in asphalt–aggregate mixes can be natural sand, manufactured sand, or a mixture of both. Natural sand is dredged from a river bed or mined from a deposit and is then processed by sieving to produce a fine aggregate having the desired particle size distribution. Manufactured sand is produced by crushing quarried stone and, like natural sand, sieving to produce the desired gradation (particle size distribution). The particles in manufactured sand tend to be more angular than those in natural sand—river sand is prone to the abrasive effects of the river flow and any particles carried by the water—and, thus, manufactured sand will produce an asphalt–aggregate mix with greater strength and rut resistance compared to the asphalt–aggregate mix composed of natural sand (Christensen et al., 2000).

Not surprisingly, specifications for aggregates that contribute to the properties of the asphalt–aggregate mix must control properties under the environmental conditions that exist at the locale where the roadway is constructed and used. Furthermore, as with the binders (see Chapter 4), test methods used to specify the properties of the aggregate and the asphalt–aggregate mix must be conducted with care and quality control (QC) (AASHTO R9; AASHTO R10; AASHTO T2; AASHTO T40; AASHTO T168; ASTM D3665), otherwise, the data will be of questionable value or even unreliable.

Asphalt–aggregate mix design is a complex process (see Chapter 9) (Newcomb, 2001a–c). The composition and properties of asphalt binders and aggregates vary considerably and the asphalt–aggregate mix manufactured for a paving project should be in accordance with a job mix formula, which is determined by a mix design process that has been standardized by the (e.g., state) agencies responsible for the specifications. The steps in a mix design are: (i) selection of the materials—the asphalt binder, the aggregate (coarse or fine), (ii) testing of the possible aggregates for size, durability, angularity, and clay content, (iii) selection of a good aggregate structure by determining the appropriate blend of sizes (gradation)—the coarse and fine aggregates should fit together with a strong stone-on-stone structure, but there should be enough empty spaces for asphalt (binder and mastic) and air voids, (iv) choice of the amount of binder in the mix and the air voids (volumetric parameters), which involves choosing the appropriate performance-graded binder for the project, and (v) choice of the mix for adhesion, cohesion, and sensitivity to water, which involves testing of proposed job mix formula samples (Kim and Burati, 1993; Kim et al., 1995; Kim and Little, 2005; Christensen and Bonaquist, 2006).

The *percent voids in the total mix* (VTM) is that part of the compacted mixture not occupied by aggregate or asphalt expressed as a percentage of the total volume. It is synonymous with air voids and is the complement of the percent density when based upon the maximum specific gravity. The VTM obtained in the Superpave mix design gives an indication of whether the mix can be compacted adequately in the field and used as high-performance asphalt (Harvey et al., 1994; Bonaquist et al., 2003; Bonaquist, 2008). In view of the excellent stabilities achieved with many mixes, the VTM is a major criterion for predicting field compaction and pavement life.

The *percent voids filled with asphalt* (VFA) is the percentage of voids in the compacted aggregate mass that are filled with asphalt cement and is synonymous with the asphalt–void ratio. The VFA property is important not only as a measure of relative durability, but also because there is an excellent correlation between it and percent density. If the VFA is too low, there is not enough asphalt to provide durability and to over-densify under traffic and bleed and the VFA becomes a very important design property.

The *voids in the mineral aggregate* (VMA) are the air-void spaces that exist between the aggregate particles in a compacted paving mixture, including spaces filled with asphalt. The VMA represent the space that is available to accommodate the asphalt and the volume of air voids necessary in the mixture. The more VMA in the dry aggregate, the more space is available for the film of asphalt. Based on the fact that the thicker the asphalt film on the aggregate particles the more durable the mix, specific minimum requirements for VMA are usually presented in specifications. Minimum values for the VMA should be adhered to so that a durable asphalt film thickness can be achieved. Increasing the density of gradation of the aggregate to a point where it is below the minimum value for the voids leads to thin films of asphalt and a dry-looking, low-durability mix.

Some aggregates are more porous than others and therefore require more asphalt binder to fill those pores (Lee et al., 1990). Based on the volumetric parameters, the job aggregate and asphalt binder are mixed at several different percentages and then compacted and blends are tested using standard test methods. Test data are used to determine the best job mix formula to meet the specified mix volumetric parameters, typically 4% (v/v) for dense-graded mixes. Mixes designed to drain away surface water (porous mixes) are designed to have 16–20% (v/v) air voids. If there is too much liquid asphalt, the hot mix asphalt might rut—aggregate particles have a tendency to *float* in the liquid asphalt during periods of hot weather or excessive road use. If there is too little asphalt, hot mix asphalt has a tendency to crack because of the insufficient amount of binder to hold the aggregate particles together.

In the laboratory, asphalt–aggregate mixtures are usually designed to contain approximately 4% (v/v) air voids, with a range on the order of 3–5% (v/v), depending on the type of mixture being designed and the design procedure being used. Typically, when first laid down and after compaction, an asphalt pavement will contain about 6–8% (v/v) air voids. However, after the flow of traffic has commenced, the asphalt mix wheel paths will gradually compact to air void levels approaching the design value of 3–5% (v/v). If the pavement is not compacted adequately during the construction phase, compaction will occur under the traffic load but the design value (3–5%, v/v, air voids) will not be achieved and, as a result, the asphalt roadway will remain permeable to air and water with ensuing shortened in-service life due to moisture damage and excessive oxidative hardening.

Typically, if the air voids are less than about 3% (v/v), the *floating of the particles in the asphalt* results in an unstable mix. On the other hand, if the air voids are in excess of approximately 8% (v/v), the air voids become interconnected, which allows water into the mix resulting in an adverse effect on long-term durability and strength.

5.2 PROPERTIES AND PERFORMANCE

The suitability of the aggregate material also depends upon the chemical and physical properties as well as the mineral type, particle size, particle shape, and interrelationship of the mineral particles. The performance of the asphalt mix (binder plus aggregate) also relies upon specifications which, in turn, depend upon the relative proportions of the components as well as the physical properties of each of the components. Thus, the strength and quality of the asphalt roadway depend upon (i) the physical properties of the mix, (ii) the design of the pavement structure, and not forgetting (iii) the quality of the construction.

5.3 GENERAL PROPERTIES

Just as the properties of the binder influence the properties of an asphalt roadway, the properties of the aggregate also play a role in determining the performance and the life of the roadway. For example, the properties of the aggregate influence the amount of binder required for satisfactory performance and the properties of interest include: (i) surface texture and shape, (ii) gradation, (iii) absorption, (iv) clay content, and (v) durability. These properties are particularly noteworthy for the specification relating to the performance grading of Superpave asphaltic roadways.

5.3.1 Surface Texture and Shape

The surface texture of the aggregate is an extremely important factor that makes a contribution to frictional resistance. The surface texture also has a strong influence on the resistance of asphalt mix to rutting—the rougher the texture of the aggregate, the better the resistance of the mix to rutting. However, during construction, a mix containing a rough-textured aggregate requires greater compaction energy to achieve the required density than hot mix asphalt containing a smooth-textured aggregate.

The rutting resistance of an asphalt mix is also influenced by the shape of the aggregate particles—angular aggregate produces a greater resistance than more rounded aggregate particles. The improved resistance of angular aggregates is a result of the increased surface roughness produced during the crushing operation as well as the ability of the aggregate particles to interlock with each other. However, as with surface texture, the more angular the aggregate, the greater the compaction energy required to produce a mix with a specified degree of density.

Two tests for objectively defining the above characteristics have been selected as a part of the Superpave system: the *coarse aggregate angularity* test (AASHTO TP56; AASHTO TP61; ASTM D5821) and the *fine aggregate angularity* test (AASHTO T304; ASTM C1252). The coarse aggregate angularity test is a method of determining the angularity of coarse aggregate. The fine aggregate angularity test is an indirect method of assessing the angularity of fine aggregate. Fine aggregate angularity is important because an excess of rounded fine aggregate (often in the form of natural sand) can lead to rutting. Generally, the acceptance criteria used for these parameters are higher as the amount of traffic increases and as the mix is placed closer to the pavement surface.

5.3.2 Particle Size Distribution

One of the important properties of aggregates for use in pavements is the distribution of particle sizes (gradation) which is generally controlled by specifications that define the distribution of particle sizes. Typically, the larger the maximum size of aggregate in a given mix type in relation to the layer thickness and the greater the amount of large aggregate in the mix, the more compaction energy that is required. Furthermore, if the nominal maximum aggregate size exceeds one-third of the compacted thickness of the pavement layer, the surface texture of the mix can be affected, and the degree of density of the mix obtained by compaction may be reduced. To improve the resistance of hot mix asphalt to rutting, both the proportion of coarse aggregate [retained on the 4.75-mm (no. 4) sieve] and the maximum particle size may be increased.

The Superpave method uses the following aggregate size definitions: (i) *maximum size*, which is one sieve size larger than the nominal maximum size, and (ii) *nominal maximum size*, which is one sieve size larger than the first sieve to retain more than 10% by weight. There is also the need to ensure that mixes have sufficient VMA to allow enough asphalt for adequate durability, since it was observed that gradations that follow the maximum density line may have, at times, lower-than-desirable VMA. Low VMA results in very little void space within which to develop sufficiently thick asphalt films for a durable mix. Another purpose of the restricted zone was to restrict the amount of natural sand in the mix—aggregate with an excessive amount of natural sand as this produces a hot mix asphalt that does not perform well in service.

The amount and size distribution of the material passing through a 0.075-mm (no. 200 sieve) is often referred to as *fines content* and influences the compactability of an asphalt–aggregate mix. Mixes with low fines content are often difficult to compact and increasing the fines content can cause the stiffness of the mix to increase, enabling the mix to become dense under the roller and resist deformation. However, too much material in this size range may also affect the compactability of the mix. The Superpave specifications place a limit on the proportion of fines (dust) or the computed ratio of the percent passing the 0.075-mm (no. 200) sieve to the effective asphalt content (expressed as a percentage of the weight of the total mix). The *effective asphalt content* is the total asphalt content less the proportion (percentage) of asphalt absorbed by the aggregate.

5.3.3 Absorption

The amount of asphalt cement that is absorbed by the aggregate can significantly affect the properties of the asphalt mixture. If the aggregate particles have high asphalt absorption, the asphalt content in the mix must be increased to compensate for binder material that is drawn into the pores of the aggregate and is unavailable as part of the film thickness around those particles. If that asphalt content adjustment is not made, the mix can be dry and stiff, the amount of compaction energy needed to achieve density in the mix will need to be increased, and the mix will have a tendency to deform in service. If absorptive aggregates that have high water content are used, extra time will be required in the production of hot mix asphalt to ensure that the moisture in the pores can evaporate or the asphalt may not be properly absorbed, leading to difficulties during the compaction process.

5.3.4 Clay Content

The presence of clay in the fine aggregate [material passing the 4.75-mm (no. 4) sieve] can have a detrimental effect on the water sensitivity of an asphalt concrete mix. For example, clay minerals coating aggregates can prevent asphalt binders from thoroughly bonding to the surface of aggregate particles, increasing the potential for water damage to the paving mixture. The sand equivalent test is used to limit the presence of clay material in the aggregate.

The sand equivalent test (AASHTO T176; ASTM D2419) is a rapid field test to show the relative proportions of fine dust or clay-like materials in fine aggregate (or granular soils). The term "sand equivalent" expresses the concept that most fine aggregates are mixtures of desirable coarse particles (e.g., sand) and generally undesirable clay or plastic fines and dust, which can coat aggregate particles and prevent proper asphalt binder–aggregate bonding.

In the test method, a sample of aggregate passing the No. 4 (4.75 mm) sieve and a small amount of flocculating solution are poured into a graduated cylinder and are agitated to loosen the clay-like coatings from the sand particles. The sample is then irrigated with additional flocculation solution forcing the clay-like material into suspension above the sand. After a prescribed sedimentation period, the height of flocculated clay and height of sand are determined and the sand equivalent is expressed as a ratio of the height of sand over the height of clay. Higher sand equivalent values indicate cleaner aggregate with a lesser amount of fine dust or clay-like materials.

5.3.5 Additional Factors Affecting Durability

In order to mitigate the degradation (production of fines) of aggregate during the production and placement of hot mix asphalt, the Los Angeles abrasion test method is used (AASHTO T96; ASTM C131). By setting a maximum abrasion loss in this test, aggregate degradation is presumed to be limited. In areas where freezing and thawing occur, the sodium or magnesium soundness test is used. By setting a maximum value in terms of aggregate degradation, the resistance of aggregate breakdown from freeze–thaw cycles is improved. In this regard, it should be noted that limits placed on the water absorption of aggregates also assist in reducing freeze–thaw damage.

Limits are also placed on the amount of deleterious materials in the aggregate—defined as the percent by weight of undesirable contaminants, such as clay lumps, soft shale, coal, wood, or mica.

5.4 SPECIFICATIONS

An aggregate, like the binder, is required to meet specifications before it can be used in road construction and it is essential that these specifications are followed. There are two general types of requirements for aggregate: (i) quality and gradation and (ii) physical quality requirements.

Typically, coarse aggregate consists of sound, durable rock, free from objectionable coatings and frozen and cemented lumps. On the other hand, fine aggregate for concrete is a fine granular material naturally produced by the disintegration of rock of a siliceous nature, or manufactured from an approved limestone or dolomite source and should be free from cemented or conglomerated lumps and not have any coating of injurious material.

Physical quality requirements are all specification provisions other than gradation or usage requirements. These requirements may be divided into distinct groups as follow: (i) absorption, (ii) abrasion resistance, (iii) soundness, and (iv) restriction on deleterious constituents. Any special requirements are subject to the regulations of the agency laying the road which, in turn, are dependent upon the environmental conditions. Furthermore, approval for use of aggregates is based upon meeting these test requirements.

The *absorption* requirement applies only to coarse aggregates, but these data are necessary on fine aggregate for other purposes, such as mix design and water/cementitious ratios. All aggregates are porous, but some are more porous than others. How porous an aggregate is determines how much liquid may be absorbed when soaked in water (AASHTO T85). Absorption is expressed as a percentage of the dry weight. The intent of using the absorption parameter is to avoid using highly porous, absorptive aggregates because extra water and cement or asphalt are needed to make a good mix. However, some aggregates, such as blast furnace slag, may be used despite their high absorptive capacity because of other characteristics that make them desirable, including skid resistance.

Abrasion resistance applies to coarse aggregates and is a measure of the resistance of the aggregate to fracturing under impact (toughness) and breaking down into smaller pieces from abrasive action (hardness). The acceptable limits are set by the Los Angeles abrasion test method (AASHTO T96), which is a measure of the degradation or loss of material as a result of impact and abrasive actions. and the limits vary from 30.0% to 50.0%, depending on the classification of the aggregate. Abrasion requirements do not apply to blast furnace slag.

Soundness of the aggregate applies to both fine and coarse aggregates. The durability of aggregates or their resistance to the forces of weathering is one of the most important considerations in the selection of a material for highway construction. Alternate freezing and thawing of the aggregates is the major concern. Three different test methods ate available to evaluate aggregate soundness: (i) the water freeze and thaw test—AASHTO T103, Procedure A, and (ii) the sodium sulfate test method—AASHTO T104. The water freeze and thaw test requires the aggregate to be sealed and totally immersed in water and then be subjected to 50 cycles of freeze and thaw. The sodium sulfate test requires the aggregate to be immersed in a sodium sulfate solution and then be subjected to five cycles of alternate immersion and drying. The freezing and thawing in water test is the method that most accurately simulates actual field conditions; however, the test requires a long period of time to conduct.

Deleterious materials in aggregates are undesirable and can affect performance. Therefore, the specifications limit the amount of deleterious constituents to a level consistent with the quality sought in the final product. Organic impurities are the only concern in fine aggregates. Generally, no restrictions are placed on organic impurities in fine aggregate for use in other types of construction. The limitations on the amount of organic impurities allowed in fine aggregates are determined by the test method for organic impurities (AASHTO T21). The results of the test are the basis for acceptance or rejection of the fine aggregate.

5.5 THE NEED FOR TESTING

The selection of an appropriate asphalt–aggregate mixture for a specific paving application is important in designing new pavements and for rehabilitation strategies for existing pavements (see Chapters 9 and 11). The type of mixture selected for the various layers of a pavement has a major effect on the long-term performance of the pavement. Asphalt–aggregate mixtures with lower binder contents and lower-quality aggregates are less expensive. To facilitate placement and compaction, thinner layers should be made with smaller nominal maximum aggregate size mixtures, while thick base layers should be made with larger nominal maximum aggregate sizes. In addition, asphalt–aggregate mixtures at the surface of a pavement should have relatively high binder content to make them more resistant to the damaging effects of traffic and the environment. Lower binder contents can be used in mixtures for intermediate and base courses because they are protected by the layers above them. Careful consideration of mix type is an important factor when staged construction is used, because the base or intermediate courses must serve temporarily as the surface during the first stages of construction.

The need for testing any product is required to ensure that the product is suitable for the claimed uses of the product and will perform when in use according to the manufacturer's claims. To do this, the product must have specified properties that are suitable to in-service use and can vary from simple properties to a more complex suite of properties that are used as a collection that is typically referred to as the product *specifications*.

Asphalt, like other petroleum products, is also required to meet specific properties (specifications) before use. In this case, the specifications will be determined by the user insofar as the specifications will vary from one region to another depending upon climatic and environmental conditions (TRB, 2004, 2010, 2012).

The in-service use of an asphalt binder depends upon the chemistry of the crude oil source and the method used to refine it, which dictate the chemical and physical properties of the binder, as any additives used. The additives, in turn, influence the properties of the binder and can also contribute to the in-service performance of the binder as well as enabling the binder to meet the required specifications. The suitability of the aggregate material for use also depends upon the chemical and physical properties as well as the mineral type, particle size, particle shape, and interrelationship of the mineral particles. The performance of the asphalt mix (binder plus aggregate) also relied upon specifications which, in turn, depend upon the relative proportions of the components as well as the physical properties of each of the components. Thus, the strength and quality of the asphalt roadway depend upon (i) the physical properties of the mix, (ii) the design of the pavement structure, and not forgetting (iii) the quality of the construction.

Specifications dictate that inferior quality materials (i.e., materials that do not meet the specifications) are not used and the physical properties (such as strength, stickiness, or elasticity) must match specifications or the result will be poor in-service performance such as rutting, cracking, and the formation of potholes (see Chapter 9). This also includes structural design—a pavement structure design and/or poor-quality construction that, for example, does not allow water to drain from the roadway or a roadway that is not strong enough to support the traffic results in a poor-quality roadway and rapid deterioration during service.

In order to assure high-quality roadways, a series of physical properties have been identified (usually by the highway agency) that will provide the best-quality roadways. Thus, a series of standard laboratory and field test methods and sampling procedures (ASTM D979) have been developed to ensure that the asphalt-binder, the aggregate, and the asphalt–aggregate mix will give the expected performance. Typically, each of the necessary properties will have upper and/or lower limits for those test results and these limits are the specifications that the binder, the aggregate, and the asphalt–aggregate mix must meet to be suitable for roadway (or other) use. Thus, for any roadway paving project to be successful and give the necessary in-service performance, the materials, structural design, and construction should meet standard specifications. Similar conditions also apply to other uses of asphalt.

The volumetric parameters (Brown et al., 2003; Spellerberg et al., 2003; Azari et al., 2006; Cross et al., 2006) include such properties as air voids, VMA, and voids filled with asphalt. Air voids are small airspaces or pockets of air that occur between the coated aggregate particles in the final compacted mix. A certain percentage of air voids is necessary in all dense-graded highway mixes to allow for some additional pavement compaction under traffic and to provide spaces into which small amounts of asphalt can flow during this subsequent compaction. The allowable percentage of air voids (in laboratory specimens) is between 2.0% and 4.0% for most surface course mixes. Furthermore, the durability of an asphalt pavement is a function of the air-void content. This is because the lower the air voids, the less permeable the mixture becomes. Too high an air-void content provides passageways through the mix for the entrance of damaging air and water. A low air-void content, on the other hand, can lead to flushing, a condition in which excess asphalt squeezes out of the mix to the surface. Density and void content are directly related—the higher the density, the lower the percentage of voids in the mix, and vice versa. Typical specifications require pavement that allows as low an air void content as is practical, approximately 8.0%.

Water sensitivity of a mix is determined by submerging the samples in water for a specified time at a specified temperature, and then the samples are retested for retained strength. The data from this test method are used to compare the strength of unconditioned mixtures with mixtures that have been conditioned under water. The asphalt binder should be sufficiently adhesive enough not to strip from the aggregate and to give a good retained strength to the mix. If the mix design testing indicates the mix may have a susceptibility to moisture, other tests can be conducted to determine the appropriate additives—such as liquid anti-strip additives for the asphalt binder or hydrated lime for the aggregate—that might be employed to enhance the performance of the asphalt–aggregate mix.

Thus, the importance of chemical and physical properties is dependent upon the purity of the petroleum or petroleum product. In the strictest sense, petroleum and petroleum products are complex chemical mixtures. These mixtures contain hydrocarbons and non-hydrocarbon compounds that confer properties on the mixture that may not be reflected in the composition. Therefore, it is necessary to apply various text methods to petroleum and petroleum products to determine whether the material is suitable for processing and (in the case of the products) for sale with a designated use in mind. The more common tests are introduced in the following sub-sections—these tests are presented in alphabetical order with no preference given to any particular test method.

5.6 AGGREGATE SUITABILITY

The mineral aggregate (i.e., any inert mineral material used for mixing in graduated particles or fragments) makes up between 80% and 90% (v/v) (94–95%, w/w) of hot mix asphalt and is an important aspect of asphalt mix production (McLeod and Davidson, 1981; Brown et al., 1989). Aggregates for asphalt concrete are generally classified according to their source or means of preparation. They include: (i) pit aggregate, such as gravel and sand, (ii) processed aggregate, which is typically natural gravel or stone that have been crushed and screened, and (iii) synthetic or artificial aggregate, which is aggregate that has resulted from the modification of materials, which may involve both physical and chemical changes—blast furnace slag is the most commonly used artificial aggregate or lightweight aggregate.

The aggregate is primarily responsible for the load-supporting capacity of a pavement. For this reason, it is important to maximize the quality of the mineral aggregates to ensure the proper performance of roadways (Cross and Brown, 1992; Kandhal and Parker, 1998). The quality of mineral aggregates for road-paving materials has been specified by the toughness, soundness (durability), cleanliness, particle shape, angularity, surface texture, and absorption. To determine aggregate suitability it is necessary to consider two categories of properties: (i) aggregate durability and (ii) the aggregate composition.

Thus, selecting an aggregate material for use in an asphalt concrete depends not only upon the availability, cost, and quality of the material, as well as the type of construction that is intended, but also upon the chemical composition of the aggregate. The suitability of aggregates for use in asphalt concrete is determined by evaluating the material in terms of: (i) size and grading, (ii) cleanliness, (iii) hardness or toughness, (iv) soundness, (v) particle shape, which can be flat and elongated, (vi) surface texture, (vii) absorption, and last but certainly not least (viii) the affinity of the aggregate for the binder.

5.7 TEST METHODS

Just as the properties of the asphalt binder contribute to the behavior of the binder in service, the physical and chemical properties of aggregates contribute to the behavior of the aggregate in service and (like the properties of the binder) also impact the adhesive bond (strength and durability) between the asphalt and the aggregate. These properties include surface free energy, chemical interaction potential, and specific surface area. The results of the various test methods can be used to estimate the role that the aggregate properties contribute to: (i) select combinations of asphalt and aggregate that are more resistant to moisture damage, (ii) select additives that can be used to improve the performance of asphalt mixtures based on the physical and chemical nature of the binder and aggregate, and (iii) predict the resistance of the mixture to moisture-induced damage (Kennedy et al., 1984a; Bhasin and Little, 2006).

In relation to aggregate durability, this generally involves two categories of tests: tests that measure aggregate abrasion resistance and aggregate attrition (aggregate breakdown) during handling, mixing, laydown, and under traffic, and tests that address aggregate weathering when exposed to freezing and thawing or wetting and drying. These tests are employed in concert to ascertain that the aggregate used in the production of asphalt mix will be suitable (Table 5.1) and also durable (Table 5.2). Deleterious materials are organic compounds or other unsuitable materials such as coal, lignite, and wood.

5.7.1 Abrasion

Aggregate abrasion resistance (sometime referred to as *aggregate toughness*) has also been shown to be related to pavement performance. Aggregate particles that are tough and resistant to abrasion will not break down during the construction process, which helps ensure that an asphalt mix can be properly constructed, placed, and compacted. Tough, abrasion-resistant aggregates also tend to produce a mix that is resistant to pop-outs and raveling. Because aggregate pop-outs and broken aggregate particles near the pavement surface make it easier for water to flow into a pavement, tough and abrasion-resistant aggregates help improve the moisture resistance of asphalt pavements. Aggregates with poor abrasion resistance can also polish under the action of traffic which can cause the pavement surface to lose skid resistance, especially when the pavement is wet.

Table 5.1 AASHTO standard test method applicable to determining aggregate suitability

Test Method	Test
AASHTO M29	Fine Aggregate for Bituminous Paving Mixtures
AASHTO M43	Standard Specification for Sizes of Aggregate for Road and Bridge Construction
AASHTO M320	Performance-Graded Asphalt Binder
AASHTO M323	Superpave Volumetric Mix Design
AASHTO M325	Standard Specification for Designing Stone Matrix Asphalt (SMA)
AASHTO R30	Mixture Conditioning of Hot-Mix Asphalt (HMA)
AASHTO R35	Superpave Volumetric Design for Hot-Mix Asphalt
AASHTO R46	Standard Practice for Designing Stone Matrix Asphalt (SMA) T 2, Sampling of Aggregates
AASHTO T27	Sieve Analysis of Fine and Coarse Aggregate
AASHTO T84	Specific Gravity and Absorption of Fine Aggregate
AASHTO T85	Specific Gravity and Absorption of Coarse Aggregate
AASHTO T166	Bulk Specific Gravity of Compacted Asphalt Mixtures Using Saturated Surface-Dry Specimens
AASHTO T209	Theoretical Maximum Specification Gravity and Density of Bituminous Paving Mixtures
AASHTO T269	Percent Air Voids in Compacted Dense and Open Asphalt Mixtures
AASHTO T283	Resistance of Compacted Asphalt Mixture to Moisture-Induced Damage
AASHTO T320	Determining the Permanent Shear Strain and Stiffness of Asphalt Mixtures Using the Superpave Shear Tester
AASHTO T321	Determining the Fatigue Life of Compacted Hot-Mix Asphalt (HMA) Subjected to Repeated Flexural Bending
AASHTO T322	Determining the Creep Compliance and Strength of Hot-Mix Asphalt (HMA) Using the Indirect Tensile Test Device
AASHTO T324	Hamburg Wheel-Track Testing of Compacted Hot-Mix Asphalt (HMA)
AASHTO TP63	Determining the Rutting Susceptibility of Asphalt Paving Mixtures Using the Asphalt Pavement Analyzer (APA)
Associated test methods	
ASTM D 3515	Hot-Mixed, Hot-Laid Bituminous Paving Mixtures
ASTM D 3665	Random Sampling of Construction Materials
ASTM D 4460	Calculating Precision Limits Where Values are Calculated from Other Test Methods

Table 5.2 Test requirement for aggregates used in asphalt mix

Test Method	Test	Requirement[a]
AASHTO T176	Sand Equivalent	45% Minimum
AASHTO T96	Los Angeles Abrasion	30% Maximum
AASHTO T182	Stripping	>95%
ASTM D5148	K-factor	Kc-2.0 Maximum Km-1.7 Maximum
ASTM D4791	Flat and elongated pieces	25% Maximum
AASHTO T11 AASHTO T27	Grading	According to job-mix requirement
AASHTO T104	Soundness	9%
AASHTO T84 AASHTO T85	Absorption	5% maximum

[a]May vary according to state or region.

The abrasion test method (Los Angeles abrasion test, LA abrasion test, ASTM C131) is a measure of degradation of mineral aggregates of standard grading resulting from a combination of actions including abrasion or attrition, impact, and grinding in a rotating steel drum containing a specified number of steel spheres. The method is a common test method used to indicate aggregate toughness and abrasion characteristics which are important because the constituent aggregate in hot mix asphalt must resist crushing, degradation, and disintegration in order to produce high-quality hot mix asphalt. In the test, the aggregate is subjected to impact and crushing. The test data are related to the expected breakdown during handling, mixing, and placement but further indications are needed to determine whether such a correlation is realistic and to what extent it is generally applicable.

Another aggregate property that is closely related to abrasion resistance is durability and soundness. Freeze–thaw cycles and alternate periods of wetting and drying in a pavement can weaken poor-quality aggregates, causing pop-outs and raveling. Aggregates that possess good durability and soundness will resist the actions of wet–dry and freeze–thaw cycles during the life of the pavement.

5.7.2 Cleanliness

Cleanliness and the presence of deleterious materials (such as clay lumps, friable particles, shale, coal, free mica, and vegetation) is an aggregate characteristic related to performance. The deleterious materials are not as strong as mineral aggregates and break down during the life of a pavement layer resulting in the occurrence of pop-outs and potholes. In addition, the term *cleanliness* is used to characterize the coatings on some aggregate particles.

The presence of clay or dust in the aggregate can prevent the asphalt binder from efficiently coating the aggregate particle with the binder. This can lead to water penetrating the asphalt binder film and, therefore, stripping of the asphalt binder from the aggregate. The *sand equivalent test method* (AASHTO T176) is used to evaluate the cleanliness of aggregates to identify when harmful clay-sized particles exist in an aggregate blend. The procedure is conducted on the aggregate fraction of the blend that passes the 4.75-mm sieve. If hydrated lime is used in the mixture, it should not be included in the fine aggregate used during the sand equivalent test. The aggregate sample is placed within a graduated, transparent cylinder that is filled with a mixture of water and flocculating agent. The combination of aggregate, water, and flocculating agent is then agitated for approximately 40–50 minutes after which the combination is allowed to sit at room temperature for 20 min. After the 20 min, the heights of the sand particles and the sand plus clay particles are measured. The sand equivalent value is then calculated as the ratio of the height of the sand to the height of sand plus clay, expressed as a percentage.

High sand equivalent values are desirable, since this indicates that the aggregate is relatively free of dust and clay particles. Therefore, minimum values for sand equivalency are specified. These minimum values do not change with depth within the pavement, but do vary somewhat with design traffic level.

Typically, a *coating* is, as the name suggest, a layer of substance covering a part or the entire surface of an aggregate particle. The coating may be of natural origin, such as mineral deposits formed in sand and gravel by groundwater, or may be artificial, such as dust formed by crushing and handling. When present, the coating can prevent a good bond from forming between the aggregate surfaces and the cementing agent and/or can even increase the quantity of bonding agent required in the mixture. If the quantity of the coating varies from batch to batch, undesirable fluctuations in the consistency of the mix may result. Deposits containing aggregates which display a history of coating problems require decantation. Generally, the aggregates are required to be washed to remove the coating (contaminant) left on the particles.

The coating is often composed of very fine clay-like materials and can affect the adhesion between the asphalt binder and aggregate particles leading to an increased potential for moisture damage. Deleterious materials are particles in an aggregate stockpile that are weak, prone to freeze–thaw damage or damage through repeated wetting and drying, or that otherwise can cause a pavement to deteriorate.

To ensure the aggregate used performs as intended for highway construction, several tests are performed to determine the physical characteristics of the material. Aggregate cleanliness typically refers to whether or not there is a coating on aggregate particles or a material fine enough to pass through a No. 200 sieve. On the other hand, *deleterious material* refers to individual particles which are made up of unsatisfactory or unsound materials. The sand-equivalent test (AASHTO T176) is a method of determining the relative proportion of detrimental fine dust or clay-like materials in the portion of aggregate passing the No. 4 (4.75 mm) sieve. Aggregate intended to perform as a drainable base or sub-base may also be adversely affected when excess amounts of clay and friable particles are present. This type of material tends to fill the void spaces intended for drainage, eventually contributing to pavement failure. Attaining a reasonably accurate determination of the amount of clay lumps and friable particles in the processed aggregate is dependent on properly obtained representative samples.

In summary, there are several test methods that are used to ensure the cleanliness of aggregates and minimize the amounts of deleterious materials (AASHTO T11; AASHTO T90; AASHTO T112; AASHTO T113; AASHTO T176; ASTM C142; ASTM D2419; ASTM D4318).

5.7.3 Composition

The chemical composition of aggregate is significant in determining the difference between limestone and dolomite. Limestone is a rock consisting mainly or wholly of calcium carbonate and has a tendency to polish smooth under traffic. Therefore, limestone is limited to use in low traffic-volume asphalt mix surface courses. Dolostone (dolomite) is a carbonate rock which consists largely of calcium magnesium carbonate ($CaCO_3 \cdot MgCO_3$), offers a higher-friction, skid-resistant surface, and is used on higher traffic-volume locations.

Some aggregates have minerals that are subject to oxidation, hydration, and carbonation. These properties are not particularly harmful, except when the aggregates are used in Portland cement concrete. However, the oxidation of the binder and the introduction of new oxygen functions can have serious effects on the bonding between the asphalt and the aggregate

(Moschopedis and Speight, 1973, 1975, 1977, 1978; Speight, 1992, 2014). As might be expected, iron sulfides, ferric and ferrous oxides, free lime, and free magnesia in industrial products and wastes are some of the common substances. Any of these substances may cause distress in Portland cement concrete and give the concrete an unsightly appearance.

5.7.4 Grading

A widely specified property of aggregate is the particle size distribution (aggregate grading or aggregate gradation), which is critical to developing an effective mix design for the production of an asphalt pavement (see Chapter 9). The maximum aggregate size in the aggregate must be matched to the lift thickness used during construction or the pavement will be difficult to place and compact properly.

Typically, the nominal aggregate size dictates lift thickness and the minimum lift thickness should be at least three times the nominal maximum aggregate size to ensure that the aggregate particles can align during compaction to achieve required density and also to ensure mix is impermeable. Therefore, the desired lift thickness can direct the decision on nominal aggregate size to use. The maximum lift thickness is dependent also upon the type of compaction equipment that is being used. For open-graded mixes, compaction is not an issue since it is intended that these types of mixes remain very open. Therefore, the maximum size aggregate can be as much as 80% of the lift thickness. Generally, as lift thickness increases, it becomes less likely that a series of interconnected voids can span the entire lift depth (Choubane et al., 1998; Huang et al., 1999).

Historically, many highway agencies have used the guideline that the minimum lift thickness of an asphalt–aggregate mix pavement should be twice the thickness of the maximum aggregate size used. With the implementation of the new Superpave mix design method in 2000, these guidelines were changed to include pavement lift-thickness values that range between 2.3 and 3.5 the nominal maximum aggregate size used. This meant that traditional designs for overlays of 1.5–2.0 inches thick would be considered *thin lifts*, which are lifts that would be less than three times the nominal maximum aggregate size. With the use of coarser materials in the mix design, thin lifts could be problematic during compaction in the field.

The distribution of particle sizes in an aggregate must have just the right density so that the resulting mix will contain the optimum amount of asphalt binder and air voids. Because the shape and texture of aggregate particles vary significantly depending on the aggregate type and the way it is mined and processed, specification limits for aggregate gradation tend to be very broad. This breadth helps technicians and engineers achieve the right blend of aggregates for different applications. The section below describes general terminology used when discussing aggregate particle size distribution and the relationship between aggregate gradation and different asphalt–aggregate mix types.

In terms of *size and grading*, the maximum size of an aggregate designates the smallest sieve size through which 100% of the material will pass—grading of an aggregate is determined by sieve analysis (AASHTO T11; AASHTO T27). In the test method, the particle size distribution of an aggregate is usually determined and specified by performing a sieve analysis in which the aggregate is passed through a stack of sieves of decreasing size. The amount of aggregate on each sieve is weighed, and the percent passing each sieve size is calculated as a percent by weight. Sometimes, for aggregates made up of different minerals or rocks having widely different specific gravities, the results of the sieve analysis are given as percent passing by volume.

Maximum size and grading are invariably controlled by specifications that prescribe the distribution of particle sizes to be used for a particular aggregate material for asphalt mixtures. The distribution of the particle sizes determines the stability and density of the asphalt mixture. Generally, there are two types of gradation specifications or requirements used in asphalt mix technology: (i) specifications for the gradation of processed aggregate, as supplied by aggregate producers, and (ii) requirements for blends of aggregate developed as part of the asphalt mix design process.

5.7.5 Hardness

The hardness of the minerals that make up the aggregate particles and the firmness with which the individual grains are cemented or interlocked control the resistance of the aggregate to abrasion and degradation. Soft aggregate particles are composed of minerals with a low degree of hardness while weak particles have poor cementation—neither type of mineral is acceptable.

Aggregates are subjected to additional crushing and abrasive wear during manufacture, placing, and compaction of asphalt concrete mixtures. Aggregates are also subjected to abrasion under traffic loads and must exhibit an ability to resist crushing, degradation, and disintegration. For an aggregate to perform satisfactorily in pavement, it must be sufficiently hard to resist the abrasive effect of traffic over a long period of time. The soft aggregates will be quickly ground to dust, whilst the hard aggregates are quite resistant to crushing effect. The test can also be used to determine the quality of the aggregate and is widely used as an indicator of the relative quality or competence of mineral aggregate. In terms of the selection of aggregate properties to mitigate rutting (Cross and Brown, 1992), the measured rut depth at each site was converted to a rutting rate by dividing the rut depth by the square root of accumulated equivalent single axle loads (ESALs). In addition, the properties of coarse aggregate (which typically consists of crushed and screened basalt that is free of soft or disintegrated pieces, clay, dirt, and other deleterious substances) can be evaluated based on test methods applied to coarse aggregate shape, angularity, and texture including: (i) the index of aggregate particle shape and texture, ASTM C1252, (ii) the flat and elongated and flat or elongated particles, ASTM D4791. Aggregate for untreated base course typically consists of crushed stone free of vegetable matter and other deleterious substances—neither reclaimed asphalt pavement (RAP) nor reclaimed concrete pavement is suitable for untreated base (AASHTO T27; AASHTO T90; AASHTO T96; AASHTO T190).

However, performance testing of RAP is a necessary part of any pavement-related program. In addition to checking the volumetric properties, it may be desirable to evaluate mixture performance of the designed asphalt mixture containing RAP, especially a high content of RAP, to assure that the mixture is able to resist low-temperature and fatigue cracking or rutting if a softer virgin binder was used in the mix design. The possible distress mechanisms that should be evaluated include permanent deformation (i.e., rutting), moisture sensitivity, fatigue, and thermal cracking and a variety of performance tests are available (Mohammad, 2006).

The moisture exudation test (AASHTO T190; ASTM D2844) for compacted soil samples is of particular interest. This test requires that the sample be compressed until either five of the six lights are illuminated or three lights and the ring light are illuminated. At this point, the load is recorded and the test is stopped. To perform this test, a compressive load is applied to a cylindrical, compacted soil sample until a certain amount of moisture is displaced from the bottom of the specimen. At this point, the load is recorded. This test requires a compression testing machine and a special exudation fixture. Older exudation fixtures contain a circle of six contact points on the base plate. When contacted by water, these points will switch on a light on the indicator box. Newer exudation fixtures contain an additional solid ring on the base plate and a corresponding light on the indicator box. This ring detects free-flowing water coming from underneath the mold.

The Los Angeles abrasion test (AASHTO T96) is used to measure the wear abrasion resistance of aggregates and another test method (AASHTO T27) is used for aggregate grading.

5.7.6 Porosity

The internal pore characteristics are very important properties of aggregates since the size, the number, and the continuity of the pores through an aggregate particle may affect the strength of the aggregate, abrasion resistance, surface texture, specific gravity, bonding capabilities, and resistance to freezing and thawing action. Absorption relates to the ability of a particle to take in a liquid. Thus, porosity is a ratio of the volume of the pores to the total volume of the particle. The related property *permeability* refers to the ability of the particle to allow liquids to pass through. If the pores in the aggregate are not connected, the result will be high porosity and low permeability.

The porosity of an aggregate is generally indicated by the amount of water *absorbed* when the aggregate is soaked in water. A certain degree of porosity is desirable, as it permits aggregates to absorb binder, which then forms a mechanical linkage between the binder film and the stone particle. Related to this the *affinity of the aggregate for the binder* is the stripping (separation) of the binder film from the aggregate through the action of water (AASHTO T283) may make an aggregate material unsuitable for asphalt concrete mixtures. Many of these materials may be used with the addition of a heat-stable additive that reduces the stripping action. Aggregates which exhibit a high degree of resistance to stripping in the presence of water are usually most suitable in asphalt concrete mixes. Such aggregates are referred to as hydrophobic (water-hating). Why hydrophobic or hydrophilic aggregates behave as they do is not completely understood. The explanation is not so as important as the ability to detect the properties and avoid the use of aggregates conducive to stripping.

5.7.7 Soundness

The *soundness* (AASHTO T104) refers to the durability of an aggregate in terms of the resistance to the action of weather and is an indication of the resistance to weathering of fine and coarse aggregates. Items for consideration under weathering action are freezing, thawing, variation in moisture content, and temperature changes. In addition, the *particle shape* of the aggregate moisture has an effect on the workability of the mix as well as the compaction energy necessary to obtain the required density. Particle shape also has an effect on the strength of the asphalt concrete mix. Irregular or angular particles tend to interlock when compacted and resist displacement. Also, like particle shape, the *surface texture* also influences the workability and strength of asphalt concrete mixtures. Surface texture has often been considered more important than shape of the aggregate particles since a rough, sandpaper-like surface texture, as opposed to a smooth surface, tends to increase the strength of the mix.

Compacted aggregates without the addition of a cementing material may be used as a base or sub-base for hot mix asphalt and Portland cement concrete pavements. Portland cement concrete pavements are rigid pavements. For these types of pavements, the purpose of the base may be to improve drainage, to prevent pumping, or to cover a material that is highly susceptible to frost. Consequently, gradation and soundness are the primary considerations in selecting or evaluating aggregates for bases under rigid pavements. The load-carrying capacity is a primary factor in the selection of aggregates for hot mix asphalt pavements. A hot mix asphalt pavement does not carry the load; help from the underlying base courses is required. In addition to grading requirements, the aggregates are required to also possess the strength to carry and transmit the applied loads. Furthermore, aggregates are sometimes used to make up the entire pavement structure. In this type of pavement, aggregates are placed on the natural soil to serve as a base course and surface course. Again, the primary requirement is the grading quality. In many instances, compacted aggregates are also used to construct roadway shoulders and berms and, in such applications, gradating and stability are important factors.

5.7.8 Specific Gravity

The specific gravity (density) of an aggregate particle is dependent upon the specific gravity of the minerals making up the particle and upon the porosity of the particle. Furthermore, specific gravity may be subdivided as: (i) bulk density or specific gravity, which includes all of the pore space, (ii) effective density or specific gravity, which includes some of the pore space, also known as the accessible pore space, and (iii) apparent density or specific gravity, which does not include any of the pore space. Determining the porosity of aggregate is often necessary; however, measuring the volume of pore space is difficult. Correlations may be made between porosity and the bulk, apparent and effective specific gravities of the aggregate.

Furthermore, the specific gravity information about a particular aggregate helps in determining the amount of asphalt needed in the hot mix asphalt. If an aggregate is highly absorptive, the aggregate continues to absorb asphalt, after initial mixing at the plant, until the mix cools down completely. This process leaves less asphalt for bonding purposes; therefore, a more porous aggregate requires more asphalt than a less porous aggregate. The porosity of the aggregate may be taken into consideration in determining the amount of asphalt required by applying the three types of specific gravity measurements.

When designing an asphalt aggregate mix, both the mass and volume of the aggregates and asphalt binder must be known (AASHTO T85). Also, determining air void content is one of the main purposes of volumetric analysis. Unfortunately, there is no simple direct way to determine the air void content of an asphalt concrete specimen. Air void content is determined by comparing the specific gravity (or density) of a compacted specimen with the maximum theoretical density of the mixture used to make that specimen.

5.8 ASPHALT CONCRETE

For the purposes of this text and many other texts, the term *asphalt concrete* is the name given to the asphalt–aggregate mix as it is laid onto the road. When set the mix approximates the properties of concrete.

Asphalt roadways (also called *asphalt concrete pavements*) are specifically engineered and consist of multiple layers or courses of asphalt mix and other materials (see Chapter 8). The structural layers are usually referred to as (i) surface course, (ii) base course, (iii) sub-base course, and (iv) subgrade course—other terminology includes the names: surface course, intermediate course (sometimes called the binder course), and base course depending on their location within the pavement structure. Asphalt roadways with higher traffic volumes may also include a wearing course composed of an open-graded friction course placed over the surface course. Each asphalt-containing layer in a pavement is composed of different materials and each layer has a specific function that affects the type of mixture that should be specified and used.

It is the purpose of this section to present the laboratory test methods that are employed for determining the properties of the asphalt concrete (binder plus aggregate). In the paving mixture, asphalt and aggregate are blended together in precise proportions and the relative proportions of these materials determine the physical properties of the mix, and ultimately, how the mix will perform as a finished pavement.

Once the mix design requirements have been established by the responsible authority, it is the responsibility of the contractor/producer and his technician to develop the mix within the framework of the specifications. It is then the moment to test the asphalt concrete in the laboratory where the data will give an indication of the probable performance of the pavement structure, as well as conformance with the relevant specifications. The laboratory tests used are typically: (i) ignition method, which is used to determine the asphalt content of the mix (ASTM D4), (ii) maximum specific gravity, which is used to determine the voids in the total mix, (iii) bulk specific gravity of mixture (AASHTO T166), which is used to determine the voids filled with binder, (iv) sieve analysis, which is used to determine the aggregate gradation and fines to asphalt (F/A) ratio, and (v) tensile strength ratio, which is used to determine the stripping potential and rutting potential (Zaniewski and Srinivasan, 2004). Other tests performed in the lab are: moisture content, particle coating, and various aggregate consensus properties tests.

5.9 TEST METHODS

Applicability of the various test methods is used to determine the quality of the asphalt concrete. However, in the formatting and application of a QC program for asphalt concrete (asphalt-aggregate mix) it is necessary to understand that asphalt concrete is, by nature, a highly variable material. Furthermore, the procedures that are used to test and determine the properties of the binder (see Chapter 4) and the aggregate are also variable. Even the procedures used to sample the materials at the asphalt plant and on the roadway may have variability and lack consistency in the test data. The three sources of variability are: (i) materials, which are the asphalt and the aggregate, (ii) sampling, and (iii) the test method used. Allowance must be made for these variable parameters and the variability must be included in an assessment of the asphalt concrete (Leahy and Witczak, 1991).

5.9.1 Aggregate Absorptivity

The *absorptivity* of the aggregate (the ability of the aggregate to absorb asphalt) used in the mix is critical in determining optimum asphalt content. Sufficient asphalt must be added to the mix to allow for absorption and still coat the particles with an adequate film. When referring to absorbed asphalt and unabsorbed asphalt, there are two distinct types of asphalt content: (i) total asphalt content and (ii) effective asphalt content.

The relationship between aggregate surface area and optimum asphalt content is most pronounced where filler material (very fine aggregate fractions which pass through the No. 200 (0.075 mm) sieve) is involved. Small increases in the amount of filler in a gradation can literally absorb much of the asphalt binder, resulting in a dry, unstable mix. Small decreases have the opposite effect: too little filler results in too rich (wet) a mixture. Variations in filler content will cause changes in mix properties, from dry to wet. If a mix contains too little or too much mineral filler, however, arbitrary adjustments to correct the situation are likely to worsen it. Instead, proper sampling and testing should be done to determine the cause of the variations and, if necessary to establish a new mix design.

The *total asphalt content* is the amount of asphalt that must be added to the mixture to produce the desired mix qualities. On the other hand, the *effective asphalt content* is the volume of asphalt not absorbed by the aggregate—the amount of asphalt that effectively forms a bonding film on the aggregate surfaces. The effective asphalt content is calculated by subtracting the amount of absorbed asphalt from the total asphalt content. The absorptivity of an aggregate is an important consideration in determining the asphalt content of an asphalt–aggregate mixture. It is generally known for established aggregate sources, but requires careful testing where new aggregate sources are being used.

5.9.2 Air Voids

By way of further explanation, the *air voids* are small airspaces or pockets of air that occur between the coated aggregate particles in the final compacted mix. A certain percentage of air voids is necessary in all dense-graded highway mixes to allow for some additional pavement compaction under traffic and to provide spaces into which small amounts of asphalt can flow during this subsequent compaction. The allowable percentage of air voids (in laboratory specimens) is between 2.0 and 4.0% (v/v) for most surface course mixes.

Air voids (VA), voids in the mineral aggregate (VMA), voids filled with asphalt (VFA), and volume of effective binder (VBE) are calculated from the following equations:

$$V_a = \left(1 - \frac{G_{mb}}{G_{mm}}\right) \times 100$$

$$\mathrm{VMA} = \left(100 - \frac{G_{mb} P_s}{G_{sb}}\right)$$

$$\mathrm{VFA} = \left(\frac{\mathrm{VMA} - V_a}{\mathrm{VMA}}\right) \times 100$$

$$V_{be} = \mathrm{VMA} - V_a$$

In the equations, G_{mb} = bulk specific gravity of the compacted sample, G_{mm} = maximum specific gravity of the asphalt mixture, P_s = percentage (by mass) of aggregate in the total mixture, and G_{sb} = aggregate bulk specific gravity.

With these equations, the effects of the specific gravity results can be analyzed more closely for approximate relationships to be determined. Thus taking each equation in turn, if G_{mb} is held constant, the following relationship between *air voids* and G_{mm} can be established—when G_{mm} changes by +0.01, V_a changes by +0.4%. Similarly, if G_{mm} is held constant, when G_{mb} changes by +0.01, V_a changes by −0.4%. In the second equation, when G_{sb} and P_s are held constant, when G_{mb} changes by +0.01, *VMA* changes by −0.4%. Furthermore, when G_{mb} and P_s are held constant, when G_{sb} changes by +0.01, *VMA* changes by +0.3%.

The *VFA* is the percentage of voids in the compacted aggregate mass that are filled with asphalt cement. It is synonymous with the asphalt–void ratio. The VFA is an important property not only as a measure of relative durability but also because there is an excellent correlation between this property and percent density. If the VFA is too low, there is insufficient asphalt to provide durability and to over-densify under traffic and bleed. Thus, the VFA is a very important design property.

The *VMA* are the air-void spaces that exist between the aggregate particles in a compacted paving mixture (AASHTO T312), including spaces filled with asphalt. The presence of VMA represents the space that is available to accommodate the asphalt and the volume of air voids necessary in the mixture. The more VMA in the dry aggregate, the more space is available for the film of the asphalt binder.

5.9.3 Anti-Stripping Additive and Moisture Content

The *boil test* is a method used to determine the effectiveness of an anti-stripping additive when used as an asphalt anti-stripping compound in asphalt mixtures (Kennedy et al., 1984b; Parker and Gharaybeh, 1987; ASTM D3625). In addition, it is important that the moisture content of the aggregate blend and asphalt concrete mixture is monitored. Many specifications require that the finished mixture have a moisture content not exceeding 1% (w/w).

In the test method, when necessary for the test to be run at the asphalt plant, the sample should be allowed to cool to 110 ± 5°C (230 ± 9°F) (Kennedy et al., 1984b). Approximately 400 g of the mixture passing the 1/2 inch (12.5 mm) sieve are used, of which approximately 200 g are placed on a paper towel before boiling. The remaining amount of the sample (approximately 200 g) is placed in boiling water and boiling continued for 10 min, after which the water is drained from the mixture and the sample placed on a paper towel and allowed to cool to room temperature. After cooling (typically an overnight procedure), the boiled and unboiled portions of the sample on the paper towels are compared. If the boiled portion shows more signs of stripping than the unboiled portion, the test fails—the test results are reported as *pass* or *fail*. The converse is true for a passing grade.

However, it is recommended (ASTM D3625) that this test method, although useful as an indicator of the relative susceptibility of bituminous-coated aggregate to water, should not be used as a measure of field performance because such correlation has not been established. If loss of adhesion due to water is indicated, testing by other procedures should be conducted to further evaluate the mixture. It is further recommended that the test method data should not be used for acceptance/rejection by owner agencies.

Other test methods include the *static immersion test method* (AASHTO T182), which has been retained as an AASHTO test method but is no longer available as an ASTM standard test method (originally ASTM D1664). In the test, the asphalt–aggregate mixture is cured for 2 h at 60°C (140°F) and cooled to room temperature. It is then placed in a glass jar and covered with 600 ml of distilled water. The jar is capped and placed in a 25°C (77°F) water bath and left undisturbed for 16–18 h. The amount of stripping is visually estimated on the basis of a subjective estimate.

The *dynamic immersion test method* is used to accelerate the stripping effect compared with the static immersion test but the test method has not been standardized and is not widely used. Samples of asphalt–aggregate mixtures are prepared in the same manner as for the static immersion test but are subjected to 4 h of agitation. As the period of agitation increases, the degree of stripping increases. Both static and dynamic immersion tests, however, fail to take into account the pore pressure effect and traffic action, as is the case for all tests on loose mixtures.

The *chemical immersion test method* covers the determination of the adhesion of asphalt to stone aggregate by means of boiling asphalt-coated aggregate successively in distilled water. Increasing concentrations of sodium carbonate (Na_2CO_3) are used, numbered 0–9 and referred to as the Riedel and Weber (R&W) number. 0 refers to distilled water, 1 implies 0.41 g of sodium carbonate in 1 l of water, and 9 refers to the highest concentration, which is 106 g of sodium carbonate (Na_2CO_3) in 1 l of water. Between 1 and 9, for every doubling of concentration the Riedel and Weber number is increased by 1. The number of the concentration at which the asphalt strips from the aggregate to such an extent that it is no longer a film but only specks or droplets is called the stripping value. An asphalt–aggregate test sample of 100 g is dried in an oven at 110°C (230°F). The aggregate is mixed with binder at high temperature and left to cool to room temperature. Solutions of sodium carbonate in distilled water are prepared at different concentrations. Approximately 50 ml of distilled water is brought to boiling in a 200-ml glass beaker. Afterwards, 10 g of the prepared aggregate–binder mix is placed into the boiling water. After boiling for 1 min, the water is drained and the sample is placed on filter paper. The sample is examined for stripping after it is dry. The stripping value of the aggregate is the lowest concentration at which stripping occurs.

5.9.4 Asphalt Content

The *asphalt content* of the mixture is critical and must be accurately determined in the laboratory and then precisely controlled on the job. The optimum asphalt content of a mix is highly dependent on aggregate characteristics such as gradation and absorptiveness. Aggregate gradation is directly related to optimum asphalt content—the finer the mix gradation, the larger the total surface area of the aggregate and the greater the amount of asphalt required to uniformly coat the particles (Goode and Lufsey, 1962). Conversely, because coarser mixes have less total aggregate surface area, they demand less asphalt.

The proportion of asphalt in the mixture is critical and must be accurately determined in the laboratory and then precisely controlled on the job (Harvey et al., 1994).

Total asphalt content is the amount of asphalt that must be added to the mixture to produce the desired mix qualities. Effective asphalt content is the volume of asphalt not absorbed by the aggregate; the amount of asphalt that effectively forms a bonding film on the aggregate surfaces. Effective asphalt content is calculated by subtracting the amount of absorbed asphalt from the total asphalt content. The absorptiveness of an aggregate is obviously an important consideration in determining the asphalt content of a mixture. It is generally known for established aggregate sources, but requires careful testing where new aggregate sources are being used.

The asphalt content of the mix will be used to determine the voids in mineral aggregate and F/A ratio of the mixture. The latter property—the F/A ratio—is an indication of the thickness of the asphalt film on the coated particles. The film thickness helps reduce premature aging and reduce moisture damage. In addition, when preparing a sample for testing by the ignition method, the quartering method may be used (AASHTO T248), which covers the reduction of large samples of aggregate to the appropriate size for testing. Techniques used should minimize variation in measured characteristics between the test samples selected and the entire sample and the end sample should be a sample representative of the source and analogous to preparing a coal sample for analysis (Speight, 2013, 2015b).

5.9.5 Bulk Specific Gravity

The *bulk specific gravity* is the ratio of the weight of a given volume of aggregate, including the permeable and impermeable voids in the particles, to the weight of an equal volume of water (Kandhal et al., 2000; Prowell and Baker, 2004; Sholar et al., 2005). Bulk specific gravity of aggregate is important information for designing hot mix asphalt because it is used to calculate voids in mineral aggregate and voids filled with asphalt. Moreover, different procedures are used to determine the *Gsb* of coarse and fine aggregate (Buchanan, 2000; Hall et al., 2001; Malpass and Khosla, 2001; Cooley et al., 2003; Williams, 2007; FHWA, 2010).

The standard test methods for the determination of specific gravity of coarse aggregate (AASHTO T85; ASTM C127) are essentially the same, except for the required time in which a sample of aggregate is submersed in water to essentially fill the pores—the AASHTO method (AASHTO T85) requires the sample be immersed for a period of 15–19h while the ASTM method (ASTM C127) specifies an immersed period of 24 ± 4h. After the specimen is removed from the water, it is rolled in an absorbent towel until all visible films of water are removed. This is defined as the saturated surface-dry (SSD) condition. Three mass measurements are obtained from a sample: (i) the SSD mass, (ii) water submerged mass, and (iii) the oven dry mass from which the value of Gsb of an aggregate can be determined.

The standard test methods for determining fine aggregate *Gsb* are also available through two standard test methods (AASHTO T84; ASTM C128) which are similar, except for the required period in which a sample of fine aggregate is submersed in water to essentially fill the pores. The AASHTO procedure requires immersion of fine aggregate in water for

15–19 h, while the ASTM method specifies a soaking period of 24 ± 4 h. For both methods, the sample is then spread on a pan and exposed to a gentle current of warm air until approaching a free-flowing condition. Periodically, the aggregate is lightly tamped into a cone-shaped mold with 25 light drops of the tamper. If the fine aggregate retains the molded shape when the mold is removed, the fine aggregate is assumed to have surface moisture, and it is dried further. When the cone of sand just begins to slump upon removal of the mold, it is assumed to have reached the SSD condition. Three masses are determined from the method using either gravimetric or volumetric methods; (i) SSD, (ii) saturated sample in water, and (iii) oven dry which are used to calculate *Gsb*.

5.9.6 Tensile Strength Ratio

With the new insight into stress-related problems of asphalt roadways, it is now recognized that the performance of an asphalt surface is dependent on the tensile strength of the surfacing (to resist the stresses). Since asphalt pavements are usually primed to provide interlock between the surfacing and the base course, the quality of the prime and the tensile strength of the surfacing/base course combination are also of importance. This explains many field observations such as rapid failure of single sand seals relative to multiple sand seals, surfacing failures over weak base courses, and better performance of thicker surfaces (Emery et al., 1992).

The *tensile strength ratio* test (AASHTO T283, which may include a freeze–thaw cycle) is a method for the measurement of the strength loss resulting from damage caused by stripping (of the asphalt from the aggregate) under laboratory-controlled accelerated water conditioning. The results may be used to predict long-term susceptibility to stripping of an asphalt concrete. To combat the effects of water damage, an anti-stripping additive is used in all asphalt mixes. To combat the effects of stripping, it is in order to use hydrated lime (on the order of 1.0%, w/w). If a freeze–thaw cycle is necessary in the test, six samples with air voids are split into two groups of three samples each. The first group of three is denoted as the control group. The second group of three samples is vacuum-saturated at a saturation level between 55% and 80% and they are placed in a freezer at −18°C (0°F) for 16 h. They are then moved to a 60°C (140°F) water bath for 24 h. All of the samples are then subjected to indirect tensile strength testing and/or resilient modulus testing.

The *indirect tensile strength test* is commonly used to predict the degree of moisture-induced damage a pavement will suffer and the test (ASTM

D4123) involves compressive loading of a cylindrical specimen of 4-inch (101.6 mm) diameter and 2.5-inch (63.5 mm) height. The single compressive load is applied vertically to the diametric plane of the cylinder. The test is performed with the specimen at 25°C (77°F) with a loading rate of 2 inches/min. Failure of the specimen is characterized by the development of a split along the vertical diameter.

The tensile strength is measured after conditioning of the core. This measurement is then compared to the tensile strength for an unconditioned core to determine the tensile strength retained (TSR) as a percentage of the original unconditioned strength:

$$\text{TSR} = \text{Tensile Strength before Conditioning}/\text{Tensile Strength after Conditioning}$$

The tensile strength ratios vary from 0 to 1—values in the range of 0.8 to 1 indicate good performance of the asphalt mix in terms of moisture susceptibility. Conversely, values below 0.7 typically represent a pavement that will likely be susceptible to moisture-induced damage. For production tests, the mixture should have a specified tensile strength ratio value not less than 0.80 for a Superpave mix.

The *thermal stress restrained specimen test* (AASHTO TP10) is used to determine the tensile strength and temperature at fracture for asphalt mixtures by measuring the tensile load in a specimen which is cooled at a constant rate while being restrained from contraction. The basic requirement for the test system is that it maintains the test specimen at constant length during cooling.

5.10 USE OF THE DATA

As already stated (see Chapter 4), the changing nature of refinery feedstocks has required that the analytical methods and interpretation of the data evolve to accommodate the new refining processes, process chemistry, and products. In this time period, the *average* quality of petroleum delivered to refineries has diminished by approximately two points on the API gravity scale. This has been accompanied by, among other things, an increase in the sulfur content. However, as a word of caution, use of the word *average* is open to definition, because some refineries accept medium- to high-quality crude oil while others accept low- to medium-quality crude oil. Nevertheless, the general quality of the crude oil accepted by refineries is lower than it was two decades ago. As a result, refineries have had to adapt processing strategies to deliver liquid fuel products from feedstocks that are deteriorating in quality. The same might be stated for products such as asphalt. The complexity of such a product renders a term such as average quality open to question and whether or not such a term is suitable for making a decision for use of the asphalt. Therefore caution is advised when considering the result(s) of analytical test method(s) and whether or not the sample represents a material suitable for the designated use.

Recognition that the behavior of the aggregate is related to composition has led to a multiplicity of attempts to establish aggregates as compositions of matter. As a result, various analytical techniques have been developed for the identification and quantification of every aggregate that is to be used for production of an asphalt mix. In fact, because of the complex nature of aggregates, there has been a need to define any aggregate in a more meaningful and understandable manner. Although this still remains to be accomplished, some success has been achieved. For example, it is now rare that aggregate deposits are referred to (geological issues notwithstanding) as gravel pits—gravel being a specific type of aggregate.

Typically, paving asphalt is specified based upon specific physical properties relating to consistency, adhesion properties, viscosity, hardness, or brittleness at a specified temperature. These attributes are important to effective asphalt pavement applications and to the resulting quality and durability of the pavement. Pavement designers select particular paving asphalts that are appropriate to the climate, traffic, and other conditions in which the pavement is used. An important physical attribute is the fact that asphalt becomes softer and more fluid when heated and hardens again when cooled. This attribute allows for the mixing, placement, and compaction of the asphalt mix to form a quality asphalt pavement that can be expediently turned over to traffic.

Asphalts are complex chemical mixtures that may be manufactured to yield very different physical and chemical attributes. For example, paving asphalt is most commonly produced through refining of crude oil using the atmospheric residuum and/or the vacuum residuum followed by, when necessary, mild oxidation (often referred to as air rectification or semi-blowing). Mildly oxidized asphalt, such as is sometimes used to produce semi-blown (air-rectified) paving asphalt, has physical properties similar to those of atmospheric or vacuum-distilled paving asphalt. Asphalt consistency has historically been measured by standardized test procedures, e.g., penetration value, ring and ball test method, softening point, and penetration index.

While these test results point to a given set of physical properties, it is the chemical constituents of asphalt that determine its consistency. For example, it is the relative proportion of the asphaltene constituents (high-molecular-weight constituents) that are responsible for strength and stiffness; resins are responsible for adhesion; and oils (low molecular weight) are responsible for viscosity and fluidity. On the other hand, oxidized asphalt is manufactured by passing air through asphalt at elevated temperatures (usually on the order of 200–250°C (390–480°F)) in order to stiffen the asphalt and/or increase the softening point (Asphalt Institute and Eurobitume, 2008). In this process, chemical reactions change the chemical nature of the asphalt. However, intense oxidation which can degrade the feedstock constituents to become an unwanted product (such as acidic by-products) is not employed in the production of paving asphalt.

One example, of data use, is the *penetration grading* system which is based on the depth a standard needle will penetrate an asphalt binder sample when placed under a 100-g load for 5 s. The test is simple and easy to perform but it does not measure any fundamental parameter and can only characterize asphalt binder at one temperature (25°C, 77°F). Penetration grades are listed as a range of penetration units (one penetration unit = 0.1 mm of penetration by the standard needle).

Similarly, the *viscosity grading* system is a measure of penetration (as in penetration grading) but also measures an asphalt binder's viscosity at 60°C (140°F) and 135°C (275°F). Testing can be done on virgin (AC) or aged (AR) asphalt binder. Grades are listed in poises (cm-g-s = dyne-second/cm^2) or poises divided by 10. Typical asphalt binders used in the United States are AC-10, AC-20, AC-30, AR-4000, and AR-8000. Viscosity grading is a better grading system but it does not test low-temperature asphalt binder rheology.

The *Superpave performance grading* system was developed as part of the Superpave research effort to more accurately and fully characterize asphalt binders for use in hot mix asphalt pavements. The system is based on the concept that the properties of a hot mix asphalt binder should be related to the conditions under which it is used. For asphalt binders, this involves expected climatic conditions as well as aging considerations. Therefore, the grading system uses a common battery of tests (as the older penetration and viscosity grading systems do) but specifies that a particular asphalt binder must pass these tests at specific temperatures that are dependent upon the specific climatic conditions in the area of intended use.

Therefore, in the United States a binder used in a southern state would be different to one used in a northern state.

Superpave performance grading is reported using two numbers: the first number is the average 7-day maximum pavement temperature (in °C) and the second number is the minimum pavement design temperature likely to be experienced (in °C). Thus, a PG 64-16 Superpave binder is intended for use where the average 7-day maximum pavement temperature is 64°C (147°F) and the expected minimum pavement temperature is −16°C (3°F). Furthermore, these numbers are pavement temperatures and not air temperatures.

The test data are only as good as the QC program allows. As stated above, in the formatting and application of a QC program for asphalt concrete (asphalt–aggregate mix) it is necessary to understand that asphalt concrete is, by nature, a highly variable material. Furthermore, the procedures that are used to test and determine the properties of the binder (see Chapter 4) and the aggregate are also variable. Even the procedures used to sample the materials at the asphalt plant and on the roadway may have variability and lack consistency in the test data. The three sources of variability are: (i) materials, which are the asphalt and the aggregate, (ii) sampling, and (iii) the test method used. Allowance must be made for these variable parameters and the variability must be included in an assessment of the asphalt concrete.

Thus, the test data are subject to the inherent variability in the individual components of the concrete (i.e., the asphalt and the aggregate) material as well as the range of asphalt content for the mix that is used during a specific project. In addition, substantial variability in test data on pavement construction projects will be from test method variability which, in turn, is subject to the sampling protocols. Even the most detailed sample history (Speight, 2015a) will not compensate for any inherent variability in the sample that is subject to any variability in the protocols used to create the mix. In fact, the actual variability in the test data can be assessed by calculating the standard deviation.

In practice, the actual precision that is represented by the standard devastation will also provide valuable information on single-operator test method precision (the variability expected for a specific given test method when repeatedly performed by the same technician in the same laboratory) and also on multi-laboratory precision (the variability expected for a given test when it is performed repeatedly on the same material by a number of different laboratories).

5.11 VALIDITY OF TEST PROCEDURES

One of the major issues in asphalt science and technology is relating the laboratory work (the science) to the field work (technology). Typically, in an industrial setting, a concept will be tested in the laboratory (laboratory scale) and, if successful (proof of concept) will be scaled up to a larger (bench scale) operation, but still in the laboratory. From the bench scale, the work is then moved to pilot plant scale which is a major test of the concept.

The pilot plant is a small (rather than full-scale) system, which is operated to generate information about the behavior of the system for use in design of larger facilities. The pilot plant is also used to reduce the risk associated with the construction of large process plants to perform: (i) computer simulations and semi-empirical methods to determine the limitations of the pilot scale system, (ii) chemical similitude studies, (iii) mathematical modeling, and (iv) risk analysis—if this has not already been accomplished. Thus, the pilot plant will provide valuable data for design of the full-scale plant. Scientific engineering data about reactions and material properties may be produced but it is difficult to predict the behavior of a process of any complexity and designers must use the data from the pilot plant to refine any existing design of the production-scale facility. All being well and in order, the concept, having been proven to be operational at the pilot plant scale, the next step is the demonstration plant. The differences between bench scale, pilot scale, and demonstration scale are strongly influenced by industry and application. Some industries use pilot plant and demonstration plant interchangeably. Some pilot plants are built as portable modules that can easily transported as a contained unit. The demonstration plant is often referred to as a semi-works plant and is used to estimate the viability of the concept (the process) on a pre-commercial scale and will generate significant quantities of product for market testing.

In the development of new processes, the design and operation of the pilot and demonstration plant will often run in parallel with the design of the future commercial plant, and the results from pilot testing programs are key to optimizing the commercial plant flow sheet. It is common in cases where process technology has been successfully implemented that the savings at the commercial scale resulting from pilot testing will significantly outweigh the cost of the pilot plant itself.

The asphalt paving industry does not always have the luxury of progressing through the sequence:

Laboratory scale → bench scale
Bench scale → pilot plant scale
Pilot plant scale → demonstration plant scale
Demonstration plant scale → commercial operation.

The laboratory data (laboratory scale or bench scale) for a roadway asphalt mix may be taken to a test strip operation and, if successful, to the asphalt paving operation. This, of course, is dependent upon the laboratory data and the validity (i.e., the reliability or usefulness) of the data.

The major issues that arise when considering the properties of the aggregate, and the asphalt–aggregate mix are (i) the validity of the testing procedure and (ii) the relationship of properties (as measured in the laboratory) to performance in terms of roadway behavior and longevity. However, there are many issues that are associated with the non-standard ways in which many materials are tested and characterized and for which relevant data or information were available. These problems make it difficult to establish relationships between properties and performance, especially identification of the trends associated with the use of modified materials (Coplantz et al., 1993). For example, while it can be reasonably concluded that modifiers have an influence on the performance-related properties of asphalt cement and, hence, on the asphalt concrete, as measured in the laboratory, they also have an influence on the performance of asphalt concrete pavements as determined from evaluation of the in-service performance of the roadways. However, whether or not the two sets of data match and any conclusions drawn can be used as a consensus, is not always the case.

Furthermore, interpretation of test data for asphalt concrete relies on site-specific factors including pavement structure characteristics, traffic, and environmental conditions. In fact, the influence of site-specific environmental factors (including traffic loads) makes it difficult to extrapolate the test data so that the data are more generally applicable to other roadways. Thus, the interpretation of test data should be considered to be valid predominantly (if not only) on a site-specific basis.

Whether or not the test methods are truly standard methods and the data can be extrapolated from the laboratory to the field is another issue that has been the subject of frequent discussion over the years. The issues related to roadway structure and the roadway environment are factors that can account for some of the differences in the test data and also in the behavior of the roadway in service. Unless standardized test methods are used, there will be many problems that are associated with the use of non-standard test methods. In addition, the repeatability and reproducibility (especially sampling of the concrete) play an important role in the believability of the data from test methods (Speight, 2015a,b).

For example, when sampling asphalt concrete mixtures (ASTM D979) the sampling is often considered to be more important that the testing, and the sampler must take every precaution to obtain samples that will yield an acceptable estimate of the nature and conditions of the materials which the samples represent (Table 5.3). In fact, if non-standard test methods are employed and three essential criteria (sampling, repeatability, and reproducibility) are not in order, the material that was tested and characterized and for which data or information were produced cannot be logically compared to the data from one laboratory to another and, thus, establishing relationships and trends associated with the use of the material is not possible. Indeed, it remains controversial as to whether any improvement in the properties of asphalt concrete, as measured by various particular non-traditional and non-standard test methods, are any type of indicator of actual performance in the field.

Furthermore, field performance of asphalt concrete is generally measured through the use of special test sections placed in a project where a control section using unmodified asphalt is compared to a *test section*. However, the test section must match the control section—for example, comparing the effect of a modified asphalt concrete section in a driving lane with the unmodified asphalt concrete section in the passing lane (and vice versa, as is also the case when the test sections of the roadway were laid in different years) is an incorrect and potentially fruitless procedure. Comparisons must be made between similar roadway sections and interpretation of the test results based on site-specific factors (including pavement structure characteristics as well as traffic loads and environmental conditions) before the data are relevant and meaningful. If the test sections are not chosen wisely, the compounding effects of the site-specific factors makes extrapolation of the test data to more generalized conditions very difficult, if not impossible. The interpretation of test results will, more than likely, be valid for only the site-specific conditions and not give a true comparison of the asphalt concrete in various roadway sections.

Table 5.3 ASTM standard methods for sampling asphalt and similar materials[a]

ASTM C702	Practice for Reducing Samples of Aggregate to Testing Size
ASTM D2041	Test Method for Theoretical Maximum Specific Gravity and Density of Bituminous Paving Mixtures
ASTM D2234	Practice for Collection of a Gross Sample of Coal
ASTM D2726	Test Method for Bulk Specific Gravity and Density of Non-Absorptive Compacted Bituminous Mixtures
ASTM D3665	Practice for Random Sampling of Construction Materials
ASTM D5361	Practice for Sampling Compacted Bituminous Mixtures for Laboratory Testing
ASTM D5444	Test Method for Mechanical Size Analysis of Extracted Aggregate
ASTM D6307	Test Method for Asphalt Content of Hot-Mix Asphalt by Ignition Method
ASTM D6925	Test Method for Preparation and Determination of the Relative Density of Hot Mix Asphalt (HMA) Specimens by Means of the Superpave Gyratory Compactor
ASTM D6926	Practice for Preparation of Bituminous Specimens Using Marshall Apparatus
ASTM D6927	Test Method for Marshall Stability and Flow of Bituminous Mixtures
ASTM E105	Practice for Probability Sampling of Materials
ASTM E122	Practice for Calculating Sample Size to Estimate, with Specified Precision, the Average for a Characteristic of a Lot or Process
ASTM E141	Practice for Acceptance of Evidence Based on the Results of Probability Sampling

[a]Listed alpha-numerically and not by preference.

However, if the test sections are chosen in a logical manner and if the standard test methods are selected with care and forethought, the data can provide meaningful correlations with the mechanical properties of the asphalt–aggregate system. This will allow the establishment of the links between laboratory data and field performance and may even point to mix properties that are likely to be required for future asphalt concrete roadways. Thus, each test method should be standardized and is evaluated on its own merits and findings from one investigation (if the method has been standardized) should be in agreement with the data from the same method applied to the same sample at a different laboratory locale.

Recognizing that problems do exist in attempting to interpret information from application of the various test methods, serious efforts should be made to extract both qualitative and quantitative information from laboratory data and field performance. In fact, in order to predict the behavior of an asphalt concrete pavement at low temperature, more than one test will be required in order to generate meaningful (and believable) data (Schmidt, 1996; Schmidt and Santucci, 1996). It is necessary to take into account all possible factors which will be affected by the composition and structure of the asphalt concrete roadway (Barth, 1962; Halstead et al., 1966; Speight, 2014, 2015a). If the test methods do not give consideration to such phenomena, the data may be of limited value. The source of the asphalt source is probably another variable that affects the glass transition temperature just as the characteristics of the monomer unit in polymers affects their glass transition temperatures (Breen and Stephens, 1967). The effects of aging on the glass transition temperature and strain energy release rate are problems that are in need of further research. Hopefully by knowing these quantities, it will be possible to select asphalt that will not be crack-susceptible at low temperature.

The need for the continuing evolution of current standard test methods and the development of new standard test methods is a necessary part of the asphalt industry, as it is for the petroleum industry and other industries (Dongre and D'Angelo, 1998; Speight, 2015a). As an example, a suite of performance-related specifications has been recently developed (within the past two decades) for asphalt binders as well as for asphalt concrete (collectively known as the Superpave pavement design system) (see Chapters 3 and 8). These new specifications were adopted by the American Association of State Highway and Transportation Officials (designated as AASHTO M-320—formerly designated as MP-1—and AASHTO MP-1a). In the former test method (AASHTO M320), empirical parameters are related to performance, while in the latter test method (AASHTO MP1a) a mechanistically based criterion to define low-temperature performance of asphalt binders is incorporated into the test protocol. In addition, property data (ASTM D6648; ASTM D6723; ASTM D6816) are used to obtain a critical cracking temperature (AASHTO MP1a), whereas other property data (ASTM D6648) are used in the Superpave pavement design system under the purchase specification (AASHTO M320; ASTM D6373).

Thus, the Superpave specification is based on test methods that were adapted and/or specifically designed to measure fundamental material properties as opposed to the empirical test methods employed by the earlier viscosity- and penetration-based specification. Test methods have been developed for use in the Superpave specification to address low-temperature cracking performance of pavements: (i) a flexural creep test called the bending beam rheometer test method to measure creep properties at low temperatures (AASHTO T313, see also ASTM D6648) and (ii) a uniaxial tension test at constant elongation rate—the direct tension test method—to measure the failure properties at low temperature. The constant elongation rate simulates the loading (in uniaxial tension) during thermal cooling of the pavement (see also ASTM D6723), and (iii) the critical cracking temperature which is calculated using an algorithm (AASHTO MP1a; ASTM D6816; Anderson et al., 1994; Roque et al., 1995; Bouldin et al., 2000).

In summary, the data produced by the various laboratory test methods must be valid and useful for commercial paving operations. QC and quality assurance (QA) programs are key components of all analytical protocols in all areas of asphalt analysis, including environmental analysis (Patnaik, 2004; Speight, 2015a). These programs mandate that the laboratories follow a set of well-defined guidelines to achieve valid analytical results to a high degree of reliability and accuracy within an acceptable range. Although such programs may vary depending on the regulatory authority, certain key features of these programs are more or less the same. The two most important components of QC are (i) determination of precision of analysis and (ii) determination of accuracy of measurement

However, in the case of asphalt science and technology, whether or not the laboratory data are sufficiently reliable to take the concept from the laboratory scale to the commercial scale (in the absence of a test strip) is always open to consideration.

REFERENCES

AASHTO M320, 2014. Standard Specification for Performance-Graded Asphalt Binder. American Association of State Highway and Transportation Officials, Washington, DC.

AASHTO MP1a, 2014. Performance Graded Binder Specification. American Association of State Highway and Transportation Officials, Washington, DC.

AASHTO R9, 2014. Acceptance Sampling Plans for Highway Construction. American Association of State Highway and Transportation Officials, Washington, DC.

AASHTO R10, 2014. Definition of Terms for Specification and Procedures. American Association of State Highway and Transportation Officials, Washington, DC.

AASHTO T2, 2014. Sampling of Aggregates. American Association of State Highway and Transportation Officials, Washington, DC.

AASHTO T11, 2014. Sieve Analysis of Fine and Coarse Aggregates. American Association of State Highway and Transportation Officials, Washington, DC.

AASHTO T21, 2014. Organic Impurities in Fine Aggregates for Concrete. American Association of State Highway and Transportation Officials, Washington, DC.

AASHTO T27, 2014. Sieve Analysis of fine and Coarse Aggregates. American Association of State Highway and Transportation Officials, Washington, DC.

AASHTO T40, 2014. Sampling Bituminous Materials. American Association of State Highway and Transportation Officials, Washington, DC.

AASHTO T84, 2014. Specific Gravity and Absorption of Fine Aggregate. American Association of State Highway and Transportation Officials, Washington, DC.

AASHTO T85, 2014. Standard Method of Test for Specific Gravity and Absorption of Coarse Aggregate. American Association of State Highway and Transportation Officials, Washington, DC.

AASHTO T90, 2014. Determining the Plastic Limit and Plasticity Index of Soils. American Association of State Highway and Transportation Officials, Washington, DC.

AASHTO T96, 2014. Standard Method of Test for Resistance to Degradation of Small-Size Coarse Aggregate by Abrasion and Impact in the Los Angeles Machine (ASTM C131). American Association of State Highway and Transportation Officials, Washington, DC.

AASHTO T103, 2014. Standard Method of Test for Soundness of Aggregates by Freezing and Thawing. American Association of State Highway and Transportation Officials, Washington, DC.

AASHTO T104, 2014. Standard Method of Test for Soundness of Aggregate by Use of Sodium Sulfate or Magnesium Sulfate. American Association of State Highway and Transportation Officials, Washington, DC.

AASHTO T112, 2014. Clay Lumps and Friable Particles in Aggregate. American Association of State Highway and Transportation Officials, Washington, DC.

AASHTO T113. Light Weight Pieces in Aggregate. American Association of State Highway and Transportation Officials, Washington, DC.

AASHTO T166, 2014. Bulk Specific Gravity of Compacted Asphalt Mixtures Using Saturated Surface-Dry Specimens. American Association of State Highway and Transportation Officials, Washington, DC.

AASHTO T168, 2014. Sampling Bituminous Paving Mixtures. American Association of State Highway and Transportation Officials, Washington, DC.

AASHTO T176, 2014. Standard Method of Test for Plastic Fines in Graded Aggregates and Soils by Use of the Sand Equivalent Test. American Association of State Highway and Transportation Officials, Washington, DC.

AASHTO T182, 2014. Static Immersion Test. American Association of State Highway and Transportation Officials, Washington, DC.

AASHTO T190, 2014. Moisture Exudation Test Compacted Soils. American Association of State Highway and Transportation Officials, Washington, DC.

AASHTO T248, 2014. Reducing Samples of Aggregate to Testing Size. American Association of State Highway and Transportation Officials, Washington, DC.

AASHTO T283, 2014. Standard Method of Test for Resistance of Compacted Asphalt Mixtures to Moisture-Induced Damage. American Association of State Highway and Transportation Officials, Washington, DC.

AASHTO T304, 2014. Uncompacted Void Content of Fine Aggregate. American Association of State Highway and Transportation Officials, Washington, DC.

AASHTO T312, 2014. Preparing and Determining Density of Hot Mix Asphalt (HMA) Specimens by Means of the Superpave Gyratory Compactor. American Association of State Highway and Transportation Officials, Washington, DC.

AASHTO T313, 2014. Determining the flexural creep stiffness of asphalt binder using the Bending Beam Rheometer (BBR) Standard Specifications for Transportation Materials and Methods of Sampling and Testing. American Association of State Highway and Transportation Officials, Washington, DC.

AASHTO TP10, 2014. Standard Test Method for Thermal Stress Restrained Specimen Tensile Strength (TSRST). American Association of State Highway and Transportation Officials, Washington, DC.

AASHTO TP56, 2015. Uncompacted Void Content of Coarse Aggregate (As influenced by Particle Shape, Surface Texture, and Grading). American Association of State Highway & Transportation Officials, Washington, DC.
AASHTO TP61, 2015. Determining the Percentage of Fracture in Coarse Aggregate. American Association of State Highway & Transportation Officials, Washington, DC.
Anderson, D.A., Christensen, D.W., Bahia, H.U., Dongre, R., Sharma, M.G., Antle, C.E., et al., 1994. Binder characterization and evaluation volume 3: physical characterization Report No. RP-A-369. Strategic Highway Research Program, National Research Council, Washington, DC.
Asphalt Institute and Eurobitume. 2008. The Bitumen Industry – A Global Perspective: Production, Chemistry, Use, Specification and Occupational Exposure. The Asphalt Institute, Lexington, Kentucky.
ASTM C127, 2014. Standard Test Method for Relative Density (Specific Gravity) and Absorption of Coarse Aggregate Annual Book of Standards. ASTM International, West Conshohocken, PA.
ASTM C128, 2014. Standard Test Method for Relative Density (Specific Gravity) and Absorption of Fine Aggregate Annual Book of Standards. ASTM International, West Conshohocken, PA.
ASTM C131, 2014. Standard Test Method for Resistance to Degradation of Small-size Coarse Aggregate by Abrasion and Impact in the Los Angeles Machine Annual Book of Standards. ASTM International, West Conshohocken, PA.
ASTM C142, 2014. Standard Test Method for Clay Lumps and Friable Particles in Aggregates Annual Book of Standards. ASTM International, West Conshohocken, PA.
ASTM C1252, 2014. Standard Test Methods for Uncompacted Void Content of Fine Aggregate (as Influenced by Particle Shape, Surface Texture, and Grading) Annual Book of Standards. ASTM International, West Conshohocken, PA.
ASTM D4, 2014. Standard Test Method for Bitumen Content Annual Book of Standards. ASTM International, West Conshohocken, PA.
ASTM D979, 2014. Standard Practice for Sampling Bituminous Paving Mixtures Annual Book of Standards. ASTM International, West Conshohocken, PA.
ASTM D1664, 1992. Test Method for Coating and Stripping of Bitumen-Aggregate Mixtures (Withdrawn 1992) Annual Book of Standards. ASTM International, West Conshohocken, PA.
ASTM D2419, 2014. Standard Test Method for Sand Equivalent Value of Soils and Fine Aggregate Annual Book of Standards. ASTM International, West Conshohocken, PA.
ASTM D2844, 2014. Standard Test Method for Resistance *R*-Value and Expansion Pressure of Compacted Soils Annual Book of Standards. ASTM International, West Conshohocken, PA.
ASTM D3625, 2014. Standard Practice for Effect of Water on Bituminous-coated Aggregate Using Boiling Water Annual Book of Standards. ASTM International, West Conshohocken, PA.
ASTM D3665, 2014. Random Sampling of Construction Materials. Section 409—Superpave Mixture Design, Standard and RPS Construction of Plant-Mixed HMA Courses. In: Publication No. 408. Pennsylvania Department of Transportation, 2007, Harrisburg, PA.
ASTM D4123, 2014. Standard Test Method for Indirect Tension Test for Resilient Modulus of Bituminous Mixtures (Withdrawn in 2003 But Still Used in Some Laboratories) Annual Book of Standards. ASTM International, West Conshohocken, PA.
ASTM D4318, 2014. Standard Test Methods for Liquid Limit, Plastic Limit, and Plasticity Index of Soils Annual Book of Standards. ASTM International, West Conshohocken, PA.
ASTM D4791, 2014. Standard Test Method for Flat Particles, Elongated Particles, or Flat and Elongated Particles in Coarse Aggregate Annual Book of Standards. ASTM International, West Conshohocken, PA.

ASTM D5821, 2014. Determining the Percentage of Fractured Particles in Coarse Aggregate Annual Book of Standards. ASTM International, West Conshohocken, PA.
ASTM D6373, 2014. Standard Specification for Performance Graded Asphalt Binder Annual Book of Standards. ASTM International, West Conshohocken, PA.
ASTM D6648, 2014. Test Method for Determining the Flexural Creep Stiffness of Asphalt Binder Using the Bending Beam Rheometer Annual Book of Standards. ASTM International, West Conshohocken, PA.
ASTM D6723, 2014. Test Method for Determining the Fracture Properties of Asphalt Binder in Direct Tension Annual Book of Standards. ASTM International, West Conshohocken, PA.
ASTM D6816, 2014. Standard Practice for Determining Low-temperature Performance Grade (PG) of Asphalt Binders Annual Book of Standards. ASTM International, West Conshohocken, PA.
Azari, H., Lutz, B., Spellerberg, P., 2006. Precision estimates of selected volumetric properties of HMA using absorptive aggregate Final Report, NCHRP Project 9-26, Phase 4.. NCHRP, Transportation Research Board, National Research Council, Washington, DC.
Barth, E.J., 1962. Asphalt: Science and Technology. Gordon and Breach Publishers, Taylor & Francis Group, Abingdon, Oxfordshire.
Bhasin, A., Little, D.N., 2006. Characterizing Surface Properties of Aggregates Used. In: Hot Mix Asphalt. Report No. ICAR/505-2. Texas Transportation Institute, Texas A&M University, College Station, TX and Aggregates Foundation for Technology, Research, and Education, Alexandria, VA.
Bonaquist, R., 2008. Ruggedness testing of the dynamic modulus and flow number tests with the simple performance tester NCHRP Report No. 629. Transportation Research Board, Washington, DC.
Bonaquist, R., Christensen, D.W., Stump, W., 2003. Simple performance tester for superpave mix design: first article development and evaluation NCHRP Report No. 513. Transportation Research Board, Washington, DC.
Bouldin, M.G., Dongre, R., Rowe, G.M., Sharrock, M.J., Anderson, D.A., 2000. Predicting thermal cracking of pavements from binder properties: theoretical basis and field validation. J. Assoc. Asphalt Paving Technol. 69, 455–496.
Breen, J.J., Stephens, J.E., 1967. Report No. JHR-67-16 Proceedings Annual Conference. Canadian Technical Asphalt Association, Halifax Nova Scotia, Canada, October 30–November 1.
Brown, E.R., McRae, J.L., Crawley, A., 1989. Effect of aggregates on performance of bituminous concrete, implication of aggregates in the design, construction, and performance of flexible pavement ASTM Special Technical Publication No. 1016. ASTM International, West Conshohocken, PA.
Brown, E.R., Hainin, M., Cooley, A., Hurley, G., 2003. Relationships of HMA in-place air voids, lift thickness, and permeability NCHRP Project No. 9-27, Vol. II, Part 3—Task 3. National Cooperative Highway Research Program. Transportation Research board, National Research Council, Washington, DC.
Buchanan, M.S., 2000. An evaluation of selected methods for measuring the bulk specific gravity of compacted hot mix asphalt (HMA) mixtures. J. Assoc. Asphalt Paving Technol. 69, 608–634.
Choubane, B., Page, G., Musselman, J., 1998. Investigation of water permeability of coarse graded superpave pavements. J. Assoc. Asphalt Paving Technol. 67, 254.
Christensen, D.W., Bonaquist, R.F., 2006. Volumetric requirements for superpave mix design NCHRP Report No. 567: Final Report for NCHRP Projects 9-25 and 9-31. Transportation Research Board, Washington, DC.
Christensen, D.W., Bonaquist, R., Jack, D.P., 2000. Evaluation of Triaxial Strength as a Simple Test for Asphalt Concrete Rut Resistance. Report No. FHWA-PA-2000-010-97-04

(19), Final Report to the Pennsylvania Department of Transportation by the Pennsylvania Transportation Institute, University Park, PA, August.
Cooley, L.A., Prowell, B.D., Hainin, M.R., 2003. Comparison of the saturated surface-dry and vacuum sealing methods for determining the bulk specific gravity of compacted HMA. J. Assoc. Asphalt Paving Technol. 72, 56–96.
Coplantz, J.S., Yapp, M.T., Finn, F.N., 1993. Review of relationships between modified asphalt properties and pavement performance Report No. SHRP-A-631. Strategic Highway Research Program, National Research Council, Washington, DC.
Cross, S.A., Brown, E.R., 1992. Selection of aggregate properties to minimize rutting of heavy duty pavements—effects of aggregates and mineral fillers on asphalt mixture performance Publication No. ASTM STP 1147. ASTM International, West Conshohocken, PA.
Cross, S.A., Mgonella, M.K., Jakatimath, Y., 2006. Evaluation of test equipment for determination of fine aggregate specific gravity and absorption Transportation Research Record: Journal of the Transportation Research Board, No. 1952. Transportation Research Board, National Research Council, Washington, DC, pp. 3–11.
Dongre, R., D'Angelo, J., 1998. Effect of new direct tension test protocol on the superpave low temperature specification for bitumen binders Proceedings 5th International Conference on Bearing Capacity of Roads and Airfields, BCRA, 2: 1035-1048. Norwegian University of Science and Technology (NTNU), Trondheim, Norway.
Emery, S.J., Van Zyl, G.D., Van Huyssteen, S., Sampson, L., 1992. Performance Limits for Bituminous Surfacings on Low Volume Road. Proceedings. Annual Transportation Conference, Pretoria, South Africa.
FHWA, 2010. A review of aggregate and asphalt mixture specific gravity measurements and their impacts on asphalt mix design properties and mix acceptance TechBrief No. FHWA-HIF-11-033. Office of Pavement Technology, Federal Highway Administration, Washington, DC, December.
Goode, J.F., Lufsey, L.A., 1962. A new graphical chart for evaluating aggregate gradations. Proc. Assoc. Asphalt Paving Technol. 31, 176–207.
Hall, K.D., Griffith, F.T., Williams, S.G., 2001. Examination of operator variability for selected methods for measuring bulk specific gravity of hot mix asphalt concrete. In: Transportation Research Record: Journal of the Transportation Research Board No. 1761, Transportation Research Board, National Research Council, Washington, DC, pp. 81–85.
Halstead, W.J., Rostler, F.S., White, R., 1966. Properties of highway Asphalts—Part III, influence of chemical composition. Proc. Assoc. Asphalt Paving Technol. 35, 91–138.
Harvey, J., Monismith, C., Sousa, J., 1994. An investigation of field- and laboratory-compacted asphalt-rubber, SMA, recycled and conventional asphalt-concrete mixes using SHRP project A-003A equipment. J. Assoc. Asphalt Paving Technol. 63, 511–548.
Huang, B., Mohammad, L., Raghavendra, A., Abadie, C., 1999. Fundamentals of permeability in Asphalt mixtures. J. Assoc. Asphalt Paving Technol. 68, 479–496.
Kandhal, P.S., Parker Jr., F., 1998. NCHRP Report 405: Aggregate Tests Related to Asphalt Concrete Performance in Pavements. Transportation Research Board, National Research Council, Washington, DC.
Kandhal, P.S., Mallick, R.B., Huner, M., 2000. Measuring bulk specific gravity of fine aggregates: development of new test method Transportation Research Record: Journal of the Transportation Research Board, No. 1721. Transportation Research Board, National Research Council, Washington, DC, pp. 81–90.
Kennedy, T.W., Roberts, F.L., Lee, K.W., 1984a. Evaluating moisture susceptibility of asphalt mixtures using the texas boiling test Transportation Research Record No. 968. Transportation Research Board, National Research Council, Washington, DC, pp. 45–54.
Kennedy, T.W., Roberts, F.L., Anagnos, J.N., 1984b. Texas Boiling Test for Evaluating Moisture Susceptibility of Asphalt Mixtures. Report No. FHWA/TX-85/63 + 253-5.

Center for Transportation Research The University of Texas at Austin, Austin, TX and Transportation Planning Division, Texas State Department of Highways and Public Transportation, Austin, TX.
Kim, K.W., Burati, J.L., 1993. Use of GPC chromatogram to characterize aged asphalt cement. J. Mater. Civil Eng. ASCE 5 (1), 41–52.
Kim, K.W., Burati, J.L., Park, J.S., 1995. Methodology for defining LMS portion in asphalt chromatogram. J. Mater. Civil Eng. ASCE 7 (1), 31–40.
Kim, Y., Little, D.N., 2005. Development of specification type tests to assess the impact of fine aggregate and mineral filler on fatigue damage Report No. FHWA/TX-05/0-1707-10. Research and Technology Implementation Office, Texas Department of Transportation, Austin, TX.
Leahy, R.B., Witczak, M.W., 1991. The influence of test conditions and asphalt concrete mix parameters on permanent deformation coefficients alpha and mu. J. Assoc. Asphalt Paving Technol. 60, 333.
Lee, D.Y., Guinn, J.A., Kandhal, P.S., Dunning, R.L., 1990. Absorption of asphalt into porous aggregates Publication No. SHRP-A/UIR-90-009. Strategic Highway Research Program, National Research Council, Washington, DC.
Malpass, G., Khosla, N.P., 2001. Evaluation of gamma ray technology for the direct measurement of bulk specific gravity of compacted asphalt concrete mixtures. J. Assoc. Asphalt Paving Technol. 70, 352–367.
McLeod, N.W., Davidson, J.K., 1981. Particle index evaluation of aggregates for asphalt paving mixtures. Proc. Assoc. Asphalt Paving Technol. 50, 251.
Mohammad, L., 2006. Performance tests for HMA including fundamental and empirical procedures Publication No. STP 1469. ASTM International, West Conshohocken, PA.
Moschopedis, S.E., Speight, J.G., 1973. The oxidation of petroleum fractions. Fuel 52, 83.
Moschopedis, S.E., Speight, J.G., 1975. Oxidation of bitumen. Fuel 54, 210.
Moschopedis, S.E., Speight, J.G., 1977. The effect of air blowing on the properties and constitution of natural bitumen. J. Mater. Sci. 12, 990.
Moschopedis, S.E., Speight, J.G., 1978. The influence of metal salts on bitumen oxidation. Fuel 57, 235.
Newcomb, D.E., 2001a. Statistical specifications for hot mix asphalt: what do we need to know? Hot Mix Asphalt Technology. National Asphalt Pavement Association, Lanham, MD, January/February: pp. 54–60.
Newcomb, D.E., 2001b. Statistical specifications for hot mix asphalt: what do we need to know? Hot Mix Asphalt Technology. National Asphalt Pavement Association, Lanham, MD, March/April: pp. 37–40.
Newcomb, D.E., 2001c. Performance related specifications developments Hot Mix Asphalt Technology. National Asphalt Pavement Association, Lanham, MD, May/June: pp. 5–37.
Parker, F., Gharaybeh, F., 1987. Evaluation of indirect tensile tests for assessing stripping of alabama asphalt concrete mixtures Transportation Research Record No. 1115. Transportation Research Board, Washington, DC.
Patnaik, P. (Ed.), 2004. Dean's Analytical Chemistry Handbook, second ed. McGraw-Hill, New York, NY.
Prowell, B.D., Baker, N.V. 2004. Evaluation of new test procedures for determining bulk specific gravity of fine aggregate by automated methods. In: Transportation Research Record: Journal of the Transportation Research Board, No. 1874, Transportation Research Board, National Research Council, Washington, DC, pp. 11–18.
Roque, R., Hiltunen, D.R., Buttlar, W.G., 1995. Thermal cracking performance and design of mixtures using superpave. J. Assoc. Asphalt Paving Technol. 64, 718–733.
Schmidt, R.J., 1996. The relationship of the low temperature properties of asphalt to the cracking of pavements. Proc. Assoc. Asphalt Paving Technol. 35, 263–269.

Schmidt, R.J., Santucci, L.E., 1996. A practical method for determining the glass transition temperature of asphalts and calculation of their low temperature viscosities. Proc. Assoc. Asphalt Paving Technol. 35, 61–90.

Sholar, G.A., Page, G.C., Musselman, J.A., Upshaw, P.B., Moseley, H.L., 2005. Investigation of the CoreLok for Maximum, Aggregate, and Bulk Specific Gravity Tests. Transportation Research Record: Journal of the Transportation Research Board, No. 1907. Transportation Research Board, National Research Council, Washington, DC, pp. 135–144.

Speight, J.G., 1978. The use of bitumen as an adhesive and building material at the ancient cities of Babylon, Calah, Nineveh, and Ur Report of Investigations. College of Science, University of Mosul, Iraq.

Speight, J.G., 1992. Asphalt. In: Kirk–Othmer's Encyclopedia of Chemical Technology fourth ed., 3: 689–724.

Speight, J.G., 2013. The Chemistry and Technology of Coal, third ed. CRC Press, Taylor & Francis Group, Boca Raton, FL.

Speight, J.G., 2014. The Chemistry and Technology of Petroleum, fifth ed. CRC Press, Taylor & Francis Group, Boca Raton., FL.

Speight, J.G., 2015a. Handbook of Petroleum Product Analysis, second ed. John Wiley & Sons Inc., Hoboken, NJ.

Speight, J.G., 2015b. Handbook of Coal Analysis, second ed. John Wiley & Sons Inc., Hoboken, NJ.

Spellerberg, P., Savage, D., Pielert, J., 2003. Precision estimates of selected volumetric properties of HMA using non-absorptive aggregate Publication No. NCHRP 54. Transportation Research Board, National Research Council, Washington, DC.

TRB, 2004. New simple performance tests for asphalt mixes Circular No. E-C068. Transportation Research Board, Washington, DC.

TRB, 2010. Development in asphalt binder specifications Transportation Research Circular E-C147. Transportation Research Board, Washington, DC.

TRB, 2012. Alternative binders for sustainable asphalt pavements Transportation Research Circular E-C165. E-C147. Transportation Research Board, Washington, DC.

White, T.D., Haddock, J.E., Rismantojo, E., 2006. Aggregate tests for hot-mix asphalt mixtures used in pavements NCHRP Report No. 557. Transportation Research Board, Washington, DC.

Williams, S.G., 2007. Bulk specific gravity of measurements of 25.0 mm and 37.5 mm Coarse-Graded superpave mixes Proceedings 86th Annual Meeting. Transportation Research Board, National Research Council, Washington, DC.

Zaniewski, J.P., Srinivasan, G., 2004. Evaluation of Indirect Tensile Strength to Identify Asphalt Concrete Rutting Potential. Department of Civil and Environmental Engineering, West Virginia University Morgantown, WV.

CHAPTER 6
Asphalt Chemistry

6.1 INTRODUCTION

The history and use of asphalt have been based on a variety of sources and most of these sources have been petroleum-based, in the form of distillation residua or natural seepages where the more volatile constituents of petroleum have evaporated over time (see Chapter 1). In the latter case and in the absence of evidence to the contrary, the natural resource is typically referred to as bitumen or in some cases (incorrectly) as natural asphalt (see Chapter 1).

Asphalt is a major product of many refineries, forms the basis of a major industry, and is one of the indispensable materials of life in many parts of the world. In fact, 94% of the paved roads in the United States are surfaced with asphalt. In order for the industry to progress and evolve, it is necessary to understand the issues associated with the technology surrounding the production and utilization of asphalt, including loading and transporting quarried materials, production of the bituminous product, and the associated health, safety, and environmental issues.

Asphalt pavements vary in longevity because of the various factors that cause deterioration of the chemical and physical association of the binder to the aggregate. As soon as the freshly laid hot asphalt pavement mix begins to cool, so also begins the aging process which will eventually initiate and enhance the deterioration of the asphalt. When oxygen and water, which constantly attack the pavement, start to combine with the asphaltic binder of the pavement, a chemical change takes place. At first, this process is necessary for the pavement to become hard and firm. Later, if this process is not arrested, a complete deterioration of the asphaltic binder will take place and reduce the pavement to a layer of loose stone. The asphalt binder is essentially what differentiates a gravel road from a paved road.

The destruction rate of an asphalt pavement varies: heat and sunlight will accelerate the deterioration process and salt will act as a catalyst. Water which penetrates the surface can get into the base course to cause trouble, and at the same time start oxidizing the asphaltic binder inside the pavement. A more dense mix will not oxidize as rapidly as a coarse,

Asphalt Materials Science and Technology.
DOI: http://dx.doi.org/10.1016/B978-0-12-800273-5.00006-4

open pavement. A thicker lift of pavement will keep its life longer than a thin one. Automobile and truck traffic create the danger of gasoline and oil spillage, which will soften the asphalt and lead to its breakup. A cold climate enhances the danger of water being caught in the pavement and freezing. This can break individual pieces of aggregate apart, or it can break out large sections.

The issue relating to pavement deterioration is that asphalt is a complex combination of organic compounds containing a high proportion of hydrocarbonaceous molecular species with high carbon numbers. Thus, asphalt is a complex material that contains a variety of functional groups containing heteroatoms (nitrogen, oxygen, and sulfur) as well as metals such as vanadium, nickel, iron, and copper. Although distributed throughout the asphalt, the heteroatoms are primarily located in the asphaltene fraction (Speight, 2014). It is these heteroatom constituents that play an important role in the development and existence of asphalt–aggregate interactions. In fact, when asphalt is brought into contact with the aggregate the oxygen-containing groups from the asphaltene constituents are preferentially adsorbed on the aggregate surface. Furthermore, understanding the chemistry and microstructure of asphalt is critical in the development and improvement of performance-based specifications—the chemistry of the asphalt–aggregate bond is an important aspect of asphalt technology.

Premature failure or poor performance of asphalt roadways results from either the weakening of the adhesive bond between the asphalt binder and the aggregate or because of mechanical stresses or aging of the asphalt binder. For example, the environmental factors that cause asphalt oxidation and moisture-associated effects vary across the United States and across many other countries. The conditioning schemes used in the specifications need to reflect the environment more accurately. Therefore, existing models must be modified and new models must be developed to reflect the effects of moisture on asphalt pavements.

Asphalt modification recently underwent a major transformation. Polymers composed of functional groups that are compatible with asphalt have been formulated. Other traditional polymers that are incompatible with asphalt are being chemically bonded with the asphalt constituents. These grafted polymers offer better storage, handling, and in-place stability characteristics than asphalt alone. Future application of polymers may entail matching polymer properties with anticipated in-service physical requirements (e.g., thin overlays, overlays on concrete, or full-depth pavements).

Most crucial is the interaction of asphalt and aggregate, particularly with the fines, which contribute the overwhelming majority of surface area in a mixture. To adequately address these issues, non-destructive techniques must be developed for evaluating binder properties in loose mix, cores, and even pavements.

6.2 COMPOSITION

Asphalt, by a very general definition, is the carbonaceous product that is produced from petroleum residua. It is unusual for a residuum produced by distillation (see Chapter 2) to meet the specifications required for asphalt (see Chapters 3 and 4). In addition, the precursor to asphalt may be blends of atmospheric residua, vacuum residua, deasphalter residua (deasphalter bottoms), and visbreaker residua (visbreaker bottoms, visbreaker tar). The disadvantage of using visbreaker residua is the presence of cracked products that are unstable and can have an adverse effect on asphalt performance. At one time, coal tar was a major constituent of road asphalt but the instability of the tar and the current environmental regulations that restrict the use of carcinogenic polynuclear aromatic compounds (in coal tar) have brought restrictions on its use.

6.2.1 General Aspects

Asphalt is the residue of mixed-base and asphalt-base crude oils and cannot be distilled even under the highest vacuum, because the temperatures required to do this promote formation of coke. Asphalt has complex chemical and physical compositions that usually vary with the source of the crude oil and are considered dispersions of particles (asphaltene constituents), in a high-boiling fluid composed of oil and resin constituents (Traxler, 1961; Barth, 1962; Hoiberg, 1964; Broome, 1973; Broome and Wadelin, 1973; Koots and Speight, 1975; Speight, 1992a; Kassir, 2009; Speight, 2014, 2015a).

On a physicochemical basis, paving asphalts are colloidal dispersions of high-molecular-weight aromatic hydrocarbonaceous species (the asphaltene constituents) dispersed in a continuous phase of lower-molecular-weight oily-resinous (paraffins and aromatic resins) material called maltenes. The asphaltenes tend to be insoluble in the oily constituent(s). They exist in asphalts as fine to coarse dispersions due to their peptization by the more aromatic resin constituent(s). The resins are considered to be part of the oily medium or in solution in it, but due to an aromaticity higher than the oily medium itself, they are adsorbed more readily by the asphaltenes than the remaining oils. In the context of asphalt being a colloidal material (Eilers, 1949), this adsorbed material is the peptizing agent or colloidal stabilizer (Koots and Speight, 1975).

The components of asphalt are classified into four general (but not chemical) classes of compounds: (i) saturate constituents, which correlate with the softening point of the material, (ii) aromatic and naphthene constituents, which consist of partially hydrogenated aromatic compounds, (iii) resin constituents, which are in some cases referred to as polar aromatic constituents, which contain a variety of functional groups, and (iv) asphaltene constituents, which consist of high-molecular-weight polar (exo-cyclic) functional groups as well as heterocyclic compounds; nickel also occur in the asphaltene fraction (ASTM D4124; Speight, 1992a,b, 2014, 2015a). Other fractionation schemes are available (Figure 6.1) (Corbett, 1969; Kassir, 2009) but it must not be forgotten that the composition nomenclature of the various fractions is based on the method of separation and typically does not refer to the chemical constituents of the fractions.

Asphalt and naturally occurring bitumen are often incorrectly named (see Chapter 1) and have been often confused with coal tar, which is a

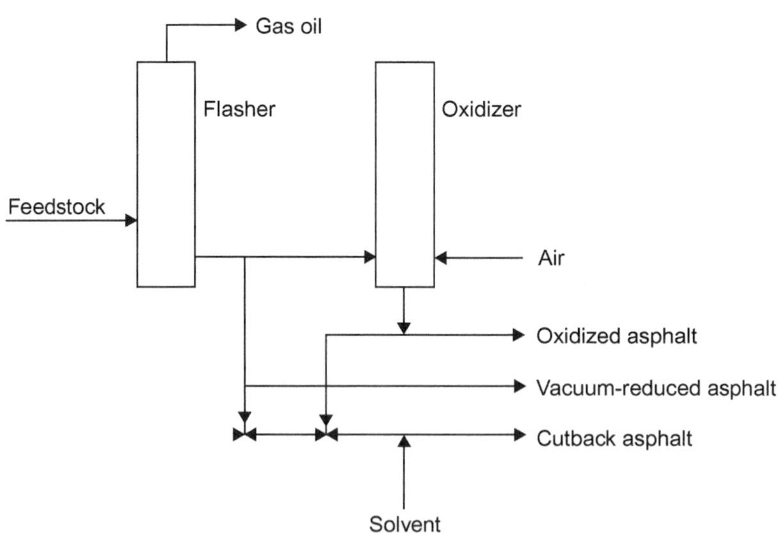

Figure 6.1 General representation of asphalt production in a refinery. The feedstock can be a residuum from the atmospheric distillation tower or from the vacuum distillation tower.

visually similar black, thermoplastic material produced by the destructive distillation of coal (see Chapter 1) (Speight, 2013, 2014). During the early-to-mid twentieth century when town gas (Speight, 2013) was produced, coal tar was a readily available by-product and extensively used as the binder for road aggregates. The addition of tar to macadam roads led to the word tarmac, which is now used in common parlance to refer to road-making materials and to airport runways when the position of an aircraft is often described as being *on the tarmac.* However, since the mid-to-late twentieth century, when the use of natural gas eventually succeeded the use of town gas, thereby reducing the need for coal as a source of gas (and the accompanying tarry by-products), petroleum-derived asphalt has completely overtaken the use of coal tar in applications related to the use of asphalt. Other examples of this confusion are inclusion of the bitumen of the La Bea Tar Pits in Los Angeles and the Canadian tar sands in Alberta within the *tar nomenclature.* In addition, pitch is another term sometimes used at times to refer to asphalt and bitumen, as in the Pitch Lake found on the island of Trinidad in the Republic of Trinidad and Tobago. Many other mislabeled examples exist through the world.

Mislabeling notwithstanding (see Chapter 1), all types of asphalt and bitumen contain at least three of the four fractions mentioned above, i.e., saturate, aromatics, resin constituents, and asphaltene constituents, which determine the properties of the material. Some asphalt products, especially the liquid products, may contain zero-to-low amounts of asphaltene constituents. Thus, the nature of the asphalt is determined by such factors as the nature of the medium (paraffinic or aromatic), as well as the nature and proportion of the asphaltenes and of the resins (Pauli and Branthaver, 1998). The asphaltenes have been suggested to be lyophobic; the resins are lyophilic, and the *interaction* of the resins with the asphaltenes is responsible for asphaltene dispersion, which seems to exercise marked control on the nature of the asphalt. The asphaltenes vary in character but are of sufficiently high molecular weight to require dispersion as micelles, which are peptized by the resins. If the asphaltenes are relatively low in molecular weight, the resins plentiful, and the medium aromatic in nature, the result may be viscous asphalt without anomalous properties. If, however, the medium is paraffinic and the resins are scarce, and the asphaltenes are high in molecular (or micellar) weight (these conditions are encouraged by vacuum, steam reduction, or air blowing), the asphalt is of the gel type and exhibits the properties that accompany such structure. A high content of resins imparts to a product desirable adhesive character and plasticity;

high asphaltene content is usually responsible for the harder, more brittle, asphalt as evidenced from the structure and rheological properties of modified asphalt (Giavarini et al., 2000).

6.2.2 Composition Maps

Asphalt is composed of a variety of hydrocarbonaceous molecular species, it is the relative proportions of the individual molecular species that dictate the character of the asphalt and present indications of the types of aggregate that are necessary to manufacture the desired asphalt mix (asphalt concrete). In addition, the composition of asphalt can be viewed as consisting of two continuous distributions, one of molecular weight and the other of molecular type. Early attempts to investigate the composition of petroleum and petroleum-derived products such as asphalt, involved subdivision into fractions conveniently named *saturates*, *aromatics*, and *resins* (sometime called *polar aromatics*) by adsorption chromatography after a preparation of the *asphaltene constituents* by use of a low-boiling liquid hydrocarbon (Speight, 2014, 2015a).

As more assiduous fractionation techniques evolved, variation of heteroatom content with the fractions allowed petroleum to be mapped on the basis of boiling point and carbon number. Composition maps became more detailed and the usefulness of the data was recognized insofar as they show that not only can every molecule in the resins and asphaltenes have more than one sulfur or nitrogen molecule but also some molecules probably contain both sulfur and nitrogen—which play an important role in asphalt–aggregate bonding and also in terms of environmental issues related to asphalt use (see Chapter 12). As the molecular weight of the aromatic fractions decreases, the sulfur and nitrogen contents of the fractions also decrease. In contrast to the sulfur-containing molecules, which appear in both the naphthene aromatics and the polar aromatics fractions, the oxygen compounds present in the heavy fractions of petroleum are normally found in the polar aromatics fraction. However, this (the oxygen distribution) can change (depending upon the source of the asphalt) through the agency of the air-blowing operation.

Fractionation work (Long and Speight, 1989, 1998; Speight, 2014, 2015a) involved the development of a compositional map (using the molecular weight distribution and the molecular type distribution as coordinates) that can also be applied to the investigation of the behavior of asphalt. The separation involved the use of an adsorbent such as clay, and the fractions were characterized by solubility parameter as a measure of the polarity of the molecular types. The molecular weight distribution can be determined by gel permeation chromatography. Using these two distributions, a map of composition can be prepared using molecular weight and solubility parameter as the coordinates for plotting the two distributions.

The molecular type was characterized by the polarity of the molecules, as measured by the increasing adsorption strength on an adsorbent. At the time of the original concept, it was unclear how to characterize the continuum in molecular type or polarity. For this reason, the molecular type coordinate of the first maps was the yield of the molecular types ranked in order of increasing polarity. The horizontal distance on the plot is a measure of the yield and there is not a continuous variation in polarity for the horizontal coordinate. It was suggested that the solubility parameter of the different fractions could be used to characterize both polarity and adsorption strength.

In order to attempt to remove some of these potential ambiguities, more recent developments of this concept have focused on the solubility parameter. The simplest map that can be derived using the solubility parameter is produced with the solubility parameters of the solvents used in solvent separation procedures, equating these parameters to the various fractions. However, the solubility parameter boundaries determined by the values for the eluting solvents that remove the fractions from the adsorbent offer a further step in the evolution of petroleum maps. Maps were obtained using three solvents, cyclohexane, toluene, and 5% water in tetrahydrofuran, that are typical solvents for chromatographic separation of saturates, aromatics, and polar constituent. Elution from the adsorbent using more complex solvent sequences yields a somewhat more detailed map in terms of the solubility parameter relationships of the various fractions. Furthermore, the solubility parameters of the individual eluted fractions in the eight-solvent case have also been measured and agree with these values, and the method gives an effective continuum in polarity for the horizontal scale.

Measuring the overall solubility parameter of a petroleum fraction is a time-consuming chore. Therefore, it is desirable to have a simpler, less time-consuming measurement that can be made on petroleum fractions which will correlate with the solubility parameter and thus give an alternative continuum in polarity. In fact, the hydrogen-to-carbon atomic ratio and other properties of petroleum fractions that can be correlated with the solubility parameter (Speight, 2014, 2015a) also provide correlations for the behavior of the feedstock under investigation.

Thus, a composition map can be used to show where a particular physical or chemical property tends to concentrate on the map. This can be followed by investigating changes in the structure of the map that are caused by the air-blowing process, as well as changes when the asphalt is mixed with the aggregate.

6.3 REACTIVITY

Asphalt is a complex mixture of hydrocarbon and heteroatom constituents (Abraham, 1945; Barth, 1962; Hanson, 1964; Speight, 1992a,b, 2014). The complex composition is reflected in the properties and behavior of asphalt due to the presence of functional groups, particularly functional groups based on basic nitrogen and acidic oxygen. The reactions of asphalt usually focus on oxidation, which serves to introduce even more oxygen-based functional groups into the asphalt (Corbett, 1975; Moschopedis and Speight, 1973, 1975, 1977, 1978; Speight, 1992a,b; WHO, 2004) and is the common reaction that causes pavement failure. The overall reaction results in increases in the softening point and asphaltene content which increase linearly with time during the blowing operation (Kleinschmidt and Snoke, 1958), leading to failure of the asphalt pavement (Abraham, 1945; Barth, 1962; Hanson, 1964; Speight, 1992a,b, 2014). The mechanism of the oxidation is partially resolved, but much about the mechanisms remains to be uncovered if oxidative failure of asphalt pavements is to be mitigated, understanding that each asphalt binder may interact differently with aerial oxygen.

6.3.1 Paving Asphalt

The constituents of asphalt concrete include the asphalt binder and the mineral aggregate, along with any mineral filler and additives. The asphalt binder is typically derived from crude oil as a by-product of fractional distillation (Figure 6.2). Due to its organic nature asphalt binder undergoes oxidative aging as time progresses, the effect of which is most prominently in the form of hardening or stiffening. The effect of aging creates graded material properties due to variation in the amount of aging across the depth of pavement. Thus, the oxidation (and, hence oxidative aging or oxidative hardening) of asphalt is of practical significance to the understanding of pavement service life (Branthaver et al., 1993).

The oxidation of the asphalt binder in pavements is an ongoing process throughout the service life of the pavement and understanding the nature

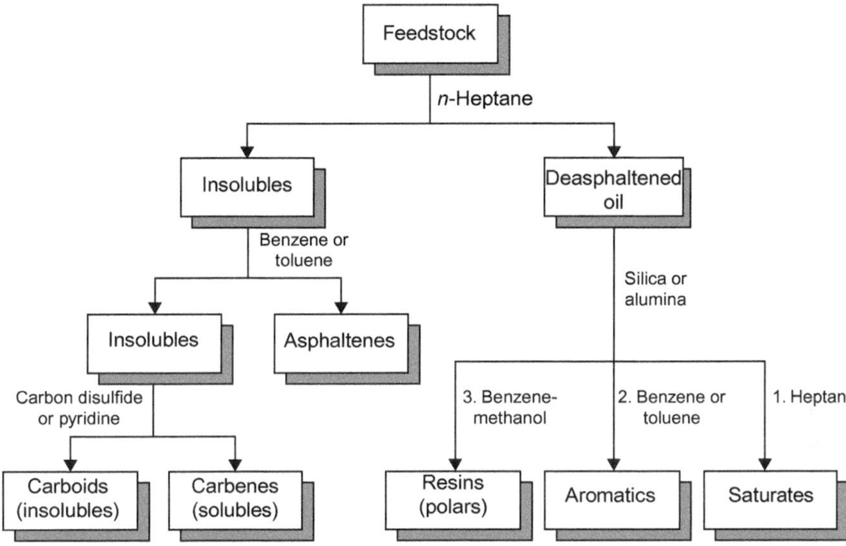

Figure 6.2 One option for the separation of asphalt into bulk fractions. Typically, the fractions named carbenes and carboids usually refer to refinery thermal products but can (in the case of asphalt) be oxidation products of the asphaltene constituents and the resin constituents.

of the oxidation process is a critical step toward better pavement design to achieve greater pavement durability.

Durability is a measure of how asphalt binder physical properties change with age (sometimes called age hardening). In general, as an asphalt binder ages, its viscosity increases and it becomes more stiff and brittle.

The main component in asphalt binder oxidation in pavements is binder oxidative hardening, which is due to oxidation of the aromatic constituents. As the aromatic compounds in asphalt binders are oxidized, more polar carbonyl compounds are created, which results in stronger intermolecular associations between the asphalt constituents and eventually leads to an increase in viscosity. Consequently, the performance of pavements is affected directly by asphalt binder hardening.

Also, low levels of accessible air voids in pavements potentially relate to binder oxidation. When the asphalt pavement has a sufficient number of highly accessible air voids, the oxidation rate is largely determined by the temperature in the pavement. On the other hand, when the percentage of accessible air voids in the pavement is considerably lower, the hardening rate of binders in pavements is reduced significantly.

Thus, oxidative behavior of asphalt is one of the critical factors contributing to the performance of asphalt pavements. Indeed, optimum performance properties are highly dependent on the flow—or more precisely the rheological properties—of the asphalt; thus, changes in flow properties with time are usually highly undesirable and often lead to reduced product performance or even failure. Because asphalt is an organic product, it is subject to chemical oxidation by reaction with atmospheric oxygen. Furthermore, asphalt hardening can be divided into two stages: (i) short-term hardening, which occurs during the mix production and (ii) construction and *in situ* field aging which occurs during the service life of a pavement (Mirza and Witczak, 1996).

Asphalt oxidation is of pragmatic importance because it leads to the hardening of the asphalt, resulting in a deterioration of desirable physical properties. In asphalt pavements, oxidative age hardening contributes significantly to pavement embrittlement, eventually resulting in excessive pavement cracking. In asphalt roofing materials, embrittlement from oxidative hardening can promote the loss of protective granules, substrate shrinkage, and cracking. Oxidative hardening is attributed primarily to the introduction of polar, oxygen-containing chemical functionalities on asphalt molecules causing increased molecular interactions. Some aromatization of asphalt molecules may also result from oxidation. These changes in chemical composition would be expected to increase asphalt hardening.

The sensitivity of asphalt to oxidative age hardening varies widely with asphalt source (chemical composition); thus, the durability and performance of asphalt are highly dependent on its chemical composition (Hveem et al., 1959). Thus, although a variety of fractionation schemes have been used to separate asphalts into generic fractions, perhaps the scheme most widely used in the past has involved separation of asphalt into four major fractions (Corbett, 1969; Speight, 2014, 2015a), which are separated into schemes in the order of their increasing molecular polarity including saturates, aromatics, resin constituents, and asphaltene constituents (Figure 6.1). Because of the relatively low chemical reactivity of the saturates fraction, it is highly resistant to ambient air oxidation. Oxidative changes in the other three fractions generally involve a movement of components from the more non-polar fractions to the more polar fractions as oxygen-containing functional groups are formed in the asphalt molecules (Corbett and Merz, 1975). Because the various fractions have different reactivity toward oxidation, the result is usually a net loss of naphthene aromatics, and possibly a net loss in polar aromatics, with a corresponding increase in the asphaltenes fraction. Relative percentage changes in only the amounts of the fractions by oxidation provide only limited understanding, and chemical changes are hard to interpret because mass changes alone provide limited information about the actual chemical changes taking place within the fractions (Petersen, 2009).

Finally there must be acknowledgment of the thermal properties of asphalt and the various constituents. The most likely constituents that exhibit temperature susceptibility are the resin constituents and the asphaltene constituents (see Chapter 7). Such temperature susceptibility is unlikely to manifest itself during service but the exposure of the asphalt to excessive heat in the asphalt mix plant (by injection of the asphalt on to the aggregate in the heated mixer may be sufficient to cause irreversible changes to the binder and the loss of necessary and desirable properties (see Chapter 11).

6.3.2 Asphalt Emulsion

The three chemical components of emulsified asphalt, that is, asphalt, water, and emulsifier, joined together by the chemical influence of a colloid mill, are the necessary ingredients in the chemistry and related production of paving asphalt emulsions (Dybalski, 1976). Furthermore, the one physicochemical property of paving asphalts that most influences the preparation of stable emulsions is the degree to which the asphaltene fraction is peptized by the maltene fraction into stable, colloidally, and uniformly dispersed micelles.

The emulsifier concentration that provides the smallest and most uniform particle size is selected as optimum for the asphalt in question. Repeating the procedure with asphalts from varying crude oil sources and processing units will allow the investigator to differentiate poor emulsion base asphalts from those that are acceptable. A typically acceptable emulsion-base asphalt using an optimum concentration of emulsifier provides an average particle size and distribution of 28% at $<1\,\mu m$, 57% at $1–5\,\mu m$, and 15% at $5–10\,\mu m$.

While water might be considered the least important of the asphalt emulsion components, water can possess chemical ionic constituents that can either be beneficial or detrimental to the formation of a stable asphalt emulsion. The presence of calcium and magnesium ions (Ca^{2+}, Mg^{2+}) can benefit the formation of a stable cationic asphalt emulsion. Calcium chloride is often added to cationic formulations to enhance storage stability. Calcium and magnesium ions could be detrimental to the formation of anionic emulsion formulations because of the formation or water-insoluble calcium and magnesium salts due to reaction with the water-soluble sodium or potassium organic acid salts normally used as anionic emulsifiers.

The presence of carbonate and bicarbonate ions (CO_3^{2-} and HCO_3^{-}) could be detrimental to the formation of a stable cationic emulsion, particularly the rapid setting type of cationic emulsion. The more water-insoluble salt will tend to form due to reaction with the water-soluble amine hydrochlorides normally used as a cationic emulsifier. By contrast the presence of carbonate and bicarbonate ions could be beneficial to the formation of an anionic emulsion because of their buffering effect. The presence of particulate matter in water could also be detrimental to the formation of a cationic emulsion due to rapid adsorption of cationic surfactants by the usually negatively charged particulate matter. This electrochemical adsorption tends to reduce the amount of cationic surfactant originally in solution, which can affect initial emulsification and performance properties of the finished emulsion.

A suitable emulsion-base asphalt has been described as a sol-type emulsion which possesses a naturally or synthetically peptized asphaltene constituent which provides colloidally and uniformly dispersed asphaltene constituents rather than the insoluble clusters that have a high propensity to separate as a distinct solid phase. Asphalt emulsions are useful and effective materials for road building when used properly. However, the user should be aware that the mechanism of use is different from hot mix asphalt and cutback asphalt. The molten asphalt in hot mix asphalt and the solvent in cutback asphalt provide the lubrication necessary for the mixing process. By contrast, emulsions are physicochemical in action and require that water is added to the aggregate prior to the addition of the emulsion to provide the necessary carrier for uniform distribution of the emulsified asphalt particles in cold mix asphalt.

6.4 ASPHALT–AGGREGATE BOND

In a road pavement, aggregate composes 94–95% (w/w) of the mix while asphalt constitutes the remaining 5–6% (w/w). Asphalt–aggregate interactions in asphalt pavements directly influence the adhesion of asphalt (the binder) to the aggregate. These interactions are directly related to the bond strength between the asphalt and the aggregate and also determine the efficiency of the asphalt–aggregate interaction in the presence of water.

Briefly, asphalt is a dark-brown to black cementitious material, solid or semisolid in consistency, in which the predominating constituents are high-molecular-weight, polar hydrocarbonaceous species (Speight, 1992a,b, 2014). Beyond that general statement, asphalt has a complex and poorly understood chemistry and structure. The chemistry and structure of asphalt depend on the source and any chemical treatment or chemical modifiers added during processing (Wolfe et al., 1986). Asphalt has a large number of heteroatomic groups with a wide range of chemical reactivity (Petersen et al., 1981; Speight, 1992a,b, 2014).

The types of functional groups that can be found in asphalt include polynuclear aromatic derivatives, phenol derivatives, 2-quinoline derivatives, pyrrole derivatives, pyridine derivatives, sulfide derivatives, sulfoxide derivatives, anhydride derivatives, carboxylic acid derivatives, and ketone derivatives. The number and distribution of these groups vary widely among different asphalt samples and are prime movers in determining the chemical behavior and performance of asphalt, such as sensitivity to moisture damage (Wolfe et al., 1986; Benedetto et al., 1970; Haxo, 1976; Petersen et al., 1982; Brule et al., 1986). Knowledge of the structure and components of asphalt which play a role in determining the strength of the asphalt–aggregate interactions is essential in understanding performance.

Asphalt–aggregate interactions are important in the adhesion of asphalt cement to aggregate because the asphalt must adhere to the aggregate for the adhesive binding action of asphalt to occur. Any additives that are added to improve the performance of the pavement are typically added to the asphalt. The aggregate is present in a multiplicity of sizes ranging from 3/4 in fraction to fines that are in the − 200 mesh range. The larger aggregate sizes are distinct entities in the asphalt pavement while the finer particles can be visualized as an extended portion of the asphalt itself.

As asphalt contacts the aggregate and forms the mix that is used for road asphalt, the asphalt molecules interact directly with the active sites on the aggregate surface. However, the surfaces of the aggregates vary considerably in their chemistry, surface area, pore-size distributions, and rugosity (having a rough, ridged, or wrinkled surface). Hence, the aggregate can be minerals such as limestone, gravel, granite, greywacke (hard gray sandstone), basalt can have markedly different active surface sites that are available for interaction with the molecular species of the asphalt. Furthermore, asphalt, being produced from petroleum residua, also varies considerably in the chemistry, because of the differences in the chemical composition of the source petroleum. Hence, the interaction between the asphalt and aggregate is dependent on the chemistry of both the asphalt and the aggregate.

Aggregate chemistry plays a key role in the adsorption of asphaltic components, the adhesion of the asphalt to the aggregate, and the retention of the asphaltic components in the presence of water. Each aggregate of a given bulk mineralogical type has a unique surface chemistry and the various active sites promote adsorption of asphaltic constituents, particularly constituents with polar functional groups. The covering of those active sites by non-polar hydrocarbons or less polar heteroatom constituents completely masks their activity, while changing the chemistry of the sites by adding amine groups, affects the adsorption behavior of asphalt. Dust coatings occurring naturally on aggregate surfaces can change the chemistry of adhesion and result in weak bonding between the dust and aggregate surface, leading to attrition of the bonding forces that help maintain the adhesion and, hence, the pavement. Thus, many factors influence the strength and longevity of the bond between asphalt and aggregate and include factors based on environmental chemistry and traffic.

For example, water, in the form of rainfall or humidity, can have a direct and insidious effect on the adhesive bond between asphalt and aggregate. Moisture damage to the asphalt–aggregate bond results in the deterioration of the bond through adhesive failure at the interface or cohesive failure within the asphalt or aggregate. These failures cause the asphalt to separate or strip from the aggregate, leaving the aggregate without a binder to keep the aggregate particles and, hence, the pavement together.

In addition, to the effects of water and traffic load, oxidative aging leads to the introduction of functional groups such as carboxylic acids, ketones, and sulfoxides (Corbett, 1975; Moschopedis and Speight, 1973, 1975, 1977, 1978). The number and type of oxidative are directly related to the compositional chemistry of the asphalt. Furthermore, the changes caused by oxidative aging can affect the nature of the chemistry of the interface since both sulfoxides and carboxylic acids have a high affinity for the aggregate surface. Hence, oxidative aging may produce substantial changes in the chemistry of the asphalt–aggregate interface, particularly with an asphalt–aggregate pair that is particularly susceptible to aging. The adhesion of the asphalt to the surface is dependent upon the molecular types at the asphalt–aggregate interface and their ability to bond strongly to the surface. The resistivity of that bond to environmental factors, particularly the intrusion of water, is a factor that determines whether or not the pavement will exhibit long pavement life. Since several of the functional groups present after oxidative aging are susceptible to water, the resistivity of the asphalt–aggregate bond will be weakened by the presence of water.

Stripping of asphalt from the aggregate arises when there is intrusion of water into the asphalt–aggregate system. The modes of failure depend upon the character of the system and include: (i) separation of the bond at the interface, (ii) molecular changes in the asphalt and aggregate due to environmental and traffic loading effects, (iii) phase separation of components due to oxidative changes in the asphalt, and, last but not least, (iv) when the presence of water increases the solubility of polar compounds through hydrogen bonding.

If the waterproofing layer of asphalt surrounding an aggregate particle is continuous, then water can penetrate the system by diffusing through the asphalt film, removing along the way those asphaltic components that are solubilized. If cracks occur in the film, then water can intrude to the asphalt–aggregate interface, causing failure at or near the interface. The failure can be interfacial or cohesive either in the asphalt or in the aggregate. A reduction in water damage can be attained by modifying the aggregate surface through chemical treatment or the addition of antistripping agents. However, complete covering of the particle by an asphalt film should decrease the quantity of water reaching the aggregate and reduce the deleterious effects of water on the aggregate. Building of roads with low air voids or good drainage may be the most influential effect in reducing water damage, by limiting the exposure of the asphalt–aggregate bond to water.

Specificity among the different asphalt–aggregate combinations was readily apparent in both the adsorption and desorption investigations as well as from the measurement of bonding energy. The interactions between asphalt and aggregate are dominated by aggregate chemistry. Asphalt chemistry also has an influence, though much smaller than that of the aggregate, on asphalt–aggregate interactions.

Asphalt–aggregate interactions are strongly influenced by the composition and surface chemistry of the aggregate. In fact, in many cases, aggregate properties may be more influential than asphalt properties in determining adsorption and stripping behavior than are asphalt properties. Furthermore, significant differences in asphalt affinity and stripping propensity occur among aggregates of different mineralogy. In fact, the polar compounds within the asphalt react with the active sites on the surface of the aggregate (i.e., those sites that contain metals or charged species). Some of the polar constituents of the asphalt, which adhere competitively to the aggregate surface, are highly susceptible to the presence of water and are readily removed from the aggregate surface by water. Changes in pH, particularly very basic pH, can be detrimental to the bond between asphalt and aggregate. Curing of the asphalt–aggregate bond can improve bonding between a particular asphalt–aggregate pair although that interaction is highly specific.

As expected, aggregate chemistry also plays a key role in adhesion. Each aggregate of a given mineralogical type with a specific history has a unique surface chemistry. The properties of the aggregate, including the electron-donating and electron-accepting ability of the aggregate, vary according to the active metal species at the aggregate surface. The active sites on the aggregate surface that have been postulated from observed behavior have been confirmed by autoradiography. These active sites promote adsorption of asphaltic components. The covering of those active sites by non-polar hydrocarbons would mask their activity. Likewise, dust coatings occurring naturally on aggregate surfaces can change the chemistry of adhesion and result in weak bonding between the dust and aggregate surface, leading to attrition of the bonding forces that help maintain the pavement (Curtis et al., 1993).

Evaluation of asphalt–aggregate interactions shows that the aggregate chemistry is much more influential than the asphalt composition for both adhesion and sensitivity to water, except where cohesive asphalt failures occur. Large differences were observed in the amount of asphalt adsorbed and the amount of asphalt retained after exposure to water with both siliceous and calcareous aggregates. Although the compositional factors of asphalt have a smaller effect, some differences in the amounts adsorbed and retained on a specific aggregate were observed.

Hence, the asphalt–aggregate mixture can be visualized as a system in which large, small, and fine aggregate particles are either coated with asphalt or suspended within the asphalt. Asphalt can penetrate into the pores, crevices, and interstices of the aggregate. The active sites on the particle attract the most polar and bondable asphaltic species on initial contact. Each asphalt molecule is contacted with an aggregate or an asphalt molecule in contact with or close to an aggregate surface. The fines that compose 5–8% of the aggregate are interspersed with the asphalt forming a mastic—a medium in which it is difficult to distinguish between asphalt and aggregate.

Thus, asphalt–aggregate interactions are strongly influenced by the composition and surface chemistry of the aggregate. Asphalt compounds with polar functional groups are highly competitive for the active sites (i.e., those sites that contain metals or charged species) on the aggregate surface. Some polar compounds that adhere competitively to the aggregate surface are highly susceptible to water and are readily removed from the aggregate surface. Changes in pH, particularly very basic pH, can be detrimental to the bond between asphalt and aggregate. Curing of the asphalt–aggregate bond can improve bonding between a particular asphalt–aggregate pair although that interaction is highly specific. Specificity among the different asphalt–aggregate combinations was readily apparent in several different sets of experimental measurements. The interactions between asphalt and aggregate are dominated by aggregate chemistry. Asphalt chemistry also has an influence, though much smaller than that of the aggregate, on asphalt–aggregate interactions. In addition, aggregate properties are much more influential in determining adsorption and stripping behavior than are asphalt properties. There are large differences in asphalt affinity and the relative stripping propensity of aggregates.

In summary, asphalt is structurally and chemically complex and dependent on the crude petroleum source and any chemical treatment or chemical modifiers added during processing. The molecular species in asphalt are comprised of a variety of functional groups whose behavior at the asphalt–aggregate interface determines the overall chemical stability of the asphalt. For the best chemical performance, the asphalt should have high contents of pyridine derivatives, phenol derivatives, and ketone derivatives, which can be achieved by selectively choosing the source material. If the situation requires special stability or redundancy, small amounts of shale oil and lime can be used as additives. Situations and conditions that favor the presence of inorganic sulfur, monovalent salts, and high-ionic-strength solutions in the asphalt should be avoided because these conditions potentially decrease the chemical stability of the asphalt cement by disrupting the functional group–aggregate bonds and increasing the overall permeability.

Four theories have been proposed to explain the adhesion between the asphalt binder and the aggregate: (i) chemical reaction, (ii) surface energy, (iii) molecular orientation, and (iv) mechanical adhesion (Terrel and Rimsritong, 1979). Realistically, it is difficult to give one theory the preference over the others and it is very likely that a combination of mechanisms occur simultaneous and synergistically to produce adhesion of the binder to the aggregate. Factors such as: (i) surface tension of the asphalt binder and aggregate, (ii) chemical composition of the asphalt and aggregate, (iii) binder viscosity, (iv) surface texture of the aggregate, (v) porosity of the aggregate, (vi) moisture content of the aggregate, and (vii) the temperature at the time of mixing with asphalt cement can exert their respective influence either collectively or individually on each of the four mechanisms.

The *chemical reaction* mechanism is based on the chemistry of the interaction of acidic and basic constituents of both the binder and the aggregate which form water-insoluble compounds that resist stripping due to the presence of selective chemical reactions or chemisorption (Rice, 1958; Thelen, 1958; Jeon et al., 1988; Curtis, 1992; Curtis et al., 1992).

The *surface energy and molecular orientation* aspects of asphalt–aggregate interactions are extremely important in terms of forming a strong (or weak) binder–aggregate bond (Lytton et al., 2005). Surface energy measurements are used to compute the adhesive bond strength between the aggregates and asphalt and cohesive bond strength in the binder. Moreover, various catalyst groups associated with the petroleum refining industry have recognized this phenomenon for several decades—without the preferred bonding and molecular orientation. In this respect, consultation of the catalyst-related literature shows that there can be vertical orientation of the asphalt constituents to the catalyst surface or horizontal orientation of the asphalt constituents to the catalyst surface (or perhaps even both forms of orientation) and if the preferred orientation is not present, the catalytic process may not produce the desired products (Speight, 2014).

A similar situation may exist in the bonding arrangement of the asphalt to the aggregate (Figures 6.3 and 6.4) and, moreover, the orientation of the molecular species of the asphalt constituents could be changed by the oxidation process in which more oxygen functions are introduced into the asphalt (Corbett, 1975; Moschopedis and Speight, 1973, 1975, 1977, 1978; Petersen, 2009). Undoubtedly, such a change would be related to the wettability of the aggregate by the asphalt and also to the potential for water displacement of the asphalt from the aggregate. In this context, the relative wettability of the aggregate surface (or molecular orientation relative to the aggregate surface) by water or asphalt can be described in terms of the surface energy phenomenon. Furthermore, in accordance with catalyst science, the orientation of the molecular species in relation to the aggregate surface can be changed by heating and may offer an insight into the

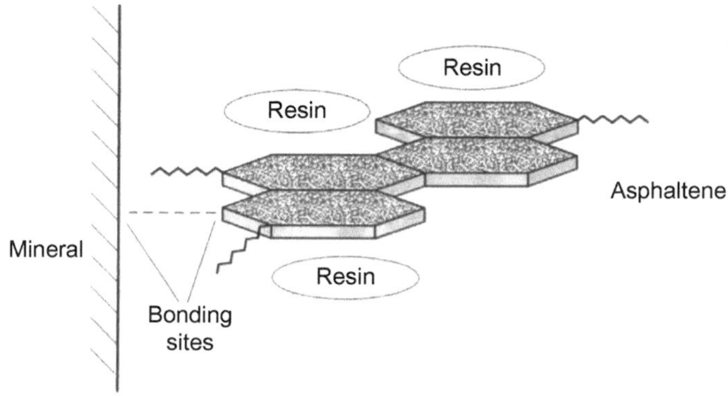

Figure 6.3 Representation of an asphalt molecule with vertical orientation to the mineral surface.

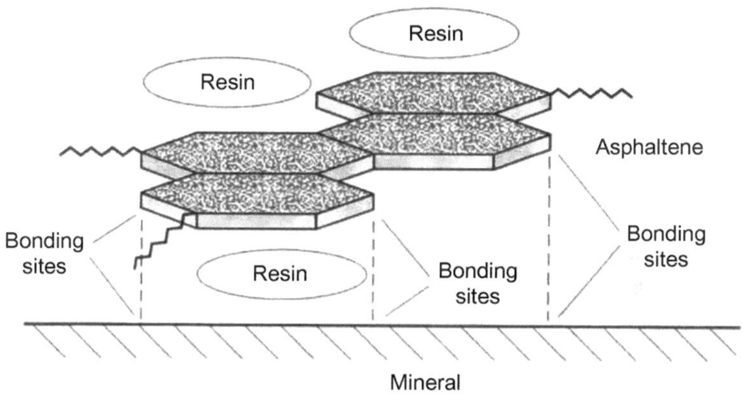

Figure 6.4 Representation of an asphalt molecule with horizontal orientation to the mineral surface.

phenomenon known as *steric hardening*, as well as a convenient resolution to this phenomenon (see Chapter 10).

Molecular orientation relates to the structuring of asphalt molecules at the asphalt–aggregate interface and assumes that adhesion between asphalt and aggregate is facilitated by a surface energy reduction at the aggregate surface as asphalt is adsorbed to the surface (Rice, 1958; Kiggundu and Roberts, 1988). The simplest example of structuring by molecular constituents of the binder on the aggregate surface is whether or not the constituent molecule is perpendicularly aligned to the aggregate surface or whether the molecule is horizontally aligned to the aggregate surface, in which case there may be more than one point of contact between the molecule and the aggregate, thereby resulting in a stronger bond (stronger adhesion).

Mechanical adhesion is a physical form of adhesion that relies on physical aggregate properties, including surface texture, porosity or absorption, surface coatings, surface area, and particle size (Terrel and Al-Swailmi, 1994). The principle of this form of adhesion involves formation of an aggregate surface that is capable of maximizing the surface area and texture to facilitate a strong physical bond which will improve the nature of the chemical bond between the asphalt and aggregate even in the presence of water and resist stripping.

Thus, damage to the asphalt–aggregate bonding association can occur within the mixture or at the binder–aggregate interface. Whether or not a cohesive or adhesive failure occurs depends on the nature of the binder as well as the relative thickness of the binder on the aggregate. In fact, damage in asphalt mixtures can occur within the binder or at the binder–aggregate interface. Whether or not failure of the pavement depends on the nature of the binder and the relative thickness of the binder around the aggregate is still open to discussion.

6.4.1 Asphalt–Aggregate Interactions

It is generally assumed, with some degree of certainty, that aggregates contain acidic and basic functions at (or near to) the surface. Furthermore, it is also generally assumed that acidic aggregates are hydrophobic (water-repellant) while basic aggregates are hydrophilic (water-attractive), but there are exceptions (Majidzadeh and Brovold, 1968). The general conclusion is that few if any aggregates can completely resist the stripping action of water and there are several factors that influence the asphalt–aggregate bond. These include surface texture, penetration of pores and cracks with asphalt, aggregate angularity, aging of the aggregate surface through environmental effects, adsorbed coatings on the surface of the aggregate, and the nature of dry aggregates versus wet aggregates.

The surface texture of the aggregate affects its ability to be properly coated, and a good initial coating is necessary to prevent stripping which relates to the importance of the interaction of the physical and the chemical bond (Cheng et al., 2002). In addition, the importance of a good mechanical bond promoted by an amenable surface texture is evident since stripping has been determined to be more severe in more angular aggregates (Gzemski et al., 1968). Angularity may promote bond rupture of the binder or mastic, leaving a point of intrusion for the water and, regardless of the strength of the bond between the asphalt and aggregate, the bond between water and aggregate is, as expected from chemical principles, considerably stronger (Cheng et al., 2002)—in fact, in many cases the aggregate–water bond can be much stronger (up to 30% stronger that the binder–aggregate bond.

When the aggregate surface is heated, the outermost adsorbed water layer is released, thereby leading to an improvement in the association between the binder and the aggregate which, in turn, leads to an improvement in the bonding association between the two. The weathering process results in a replacement of the adsorbed water layer with organic fatty acids from the air. This results in an improved asphalt–aggregate bond (Fromm, 1974), not forgetting the introduction of additional functional group into the binder that can lead to multiple association (on a molecular basis) between the binder and the aggregate—i.e., horizontal bonding of the binder molecule to the aggregate leading to the potential of multiple bonding contact that was not available in the unoxidized binder. Furthermore, a dust coating on the aggregate surface can interfere with binder–aggregate contact by preventing intimate contact (whether it is single or multiple contact points) between the asphalt and aggregate and by creating channels through which water can penetrate (Castan, 1968).

6.4.2 Effect of Asphalt Composition on Adhesion

The molecular species within asphalt are comprised primarily of carbon and hydrogen (between 90% and 95%, w/w) and polar constituents—which contain heteroatoms (nitrogen, oxygen, sulfur, nickel, vanadium, and iron)—that are extremely important to the interaction of asphalt with the aggregate.

The polar constituents have the ability to form molecules from molecular networks or clusters through hydrogen and π–π bonds that confer the elastic properties on asphalt. The non-polar constituents form the body of the material in which the network is formed and contribute to the viscous properties of asphalt (Jones, 1992; Speight, 1992a,b, 2014). The highly polar and highly aromatic molecules form the most interactive and strongest molecular networks. Non-polar molecules do not interact strongly enough to form networks, but they do substantially influence asphalt performance. A preponderance of high-molecular-weight non-polar molecules will lead to asphalt that hardens and performs poorly at low service temperatures. Furthermore, if the non-polar constituents are waxy, they will crystallize at low temperatures and become susceptible to cracking.

However, non-polar and polar molecules must interact in an acceptable manner or be compatible or phase separation (incompatibility) will ensue, thereby destroying the effectiveness of the asphalt for its designated use.

The affinity of the various functional groups to bond to the aggregate has been ranked (Curtis, 1992). In general, acidic groups, carboxylic acids, and sulfoxides have the highest propensity for adsorption while ketones and nonbasic nitrogen groups have the least propensity for adsorption on the aggregate. However, the sulfoxide and carboxylic acids were more susceptible to desorption in the presence of water and the general trend of desorption potential of polar groups from aggregate surfaces is sulfoxide > carboxylic acid > non-basic nitrogen ≥ ketone > basic nitrogen > phenol (Curtis, 1992).

6.4.3 Effect of Aggregate Properties on Adhesion

The adhesive bond between asphalt and aggregate is affected by a number of properties, such as: (i) size and shape of aggregate, (ii) pore volume and size, (iii) surface area, (iv) chemical constituents at the surface, (v) acidity and alkalinity, (vi) adsorption size surface density, and (vii) surface charge or polarity.

The efficiency of the bond between the asphalt and the aggregate is ultimately defined by the *pore volume and surface area*, which are indicative of stripping resistance (Yoon and Tarrer, 1988). The optimal resistance to stripping was developed in aggregates that provide a large surface area for bonding as well as a favorable pore size for adequate (deep) asphalt penetration. However, the penetration of asphalt cement into pores is not only dependent on the pore structure but also on the viscosity of the asphalt cement at the mixing temperatures. Furthermore, and not surprisingly when compared to refining catalysts, aggregates with approximately equal physical properties (e.g., pore volume and structure and surface area) can have very different properties depending on their basic chemistry and mineralogy, which define surface activity. For example, limestone ($CaCO_3$) has a substantially higher bonding power for asphalt than gravel, which is typically a quartz mineral (SiO_2), even though both have similar physical surface structures (Yoon and Tarrer, 1988; Cheng et al., 2002).

Another factor to be considered is the pH of contacting water which has a tendency to weaken the asphalt–aggregate bond when the pH lies on the acid side of the pH scale (Scott, 1978; Yoon and Tarrer, 1988). It is believed, with some justification, that when an aggregate is being coated with asphalt, the aggregate selectively adsorbs some components of the asphalt. However, the presence of ketone derivatives and phenol derivatives is thought to improve stripping resistance, whereas carboxylic acid derivatives, anhydride derivatives, and 2-quinoline derivatives increase stripping sensitivity because of the substantial water susceptibility of the associated bonds.

The addition of hydrated lime offers a mechanism to fix carboxylic acid derivatives and 2-quinoline acid derivatives so they cannot interact with hydrogen-bonding functions on the aggregate surface to produce moisture-sensitive bonds. In fact, it is considered likely that the interaction of lime with components in the asphalt not only prevents the formation of moisture-sensitive bonds but also subsequently allows more resistant bonds (such as nitrogen compounds from the asphalt) to proliferate. An additional benefit of the use of lime is to react with or adsorb compounds that can be further oxidized and enhance the increase in viscosity as a result of oxidation.

In fact, there is a substantial improvement in moisture resistance after severe freeze–thaw experiments when lime is added directly to the asphalt and before the asphalt is coated on the aggregate (Plancher et al., 1977; Petersen et al., 1987). In a manner similar to the reaction between acidic compounds, such as carboxylic acid derivatives in asphalt and alkaline aggregate or with lime, an amine compound either if present in asphalt or added in the form of an antistripping additive will react with acidic surfaces as in the case of siliceous aggregates to form a surface compound (Titova et al., 1987). When lime is added, some dissociation of the $Ca(OH)_2$ molecule occurs, resulting in calcium ions (Ca^{2+}) which interact with the carboxylic acid derivatives and 2-quinoline derivatives (Petersen et al., 1987) to form insoluble calcium organic salts. This leaves the SiOH molecule free to bond with nitrogen groups in the asphalt to from strong bonds that contribute to adhesion (Schmidt and Graf, 1972; Petersen et al., 1987).

6.5 MOISTURE DAMAGE AND AGING

Asphalt pavement is susceptible to a variety of forces that cause *distress*, which necessitates that tests to determine and correct discovered deficiencies be applied. The increasing use of asphalt-type pavement in new roadway construction also requires large quantities of materials.

6.5.1 Moisture Damage

The moisture susceptibility of an asphalt mix is the propensity for the mix to allow the infiltration of water into the matrix to form a boundary layer between asphalt cement and aggregate. This separation of the asphalt cement from the aggregate is known as stripping (moisture damage), which is the loss or weakening of the cohesive bond between asphalt cement and aggregate. Once the bond begins to lose cohesion, the decrease in the asphalt mixes' strength is often great. The phenomenon typically occurs first at the bottom of the asphalt layer due to water running downward and residing in the lowest pores in the pavement.

Various modes of infiltration will allow water to reach the bottom of the asphalt layer, which are dependent upon the physical and/or chemical properties of the aggregate used in the mix (Yoon and Tarrer, 1988). Commonly, asphalt can become saturated if water is left standing on the pavement surface due to inadequate drainage. Secondly, poor local subsurface drainage will allow the local water table to rise which can encourage a wicking action, which allows the water to enter the asphalt from below. Also, due to nearby bodies of water, a locally elevated water table will allow lateral flow of water from the phreatic surface into the asphalt. The internal pore pressure then leads to a weakening of the bond. As the bottom of the pavement gradually weakens, more water is allowed into the asphalt which lets the water level rise. The higher water level promotes further stripping in the lowest reaches of the pavement as well as beginning the stripping process further up into the matrix. In this manner, stripping travels upward through the pavement until it reaches the surface. At the surface, the weakened asphalt may exhibit rutting in the travel lanes or shoving where vehicles are required to make repeated stops.

Thus, the infiltration of moisture into hot mix asphalt pavements is one of the most common causes of degradation in pavement structures. When moisture enters the pavement structure, it can find its way between the aggregate and the asphalt cement, leading to a loss of cohesion within the pavement. This separation of aggregate from the asphalt cement by an aqueous boundary layer is commonly referred to as stripping. Stripping is a failure mode that can be manifested in many different ways. The effects of stripping are most often very evident in surface features such as ruts or cracks in the asphalt (Lottman and Frith, 1989). However, the effects of stripping can also be seen in situations where shoving of the asphalt matrix has taken place. Many factors in addition to the presence of water can contribute to stripping. Relevant factors include, but are not limited to: asphalt mix characteristics, traffic loading, and climate. However, it is believed that pavement experiences the greatest destructive effect when an interaction occurs between one or more of the aforementioned effects while moisture is present. Several laboratory and field tests have been established to evaluate how moisture-susceptible a paving mix is. The tests performed have been both quantitative and qualitative in nature.

Moisture damage (water-induced damage) of asphalt mixtures has produced serious distress, reduced performance, and increased maintenance for pavements in many areas of the United States as well as in other countries. This damage occurs due to stripping of asphalt from aggregate and in some cases possibly due to softening of the asphalt matrix (Kennedy et al., 1984a). Moisture-induced damage produces several forms of distress including localized bleeding, rutting, shoving, and ultimately complete failure due to permanent deformations and cracking. The two basic forms of moisture-related distress are stripping and softening. Stripping, which is of primary concern, is the physical separation of the asphalt cement and aggregate produced by the loss of adhesion between the asphalt cement and the aggregate surface primarily due to the action of water or water vapor. Stripping is accentuated by the presence of aggregate surface coatings and by smooth surface textured aggregates. Softening is a general loss of stability, strength, and stiffness of a mixture that is caused by a reduction in cohesion due to the action of moisture within the asphalt matrix.

This moisture damage has a major effect on roadway performance. The term *performance*, as applied in this text, generally means the time to pavement failure (service life), which is determined by numerous other factors. The performance or service lifetime of an asphalt pavement can be on the order of months to years and is due to the occurrence of one or more modes of pavement failure, such as: (i) permanent deformation, (ii) rutting, (iii) fatigue cracking, (iv) low-temperature cracking, (v) moisture damage, and (vi) total loss of adhesion. Aging of the asphalt binder is also a factor but is sometimes regarded as being beneficial or (for the most part) detrimental. Aging is detrimental when excessive hardening or stiffening is observed in an already adequate pavement, but may be considered to be beneficial when a soft mixture hardens into an adequate pavement. Thus, a simple view of performance is that a pavement may be too soft and therefore rut and deform or the pavement may be too stiff and brittle and subject to cracking, either under traffic load or under thermal stress.

Moisture damage is the loss of strength and durability in asphalt mixtures due to the effects of moisture which occurs when there is a loss of bond between the asphalt cement or the mastic and the aggregate. The mechanisms associated with moisture damage are varied and include: (i) detachment, (ii) displacement, (iii) spontaneous emulsification, (iv) pore pressure, (v) hydraulic scour, (vi) pH instability, and (vii) the effects of environmental and/or climatic conditions on asphalt–aggregate systems (Taylor and Khosla, 1983; Kiggundu and Roberts, 1988; Terrel and Al-Swailmi, 1994).

By way of explanation, *detachment* is the separation of asphalt binder from the aggregate surface by a thin film of water without a break in the film (Majidzadeh and Brovold, 1968). If a strong bond between the asphalt and the aggregate is nonexistent, wettability of the rock can occur, thereby further decreasing the ability of the asphalt binder to wet the aggregate and the energy of adhesion decreases (Majidzadeh and Brovold, 1968). Furthermore, if a three-phase interface consisting of asphalt, aggregate, and water exists, water reduces the free energy of the system more than asphalt to form a thermodynamically stable condition of minimum surface energy and the aggregate has a strong preference for water over asphalt. In some cases, where the asphalt binder has a low polar activity and the bond that develops between the aggregate and asphalt is chiefly due to relatively weak dispersion forces, the highly polar water molecules will replace the asphalt at the asphalt–aggregate interface (Cheng et al., 2002).

Displacement is the removal of the asphalt binder at the aggregate surface through a break in the asphalt film when chemical forces take over and complete the break between the binder and the rock (Fromm, 1974; Scott, 1978). Displacement can proceed through changes in the pH of the water at the aggregate surface that enters through the point of disruption, which changes the types of polar groups adsorbed, leading to physical separation of the asphalt from the aggregate (Scott, 1978).

Spontaneous emulsification is an inverted emulsion of water droplets in asphalt cement, especially when the emulsifiers such as clays and asphalt additives are present (Fromm, 1974; Scott, 1978). Furthermore, spontaneous emulsification occurs when asphalt films are immersed in water and, as might be expected, the rate of emulsification depends on the nature of the asphalt and the presence of additives. In addition, organic amines (basic nitrogen compounds), that are present in the residuum from which the asphalt is produced, bond strongly to aggregates and the rate of emulsification is dependent on the nature and viscosity of the asphalt.

Pore pressure is the development of pressure through water trapped within the pores of the aggregate which disrupts the asphalt film from the aggregate surface or can cause the growth of microcracks in the asphalt mastic (Bhairampally et al., 2000). Microcracking, which occurs under heavy repeated loading, results in cohesive failure, adhesive failure, or both, occurs and increases in the presence of water as the pore pressure developed in the microcrack voids. This increases the rate of crack growth and damage through the development of higher pressures and weakening of the mastic and of the adhesive bond between the mastic and the aggregate.

Hydraulic scour occurs at the pavement surface where stripping of the binder from the aggregate is the result of the action of tires on a saturated surface (Fromm, 1974). Water is sucked under the tire into the pavement by the tire action—osmosis and pullback have been suggested as possible mechanisms of scour. Osmosis occurs in the presence of salts or salt solutions in the pores of the aggregate which creates an osmotic pressure gradient that actually encourages passage of water through the asphalt film and the diffusion of water vapor through asphalt cement occurs and can be considerable (Cheng et al., 2002).

The *instability of the pH* refers to change in the acidity–alkalinity of the system due to a variety of factors and that asphalt–aggregate adhesion is strongly influenced by the pH of the contact water (Scott, 1978; Kennedy et al., 1984b). In fact, as the pH is increased, the value of interfacial tension between asphalt and glass peaks at intermediate pH values, up to about 9, and then decreases as the pH increases to higher values (Scott, 1978). This indicates that stabilization of the pH sensitivity at the asphalt–aggregate interface can (i) minimize the potential for bond breakage, (ii) provide strong, durable bonds, (iii) thereby reducing the tendency for stripping of the binder from the aggregate (Kiggundu and Roberts, 1988; Tarrer, 1996).

Environmental and climatic conditions such as temperature, air, and water also have an effect on the durability of asphalt concrete mixtures. However, under mild climate conditions where good-quality aggregates and good-quality asphalt binders are available, the major contribution to deterioration is traffic loading. Premature failure will result when poor materials and traffic are coupled with severe weather when factors such as: (i) water from precipitation of groundwater sources, (ii) temperature fluctuations, including freeze–thaw conditions, and (iii) aging of the asphalt are operative (Terrel and Al-Swailmi, 1994).

Although several separate chemical and physical mechanisms have been identified to explain the process of moisture damage in asphalt pavements, it is more likely that most asphalt pavements suffer moisture damage as a result of the occurrence of several simultaneous processes. From a chemical standpoint, certain polar constituents of the asphalt (particularly the phenol derivative and the basic nitrogen derivatives) form strong moisture-resistant bonds with the aggregate and the development of these bonds can be enhanced by treatment of the asphalt mixtures with additives. While sulfoxide derivatives and carboxylic acid derivatives have a greater affinity for the aggregate surfaces, they are also most susceptilbe to destruction by water. Also, the asphalt–aggregate bond is affected by aggregate mineralogy and asphalt rheology at the temperature at which the binder and aggregate are mixed as well as the nature of the aggregate surface, pore size, and pore size/shape.

By means of explanation, rheology is the investigation of the deformation and flow of matter, such as asphalt. Deformation and flow of the asphalt binder in hot mix asphalt is important in hot mix asphalt pavement performance. Hot mix asphalt pavements that deform and flow too much may be susceptible to rutting and bleeding, while those that are too stiff may be susceptible to fatigue cracking. Hot mix asphalt pavement deformation is closely related to asphalt binder rheology. Rheological properties of asphalt binder vary with temperature, so rheological characterization involves two key considerations. First, to fully characterize an asphalt binder, its rheological properties must be examined over the range of temperatures that it may encounter during its life. Second, to compare different asphalt binders, their rheological properties must be measured at some common reference temperature.

In addition, the bonding of the asphalt to the aggregate dynamic changes with time, which is predominantly affected by the shift in pH at the aggregate–water interface, which can be due to dissociation of aggregate minerals or by the nature of the pore water (cation type and concentration). Finally, moisture damage is not only limited to adhesive failure but also to weakening of the cohesive strength of the mastic due to moisture infiltration which is equally important.

In order to mitigate the damage caused by stripping, it is essential to have the correct mix design for the paving application. However, even with the correct mix design, stripping can still occur due to improper compaction of the asphalt during placement at the job site. An under-compacted mix will typically exhibit a higher percentage of air voids than is desirable—the voids provide a place for water to enter the pavement and to reside. Properly compacted mixes will have fewer voids where water can reside—for example, asphalt mixes prepared with air voids in the range of $\leq$ 5% will significantly reduce moisture infiltration into the asphalt mix, even to the point of becoming nearly impervious. Conversely, asphalt mixes exhibiting air voids of $\geq$ 8% will allow an unimpeded influx of water into the asphalt matrix. For this reason, many test procedures used to predict the moisture susceptibility of an asphalt mix specify an air-void content of 7 ± 1%—such as the modified Lottman test (AASHTO T283) (see Chapter 4).

Several other factors can have contributing effects to the stripping process. Open-graded aggregates, wet aggregates, or poor-quality aggregates with low tensile strength can increase the amount and magnitude of stripping. Use of antistripping additives can sometimes decrease susceptibility to stripping (Tunnicliff and Root, 1984). However, if an antistripping additive approximates the action of a detergent, as is the case when certain chemicals are used for enhanced oil recovery (Speight, 2009, 2014), the so-called antistripping additive can actually have the reverse effect and promote stripping of the asphalt from the aggregate.

6.5.2 Asphalt Aging

The destruction of an asphalt pavement during the in-service period is due, in large part, to the aging of the asphalt binder, which made a contribution to bring down the durability and service life of asphalt pavements. The factors affecting asphalt aging included (i) characteristics of the asphalt and its content in the mix, (ii) the nature of aggregate and particle size distribution, (iii) the void content of the mix, and (iv) production-related factors such as temperature and time. The most important aging-related modes of failure were traffic and thermally induced cracking, and raveling. Furthermore, asphalts from different sources age at different rates and show differing rheological sensitivity to oxidative aging in products depending on their chemical compositions as well as the influence of water (Nourendin and Wood, 1989; Lau et al., 1992; Petersen et al., 1993; Thomas, 2002; Kim et al., 2004).

The oxidative hardening of asphalt (sometime referred to as oxidative hardening) is a result of the formation of oxygen functions (such as ketones and sulfoxides) during the oxidation process (Moschopedis and Speight, 1973; Dorrence et al., 1974; Moschopedis and Speight, 1975, 1977, 1978; Petersen et al., 1982; Speight, 1992a,b; Petersen et al., 1993; Speight, 2014). The relative amount of hardening that the functions produce depends upon the state of dispersion of the molecules on which they are formed. The main aging mechanism is irreversible and is characterized by chemical changes of the binder which, in turn, has an impact on the rheological properties. The processes contributing to this type of aging include oxidation, loss of volatile components, and exudation (migration of oily components from the asphalt into the aggregate). Asphalt aging occurred during the mixing and construction process as well as during long-term service in the road. However, the circumstances at different aging stages vary considerably.

In order to evaluate the aging properties of asphalt, a number of laboratory test methods are available (see Chapter 4), such as (i) the thin-film oven test (TFOT), (ii) the rolling thin-film oven test (RTFOT), (iii) the pressure aging vessel (PAV) test method, and (iv) the ultraviolet (UV) light aging test method. In these tests, asphalt aging is accelerated by increasing temperature, decreasing asphalt film thickness, increasing oxygen pressure, or applying various combinations of these factors.

The RTFOT is meant to simulate asphalt binder age hardening as it occurs during mixing, transport, and placement. In this procedure, 35 g of asphalt are carefully weighed into glass bottles, which are placed in a circular rack in a specially designed oven. The rack slowly rotates the bottles while the oven maintains the specified test temperature (AASHTO T240, ASTM D2872). During the test, a jet of air is blown into the bottles for a few seconds once every rotation. The test is continued for 75 min; the bottles are then removed from the oven, cooled, and weighed. The percent mass loss is calculated from the initial and final weight of the asphalt binder in the bottle. High values of mass loss mean that a significant amount of light oils have volatized during aging and, as a result, the asphalt binder might be prone to excessive age hardening, shrinkage, and cracking. Current performance grading standards require that mass loss during aging (in the RTFOT) be no more than 1.0%. After mass loss determination, the bottles are heated, and the asphalt is poured either into a tin for further testing, or into PAV pans for additional aging.

In the PAV test, 125-mm-diameter stainless steel pans are filled with asphalt that has already been aged in the RTFOT. Six of these pans are placed in a vertical rack, which is then placed in the pressure vessel, which in turn is placed inside an oven. The pressure vessel is a heavily constructed steel chamber, designed to withstand the high pressure and temperature used in the RTFOT. These high temperatures and pressures help accelerate aging of the asphalt binder. At the end of the RTFOT, the asphalt binder is assumed to have aged approximately as much as it would in a pavement after several years of service.

The RTFOT was developed as an improvement to the TFOT for short-term asphalt binder aging, which placed asphalt binder samples in shallow pans (of the same dimensions as those used for the PAV) and then heated them in an oven for an extended period of time to accomplish simulated aging. The RTFOT is an improvement over the TFOT because fresh asphalt binder is continuously exposed to heat and air flow due to the rolling action of the carousel. In addition, asphalt binder modifiers, when used, usually remain dispersed in the asphalt binder due to the rolling action of the carousel. Also, no surface skin, which inhibits aging, forms on the asphalt binder because of the rolling action of the carousel.

However, there is a basic assumption built into the RTFOT and PAV procedures that asphalt will, within a narrow range, exhibit consistent aging behavior and that that aging will have a predictable effect on the performance in mixtures and pavements. An alternative binder may have significantly different aging characteristics. Consequently, the rolling thin-film oven and PAV procedures may not adequately represent plant and field aging. It is suggested that experimental considerations start with determining an aging index to compare against conventional asphalt. The rolling thin film oven procedure should be run over a range of times and temperatures to see whether normal time–temperature correlations still hold. The same applies for the PAV procedure which should be run at 60°C (140°F) for extended times and the resulting behavior compared to the results at standard conditions (Kluttz, 2012).

6.6 STABILITY AND INSTABILITY

The instability of asphalt (usually reflected in poor performance and aggregate stripping) occurs when external effects perturb the system. When such disturbances occur to the organic constituents, it is the higher-molecular-weight constituents that are most seriously affected, eventually leading to incompatibility (precipitation, sediment formation, sludge formation) depending upon the circumstances. Thus, the dispersability of the higher-molecular-weight constituents (the asphaltene constituents and, in many instances, the resin constituents) becomes an issue that needs attention. And one of the ways by which this issue can be understood is to be aware of the chemical and physical character of the higher-molecular-weight constituents. By such means, the issue of dispersability, and the attending issue of incompatibility can be understood and even predicted (see Chapter 7).

This, an important fundamental molecular property of asphalt is the physical–chemical structure that is associated with the asphaltene constituents and the resin constituents. It is due to this association that asphalt can be stated to have a structure which dictates its temperature-dependent performance. However, the extent of molecular motions depends on the intramolecular configuration of the various asphalt components and the manner in which they interact by intermolecular association will depend upon the temperature and other environmental conditions, not least of which is the association of the asphalt with the aggregate.

Any changes in structure and composition brought about by association with the aggregate can affect the overall long-term road performance behavior of asphalts. On a macro-scale, the performance behavior of asphalts is manifested in changes in the rheological properties (such as the viscoelastic nature of the asphalt). Furthermore, the stability of asphalt in service (durability) relates to changes with time in the original properties of the asphalt in general, while the latter refers rather to an age hardening process, however, both these phenomena are related to each other and are associated with the presence of resin and asphaltene constituents which can cause serious instability problems (see Chapter 7). Asphalt produced by vacuum distillation tends to exhibit instability, manifested by a drop in the penetration and ductility and an increase in viscosity during storage and handling.

After application of the asphalt in road pavements, it is exposed to extremes of environmental conditions: high temperatures (especially in warm regions), subzero temperatures in near-Arctic regions, atmospheric radiation, and mechanical stress. Rheological as well as chemical considerations have therefore to be taken into account. Furthermore, the chemical nature of asphalt, which is influenced by the production methods, affects its behavior in this respect. And many of the general properties of asphalt, which also serve as methods for routine inspection, are actually stability estimates.

The stability (and instability) of asphalt is associated with the durability of the asphalt in service as, for example, a paving asphalt. This relates to changes with time in the original properties of the asphalt in general, while the latter refers rather to an age hardening process, but both these phenomena are related to each other.

6.6.1 Stability

The stability of the asphalt–aggregate mix is essential for roadway performance and longevity. However, stabilization is where chemical bonding techniques render a binder less toxic and less harmful to the environment. Examples of such stabilization include (i) ion exchange of heavy metals in an alumina silicate matrix of a cementitious agent or (ii) sorption of heavy metals on fly ash in an aqueous system. Stabilization of the asphalt–aggregate mix can be functionally described in terms of the final product, where the binder and the aggregate are transformed into a stable and durable matrix that is more compatible for roadway use. The chemical stability and environmental performance of asphalt are dependent on the nature and extent of the asphalt–aggregate bonds. Furthermore, much of the information concerning these bonds comes from investigations of the sensitivity of the asphalt pavement to moisture-induced damage (Petersen et al., 1982).

Because of the polar/non-polar and hydrophobic/hydrophilic interactions among the various phases in the system, the functional groups in asphalt, as determined by a concentration effect as a result of the distillation process (Figure 6.5), will be concentrated at the interfaces between phases, such as the asphalt–aggregate interface, the polar micelle–asphalt oil interface, and the asphalt–water interface. Thus, the stability of the asphalt–aggregate mix is addressed on the basis of the relative tendency of the functional groups to be concentrated at the asphalt–aggregate interface, sorption affinity, and ability to be displaced by water.

In fact, the behavior of functional groups at the asphalt–aggregate interface determines the chemical stability of the asphalt to a large degree. At the asphalt–aggregate interface, these groups are susceptible to interactions with the aqueous phases and any contaminants present in the

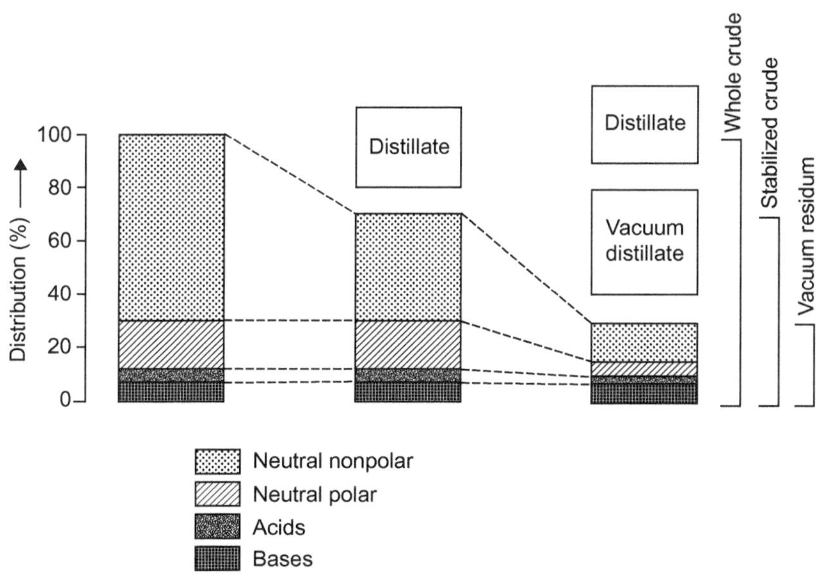

Figure 6.5 Accumulation of polar functions in residua during distillation.

affected soil. The relative tendency of the asphalt functional groups to be concentrated at the asphalt–aggregate interface has been determined (Petersen et al., 1981), in order of decreasing tendency, as:

Carboxylic acid > anhydride >> quinoline, phenolic > sulfoxide, ketone > pyrrole

The sorption affinity of the functional groups with the aggregate surface has also been investigated (Petersen et al., 1982) and is given, in order of decreasing tendency, as:

Pyridine, carboxylic acid ≫ anhydride > quinoline, phenol > sulfoxide > ketone ≫ pyrrole, polynuclear, aromatic

The third important tendency of the functional groups affecting stability is the ability of the functional group–aggregate bond to be displaced by water. This tendency, listed in order of decreasing tendency, is given as:

Anhydride derivatives > quinoline derivatives
Quinoline derivatives > carboxylic acid derivatives
Carboxylic acid derivatives > sulfoxide derivatives
Sulfoxide derivatives > ketone derivatives
Ketone derivative > pyrrole derivatives
Pyrrole derivatives > phenol derivatives

Combining these tendencies provides some indications of performance—for the best chemical performance, the asphalt should have high contents of pyridine, phenol, and ketone groups, which can be achieved by selectively choosing the source material or using additives, such as shale oil that is rich in pyridine-type basic nitrogen. Also, the presence of inorganic sulfur, monovalent salts, and high-ionic-strength solutions in the asphalt decreases the chemical stability of the asphalt cement by disruption of the functional group–aggregate bonds and increases the system's permeability (Petersen et al., 1982). Addition of lime to the aggregate can thus be used, and often is, to counter this effect.

In addition, asphalt can change during storage due to oxidation (oxidative fouling) (Speight, 2015b), resulting in degradation of the product which can be determined from changes in the penetration (ASTM D5) or in viscosity (ASTM D2170).

A penetration drop below the required limit or a viscosity rise above the specification limit have been used to assess the suitability of the asphalt for designated uses. In addition to these two properties, two additional rheological estimates are included in most of the asphalt specifications:

the Ring and Ball softening point (ASTM D36) and the ductility (ASTM D113). All four properties are subject to changes during storage and handling and are affected by environmental conditions: temperature, exposure to air (oxygen), and atmospheric radiation. Another test method, also in current use is the TFOT (ASTM D1754) which can be used to indicate the rate of change of penetration, viscosity, and ductility in a film of asphalt heated in an oven for 5 h at 165°C (325°F) on a rotating plate. This test establishes the effects of heat and air on the basis of changes incurred in the above physical properties measured before and after the oven test. The allowed rate of change in the various properties of the asphalt, after the exposure of the tested sample to the oven test, is specified in the relevant specifications (ASTM D3381).

Some of the other inspection data relating to asphalt quality and changes during storage are: (i) the Cleveland open cup flash point, ASTM D92, which relates to the safety in handling the asphalt (ii) solubility in various solvents, which is an estimate pertaining to the required absence in the asphalt of insoluble impurities, and (iii) the asphaltene content (ASTM D893, ASTM D2007, ASTM D3279, ASTM D4124), which is an indication as to the bulk composition of the asphalt and also the stability and incompatibility of blended asphalt (i.e., asphalt that is a blend of several asphalts from different sources).

An appropriate standard test method (ASTM D6703) can be applied as a laboratory diagnostic tool for estimating the colloidal stability of asphalt, asphalt cross blends, aged asphalt, and heavy oil residuum. Historically, asphalt and heavy oil residua have been modeled as colloidal suspensions in which a polar associated asphaltene moiety (the dispersed phase) is suspended in a maltene solvent moiety (the dispersing medium) (ASTM D3279; ASTM D4124; ASTM D5546). The extent to which these two moieties remain in state of peptization is a measure of the compatibility (colloidal stability) of the suspension. Compatibility influences the physical properties of these materials, including rheological properties, for example, phase angle and viscosity. This test method and other similar test methods, along with the classical Heithaus test, measure the overall compatibility of a colloidal system by determining a parameter referred to as the state of peptization, P. The value of P commonly varies between 2.5 and 10 for unmodified or neat asphalt samples. Materials calculated to have low values of P are designated incompatible. Materials calculated to have high P values are designated compatible. Values in P are calculated as a function of two parameters that relate to the peptizability of the asphaltene moiety (the asphaltene peptizability parameter, p_a) and the solvent power of the maltene moiety (the maltene peptizing power parameter, p_o). Values of p_a and p_o are calculated as functions of the quantities C_{min} and FR_{max}, which are determined from experimental variables, the weight of asphalt (W_a), the volume of solvent (V_S) to dissolve the weight of asphalt, and the volume of titrant (V_T) added to initiate flocculation.

6.6.2 Instability

Asphalt concrete pavements are inherently nonhomogeneous in nature and exhibit instability (incompatibility) which can have serious consequences for mechanical properties of a roadway asphalt mix (Masad et al., 2009). Property gradients are most severe across the thickness of the asphalt concrete layers. The main sources of nonhomogeneity (and instability) are (i) aging and (ii) varying temperature profile (Dave et al., 2010). Thus, the term *long-term stability* refers to the durability of asphalt not only in its original form, but also in the form of asphalt paving cements. After application of the asphalt as a paving asphalt, it is exposed to extremes of environmental conditions: (i) high temperatures, especially in the southern regions of the United States, (ii) subzero temperatures, especially in the northern states of the United States, (iii) atmospheric radiation, and (iv) mechanical stress. Therefore, potential rheological issues as well as chemical issues have to be taken into consideration.

Asphaltenes and polar-aromatics play a fundamental role in determining the mechanical and rheological properties of asphalts (penetration index (PI) and kinematic viscosity with aging time and temperatures). Other important asphalt properties, such as temperature susceptibility coefficient and characterizing factor, were dependent on the quantity of the asphaltenes and polar-aromatics. The asphalt separation procedure is based on solubility in normal heptane, followed by adsorption chromatography of the soluble portion (Speight, 1992a; Mohammed and Morshed, 2008; Speight, 2014).

The term age hardening is used to describe the phenomenon of hardening. Hardening is primarily associated with loss of volatile components in asphalt aging during the services. This factor causes an increase in viscosity of the asphalt and an increase in stability. In addition, the durability of asphalt is a prime consideration in the economics of all asphalt contractions. Asphalt highway systems present the more obvious and perhaps the most important area in which asphalt performance is observed. Hardening, as occurs in asphalt under service conditions, was for a long time taken as the best measure of its economic value.

Asphalt composition is largely dependent upon the crude from which it was derived and is typically based upon the qualitative determination of the four generic fractions found in all asphalts (Speight, 1992a, 2014). Each fraction differs considerably in color, density, and aromatic carbon content. An asphalt meeting a given specification is thus a composite of these quantities.

The asphaltene constituents are isolated as a fraction that is insoluble in low-molecular-weight paraffin solvents such as *n*-pentane or *n*-heptane, but soluble in aromatic solvents—maltenes are defined as components soluble in paraffin solvents (Figure 6.2) (ASTM D893, ASTM D2007, ASTM D3279, ASTM D4124; Speight, 1992a,b, 1994, 2014, 2015a). Carbenes are insoluble in aromatic solvents but are soluble in carbon tetrachloride or trichloroethylene. Carboids are insoluble in all of the solvents that dissolve asphaltenes and carbenes.

In terms of oxidation and the onset of instability, the more polar species in the asphalt (i.e., the resin and asphaltene constituents) will oxidize first during air blowing. After incorporation of oxygen to a limit, significant changes can occur in the asphaltene structure, especially in terms of the incorporation of polar oxygen, which can influence molecular weight. Thus, the change in the character of the asphalt may not be so much due to oxidative degradation but to the incorporation of oxygen functions that interfere with the natural order of intramolecular structuring. There is the potential that the incorporation of oxygen functions enhances the ability of the asphalt to bond to the aggregate. Uncontrolled introduction of oxygen functions could result in the production of a poor-grade asphalt, where phase separation of the oxidized asphaltene may already have occurred, or, should it occur in the product, the result can be pavement failure due to a weakening of the asphalt–aggregate interactions.

Although not defined as a stability property of asphalt (since it measures decreases in penetration and ductility and increases in viscosity), when a thin layer is exposed to heat and air a thin film forms which would support formation of oxygen-containing polymerization products. This, in turn, would reduce the penetration of the asphalt (increase its hardness), reduce its ductility (make it more brittle or less elastic), and increase its viscosity. Stable asphalt would be affected much less by this exposure to heat and oxygen. There is a direct relation between the results of this test and the changes in the original properties of the asphalt during handling, storage, and use.

Since asphalt is considered, with some justification, as a colloidal system, the nature of this system will determine the rheological properties of the asphalt, defined by its penetration, softening point, ductility, and viscosity at given temperatures. At this point, it is worth considering the approach applied to the instability/incompatibility of heavy residual fuels and its potential application to asphalt (Speight, 1992a,b, 2014). This concept bases the instability/incompatibility on the chemical composition as well as on the internal colloidal structure (Por, 1992) by defining a colloidal instability index, which is the ratio of the sum of asphaltene constituents and saturated oils to the sum of the resins and aromatic solvents:

$$\text{CII} = (\text{Asphaltenes} + \text{Saturates})/(\text{Aromatics} + \text{Resins})$$

The equilibrium of a well-peptized asphaltene system, such as asphalt, can be easily disturbed by (i) application of heat during service on days of extremely high temperatures and/or friction from vehicle tires; (ii) oxidation due to constant exposure to air; (iii) UV irradiation by prolonged exposure to sunlight; and (iv) addition of a paraffinic diluent. In each case, the chemical composition is altered and the aromaticity is affected, thereby causing a disruption of the equilibrium of the colloidal system (Moschopedis and Speight, 1973, 1975, 1977, 1978; Speight, 1992a,b, 2014).

As a result, asphaltene particles are deprived of their enveloping layers, which previously merged continuously with the consecutive layers. The micelle system becomes non-continuous, and the asphaltene cores are prone to agglomeration. Such a process leads to instability of the asphalt, perhaps resulting in phase separation of the asphaltenes from the asphalt, thereby causing a loss of the asphalt–binder interactions. Pavement failure is the result.

Another of the estimates regarding the rheological properties is the PI. The logarithm of the penetration is in a linear relation to the temperature:

$$\log \text{pen} = AT + K$$

In this equation, A is the slope of the temperature sensitivity as related to the logarithm of the penetration. By extrapolation to the softening point temperature, a penetration of approximately 800 is obtained. The slope, A, can be obtained by measurement of the penetration at two different temperatures or by the relation of the penetration and the softening point temperature.

A PI can be obtained by:

$$d\log \text{pen} / dt = (20 - PI)/(10 + PI) \times 0.02$$

$$A = d\log \text{pen} / dt$$

$$PI = 10 \times (2 - 50A)/(1 + 50A)$$

A PI in excess of + 2 would indicate a gel structure of elastic properties and a thixotropic nature, while a PI of below − 2 would indicate a sol structure of Newtonian properties, whereas an asphalt exhibiting satisfactory rheological properties should have a PI between + 1 and − 1. After an initial deformation, with a certain elasticity, Newtonian properties should prevail, with a proportionality between the deformation rate and the applied stress. Curves above these regions indicate gel-structured asphalt, and curves below these regions indicate sol-structured asphalt.

The presence of propane-precipitated asphalt in the asphalt mix improves the stability properties of such blends because of both the rheological and the chemical natures of the propane-precipitated asphalt. The improved stability properties of such blends can be seen from the decreasing differences in the viscosities, penetrations, and ductility after exposure to elevated temperatures and oxygen (as, e.g., in the TFOT).

The beneficial effect of the propane-precipitated asphalt is subject to limitations in their proportions—for example, up to 35% (v/v) of the propane-precipitated asphalt in blends with vacuum residua might be the upper limit, depending on the nature of the vacuum residues as well as on the nature and proportions of other components, as for example, lubricating oil extracts, which are sometimes used in such blends (Ishai et al., 1988).

The indications are that long-term asphalt stability is related to basic rheological and physicochemical characteristics of original as well as aged asphalt samples of varying compositions. It is also believed (Ishai et al., 1988) that the understanding of the relations of these characteristics on the properties of asphalt (paving asphalt cement) makes possible the prediction of asphalt durability performance in the field, as well as the prediction of other relevant properties (Mohammed and Eweed, 2012). The stability is indicated in this case by aging indices: viscosity and softening point ratios and retained penetration percentage before and after exposure to the TFOT (Por, 1992).

Finally, the consequence of particle fouling of the residuum during distillation or the visbreaker residuum (also called *visbreaker tar*)

(Speight, 2014, 2105b) can have serious consequences for performance of the asphalt in service. If particle-fouled resid or visbreaker tar is used for asphalt production, the result is a poor-grade asphalt (more suitable as parking-lot asphalt) that would be priced at a much lower level than good-grade (particle-free) roadway asphalt.

6.7 CHEMICAL COMPOSITION AND PERFORMANCE

There have been many attempts to relate asphalt performance to composition (Corbett, 1969, 1970; Hattingh, 1984; Petersen, 1984; Speight et al., 1984; Jennings, 1985; Thenoux et al., 1985; Goodrich et al., 1986; Kim et al., 1986; Thenoux et al., 1986; Speight, 1994, 2014, 2015a), with some showing a modicum of success. However, asphalt performance (frequently referred to as pavement performance) is a complex term that refers to the quality of service that a pavement structure gives during its service life. There are numerous factors that contribute to the deterioration of a pavement structure and the subsequent reduction in its performance, such as (i) materials, (ii) the type of mix and the mixing procedure, (iii) construction, (iv) mix design, (v) pavement structure, (vi) traffic load, and (vii) the environment (Table 6.1). Two other factors: time and maintenance, can also be included.

Any project should include or combine many of the above factors in a rather random arrangement where it is almost impossible to isolate any detrimental element and subsequently to relate performance to the composition of the asphalt used. Further, a pavement structure may show different types of distress at the same time and each distress mode may be caused by the same or a different factor. Another problem in relating pavement performance to any of the above factors is the difficulty in compiling all the construction and field data, particularly when an asphalt pavement has been in service for a long period of time and it has been exposed to a wide range of traffic and environmental conditions.

Most types of distress that shorten the life of an asphalt pavement are related to an increase in hardening of the asphalt material, reducing its ability to absorb energy (brittleness) and causing it to become more sensitive to failure at low temperatures and/or normal highway strains and stresses (aging). Generally, an increase in hardening is associated with an increase in asphaltene content and the asphaltene content is the only chemical component that consistently relates to pavement performance. Another reason for choosing the asphaltene fraction as the prime focus of

Table 6.1 Factors that influence pavement performance

Materials	Mixing	Construction	Mix design	Pavement structure	Traffic load	Environment
Asphalt type	Cold or hot	Transportation	Dense-graded	Thickness	Types	Temperature
Asphalt properties	Mixing plant	Compaction	Open graded	Base strength	Load repetitions	Water
Aggregates		Quality control	Asphalt content	Subbase strength	Tire properties	Oxidation
Asphalt–aggregate affinity			Film thickness	Subgrade strength		Spring thaw
Additives			Air voids			

N.B.: The *Time Factor* refers to the duration of the pavement to any of the other effects.
The *Maintenance Factor* refers to the maintenance techniques employed and the frequency of maintenance.

properties and performance is that there has been a stronger focus on the characteristics of this generic fraction than for the other three fractions (Speight et al., 1984; Goodrich et al., 1986; Speight, 1994, 2014, 2015a).

The analysis of relationships between chemical components showed that the recovered asphalt did not have the same profile relationships as original asphalt, indicating that recovered asphalt, after going through the extraction and recovery procedure may be chemically altered and no longer represents the in-place asphalt and/or aged asphalt. Indeed, it is not certain that some of the test methods used to age asphalt actually duplicate the chemical changes made by the asphalt under natural weathering and in contact with mineral aggregates. In fact, the bulk fractional composition analysis of original and laboratory-aged asphalt should be analyzed separately from recovered asphalt when investigating the relationships between chemical composition and the physical properties as well as the behavior of asphalt.

For the most part, there will be a correlation between one or more selected physical properties with the bulk composition. These correlations will vary with higher temperatures and whether or not this is due to the effect of molecular shape and geometry which are minimized in the higher-temperature range. But, overall, it is difficult to arrive at recommendations with regard to the relationship between chemical composition and asphalt physical properties. Furthermore, the interaction of two or more components of the asphalt (as for heavy oil), as well as other molecular properties, can greatly influenced rheological behavior (Speight, 2014).

Moreover, a certain level of generalization of rheological and chemical behavior on original and aged asphalt may be made by investigating a group of different asphalts that are known to vary in performance when placed in service. In addition, different asphalts do age differently and the varied behavior patterns when comparing the original to the aged asphalt suggest that, to better characterize asphalt properties after aging, a thorough investigation of the aging process is necessary. In fact, measuring absolute changes of asphalt properties based on one aging condition will not reflect the overall aging behavior of the asphalt. The conditions under which aging occurs should be considered, as should the rate of aging and the chemistry of aging.

Furthermore, drawing meaningful conclusions regarding the low-temperature behavior of asphalt and the relationship to asphalt is also difficult. There are suggestions that other molecular properties (such as molecular size, molecular structuring, and molecule geometry) may be more important than fraction composition as related to low-temperature behavior.

On the positive side, it is known that asphalt samples stored in sealed cans did not show significant variations in their physical properties. Minor variations do not give a clear indication that physical changes were due to aging and these variations were attributed to the reproducibility of the test results. Physical properties did show some correlation with all four fractions.

To recap, it is not surprising that different asphalts behave differently and age differently. The different types of behavior when changing from original to aged materials suggest that more than one aging condition must be investigated. As a result, measuring absolute changes of asphalt properties based on one aging condition may not reflect the overall aging behavior.

In summary, there is not a strong correlation of pavement performance with an individual property. Rather, pavement performance statistically relates to many physical and chemical properties of asphalt. However, the physical and chemical properties that may enter into the correlation depend on which parameters are used to evaluate pavement conditions.

REFERENCES

AASHTO T240, 2014. Standard Method of Test for Effect of Heat and Air on a Moving Film of Asphalt (Rolling Thin-Film Oven Test) (ASTM D2872). American Association of State Highway and Transportation Officials, Washington, DC.

AASHTO T283, 2014. Standard Method of Test for Resistance of Compacted Asphalt Mixtures to Moisture-Induced Damage. American Association of State Highway and Transportation Officials, Washington, DC.

Abraham, H., 1945. Asphalts and Allied Substances: Their Occurrence, Modes of Production, Uses in the Arts and Methods of Testing, fifth ed. Van Nostrand, New York, NY.

ASTM D5, 2014. Standard Test Method for Penetration of Bituminous Materials Annual Book of Standards. ASTM International, West Conshohocken, PA.

ASTM D36, 2014. Standard Test Method for Softening Point of Bitumen (Ring-and-Ball Apparatus) Annual Book of Standards. ASTM International, West Conshohocken, PA.

ASTM D92, 2014. Standard Test Method for Flash and Fire Points by Cleveland Open Cup Tester Annual Book of Standards. ASTM International, West Conshohocken, PA.

ASTM D113, 2014. Standard Test Method for Ductility of Bituminous Materials Annual Book of Standards. ASTM International, West Conshohocken, PA.

ASTM D893, 2014. Standard Test Method for Insolubles in Used Lubricating Oils. Annual Book of Standards. ASTM International, West Conshohocken, PA.

ASTM D1754, 2014. Standard Test Method for Effects of Heat and Air on Asphaltic Materials (Thin-Film Oven Test) Annual Book of Standards. ASTM International, West Conshohocken, PA.

ASTM D2007, 2014. Standard Test Method for Characteristic Groups in Rubber Extender and Processing Oils and Other Petroleum-Derived Oils by the Clay-Gel Absorption

Chromatographic Method Annual Book of Standards. ASTM International, West Conshohocken, PA.
ASTM D2170, 2014. Standard Test Method for Kinematic Viscosity of Asphalts (Bitumens) Annual Book of Standards. ASTM International, West Conshohocken, PA.
ASTM D2872, 2014. Standard Test Method for Effect of Heat and Air on a Moving Film of Asphalt (Rolling Thin-Film Oven Test) Annual Book of Standards. ASTM International, West Conshohocken, PA.
ASTM D3279, 2014. Standard Test Method for n-Heptane Insolubles Annual Book of Standards. ASTM International, West Conshohocken, PA.
ASTM D3381, 2014. Standard Specification for Viscosity-Graded Asphalt Cement for Use in Pavement Construction Annual Book of Standards. ASTM International, West Conshohocken, PA.
ASTM D4124, 2014. Standard Test Method for Separation of Asphalt into Four Fractions Annual Book of Standards. ASTM International, West Conshohocken, PA.
ASTM D5546, 2014. Standard Test Method for Solubility of Asphalt Binders in Toluene by Centrifuge Annual Book of Standards. ASTM International, West Conshohocken, PA.
ASTM D6703, 2014. Standard Test Method for Automated Heithaus Titrimetry Annual Book of Standards. ASTM International, West Conshohocken, PA.
Barth, E.J., 1962. Asphalt: Science and Technology. Gordon and Breach Publishes, Taylor & Francis Group, Abingdon, Oxfordshire, UK.
Benedetto, A.T., Lottman, L.P., Cratin, P.D., Ensley, E.K., 1970. Asphalt Adhesion and Interfacial Phenomena. Highway Research Record No. 340. Highway Research Board, National Research Council, Washington, DC.
Bhairampally, R.K., Lytton, R.L., Little, D.N., 2000. Numerical and Graphical Method to Assess Permanent Deformation Potential for Repeated Compressive Loading of Asphalt Mixtures. Transportation Research Record: Journal of the Transportation Research Board, No. 1723. T Transportation Research Board, National Research Council, Washington, DC, pp. 150–158.
Branthaver, J.F., Petersen, J.C., Robertson, R.E., Duvall, J.J., Kim, S.S., Harnsberger, P.M., et al., 1993. Binder Characterization and Evaluation, Volume 2: Chemistry Final Report, SHRP A-368. Strategic Highway Research Program, National Research Council, Washington, DC.
Broome, D.C., 1973. In: Hobson, G.D., Pohl, W. (Eds.), Modern Petroleum Technology Applied Science Publishers Inc., Barking, Essex, UK. (Chapter 23).
Broome, D.C., Wadelin, F.A., 1973. In: Allinson, J.P. (Ed.), Criteria for Quality of Petroleum Products Halsted Press, Toronto, ON. (Chapter 13).
Brule, B., Ramond, G., Such, C., 1986. Relationships Between Composition, Structure and Properties of Road Asphalts: State of Research at the French Public Works Central Laboratory: Transportation Research Record No. 1096. Transportation Research Board, National Research Council, Washington, DC, pp. 22–24.
Castan, M., 1968. Rising of binder to the surface of an open-graded bituminous mix. Bull. Liaison Lab. Rout. 33, 77–84.
Cheng, D., Little, D.N., Lytton, R.L., Holste, J.C., 2002. Surface Energy Measurement of Asphalt and Its Application to Predicting Fatigue and Healing in Asphalt Mixtures. Transportation Research Record: Journal of the Transportation Research Board, No. 1810. Transportation Research Board, National Research Council, Washington, DC, p. 4453.
Corbett, L.W., 1969. Composition of asphalt based on generic fractionation, using solvent deasphaltening, elution-adsorption chromatography, and densimetric characterization. Anal. Chem. 41 (4), 576–579.

Corbett, L.W., 1970. Relationship between composition and physical properties of asphalt. Proc. Assoc. Asphalt Paving Technol. 39, 481–491.

Corbett, L.W., 1975. Reaction variables in the air blowing of asphalt. Ind. Eng. Chem. Process Des. Dev. 14 (2), 181–187.

Corbett, L.W., Merz, R.E., 1975. Asphalt Binder Hardening in the Michigan Test Road After 18 Years of Service. Transportation Research Record 544. TRB, National Research Council, Washington, DC, pp. 27–34.

Curtis, C.W., 1992. Fundamental Properties on Asphalt Aggregate Interactions Adhesion and Adsorption Final Report on Contract A-003B. Strategic Highway Research Program, National Research Council, Washington, DC.

Curtis, C.W., Lytton, R.L., Brannan, C.J., 1992. Influence of Aggregate Chemistry on the Adsorption and Desorption of Asphalt. Transportation Research Record No. 1362. Transportation Research Board, National Research Council, Washington, DC, pp. 1–9.

Curtis, C.W., Ensley, K., Epps, J., 1993. Fundamental Properties of Asphalt-Aggregate Interactions Including Adhesion and Absorption Report No. SHRP-A-341. Strategic Highway Research Program, National Academy of Sciences, National Research Council, Washington, DC, December.

Dave, E.V., Buttlar, W.G., Paulino, G.H., 2010. Asphalt pavement aging and temperature dependent properties through a functionally graded viscoelastic model, part-II: applications. Mater. Sci. Forum 631–632, 53–58. Trans Tech Publications, Pfaffikon, Switzerland.

Dorrence, S.M., Barbour, F.A., Petersen, J.C., 1974. Direct evidence of ketones in oxidized asphalts. Anal. Chem. 46, 2242–2244.

Dybalski, J.N., 1976. The chemistry of asphalt emulsions Proceedings. 55th Annual Meeting. Transportation Research Board, Washington, DC, January.

Eilers, H., 1949. The colloidal structure of asphalt. J. Phys. Chem. 53 (8), 1195–1211.

Fromm, H.J., 1974. The mechanisms of asphalt stripping from aggregate surfaces. Proc. Assoc. Asphalt Paving Technol. 43, 191–223.

Giavarini, C., Mastrofini, D., Scarsella, M., 2000. Macrostructure and rheological properties of chemically modified residues and bitumens. Energy Fuels 14, 495–502.

Goodrich, J.G., Goodrich, J.E., Kari, W.J., 1986. Asphalt composition tests: their application and relation to field performance. In: Proceedings. Annual Meeting. Transportation Research Board Meeting, Washington, DC. December 15.

Gzemski, F.C., McGlashan, D.W., Dolch, W.L., 1968. Highway Research Circular No. 78. Thermodynamic Aspects of the Stripping Problem. Highway Research Board, National Research Council, Washington, DC.

Hanson, W.E., 1964. In: Hoiberg, A.J. (Ed.), Bituminous Materials: Asphalts, Tars, and Pitches. Volume I Interscience Publishers, New York, NY. (Chapter 1).

Hattingh, M.M., 1984. The fractionation of asphalt. Proc. Assoc. Asphalt Paving Technol. 53, 197–215.

Haxo Jr., J.E., 1976. Assessing synthetic and admixed materials for liner landfills, gas and leachate from landfills; formation, collection and treatment. In: Genetelli, E.J., Circello, J. (Eds.), EPA Report No. 600/9-76-004; NTIS Report No. PB 251161 United States Environmental Protection Agency, Cincinnati, OH, pp. 130–158.

Hoiberg, A.J., 1964. Bituminous Materials: Asphalts, Tar, and Pitches. Interscience Publishers, New York, NY.

Hveem, F.N., Zube, E., Skog, J., 1959. Progress Report. The Zaca-Wigmore Experimental Asphalt Test Project Special Technical Publication No. 277. American Society for Testing and Materials (ASTM International)., pp. 1–45.

Ishai, I., Brulé, B., Vaniscote, J.C., Raimond, G., 1988. Proceedings. Annual Meeting Association of Asphalt and Paving Technologists, Williamsburg, VA. March 1–2.

Jennings, P.W., 1985. Prediction of asphalt performance by HP-GPC. Proc. Assoc. Asphalt Paving Technol. 54, 635–651.

Jeon, W.Y., Curtis, C.W., Kiggundu, B.M., 1988. Adsorption Behavior of Asphalt Functionalities on Dry and Moist Silica. Transportation Research Board, National Research Council, Washington, DC.

Jones, D.R., 1992. An Asphalt Primer: Understand How the Origin and Composition of Paving-Grade Asphalt Cements Affect Their Performance. SHRP Technical Memorandum No. 4. Strategic Highway Research Program, National Research Council, Washington, DC.

Kassir, A.F., 2009. Fractional chemical composition of asphalt as a function of its durability. Ibn Al-Haitham J. Pure Appl. Sci. 22 (4), 1–6.

Kennedy, T.W., Roberts, F.L., Anagnos, J.N., 1984a. Texas Boiling Test for Evaluating Moisture Susceptibility of Asphalt Mixtures Report No. FHWA/TX-85/63 + 253-5. Center for Transportation Research The University of Texas at Austin/Transportation Planning Division, Texas State Department of Highways and Public Transportation, Austin, Texas.

Kennedy, T.W., Roberts, F.L., Lee, K.W., 1984b. Evaluating Moisture Susceptibility of Asphalt Mixtures Using the Texas Boiling Test. Transportation Research Record No. 968. Transportation Research Board, National Research Council, Washington, DC, pp. 45–54.

Kiggundu, B.M., Roberts, F.L., 1988. The Success/Failure of Methods Used to Predict the Stripping Potential in the Performance of Bituminous Pavement Mixtures. Transportation Research Board, National Research Council, Washington, DC.

Kim, O.K., Bell, C.A., Wilson, J.E., Boyle, G., 1986. Effect of Moisture and Aging on Asphalt Pavement Life—Part 2—Effect of Aging Report No. FHWA-OR-RD-86-01-2. Oregon Department of Transportation, Salem, OR, January.

Kim, K.W., Doh, Y.S., Amirkhanian, S.N., 2004. Evaluation of aging characteristics of selected PMA using HP-GPC. J. Korean Soc. Pavement Eng. 6 (2), 15–24.

Kleinschmidt, R., Snoke, H.R., 1958. Changes in the properties of an asphalt during the blowing operation. Research paper 2835. J. Res. Natl. Bur. Stand. 60 (3), 169–172.

Kluttz, R., 2012. Considerations for use of alternative binders in asphalt pavements: material characteristics Proceedings. Alternative Binders for Sustainable Asphalt Pavements. Transportation Research Circular E-C165. E-C147. Transportation Research Board, National Research Council, Washington, DC, pp. 2–6. December.

Koots, J.A., Speight, J.G., 1975. The relation of petroleum resins to asphaltenes. Fuel 54, 179.

Lau, C.K., Lunsford, K.M., Glover, C.J., Davison, R.R., Bullin, J.A., 1992. Reaction Rates and Hardening Susceptibilities as Determined from Pressure Oxygen Vessel Aging of Asphalts. Transportation Research Record No. 1342. Transportation Research Board, National Research Council., pp. 50–57.

Long, R.B., Speight, J.G., 1989. Studies in petroleum composition. I: Development of a compositional map for various feedstocks. Rev. Inst. Français du Pétrole. 44, 205.

Long, R.B., Speight, J.G., 1998. The composition of petroleum. In: Speight, J.G. (Ed.), Petroleum Chemistry and Refining Taylor & Francis, Philadelphia, PA, pp. 1–38. (Chapter 1).

Lottman, R.P., Frith, D.J., 1989. Predicting the Effects of Moisture on Wheelpath Rutting in Asphalt Concrete. Transportation Research Record No. 1228. Transportation Research Board, National Research Council, Washington, DC.

Lytton, R.L., Masad, E.A., Zollinger, C., Bulut, R., Little, D., 2005. Measurements of Surface Energy and Its Relationship to Moisture Damage Report No. FHWA/TX-05/0-4524-2. Research and Technology Implementation Office, Texas Department of Transportation, Austin, TX.

Majidzadeh, K., Brovold, F.N., 1968. State of the Art: Effect of Water on Bitumen–Aggregate Mixtures Special Report No. 98. HRB, National Research Council, Washington, DC.

Masad, E., Kassem, E., Chowdhury, A., You, Z., 2009. A Method for Predicting Asphalt Mixture Compactability and Its Influence on Mechanical Properties Report No. 0-5261-2. Texas Transportation Institute, Texas A&M University System, College Station, TX.
Mirza, M.W., Witczak, M.W., 1996. Development of a global aging system for short and long term aging of asphalt cements Proceedings. Annual Meeting, vol. 64. Association of Asphalt Paving Technologists., pp. 393–430.
Mohammed, A.-H.A.-K., Eweed, K.M., 2012. Prediction of the chemical composition and physical properties of aged asphalt cement. Iraqi J. Chem. Pet. Eng. 13 (4), 1–12.
Mohammed, A.-H.A.-K., Morshed, K., 2008. The effect of aging on physical and chemical properties of asphalt cement. Iraqi J. Chem. Pet. Eng. 9 (2), 9–15.
Moschopedis, S.E., Speight, J.G., 1973. The oxidation of petroleum fractions. Fuel 52, 83.
Moschopedis, S.E., Speight, J.G., 1975. Oxidation of bitumen. Fuel 54, 210.
Moschopedis, S.E., Speight, J.G., 1977. The effect of air blowing on the properties and constitution of natural bitumen. J. Mater. Sci. 12, 990.
Moschopedis, S.E., Speight, J.G., 1978. The influence of metal salts on bitumen oxidation. Fuel 57, 235.
Nourendin, A.S., Wood, L.E., 1989. Variation in Molecular Size Distribution Virgin and Recycled Binders Associated with Aging. Transportation Research Record: Journal of the Transportation Research Board, No. 1228. Transportation Research Board, National Research Council, Washington, DC, pp. 191–197.
Pauli, A.T., Branthaver, J.F., 1998. Relationship between asphaltenes, heithaus compatibility parameters, and asphalt viscosity. Pet. Sci. Technol. 16 (9–10), 1125–1147.
Petersen, J.C., 1984. Chemical composition of asphalt as related to asphalt durability. Transportation Research Record No. 999 Proceedings. 63rd Annual Meeting. Transportation Research Board, National Research Council, Washington, DC.
Petersen, J.C., 2009. A Review of the Fundamentals of Asphalt Oxidation: Chemical, Physicochemical, Physical Property, and Durability Relationships. Transportation Research Circular E-C140. Transportation Research Board, Washington, DC.
Petersen, J.C., Dorrence, S.M., Nazir, M., Plancher, H., Barbour, F.A., 1981. Oxidation of sulfur compounds in petroleum residues: reactivity-structural relationships Preprints. Div. Petrol. Chem. A. Chewm. Soc. 26 (4), 898–906.
Petersen, J.C., Plancher, H., Ensley, E.K., Venable, R.L., Miyake, G., 1982. Chemistry of Asphalt-Aggregate Interactions: Relationship with Pavement Moisture Damage Prediction Test. Transportation Research Record No. 843. Transportation Research Board, National Research Council, Washington, DC, pp. 95–104.
Petersen, J.C., Plancher, H., Harnsberger, P.M., 1987. Lime Treatment of Asphalt. Final Report. National Lime Association, Arlington, VA.
Petersen, J.C., Branthaver, J.F., Robertson, R.E., Harnsberger, P.M., Duvall, J.J., Ensley, E.K., 1993. Effects of Physicochemical Factors on Asphalt Oxidation Kinetics. Transportation Research Record no. 1391. Transportation Research Board, National Research Council., pp. 1–10.
Plancher, H., Dorrence, S.M., Petersen, J.C., 1977. Identification of chemical types in asphalts strongly absorbed at the asphalt-aggregate interface and their relative displacement by water. Proc. Assoc. Asphalt Paving Technol. 46, 151–175.
Por, N., 1992. Stability Properties of Petroleum Products. The Israel Institute of Petroleum and Energy, Ministry of National infrastructures, Energy, and Water Resources, Haifa, Israel.
Rice, J.M., 1958. Relationship of aggregate characteristics to the effect of water on bituminous paving mixtures Proceedings. Symposium on Effect of Water on Bituminous Paving Mixtures ASTM STP No. 240. ASTM International, West Conshohocken, PA, pp. 17–34.

Schmidt, R.J., Graf, P.E., 1972. The effect of water on the resilient modulus of asphalt treated mixes. Proc. Assoc. Asphalt Paving Technol. 41, 118–162.

Scott, J.A.N., 1978. Adhesion and disbonding mechanisms of asphalt used in highway construction and maintenance. Proc. Assoc. Asphalt Paving Technol. 47, 19–48.

Speight, J.G., 1992a. Asphalt. In: Kirk-Othmer's Encyclopedia of Chemical Technology, fourth ed., vol. 3, pp. 689–724.

Speight, J.G., 1992b. A chemical and physical explanation of incompatibility during refining operations. In: Proceedings. 4th International Conference on the Stability and Handling of Liquid Fuels. U.S. Department of Energy (DOE/CONF-911102), p. 169.

Speight, J.G., 1994. Chemical and physical studies of petroleum asphaltenes In: Yen, T.F. Chilingarian, G.V. (Eds.), Asphaltenes and Asphalts, I. Developments in Petroleum Science, 40 Elsevier, Amsterdam. (Chapter 2).

Speight, J.G., 2009. Enhanced Recovery Methods for Heavy Oil and Tar Sands. Gulf Publishing Company, Houston, TX.

Speight, J.G., 2013. The Chemistry and Technology of Coal, third ed. CRC Press, Taylor & Francis Group, Boca Raton, FL.

Speight, J.G., 2014. The Chemistry and Technology of Petroleum, fifth ed. CRC Press, Taylor & Francis Group, Boca Raton, FL.

Speight, J.G., 2015a. Handbook of Petroleum Product Analysis, second ed. John Wiley & Sons Inc., Hoboken, NJ.

Speight, J.G., 2015b. Fouling in Refineries. Gulf Professional Publishing, Elsevier, Oxford, UK.

Speight, J.G., Long, R.B., Trowbridge, T.D., 1984. Factors influencing the separation of asphaltenes from heavy petroleum feedstock. Fuel 63, 616–620.

Tarrer, A.R., 1996. Use of hydrated lime to reduce hardening and stripping in asphalt mixtures. In: Proceedings. 4th Annual International Center for Aggregate Research Symposium, Atlanta, GA.

Taylor, M.A., Khosla, N.P., 1983. Stripping of Asphalt Pavements: State of the Art. Transportation Research Record No. 911. Transportation Research Board, National Research Council, Washington, DC, pp. 150–158.

Terrel, R.L., Al-Swailmi, S., 1994. Water Sensitivity of Asphalt–Aggregate Mixes: Test Selection SHRP Report A-403. Strategic Highway Research Program, National Research Council, Washington, DC.

Terrel, R.L., Rimsritong, S., 1979. Proc. Assoc. Asphalt Paving Technol. 48, 111.

Thelen, E., 1958. Surface Energy and Adhesion Properties in Asphalt-Aggregate Systems. Bulletin No. 192. Highway Research Board, National Research Council, Washington, DC, pp. 63–74.

Thenoux, G., Lees, G., Bell, C.A., 1985. Laboratory investigation of the fraass brittle test. Proc. Assoc. Asphalt Paving Technol. 54, 529–550.

Thenoux, G., Bell, C.A., Wilson, J.E., Eakin, D., Schroeder, M., 1986. Evaluation of Asphalt Properties and Their Relationship to Pavement Performance—Part 1 Interim Report No. FHWA-OR-86-02. Oregon Department of Transportation, Salem, OR, February 1986.

Thomas, K.P., 2002. Impact of water during the laboratory aging of asphalt. Road Mater. Pavement Des. 3 (3), 299–315.

Titova, T.I., Kosheleva, L.S., Zhdanov, S.P., 1987. IR study of hydroxylated silica. Langmuir 3 (6), 960–967.

Traxler, R.N., 1961. Asphalt: Its Composition, Properties, and Uses. Reinhold Publishing Corp., New York, NY.

Tunnicliff, D., Root, R., 1984. Use of Antistripping Additives in Asphaltic Concrete Mixtures NCHRP Report No. 274. Transportation Research Board, National Research Council, Washington, DC.

WHO, 2004. Asphalt (Bitumen). Concise International Chemical Assessment Document No. 59. World Health Organization, Geneva, Switzerland.
Wolfe, D.L., Armentrout, D., Arends, C.B., Baker, H.M., Petersen, J.C., 1986. Crude Source Effects on the Chemical, Morphological, and Viscoelastic Properties of Styrene/ Butadiene Latex Modified Asphalt Cements. Transportation Research Record No. 1096. Transportation Research Board, National Research Council, Washington, DC, pp. 12–21.
Yoon, H.H., Tarrer, A.R., 1988. Effect of aggregate properties on stripping Transportation Research Record No. 1171. Transportation Research Board, National Research Council, Washington, DC, pp. 37–43.

CHAPTER 7

Asphaltene Deposition

7.1 INTRODUCTION

The term *asphalt* as used in this book and elsewhere (Speight, 2014) is used to indicate that the product is a manufactured product. It is not a residuum or tar (the volatile high-boiling product from the thermal decomposition of carbonaceous material such as coal and biomass), or pitch (the non-volatile residue from the thermal decomposition of carbonaceous material such as coal and biomass), or naturally occurring bitumen. Using this nomenclature and understanding the terminology, convention will alleviate some of this. Thus, the term *bitumen* is used to designate the naturally occurring material, and the term *asphalt* is used to designate the petroleum product.

Asphalt manufacture is, in essence, a matter of distilling everything possible from crude petroleum until a residue with the desired properties is obtained (Speight, 2014). This is usually done by stages: petroleum distillation at atmospheric pressure removes the lower-boiling fractions and yields reduced crude that may contain higher-boiling (lubricating) oils, asphalt, and even wax. Distillation of the reduced crude under vacuum removes the oils (and wax) as overhead products and the asphalt remains as a bottom (or residual) product. The majority of the polar functionality in the original crude oil end to in non-volatile constituents and concentrate in the vacuum residuum. It is this concentration effect that confers upon asphalt some of its unique properties. At this stage the asphalt is frequently and incorrectly referred to as pitch and has a softening point related to the amount of oil removed: the more oil distilled from the residue, the higher the softening point.

However, since there are wide variations in refinery operations and the properties of the crude oil (Speight, 2014), asphalt with softening points ranging from 25°C to 55°C (80°F to 130°F) may be produced. Blending with higher and lower softening point asphalt may make asphalt of intermediate softening points. If lubricating oils are not required, the reduced

Asphalt Materials Science and Technology.
DOI: http://dx.doi.org/10.1016/B978-0-12-800273-5.00007-6

crude may be distilled in a flash drum, which is similar to a bubble tower but has few, if any, trays. Asphalt descends to the base of the drum as the oil components pass out of the top of the drum. If the asphalt has a relatively low softening point, it can be hardened by further distillation with steam or by oxidation. Asphalt is also produced by propane deasphalting (see Chapter 2) and the asphalt so produced may have a softening point of about 95°C (200°F). Softer grades are made by blending *hard asphalt* with the extract obtained in the solvent treatment of lubricating oils.

If the asphalt requires a change in properties to a harder material, such as is required for roofing, pipe coating, or as an under-sealant or waterproofing material, *soft asphalt* can be converted into harder asphalt by oxidation. This process produces a material that softens at a higher temperature than paving asphalts. It may be air blown at the refinery, at an asphalt processing plant, or at a roofing material plant (Speight, 2014). The asphalt is heated to approximately 260°C (500°F) and air is bubbled through the hot mix for 1–5 h. The product, when cooled, remains in the liquid phase.

Asphalt can be represented as a compendium of four major fractions: (i) saturate constituents, (ii) aromatic constituents, (iii) resin constituents, and (iv) asphaltene constituents (Figure 7.1). Two other fractions,

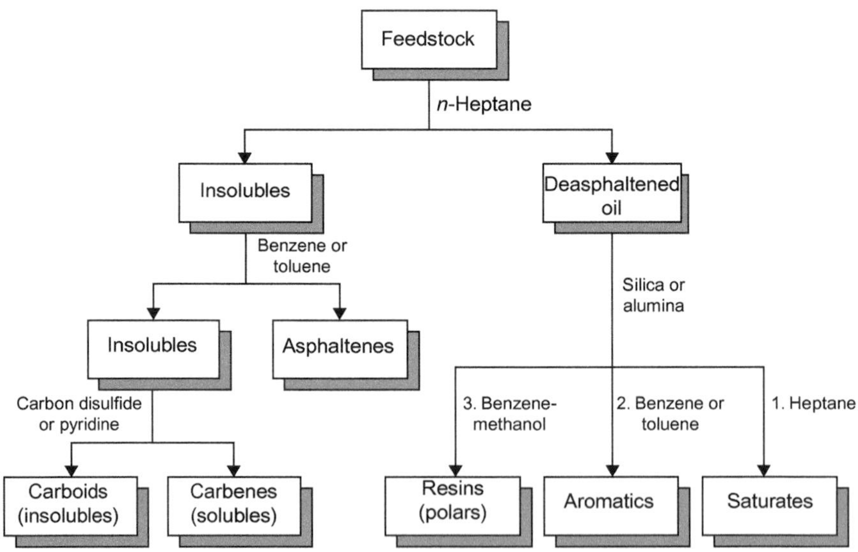

Figure 7.1 General fractionation scheme and nomenclature of petroleum fraction; carbenes and carboids are generally recognized to be thermally generated refinery product fractions but can be generated by oxidation during asphalt service.

conventionally named as carbenes and carboids, are generally recognized as thermal products, although the name can also be used to apply to product formed from other reactions, such as the oxidation of asphaltene constituents. The presence of asphaltene constituents in asphalt is particularly acute when the circumstances permit their separation (precipitation) from the feedstock in such a manner that the phenomenon of phase separation causes asphalt instability and pavement deterioration during service. This is analogous to the phenomenon known as fouling (the formation of a separate phase) that occurs during petroleum recovery and refining when external forces are applied and many parallels can be drawn by a combination of the two (Speight, 2015b).

It will become immediately apparent from the contents of this chapter that the majority of the work carried out on the higher-boiling constituents of petroleum concentrated on the asphaltene fraction. Parallel work on the resin and oil fractions is extremely sketchy and has not been considered worthwhile, insofar as it is generally believed that the constituents of the asphaltene fraction have a markedly adverse effect on the processability of crude oils. Furthermore, it is the asphaltene constituents that also play a role in the rheological properties of asphalt (Nellensteyn, 1924, 1938; Katz and Beu, 1945; Speight, 2014; Ancheyta et al., 2002; Sheu, 2002).

Briefly, the asphaltene fraction is the most refractory or the highest-molecular-weight fraction of petroleum (Speight, 1994, 2014). A generally accepted definition of the asphaltene constituents (asphaltene fraction) is in terms of their solubility in low-boiling liquid hydrocarbons (Speight, 1994, 2001, 2014, 2015a). Thus, in many vacuum residua, where the asphaltene fraction is a major component, it is the propensity of the asphaltene constituents to produce a separate phase when reacted under specified conditions, such as oxidation, in the current context of asphalt. In this context, the ultimate goal is to prevent the formation of a separate asphalt phase and prevent deterioration of the asphalt mix (and, hence, the asphalt roadway) through detachment of the binder from aggregate. In order to accomplish this goal, an understanding of the fundamental chemistry of the asphaltene constituents can be rewarding. Also, the complex nature of the asphaltene fraction can be taken several steps further and made even more formidable when it is acknowledged that the constituents are *multidimensional in space* in contrast to the two-dimensional aspects often represented, for convenience, *on paper* (Speight, 1994, 2014). Such knowledge will also offer new insights into the thermal chemistry of the asphaltene constituents and their role in the structure and behavior of asphalt.

In addition to understanding the chemical behavior of the asphaltene constituents, it is also necessary to understand the behavior of the resin constituents as a means of truly understanding asphalt behavior. The resin constituents are also capable of playing a role in the phase separation process and the means by which these constituents can also react and phase separate. Further understanding might be added through investigations of the relationships between the asphaltene constituents and the resin constituents (as well as the other constituents) of asphalt. Such investigations will help in developing behavioral traits and afford some degree of predictability of the outcome of asphalt deterioration in-service. An example, of asphaltene behavior and resin behavior (or misbehavior) is the asphalt air-blowing (oxidation) process.

The term *resin* generally implies material that has been eluted from various solid adsorbents, whereas the term *maltenes* (or *petrolenes*) (Speight, 2014) indicates a mixture of the resin constituents and oil constituents obtained in the filtrates from the asphaltene precipitation. Thus, after the asphaltenes are precipitated, adsorbents are added to the *n*-pentane solutions of the resins and oils, by which process the resins are adsorbed and subsequently recovered by use of a more polar solvent and the oils remain in solution (Speight, 1994, 2001, 2014, 2015a).

Thus, this chapter summarizes the data related to the character of the asphaltene and resin constituents and the interrelationships of these constituents in asphalt. This can then be used to explain various behavioral aspects, including the chemical and physical events that lead to sediment formation within the asphalt and related phenomena that contribute to diminished performance of the asphalt.

7.2 ASPHALTENE AND RESIN CONSTITUENTS

Under typical circumstances when external influences have not had any effect, asphalt is stable. However, when such disturbances occur to the organic constituents, it is the higher-molecular-weight constituents (the asphaltene constituents and the resin constituents) that are most seriously affected, eventually leading to incompatibility (precipitation, sediment formation, sludge formation) depending upon the external forces and circumstances. Thus, the dispersability of the higher-molecular-weight constituents in asphalt becomes an issue that needs attention. And one of the ways by which this issue can be understood is to be aware of the chemical and physical character of the higher-molecular-weight constituents. By such means, the issue of stability (or dispersability) and the attending issue of incompatibility can be understood and even predicted.

Asphaltene constituents are best known for the problems they cause as solid deposits that obstruct flow in the petroleum production systems as well as the formation of coke during processing. Asphaltene separation (phase-separation) can also be a cause of asphalt instability. A better understanding of the effect of asphaltene constituents and resin constituents is necessary to prevent the formation of deposits during production and refining and mitigate the deleterious effects of these petroleum constituents (Speight, 2014, 2015a,b). In addition, the resin constituents of petroleum and their function within the asphalt system also should be considered because the influence of these constituents on the properties and performance of asphalt is of unquestionable importance.

For the purposes of this chapter, the *resin constituents* are those materials soluble in *n*-pentane or *n*-heptane (i.e., whichever hydrocarbon is used for the separation of asphaltene constituents) but insoluble in liquid propane. The term *resin* generally implies material that has been eluted from various solid adsorbents (Koots and Speight, 1975; Andersen and Speight, 2001; Speight, 2001, 2014, 2015a). Thus, after the asphaltene constituents are precipitated, adsorbents are added to the *n*-pentane or *n*-heptane solutions of the resin constituents and oils, by which process the resin constituents are adsorbed and subsequently recovered by use of a more polar solvent and the oils remain in solution.

The *oils fraction* (comprising the *saturates fraction* plus the *aromatics fraction*) is, therefore, that fraction of crude oil soluble in *n*-pentane or in *n*-heptane that is extractable from an earth by *n*-pentane or by *n*-heptane (Figure 7.1). The term *maltenes* (sometime called *petrolenes*) indicates a mixture of the resin constituents and oil constituents obtained as the filtrate from the asphaltene precipitation (Speight, 2014, 2015a). However, the resin constituents cannot be extracted from an adsorbent such as clay or alumina by *n*-pentane or *n*-heptane. Thus, resin constituents are soluble in the liquids that precipitate asphaltene constituents and are usually soluble in most organic liquids, except in the lower alcohols and acetone, but they are precipitated by liquid propane and liquid butanes. The resin constituents often co-precipitate with the asphaltenes in controlled propane deasphalting procedures, and the product, called *propane asphalt*, contains appreciable amounts of adsorbed resins and has the properties of a low-melting-point asphalt. The resins are dark, semisolid or solid, very adhesive materials of high molecular weight. Their composition can vary depending on the kind of precipitating liquid and on the temperature of the liquid system. They become quite fluid on heating but often show pronounced brittleness when cold.

It is generally recognized and accepted that the asphalt system is a colloid-type system comprising fractions of saturate constituents, aromatic constituents, resin constituents, and asphaltene constituents (Figure 7.1) (Speight, 1992a, 2014)—the names of the respective fractions are a matter of convenience rather than a description of the chemical types. The asphaltene constituents are stabilized by the resin constituents but the asphaltene constituents can be naturally or artificially precipitated if the molecular association with the resin constituents is disturbed. As a result, destabilization (flocculation) and deposition of the asphaltene constituents can occur from asphalt under different circumstances, depending upon the condition and use of the asphalt which causes rupture of the attraction forces between the associated resin constituents and the asphaltene constituents (Koots and Speight, 1975; Branco et al., 2001; Speight, 2014). Thus, the source of the deposited material foulant could be (i) resin or asphaltene constituents, (ii) thermally altered resin or asphaltene constituents, or (iii) a mixture of the former. Careful analysis, with the options to detect specific metal constituents, can give an indication of whether the major mechanism disposition is due to is inorganic constituents, organic constituents, or a combination of the two.

Indeed, although these four fractions are chemically complex, the methods of separation have undergone several modifications to such an extent that the evolution of the separation techniques is a separate subject in itself (Speight, 2001, 2014, 2015a). Furthermore, because the fractions are in a balanced relationship, the chemical and physical character of these constituents needs reference in this chapter.

7.3 SEPARATION

The separation of the asphalt portion from petroleum can be accomplished by a variety of methods and has been presented in detail elsewhere (Speight and Ozum, 2002; Hsu and Robinson, 2006; Gary et al., 2007; Speight, 2014, 2015a), so further discussion about these techniques is not warranted here. However, it should be noted that, unless otherwise stated, the asphalt materials described in this chapter are those materials separated

Table 7.1 Examples of Different Methods for Asphaltene Separation

Precipitant	Volume Precipitant per gram of Sample	Method
n-Pentane	10 ml	ASTM D893
n-Pentane	10 ml	ASTM D2007
n-Heptane	100 ml	ASTM D3279
n-Heptane	100 ml	ASTM D4124

by non-destructive distillation or by solvent-adsorbent treatment; materials such as cracked residua and visbreaker bottoms as well as materials separated by chemical methods or by other separation (or production) methods that could involve substantial structural changes within the molecular framework of the constituents are not considered here.

Contrary to the belief of many investigators, the asphaltene fraction and the resin fraction are not chemical classes. By definition, each fraction is a *separation class* and the composition of each class is dictated by the method of separation. In the case of the asphaltene fraction, it is the collection of constituents that is precipitated from feedstocks by the addition of 40 volumes of the liquid low-boiling hydrocarbon (Table 7.1) (Girdler, 1965; Mitchell and Speight, 1973; Speight et al., 1982, 1984; Andersen and Birdi, 1990, 1991; Andersen, 1994; Speight, 1994, 2001, 2014, 2015a; Gholoum et al., 2003; Gholoum and Oskui, 2004).

As a recap, the asphaltene fraction is, by definition, a *solubility class* and is the fraction precipitated from feedstocks by the addition of 40 volumes of the liquid hydrocarbon (Girdler, 1965; Mitchell and Speight, 1973; Speight et al., 1982, 1984; Andersen and Birdi, 1990; Andersen, 1994; Speight, 1994; Cimino et al., 1995). The fraction is separated as a dark-brown to black friable solid that has no definite melting point and usually foams and swells on heating to leave a carbonaceous residue. They are obtained from petroleum by the addition of a non-polar solvent (such as a hydrocarbon) with a surface tension lower than that of 25 dyne cm^{-1} at 25°C (77°F), such as low-boiling petroleum naphtha, petroleum ether, *n*-pentane, *iso*-pentane, *n*-heptane, liquefied petroleum gases and the like and, in fact, propane is used commercially in processing petroleum residua for asphalt production (Girdler, 1965; Mitchell and Speight, 1973; Speight, 2014, 2015a). On the other hand, the asphaltene constituents are soluble in liquids with a surface tension above 25 dyne cm^{-1}, such as pyridine, carbon disulfide, carbon tetrachloride, and benzene.

Thus, the nomenclature of the petroleum constituents is an *operational aid* and is not usually based on chemical or structural features. Furthermore, there is no one parameter that is operational in the separation of asphaltenes. The relevant parameters for asphaltene separation are physical and chemical in nature and include:

1. Polarity (the presence of functional groups derived from the presence of heteroatoms in the asphaltenes).
2. Aromaticity (the presence of polynuclear aromatic systems in the asphaltenes).
3. Molecular weight (molecular size).
4. Three-dimensional structure (the micelle) of the asphaltene constituents as they exist in relationship with the other constituents of crude oil.
5. Solvent power of the precipitating/extracting liquid used for the separation.
6. Time required to allow the precipitating/extracting liquid to penetrate the micelle, which is dependent upon the ability of the hydrocarbon liquid to penetrate the micelle, indicating that the process is diffusion-controlled.
7. Ratio of the precipitating/extracting liquid to crude oil, which dictates the yield and character of the asphaltene product.
8. Temperature, which may reduce the induction period that is a requirement of diffusion-controlled processes.
9. Pressure, employed in several refinery processes as a means of maintaining the low-boiling liquid hydrocarbon in the liquid phase.

Other parameters may be defined as subsets of those enumerated above, and it is worthy of note that, in order to remove entrained resin constituents, reprecipitation of the asphaltene constituents from toluene (or a similar suitable solvent) and a hydrocarbon non-solvent may be necessary (Speight et al., 1982, 1984; Ali et al., 1985).

Despite the reported need for an excess (40 volumes per volume of feedstock) of the hydrocarbon liquid, there are still reports of *asphaltenes* being isolated from crude oil by much lower proportions of the precipitating medium, which can lead to errors not only in the determination of the amount of asphaltenes in the crude oil but also in the determination of the compound type. For example, when insufficient proportions of the precipitating medium are used, resins may also appear within the asphaltene fraction by adsorption onto the asphaltenes from the supernatant liquid and can be released by reprecipitation (Speight et al., 1984; Ali et al., 1985). These resins would be isolated at a later stage of the separation procedure by adsorption chromatography if standardized parameters were employed for the separation of the asphaltenes (Speight et al., 1984; ASTM D3279; ASTM D4124). Questionable isolation techniques throw serious doubt on any conclusions drawn from subsequent work on the isolated material (Andersen and Speight, 2001, 2002).

In fact, it is now accepted that to ensure *stable* asphaltene yields it is necessary to employ the following parameters: (i) an excess of the liquid hydrocarbon, >40 ml hydrocarbon per g or per ml feedstock, (ii) use of *n*-pentane or *n*-heptane as the liquid hydrocarbon; volatility constraints and consistency of the asphaltene fraction favor the use of *n*-heptane, (iii) a prolonged contact time on the order of 8–10 h is preferable, and (iv) a precipitation sequence to remove any constituents that separate with the asphaltene constituents. The precipitation sequence involves dissolution of the asphaltenes in benzene or toluene (10 ml per g asphaltene) followed by the addition of the hydrocarbon (50 ml precipitant per ml toluene or benzene) to the solution. This sequence should be repeated three times to remove adsorbed lower-molecular-weight resin material and to provide consistency of the asphaltene fraction.

Briefly, the precipitation of asphaltenes can be ascribed to changes in crude oil composition caused by the addition of lower-boiling components than the complex equilibrium keeping the asphaltenes in solution or in a *peptized state*. In the case of addition of low-boiling liquid hydrocarbons to the feedstock, the hydrocarbon causes a change in the solubility parameter of the oil medium that, in turn, changes the tolerance of the medium for the complex micelle structure. As this occurs, the lower-molecular-weight and less polar constituents of the micelle are extracted into the liquid, leaving the asphaltene constituents without any surrounding (dispersing) sheath. Separation then ensues. If the process is chemical, i.e., due to oxidation, a different sequence of events ensues leading to the formation of an insoluble phase dues to changes in the chemical functionality of the resin-type constituents (reacted resin constituents) and asphaltene-type constituents (reacted asphaltene constituents) (Speight, 1994, 2014).

The focal point of the asphaltene separation method is the isolation of the asphaltene constituents as a discrete fraction. However, a question that often arises is related to whether or not the asphaltenes self-associate in the crude oil or whether the asphaltene constituents interact directly with the other crude oil constituents. In addition, there is also a question related to (i) precipitation of the asphaltenes or (ii) extraction of the constituents from the micelle that is the prime means of the phase separation of the asphaltene constituents from the other components of asphalt.

In the former case (i.e., precipitation of the asphaltene constituents), it is assumed to be the means by which asphaltene constituents are separated from light petroleum and medium petroleum, whereas extraction is the means by which the asphaltene constituents are separated from asphalt—for clarification, the lower-molecular-weight constituents in asphalt feedstocks are presumed to be extracted by the liquid hydrocarbon used for the separation. If the micelle concept is true, extraction of the lower-molecular-weight constituents from the micelle should occur whatever the specific gravity or viscosity of the feedstock, in this case asphalt. The precipitation of asphaltene constituents can be ascribed to changes in asphalt (or petroleum) composition that are caused by the addition of lower-boiling components that alter the complex equilibrium keeping the asphaltenes in solution or in a *peptized state.* The extraction concept is most likely.

In the case of addition of a low-boiling liquid hydrocarbon to the asphalt, the hydrocarbon causes a change in the solubility parameter of the oil medium that, in turn, changes the ability of the medium to hold the complex micelle structure as a dispersed phase. As this occurs, the lower-molecular-weight and less polar constituents of the micelle are extracted into the liquid, leaving the asphaltene constituents without any surrounding (dispersing) sheath. Separation of the asphaltene constituents then ensues.

The fraction is separated as a dark-brown to black friable solid that has no definite melting point and usually foams and swells on heating to leave a carbonaceous residue. The composition of the asphaltene fraction varies with the hydrocarbon employed for the separation (Girdler, 1965; Mitchell and Speight, 1973; Speight, 2014, 2015a) and, in fact, propane is used commercially in processing petroleum residua for asphalt production—the asphalt is a mixture of resin and asphaltene constituents with some heavy gas oil constituents (Speight and Ozum, 2002; Hsu and Robinson, 2006; Gary et al., 2007; Speight, 2014).

The complexity of the asphaltene fraction is reflected in the separation of the fraction by functional group composition (Figure 7.2) (Francisco and Speight, 1984; Andersen et al., 1997) but it cannot be defined by the separation method and the solvent employed. In fact, there is no one parameter that is operational in the separation of the asphaltene fraction (Speight, 1994, 2014, 2015a; Chukwu et al., 2011). The relevant

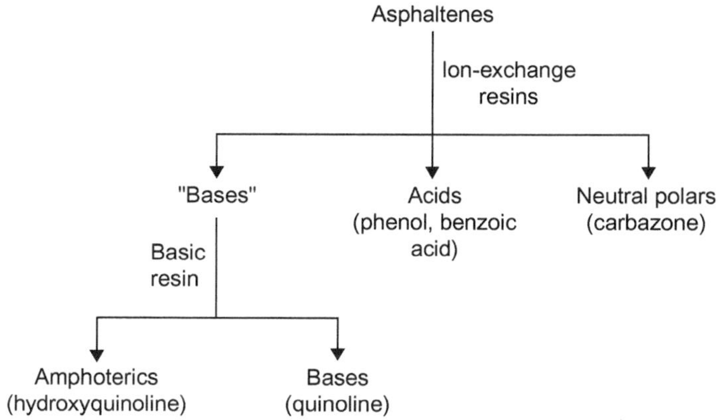

Figure 7.2 Separation of asphaltene constituents based on functionality (polarity).

parameters for asphaltene separation are physical rather than chemical in nature (Girdler, 1965; Mitchell and Speight, 1973; Long, 1979, 1981; Speight and Long, 1981; Hirschberg et al., 1984; Speight et al., 1984; Andersen and Birdi, 1990; Speight, 1992b, 1994; Andersen and Speight, 1992, 1994; Hassanvand et al., 2012). It is also worthy of note that, in order to remove entrained resin material, precipitation of the asphaltene constituents from benzene or toluene is often necessary (Speight et al., 1984; Ali et al., 1985).

In addition to the complexity of the asphalt (especially the complexity of the resin and asphaltene fractions), the control and mitigation of asphaltene disposition requires that it is necessary to consider the roadway conditions such as temperature range, fluid flow rate, as well as chemical composition and, where possible, make such changes as are required by the severity and type of the deposition. This is especially true if the asphalt contains foreign material, such as moisture, that is capable of interfering with the asphalt–aggregate bond.

Finally, another aspect of deposition mitigation is the use of additives. Additives also help control deposition phenomena within critical locations of the process and improve the stability of the asphalt. The additives (the composition of which will be asphalt-dependent) can enable the roadway to achieve maximum, while monitoring helps to determine and control the deposition process.

Briefly, the separation of the asphaltene fraction can be ascribed to changes in asphalt composition and chemistry that occurs during use. In the case of asphalt oxidation (the most common form of asphalt deterioration), the change in the functional groups distribution has a significant effect on resin–asphaltene interaction. If the process is thermal (i.e., at a temperature in excess of the thermal decomposition temperature) a different sequence of events ensues, leading to the formation of an insoluble phase and ultimately to coke (Speight, 2014).

7.4 COMPOSITION

In attempts to define and understand the nature of destabilization of asphaltene constituents, many models have been proposed but few, if any, consider the complexity of the asphaltene fraction (as demonstrated by Bestougeff and Darmois, 1947; Bestougeff and Byramjee, 1994; Speight, 1994, 2014) preferring instead to use *average parameters*, which are followed by so-called *fine tuning* of the model. Furthermore, the effect of resins in the feedstock has been largely ignored in such models and there has been a general failure of the model proponents to recognize that the resin constituents also play a major role in crude oil chemistry and physics insofar as the resin constituents stabilize the asphaltene constituents (Koots and Speight, 1975; Hammami and Freworn, 1998; Carnahan et al., 1999; Gawrys et al., 2003; Speight, 2014, 2015b). For the remainder of this chapter, although the focus may appear to be on the asphaltene constituents, the resin constituents are also included by inference.

However, these investigations ignore the fact that the asphaltene fraction can be separated into various sub-fractions on the basis of polarity (functionality) and molecular weight (using different hydrocarbon solvents) by a variety of techniques (Bestougeff and Darmois, 1947, 1948; Bestougeff and Mouton, 1977; Francisco and Speight, 1984; Speight, 1994). Of specific interest is the observation that when asphaltenes are fractionated on the basis of aromaticity and polarity, it appears that the more aromatic species contain higher amounts of nitrogen (Speight, 1984, 1994, 2014), suggesting that the nitrogen species are located predominantly in aromatic systems.

The solvent-based fractionation of the asphaltene component of petroleum showed that it is possible to obtain asphaltene fractions characterized by different degrees of aromaticity or heteroatom content by using benzene/pentane or benzene/methanol mixtures in variable ratios (Speight, 1979; Andersen et al., 1997; Miller et al., 1998). The use of mixtures of a polar and a non-polar solvent in order to fractionate an asphaltene sample will tend to direct the fractionation by introducing polar forces and hydrogen bonding, as well as dispersion forces, as factors determining which components of the asphaltene sample are soluble in the mixture.

The fractionation of asphaltenes into a variety of functional (and polar) types (Figure 7.2) (Francisco and Speight, 1984) has confirmed the complexity of the asphaltene fraction. High-performance liquid chromatography (HPLC) has also been used to illustrate the diversity of the structural and functional types in asphaltenes to the extent that different HPLC profiles can be expected for different asphaltenes (Speight, 1986, 1991). In summary, fractionation data lend support to, and reinforce, the concept that asphaltenes are complex mixtures of molecular sizes and various functional types (Speight, 1994).

On a molecular basis, the nature and location of the nitrogen, oxygen, and sulfur atoms in the asphaltene structure offer some indication of the behavior of the asphaltene in asphalt. For example, investigations on the disposition of *nitrogen* in petroleum asphaltenes indicated the existence of nitrogen as various heterocyclic types (Clerc and O'Neal, 1961; Nicksic and Jeffries-Harris, 1968; Moschopedis and Speight, 1976b, 1979; Jacobson and Gray, 1987). The more conventional (from the organic chemist's viewpoint) nitrogen species, i.e., primary, secondary, and tertiary aromatic amines have not been established as being present in petroleum asphaltenes. There are also reports in which the organic nitrogen in petroleum asphaltenes has been defined in terms of basic and non-basic types (Nicksic and Jeffries-Harris, 1968). Spectroscopic investigations (Moschopedis and Speight, 1979) suggest that carbazole derivatives occur in asphaltenes that support earlier mass spectroscopic evidence (Clerc and O'Neal, 1961) for the occurrence of carbazole nitrogen in asphaltenes. The application of x-ray absorption near-edge structure (XANES) spectroscopy to the investigation of the asphaltene constituents has led to the conclusion that a large portion of the nitrogen present is in the pyrrole rather than pyridine form (Mitra-Kirtley et al., 1993), although it is evident that the location of nitrogen in the asphaltene constituents will vary according to the source and maturation paths of the crude oil.

In addition, investigations of the thermal behavior of the asphaltene constituents have shown that only 1% (w/w) of the nitrogen is lost during thermal decomposition, but substantially more sulfur (23%, w/w) and almost all the oxygen (81%, w/w) is lost as a result of this treatment. The tendency for nitrogen and sulfur to remain in the non-volatile residue produced during thermal decomposition, as opposed to the relatively facile elimination of oxygen, supports the concept that nitrogen and sulfur have stability because of their location in ring systems. These types of nitrogen will (for example) readily form bonding arrangements with aggregates that are acidic in nature.

Oxygen has been identified in carboxylic, phenol, and ketone function (Nicksic and Jeffries-Harris, 1968; Petersen et al., 1974; Moschopedis and Speight, 1976a,b; Ritchie et al., 1979; Speight and Moschopedis, 1981; Rose and Francisco, 1987) locations but is not usually regarded as located primarily in heteroaromatic ring systems. Furthermore, evidence for the location of oxygen within the asphaltene fraction has been obtained by infrared spectroscopy. Examination of dilute solutions of the asphaltenes in carbon tetrachloride shows that at low concentration (0.01%, w/w) of asphaltenes a band occurs at 3585 cm^{-1}, which is within the range anticipated for free non-hydrogen-bonded phenolic hydroxyl groups. In keeping with the concept of hydrogen bonding, this band becomes barely perceptible, and the appearance of the broad absorption in the range 3200–3450 cm^{-1} becomes evident at concentrations above 1% (w/w). Similar to the nitrogen content, these types of oxygen functions will readily form (for example) hydrogen-bonding arrangements, with aggregates containing hydroxyl and similar functions.

Other evidence for the presence and nature of oxygen functions in asphaltenes has been derived from infrared spectroscopic examination of the products after interaction of the asphaltenes with acetic anhydride. Thus, when asphaltenes are heated with acetic anhydride in the presence of pyridine, the infrared spectrum of the product exhibits prominent absorptions at 1680, 1730, and 1760 cm^{-1}. These changes in the infrared spectrum of the asphaltene fraction as a result of treatment with refluxing acetic anhydride suggest acetylation of free and hydrogen-bonded phenolic hydroxyl groups present in the asphaltene constituents (Moschopedis and Speight, 1976a,b).

The absorption bands at 1680 and 1760 cm^{-1} are attributed to a non-hydrogen-bonded carbonyl of a ketone, for example, a diaryl ketone and/or a quinone, and to the carbonyl of a phenolic acetate, respectively. If the 1680 cm^{-1} band is in fact the result of the presence of non-hydrogen-bonded ketone derivatives or quinone derivatives, the appearance of this band in the infrared spectra of the products could conceivably arise by acetylation of a nearby hydroxyl function. This hydroxyl function may have served as a hydrogen-bonding partner to the ketone (or quinone), thereby releasing this function and causing a shift from about 1600 cm^{-1} to higher frequencies.

The 1730 cm^{-1} band is the third prominent feature in the spectrum of the acetylated products. It is also ascribed to phenolic acetates and, as with the 1760 cm^{-1} band, falls within the range 1725–1760 cm^{-1} assigned to esters of polyfunctional phenols. This suggests that a considerable portion of the hydroxyl groups present in the asphaltenes may occur as collections of two or more hydroxyl functions on the same aromatic ring. Alternate sites include adjacent peripheral sites on a condensed ring system or sites adjacent to a carbonyl function in a condensed ring system. In the context of polyhydroxyaromatic nuclei existing in asphaltenes, it is of interest to note that pyrolysis at 800C (1470°F) results in the formation of resorcinol derivatives (Ritchie et al., 1979), implying that such functions may indeed exist in the asphaltenes.

Sulfur occurs as benzothiophene derivatives, dibenzothiophene derivatives, and naphthene benzothiophene derivatives (Clerc and O'Neal, 1961; Nicksic and Jeffries-Harris, 1968; Drushel, 1970; Yen, 1974; Speight and Pancirov, 1984; Rose and Francisco, 1988; Kelemen et al., 1990); more highly condensed thiophene types may also exist but are precluded from identification by low volatility. Other forms of sulfur that occur in asphaltene constituents include the alkyl–alkyl sulfides, alkyl–aryl sulfides, and aryl–aryl sulfides (Yen, 1974).

Metals (i.e., nickel and vanadium) are much more difficult to integrate into the asphaltene system. Nickel and vanadium occur as porphyrins (Baker, 1969; Yen, 1975, 1994), but whether these are an integral part of the asphaltene structure is not known. Some of the porphyrins can be isolated as a separate stream from petroleum (Branthaver, 1990; Reynolds, 1998).

Amide and carbazole types have also been recognized in natural product systems (Weiss and Edwards, 1980). Sulfur (as thiophene sulfur) is more difficult to identify, and its presence in biological systems, other than the sulfur-containing amino acids (cystine, cysteine, and methionine) is still speculative, although there are means by which inorganic sulfur can be incorporated into petroleum (Hobson and Pohl, 1973; Orr, 1977). It is also worthy of note here that asphaltenes have been postulated to contain charged species within their structure (Nicksic and Jeffries-Harris, 1968). This, of course, is analogous to the occurrence of zwitterion (dipolar ion) structures in proteins and peptides (Lehninger, 1970).

In summary, asphaltenes contain similar functionalities but with some variation in degree. For example, asphaltenes from the more paraffinic crude oils appeared (through this cursory examination) to contain less quinone and amide systems than those from the heavy feedstocks.

7.5 REACTIONS

The complexity of the asphaltene fraction, which can be separated by solvents or adsorbents into a variety of sub-fractions, each with different properties (Bestougeff and Darmois, 1947, 1948; Bestougeff and Mouton, 1977; Bestougeff and Byramjee, 1994; Speight, 1994, 2014; Andersen et al., 1997), dictates the complexity of the various reactions of the asphaltene constituents.

7.5.1 Thermal Reactions

The thermal reactions of the asphaltene fraction (Speight, 1970; Moschopedis et al., 1978) and the resin fraction (Parkash et al., 1980) may appear to be of little consequence to the investigation of asphalt aging and deterioration (because of the higher-than-normal temperatures used for thermal decomposition) except to confirm the complexity of the fraction. However, these investigations do emphasize that the relative reactivity of resin constituents and asphaltene constituents can be assessed on the basis of bond energies. Whilst the use of bond energy data is a method for predicting the reactivity or the stability of specific bonds under designed conditions, it must be remembered that the reactivity of a particular bond is also subject to the molecular environment surrounding the bond. In addition, the application of excessive heat in the asphalt mix plant may be sufficient to cause irreversible changes to the binder and the loss of necessary and desirable properties (see Chapters 5 and 11).

Thus, there must be acknowledgment of the thermal properties of asphalt and the various constituents. The most likely constituents that exhibit temperature susceptibility are the resin constituents and the asphaltene constituents (see Chapter 6) (Speight, 1970; Moschopedis et al., 1978; Parkash et al., 1980). Such temperature susceptibility is unlikely to manifest itself during roadway service (with the exception of the ignition of spilled oil on the roadway) but the exposure of the asphalt to excessive heat in the asphalt mix plant (by injection of the asphalt on to the aggregate in the heated mixer) may be sufficient to cause irreversible chemical and physical changes to the binder and the loss of necessary and desirable properties (see Chapter 11).

Asphaltene constituents and resin constituents can be *thermally decomposed* under conditions similar to those employed for visbreaking (viscosity breaking); approximately 470°C (880°F) to afford, on the one hand, light oils that contain higher (to $>C_{40}$) paraffins and, on the other hand, coke:

Asphaltene fraction $\rightarrow$ H_2,CO,CO_2,H_2S,SO_2,H_2O + CH_4,$CH_3(CH_2)_nCH_3$, + olefins

The application of thermal techniques to investigation of the nature of the volatile thermal fragments from petroleum asphaltene constituents has produced some interesting data relating to the polynuclear aromatic systems (Speight, 1971, 1972; Ritchie et al., 1979; Schucker and Keweshan, 1980; Gallegos, 1981). These thermal techniques have produced strong evidence for the presence of small (one-ring to four-ring) polynuclear aromatic systems (Speight and Pancirov, 1984), and now, application of the technique to the various functional fractions confirmed the general but unequal distribution of these systems throughout asphaltene constituents.

Paraffins are not the only hydrocarbon products of the thermal reactions of asphaltene constituents. The reaction paths are extremely complex; spectroscopic investigations indicate an overall dealkylation of the aromatics to methyl (predominantly) or ethyl (minority) groups. This is in keeping with a mass spectroscopic examination of asphaltene fractions (by direct introduction into the ionization chamber), which indicates a progressive increase with increasing temperature (50–350C, 120–660F) of ions attributable to low-molecular-weight hydrocarbons. Higher temperatures (500C, 932F) promote the formation of benzene and naphthalene nuclei as the predominant aromatics in the light oil, but unfortunately an increase in coke production also occurs.

In conclusion, thermal decomposition of asphaltene constituents affords light oil having a similar composition to the crude oil from which the asphaltene fraction was produced and a hydrocarbon gas composed of the lower paraffin derivatives. The thermal decompositon of the constituents of the resin fraction follows similar pathways but there are differences that can be assigned to the molecular composition of the resin constituents and the molecular weight of the various species.

7.6 OXIDATION

Oxidation of resin and asphaltene constituents is an important aspect of the contribution of the various molecular species to asphalt stability and instability. Oxidation of the resin and asphaltene constituents introduces oxygen-containing functional groups in to the molecular structures of the constituents, which disturbs the resin–asphaltene relationships, leading to phase separation of the oxidized products (Speight, 1994, 2014). The oxidation is exemplified by the use of chemical reagents and aerial oxygen – the latter form of oxidation is directly related to oxidation during the storage of asphalt.

Oxidation of asphaltene constituents in solution, by air, and in either the presence or absence of a metal salt is also possible (Moschopedis and Speight, 1978). There is some oxygen uptake, as can be seen from the increased O/C atomic ratios, but the most obvious effect is the increase in the amount of *n*-heptane-insoluble material. And analysis of the data show that it is the higher heteroatoms (more polar constituents) of the asphaltene constituents that are more susceptible to oxidation, leaving the suggestion that the polarity of the constituents may be determined by the incorporation of the heteroatoms into ring systems.

Air-blowing of asphaltene constituents at various temperatures brings about significant oxygen uptake. This is accompanied by a marked decrease in the molecular weight (vapor pressure osmometry, benzene solution) of the product. This indicates that intermolecular hydrogen bonding of oxygen functionality may play a part in the observed *high* molecular weights and physical structure of petroleum (Moschopedis and Speight, 1978; see also Taft et al., 1996).

The chemistry of the oxidation process, although more complex than a simple representation, promotes the formation of additional resin constituents and additional asphaltene constituents from the polar and/or higher-molecular-weight oil non-resin non-asphaltene constituents:

Oil (aromatic) constituents→resin constituents
Resin constituents→asphaltene constituents
Resin constituents→[asphaltene constituents]→carbenes/carboids
Asphaltene constituents→carbenes/carboids.

The move to the right in the above equations signifies an increase in viscosity (Speight, 1992, 2014; Petersen et al., 1993).

The increase in the proportion of semisolid (resin) and solid (asphaltene) constituents as a result of air blowing is accompanied by an increase in softening point with only a small loss in volume. A similar increase in softening point by removing oily constituents would cause a considerable decrease in volume. Oxidation is carried out by blowing air through asphalt heated to about 260°C (500°F) and is usually done in a tower (an oxidizer) equipped with a perforated pipe at the bottom through which the air is blown. The asphalt, in the batch mode or continuous mode, is heated until the oxidation reaction starts, but the reaction is exothermic, and the temperature is controlled by regulating the amount of air and by circulating oil or water through cooling coils within the oxidizer. Asphalt products with softening points as high as 180°C (350°F) may be produced.

7.7 SOLUBILITY PARAMETER

The solubility parameter (Table 7.2) is a measure of the ability of the compound to act as a solvent or non-solvent and as well as a measure of the intermolecular and intramolecular forces between the solvent molecule and between the solvent and solute molecules (Barton, 1991). The solubility parameter is also a major parameter in certain petroleum products, especially in material such as residua and asphalt (Redelius, 2000, 2004).

The most prevalent thermodynamic approach to describing asphaltene solubility has been the application of the solubility parameter or the concept of cohesive energy density. The application of solubility parameter data to correlate asphaltene precipitation, and, hence, crude oil–solvent interaction, has been used on prior occasions (Mitchell and Speight, 1973; Speight, 2014). The solubility parameters of asphaltene constituents can be estimated from the properties of the solvent used for separation (Long and Speight, 1989) or be measured by the titration method (Andersen and Speight, 1992) or even from the atomic hydrogen–carbon ratio (Figure 7.3).

Table 7.2 Hansen Solubility Parameters for Selected Hydrocarbons

Compound	dD	dP	dH
Toluene	18.0	1.4	2.0
Cyclopentane	16.4	0.0	1.8
Cyclohexane	16.8	0.0	0.2
o-Xylene	17.8	1.0	3.1
Trans-decalin	18.0	0.0	0.6
p-Diethylbenzene	18.0	0.0	0.6
Mesitylene	18.0	0.0	0.6
Benzene	18.4	0.0	2.0
Cis-decalin	18.8	0.0	0.0
Naphthalene	19.2	2.0	5.9
Tetrahydronaphthalene	19.6	2.0	2.9
1-Methylnaphthalene	20.6	0.8	4.7
Biphenyl	21.4	1.0	2.0

dD: the energy from dispersion forces between molecules.dP: the energy from dipolar intermolecular forces between molecules.dH: the energy from bonds between molecules.

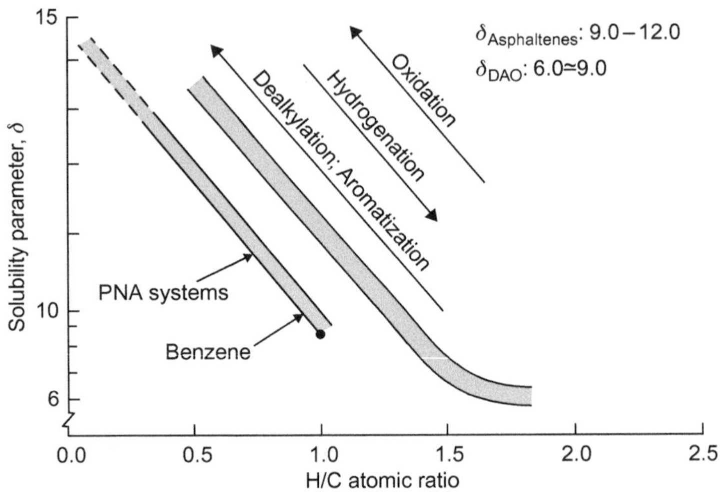

Figure 7.3 Estimation of the solubility parameters of asphaltene constituents from the atomic hydrogen–carbon ratio (compared to polynuclear aromatic systems) and changes caused by chemical-thermal effects.

The solubility parameter difference that results in a phase separation of two materials, such as asphaltene constituents in a solvent, can be estimated using the Scatchard–Hildebrand equation:

$$\ln a_a = \ln x_a + M_a/\{RT\rho_a[\Phi_s^2(\delta_s - \delta_a)^2]\}$$

In the equation, a_a is the activity of the solute a, x_a is the mole fraction solubility of a, M_a is the molecular weight of a, ρ_a is the density of a, Φ_s is the volume fraction of solvent, and $(\delta_s - \delta_a)$ is the difference between the solubility parameters of the solute a and the solvent *s*. Assuming that the activity of the asphaltene constituents a_a is 1 (solid asphaltene constituents in equilibrium with dissolved asphaltene constituents) and the volume fraction of an excess of solvent is essentially 1, the equation can be rearranged into a form that can be used to gain insight into the solubility of asphaltene constituents:

$$\ln x_a = -M_a/RT\Delta_a[(\delta_s - \delta_a)^2]$$

Assuming a density for asphaltene constituents of 1.28 g/cc and a molecular weight of 1000 g/mol, the solubility of asphaltene constituents as a function of the differences between solubility parameters of the asphaltene constituents and precipitating solvent can be calculated. Thus, the solubility of asphaltene constituents can be shown to decrease as the difference between solubility parameters increases, with the limit of solubility attained at a difference of (approximately) 3. Thus, if the asphaltene constituents are part of a polarity and molecular weight continuum in a crude oil, their precipitation is not as straightforward as it would be for a single species with a single solubility parameter, rather than a range of solubility parameters.

Incompatibility phenomena can be explained by the use of the solubility parameter for asphaltene constituents and other petroleum fractions (Speight, 1992b, 1994, 2014). As an extension of this concept, there are sufficient data to draw an approximate correlation between hydrogen-to-carbon atomic ratio and the solubility parameter, δ, for hydrocarbon solvents and petroleum constituents (Speight, 1994). By this means, the solubility parameter of asphaltene constituents can be estimated to be a range of values that is also in keeping with the asphaltene constituents being composed of a mixture of different compound types with the accompanying variation in polarity

7.8 THE STRUCTURE OF ASPHALT

The structure of petroleum has long been a topic for investigation and discussion but the structure of asphalt has received less attention. Nevertheless the same species are present in petroleum and asphalt, assuming that the asphalt is not produced from cracked residua in which molecular alteration has occurred due to the cracking process. The structural aficionados might disagree with this concept but the evidence is not in favor of widely different structures between the original petroleum and the non-thermal asphalt but this does not refer to asphalt produced by oxidation. With this caveat included in any discussion, it is possible to formulate a structure for asphalt using the native petroleum as the starting point.

As a starting point, the pronounced tendency of asphaltene constituents to form aggregates in hydrocarbon solution is one of their most characteristic traits, while the resin constituents are less likely to form such aggregates. Thus, while a number of experimental methods indicate that isolated asphaltene monomers have an *average* molecular weight on the order of 1500–2500 (vapor pressure osmometry in polar solvents), numerous methods show that asphaltenes associate spontaneously in most hydrocarbon media, resulting in observed molecular weights of 10,000 or more (Speight et al., 1985). Thus, it is not surprising that this tendency for asphaltenes to aggregate in hydrocarbon solvents has been applied to the existence of asphaltenes in crude oil. The debate is still active as to whether or not this line of thinking is correct.

An early hypothesis of the physical structure of petroleum (Pfeiffer and Saal, 1940a, b) indicated that asphaltenes are the centers of micelles formed by adsorption, or even by absorption of part of the maltenes, that is, resin material, on the surfaces or in the interiors of the asphaltene particles. On the basis of this hypothetical picture, resins or maltenes are necessary in order to suspend the asphaltene constituents and, thus, the structure of the micelle is an asphaltene–resin complex rather than an asphaltene–asphaltene micelle (Koots and Speight, 1975).

A later model of asphaltene micelle structure was proposed in which the asphaltenes, having a strong propensity for self-association, formed self-associated cores as the central unit of the micelle (Dickie and Yen, 1967). Within a unit, the polynuclear aromatics of the asphaltene molecules stack to form local discotic structures. However, the vast differences in molecular structure among individual asphaltene molecules prevent these agglomerates from forming a discotic phase, which requires the monomers to have identical molecular structures (Speight, 1994, 2014). Instead, the asphaltenes randomly stack to form globular micelles as proposed in this model. It was also proposed that the elementary micelle units can further cluster into larger particles similar to the physical structure of a floc.

Acceptance of these models and the concept of a micelle is easier to understand if the constituents of the resin and asphaltene fractions are viewed as two adjacent portions of a single broad compositional continuum that contains the polar, aromatic, and higher-molecular-weight components of petroleum (Speight, 1991). This allows the further acceptance of structural similarities in the asphaltenes and resins that facilitate formation of the micelle(s) (Speight, 1994, 2014).

The means by which asphaltene constituents associate in solution and in crude oil has been the subject of many investigations (Speight, 1994, 2014). Much of the work related to asphaltene association has arisen from molecular weight measurements from which deductions have been made about the asphaltene micelle, but there is still considerable conjecture about the true nature of the micelle. Specifically, there is some debate about whether the micelle in petroleum is composed of homogeneous material, insofar as it is composed only of asphaltene molecules, or is composed of asphaltene and resin molecules (Pfeiffer and Saal, 1940a,b; Dickie and Yen, 1967; Koots and Speight, 1975; Speight, 1994, 2014).

If the composition and properties of the precipitated asphaltene constituents reflect those of the micelles in solution, the latter should be considered as mixed micelles. This is a reasonable assumption in view of the large quantities of soluble resins found in the precipitated solid. It should be noted that if mole fractions are used as a composition measure, the one for resins in the precipitates will be greater than that for asphaltenes.

Considering the micelles to be composed of asphaltene species that are *peptized* by resins and asphaltenes is a more reasonable approach. An important corollary of petroleum composition is that the mole fraction of resins is always larger than that of asphaltenes and hence the micelles are expected to be richer in resins (Speight, 1998). The center or core of the micelle would be occupied by highly polar asphaltene molecules that are surrounded by other more soluble asphaltene molecules, which would be placed between the center and the periphery. The inclusion of other asphaltene species is not seen as a graphite-type stack, as has been proposed previously. It is more likely an association of convenience that facilitates the association of the central asphaltene with resin species. In many cases, more than one asphaltene molecule per micelle is considered unlikely.

However, in such a model, the substances with higher molecular weights and with the most pronounced aromatic nature are situated closest to the nucleus and are surrounded by lighter constituents of less aromatic nature. The transition of the inter-micellar (dispersed or oil) phase is gradual and almost continuous. Furthermore, since asphaltenes are incompatible with the oil fraction (Swanson, 1942; Koots and Speight, 1975), asphaltene dispersion is attributable mainly to the resins (polar aromatics).

Empirical observations indicate that the resins play an important role in stabilizing asphaltenes in crude oil (Koots and Speight, 1975), and under unfavorable solvent conditions the asphaltene species are prone to further aggregation into clusters that are unstable and precipitate from the crude oil (Mitchell and Speight, 1973; Koots and Speight, 1975; Speight, 1994).

$$\text{Asphaltene molecule} \leftrightarrow \text{micelle} \leftrightarrow \text{Micelle} \leftrightarrow \text{micelle cluster/floc}$$

This model requires that the asphaltene micelles are composed of an insoluble molecular core that associates with the resins, thereby providing steric stabilization against flocculation and precipitation. Thus, the phase separation of asphaltenes upon the addition of non-polar solvents to the crude oil can be rationalized in terms of reduction of the solubility parameter or polarity of the hydrocarbon. Thus, the following events ensue: (i) solubilization of the resin constituents, (ii) dissociation of the resin–asphaltene complexes, (iii) destabilization of the soon-to-be-unattended asphaltenes, and (iv) flocculation, precipitation, and phase separation.

An alternative explanation for the stabilizing influence of the resins, and for the stability of asphaltenes in hydrocarbon media in general, is that individual asphaltene species and micelles are in thermodynamic equilibrium (Cimino et al., 1995). Thus, the degree of association is determined by the relative energies of solvation of the monomers and the micelles and the entropy and enthalpy changes associated with the self-association of the asphaltene monomers. According to this model, the stabilization of the asphaltene micelles by the resin fraction derives from the contribution of the latter to the solvent power (polarity and aromaticity) of the medium. Destabilization of the asphaltenes, leading to phase separation, occurs when the solvating power of the medium toward the asphaltene monomers and micelles is reduced to the point at which they are no longer fully soluble.

The size of the asphaltene micelles varies as a function of the temperature, the asphaltene concentration, and the identity of the solvent, thereby establishing that the self-association of the asphaltenes is reversible and that the sizes in the solution state result from a true thermodynamic equilibrium between primary particles and reversible aggregates, the equilibrium constant depending on some property of the solvent. But, again, the relationship of the behavior of asphaltenes in hydrocarbon media to their behavior in crude oil is still speculative.

Thus, the means by which asphaltenes exist in crude oil and in asphalt is assumed to involve asphaltene–resin interactions leading to the formation of micelles (Koots and Speight, 1975). On this basis, the stability of asphalt can be represented by a three-phase system in which the asphaltenes, the aromatics (including the resins), and the saturates are delicately balanced. Various factors, such as oxidation, can have an adverse effect on the system, leading to instability or incompatibility as a result of changing the polarity, and bonding arrangements, of the molecular species in asphalt.

Following from this, the structure and performance of asphalt concrete (asphalt–aggregate) mixtures will be influenced by: (i) the structure of the asphalt, (ii) the arrangement of aggregates, and (iii) the associated air voids within the asphalt concrete (Masad et al., 1999). As a result, the asphalt, by association with the aggregate, can form stable chemical associations in which the asphaltene constituents and the resin constituents play a major role.

7.9 EFFECT ON ASPHALT PERFORMANCE

Under typical ambient conditions, the asphaltene constituents, resin constituents, maltenes, and the oil phase are in thermodynamic equilibrium. This equilibrium can be disturbed by a number of factors such as pressure reductions, change in temperature, and change in asphalt composition (through blending with asphalt samples from different crude oil sources). Phase separation leading to solids deposition from asphalt during production is a serious problem as it can cause serious problems in roadway performance and longevity. However, even when asphalt has high resin content and high asphaltene content, the asphalt may not necessarily produce deposition under general in-service conditions. Even when typical roadway conditions are known, deposition may have been caused by an induced physical effect which caused a momentary or semi-permanent or permanent change to the roadway conditions after which the resin–asphaltene chemistry changes and deposition ensued.

The stability of asphalt in service (durability) relates to changes with time in the original properties of the asphalt in general, while the latter refers rather to an age hardening process, but both these phenomena are related to each other and are associated with the presence of resin and asphaltene constituents which can cause deposition (see Chapter 6). Asphalt produced by vacuum distillation tends to exhibit instability, manifested by a drop in the penetration and ductility and an increase in viscosity during storage and handling. Furthermore, in the context of production, transportation, and storage of asphalt, the presence and separation of high-molecular-weight constituents (such as resin and asphaltene constituents) from the liquid phase, it is valuable (even necessary) to know the point (related to the operating conditions) at which the separation will occur and the quantity of material that will separate. Thus, accurate determination of the threshold points for phase separation (Figures 7.4 and 7.5) from asphalt is required in order to (i) characterize crude oil behavior, (ii) predict phase separation, and (iii) develop preventive measures for phase separation (Mitchell and Speight, 1973; Kokal et al., 1992; Dandekar et al., 2004; Mousavi-Dehghania et al., 2004; Udourioh et al., 2014).

While the separation of asphaltene constituents has been discussed in terms of the asphaltene separation procedure, the same lines of thinking can also be applied to the phase separation of asphaltene constituents in asphalt. However, in the case of asphalt, it is not the change in the medium that causes the asphaltene constituents to separate but a

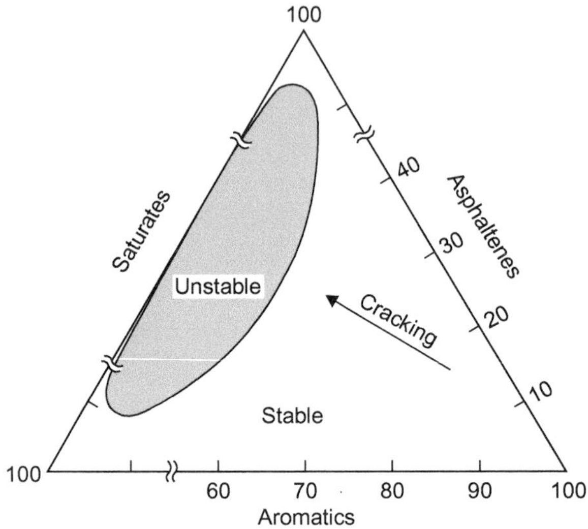

Figure 7.4 Schematic representation of phase relationships during feedstock cracking.

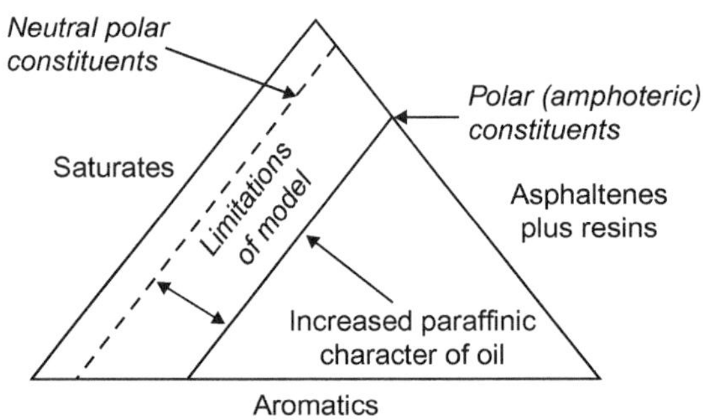

Figure 7.5 Different asphaltene constituents will phase separate at different times depending upon the composition of the feedstock.

change in the chemical character of the asphaltene constituents that can be ascribed to the inclusion of additional oxygen functions as a result of aerial oxidation. These additional functions disturb the delicate balance of the interrelationship of the constituents of asphalt leaving the asphaltene constituents as a separate phase. Furthermore it has been shown that when oxidation occurs, the most polar compounds are more prone to oxidation (Moschopedis and Speight, 1976a,b). As a result, the threshold of the medium for toluene of the asphaltene constituents varies with the polarity, and the more polar constituents will phase separate more rapidly than the less polar constituents.

After the application of asphalt in road pavements, it is exposed to extremes of environmental conditions: high temperatures (especially in warm regions), subzero temperatures in near-Arctic regions, atmospheric radiation, and mechanical stress. Rheological as well as chemical considerations have therefore to be taken into account. Furthermore, the chemical nature of asphalt, which is influenced by the production methods, affects its behavior in this respect. And many of the general properties of asphalt, which also serve as methods for routine inspection, are actually stability estimates.

In terms of oxidation and the onset of deposition, the more polar species in the asphalt (i.e., the resin and asphaltene constituents) will oxidize first during air blowing. After incorporation of oxygen to a limit, significant changes can occur in the asphaltene structure, especially in terms of the incorporation of polar oxygen, which can influence molecular weight. Thus, the change in the character of the asphalt may not be so much due to oxidative degradation but to the incorporation of oxygen functions that interfere with the natural order of intramolecular structuring. There is the potential that the incorporation of oxygen functions enhances the ability of the asphalt to bond to the aggregate. Uncontrolled introduction of oxygen functions could result in the production of a poor-grade asphalt, where phase separation of the oxidized asphaltene may already have occurred, or, should it occur in the product, the result can be pavement failure due to a weakening of the asphalt–aggregate interactions.

Although not defined as a stability property of asphalt (since it measures decreases in penetration and ductility and increases in viscosity), a thin layer of which is exposed to heat and air, this exposure to heat and air of a thin film of an unstable asphalt would support formation of oxygen-containing polymerization products, which, in turn, would reduce the penetration of the asphalt (increase its hardness), reduce its ductility (make it more brittle or less elastic) and increase its viscosity. Stable asphalt would be affected much less by this exposure to heat and oxygen. There is a direct relation between the results of this test and the changes in the original properties of the asphalt during handling, storage, and use.

Disturbance of the asphalt system can result in irreversible phase separation of asphaltene constituents, which can severely reduce the bonding arrangement of the asphalt binder to the aggregate. Thus, any factor such as changes in pressure, temperature, or composition that disrupts this

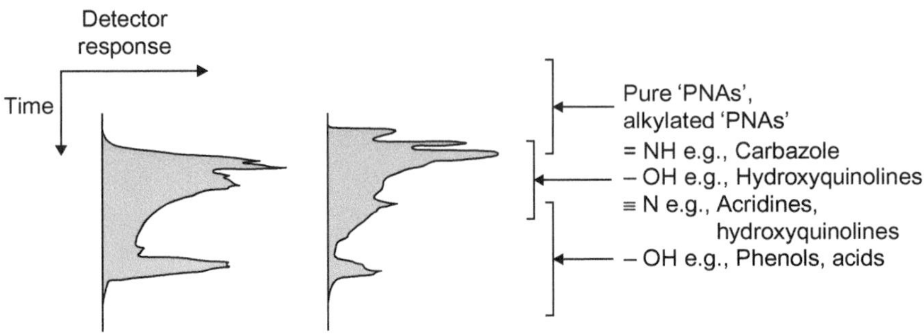

Figure 7.6 The complexity of the asphaltene fraction as shown by HPLC.

asphalt–aggregate interaction can cause asphaltene separation and deposition. In fact, as asphalt is a system in which the various constituents are in balance, any change in the physical and chemical properties of the system is a major factor in asphalt stability (Koots and Speight, 1975; Speight, 1994, 2014; Otzisk and Kempen, 2009).

Association of the asphaltene constituents and the appearance of a separate phase (often referred to as flocculation) is often irreversible, even when the conditions are returned to the pre-flocculation point (Koots and Speight, 1975; Fuhr et al., 1991; Speight, 1994; Abebi et al., 1998; Acevedo et al., 1995; Hammami et al., 2000; Karan et al., 2003; Gharfeh et al., 2004). This is the major cause of unrepairable arterial blockage damage to the behavior patterns of asphalt. Due to their large size and the adsorption affinity to solid surfaces, flocculated asphaltene constituents can cause irreversible deposition on the aggregate, thereby disturbing any other asphalt–aggregate interactions. In addition, the complexity and variation in the properties of sub-fractions of the resin and asphaltene fraction (Figures 7.6 and 7.7) detracts from the concept of any form of meaningful average structure for these fractions. Thus, the mechanism of asphaltene deposition is a multi-path approach where each constituent (chemical type) must be considered on the basis of individual chemistry and chemical properties.

Thus, although the separation of asphaltene-type material during asphalt in-service operations can now be chemically and physically rationalized, there is some degree of predictability of phase separation (Speight, 2014). There is also the suggestion (not always real) that the sediments and deposits originate from the inorganic constituents that are associated with the asphaltene constituents. These inorganic constituents may be formed from the inherent components of crude oil (i.e., the metallo-porphyrin

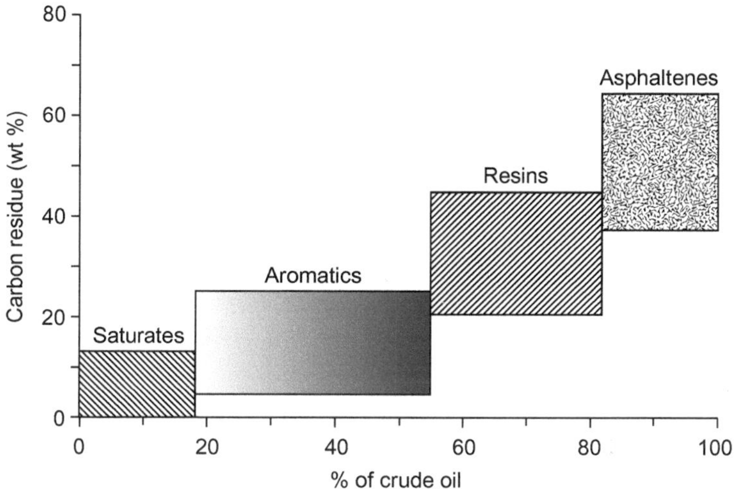

Figure 7.7 Sub-fractionation of the various fractions of crude oil yield sub-fractions with different carbon residues.

constituents) or from the ingestion of contaminants by crude oil (and, hence, by the asphalt) during the initial processing operations. For example, crude oil and fractions are known to *pick up* iron and other metal contaminants from contact with pipelines and pumps.

The presence of propane-precipitated asphalt in the asphalt mix improves the stability properties of such blends because of both the rheological and the chemical nature of the propane-precipitated asphalt. The improved stability properties of such blends can be seen from the decreasing differences in the viscosities, penetrations, and ductility after exposure to elevated temperatures and oxygen. The beneficial effect of the propane-precipitated asphalts is, of course, subject to limitations in their proportions; up to 35% of the propane-precipitated asphalt in blends with vacuum residua would be the upper limit, much depending on the nature of the vacuum residues as well as on the nature and proportions of other components, as, for example, lubricating oil extracts, which are sometimes used in such blends.

The indications are that long-term asphalt stability is related to basic rheological and physicochemical characteristics of original as well as aged asphalt samples of varying compositions. Also, understanding of the relations of these characteristics on the properties of asphalt (paving asphalt cement) makes possible the prediction of asphalt durability performance in the field. The stability is indicated in this case by aging indices: viscosity and softening point ratios and retained penetration percentage before and after exposure to the oxygen and which can be ascribed to the presence of asphaltene constituents (Speight, 1992a; Choquet and Verhasselt, 1994; Speight, 1994, 2014).

Finally, the consequence of particle deposition of the residuum during distillation or the visbreaker tar (Speight, 2014) can have serious consequences for performance of the asphalt in service. If particle-fouled resid or visbreaker tar is used for asphalt production, the result is a poor-grade asphalt would that would be priced at a much lower level than good-grade (particle-free) asphalt.

REFERENCES

Abebi, S.J., Seifried, S., Shaw, J.M., 1998. Unusual retrograde condensation and asphaltene precipitation in a model heavy oil system. J. Pet. Sci. Technol. 16 (3/4), 209–226.

Acevedo, S., Ranaudo, M.A., Escobar, G., Gutierrez, L., Ortega, P., 1995. Adsorption of asphaltene constituents and resins on organic and inorganic substrates and their correlation with precipitation problems in production well tubing. Fuel 74, 595–598.

Ali, L.H., Al-Ghannan, A., Speight, J.G., 1985. Investigations on asphaltenes in heavy crude oils: part II. Determination of resins occlusion through swelling by organic solvents. Iraqi J. Sci. 26, 41.

Ancheyta, J., Centeno, G., Trejo, F., Marroquin, G., Garcia, J.A., Tenorio, E., et al., 2002. Extraction and characterization of asphaltenes from different crude oils and solvents. Energy Fuels 16, 1121–1127.

Andersen, S.I., 1994. Dissolution of solid boscan asphaltene constituents in mixed solvents. Fuel Sci. Technol. Int. 12, 1551–1577.

Andersen, S.I., Birdi, K.S., 1990. Influence of temperature and solvent on the precipitation of asphaltene constituents. Fuel Sci. Technol. Int. 8, 593–615.

Andersen, S.I., Birdi, K.S., 1991. Aggregation of asphaltenes as determined by calorimetry. J. Colloid Interface Sci. 142, 497.

Andersen, S.I., Speight, J., 1994. Observations on the critical micelle concentration of asphaltene constituents. Fuel 72, 1343–1344.

Andersen, S.I., Speight, J.G., 1992. Asphaltene precipitation and incipient flocculation in mixed solvents Preprints. Div. Fuel Chem. Am. Chem. Soc. 37 (3), 1335.

Andersen, S.I., Speight, J.G., 2001. Petroleum reins: separation, character, and role in petroleum. Petrol. Sci. Technol. 19, 1.

Andersen, S.I., Keul, A., Stenby, E., 1997. Variation in composition of the sub-fractions of petroleum asphaltene constituents. Petrol. Sci. Technol. 15, 611–645.

ASTM D893, 2014. Standard Test Method for Insolubles in Used Lubricating Oils Annual Book of Standards. ASTM International, West Conshohocken, PA.

ASTM D2007, 2014. Standard Test Method for Characteristic Groups in Rubber Extender and Processing Oils and Other Petroleum-Derived Oils by the Clay-Gel Absorption Chromatographic Method Annual Book of Standards. ASTM International, West Conshohocken, PA.

ASTM D3279, 2014. Standard Test Method for *n*-Heptane Insolubles Annual Book of Standards. ASTM International, West Conshohocken, PA.

ASTM D4124, 2014. Standard Test Method for Separation of Asphalt into Four Fractions Annual Book of Standards. ASTM International, West Conshohocken, PA.
Baker, E.W., 1969.. In: Eglinton, G., Murphy, M.T.J. (Eds.), Organic Geochemistry Springer-Verlag, New York, NY.
Barton, A.F.M., 1991. CRC Handbook of Solubility Parameters and Other Cohesion Parameters, second ed. CRC Press, Boca Raton, FL.
Bestougeff, M.A., Byramjee, R.J., 1994. Chemical constitution of asphaltene constituents. In: Yen, T.F. Chilingarian, G.V. (Eds.), Asphaltene Constituents and Asphalts. Developments in Petroleum Science, vol. 40 Elsevier, Amsterdam, Netherlands. (Chapter 3).
Bestougeff, M.A., Darmois, R., 1947. Détermination des Points de Fusion et des Poids Moléculaires des Asphaltènes. Comptes Rend. 224, 1365.
Bestougeff, M.A., Darmois, R., 1948. Études des Asphaltènes. Comptes Rend. 227, 129.
Bestougeff, M.A., Mouton, Y., 1977. Bull. Liason Lab. Pont. Chaus., 79. Special Volume.
Branco, V.A.M., Mansoori, G.A., De Almeida Xavier, L.C., Park, S.J., Manafi, H., 2001. Asphaltene flocculation and collapse from petroleum fluids. J. Petrol. Sci. Eng. 32, 217–230.
Branthaver, J.F., 1990. Porphyrins. In: Speight, J.G. (Ed.), Fuel Science and Technology Handbook Marcel Dekker Inc., New York, NY.
Carnahan, N.J., Salager, R., Davila, A., 1999. Properties of resins extracted from boscan crude oil and their effect on the stability of asphaltene constituents in Boscan and Hamaca crude oils. Energy Fuels 13, 309.
Choquet, F.S., Verhasselt, A.F., 1994. Natural and accelerated aging of bitumens. In: Sharma, M.K., Yen, T.F. (Eds.), Asphaltene Particles in Fossil Fuel Exploration, Recovery, Refining and Production Plenum Press, New York, NY.
Chukwu, O.C., Ofodile, S.E., Okoye, I.P., 2011. The effect of different solvent polarity on the precipitation of heavy organics from a crude oil deposit in the Niger Delta. Sci. Afr. 10 (2), 9–15.
Cimino, R., Correrra, S., Del Bianco, A., Lockhart, T.P., 1995.. In: Sheu, E.Y., Mullins, O.C. (Eds.), In Asphaltenes: Fundamentals and Applications Plenum Press, New York, NY. (Chapter III).
Clerc, R.J., O'Neal, M.J., 1961. The mass spectrometric analysis of asphalt. A preliminary investigation. Anal. Chem. 33, 380–382.
Dandekar, A.Y., Andersen, S.I., Stenby, E.H., 2004. Solid organic deposition during gas injection studies. Petrol. Sci. Technol. 22 (7,8), 1209–1229.
Dickie, J.P., Yen, T.F., 1967. Macrostructures of asphaltic fractions by various instrumental methods. Anal. Chem. 39, 1847–1852.
Drushel, H.V., 1970. Sulfur compounds in petroleum Preprints. Div. Petrol. Chem. Am. Chem. Soc. 15.
Francisco, M.A., Speight, J.G., 1984. Asphaltene characterization by a non-spectroscopic method Preprints. Div. Fuel. Chem. Am. Chem. Soc. 29 (1), 36.
Fuhr, B.J., Cathrea, L., Coates, H., Kalra, H., Majeed, A.I., 1991. Properties of asphaltene constituents from waxy crude. Fuel 70, 1293.
Gallegos, E.J.J., 1981. Alkylbenzenes derived from carotenes in coals by GC/MS. Chromatograph. Sci. 19, 177–182.
Gary, J.G., Handwerk, G.E., Kaiser, M.J., 2007. Petroleum Refining: Technology and Economics, fifth ed. CRC Press, Taylor & Francis Group, Boca Raton, FL.
Gawrys, K., Spiecker, L., Matthew, P., Kilpatrick, P.K., 2003. The role of asphaltene solubility and chemical composition on asphaltene aggregation. Petrol. Sci. Technol. 21 (3,4), 461–489.
Gharfeh, S., Yen, A., Asomaning, S., Blumer, D., 2004. Asphaltene flocculation onset determinations for heavy crude oil and its implications. Petrol. Sci. Technol. 22 (7,8), 1055–1075.

Gholoum, E.F., Oskui, G.P., 2004. Investigation of the asphaltene precipitation process for Kuwaiti reservoirs. Petrol. Sci. Technol. 22 (7,8), 1165–1185.
Gholoum, E.F., Oskui, G.P., Salman, M., 2003. Investigation of Asphaltene Onset Conditions for Kuwaiti Reservoirs. SPE Paper No. 81571. Proceedings SPE 13th Middle East Oil Show Conference, Bahrain April 5–8. Society of Petroleum Engineers, Richardson, TX.
Girdler, R.B., 1965. Constitution of asphaltenes and related studies. Proc. Assoc. Asphalt Paving Technol. 34, 45.
Hammami, A., Freworn, K.A., 1998. Asphaltic crude oil characterization: an experimental investigation of the effect of resins on the stability of asphaltene constituents. Petrol. Sci. Technol. 16 (3,4), 227–249.
Hammami, A., Phelps, H., Little, T.M., 2000. Asphaltene precipitation from live oils: an experimental investigation of the onset conditions and reversibility. Energy Fuels 14, 14–18.
Hassanvand, M., Shahsavani, B., Anooshe, A., 2012. Study of temperature effect on asphaltene precipitation by visual and quantitative methods. J. Petrol. Technol. Alt. Fuels 3 (2), 8–18.
Hirschberg, A., deJong, L., Schipper, B., Maijer, J., 1984. Influence of temperature and pressure on asphaltene flocculation. SPE J., 283–293. June.
Hobson, G.D., Pohl, W., 1973. Modern Petroleum Technology. Applied Science Publishers, London.
Hsu, C.S., Robinson, P.R., 2006. Practical Advances in Petroleum Processing, Volumes 1 and 2. Springer, New York, NY.
Jacobson, J.M., Gray, M.R., 1987. The use of infrared spectroscopy and nitrogen titration data in the structural group analysis of bitumen. Fuel 66, 749–752.
Karan, K., Hammami, A., Flannery, M., Stankiewicz, B.A., 2003. Evaluation of asphaltene instability and a chemical control during production of live oils. Petrol. Sci. Technol. 21 (3,4), 629–645.
Katz, D.H., Beu, K.E., 1945. Nature of asphaltic substances. Ind. Eng. Chem. 37, 195–200.
Kelemen, S.R., George, G.N., Gorbaty, M.L., 1990. Speciation of sulfur forms. Fuel 69, 939.
Kokal, S.L., Najman, J., Sayegh, S.G., George, A.E., 1992. Measurement and correlation of asphaltene precipitation from heavy oils by gas injection. J. Can. Petrol. Technol. 31, 24.
Koots, J.A., Speight, J.G., 1975. The relation of petroleum resins to asphaltene constituents. Fuel 54, 179.
Lehninger, A.L., 1970. Principles of biochemistry. Worth Publishers, MacMillan, New York, NY.
Long, R.B., 1979. The concept of asphaltene constituents Preprints. Div. Petrol. Chem. Am. Chem. Soc. 24 (4), 891.
Long, R.B., 1981. The concept of asphaltene constituents. In: Bunger, J.W., Li, N. (Eds.), The Chemistry of Asphaltene Constituents American Chemical Society, Washington, DC. Advances in Chemistry Series No. 195.
Long, R.B., Speight, J.G., 1989. Studies in petroleum composition. I: Development of a compositional map for various feedstocks. Rév. Inst. Franç. Petrol. 44, 205.
Masad, E., Muhunthan, B., Shashidhar, N., Harman, T., 1999. Internal structure characterization of asphalt concrete using image analysis. J. Comput. Civ. Eng. 13 (2), 88–95.
Miller, J.T., Fisher, R.B., Thiyagarajan, P., Winans, R.E., Hunt, J.E., 1998. Sub-fraction and characterization of Mayan asphaltenes. Energy Fuels 12 (6), 1290–1298.
Mitchell, D.L., Speight, J.G., 1973. The solubility of asphaltene constituents in hydrocarbon solvents. Fuel 52, 149.
Mitra-Kirtley, S., Mullins, O.C., van Elp, J., George, S.J., Chen, J., Cramer, S.P., 1993. Determination of the nitrogen chemical structures in petroleum asphaltenes using XANES spectroscopy. J. Am. Chem. Soc. 115, 252–258.

Moschopedis, S.E., Speight, J.G., 1976a. An investigation of hydrogen bonding by oxygen functions in athabasca bitumen. Fuel 55, 187.
Moschopedis, S.E., Speight, J.G., 1976b. Oxygen functions in asphaltene constituents. Fuel 55, 334.
Moschopedis, S.E., Speight, J.G., 1978. The influence of metal salts on bitumen oxidation. Fuel 57, 235.
Moschopedis, S.E., Speight, J.G., 1979. Investigation of nitrogen types in athabasca bitumen Preprints. Div. Petrol. Chem. Am. Chem. Soc. 24 (4), 1007.
Moschopedis, S.E., Parkash, S., Speight, J.G., 1978. The thermal decomposition of asphaltenes. Fuel 57, 431.
Mousavi-Dehghania, S.A., Riazi, M.R., Vafaie-Sefti, M., Mansoori, G.A., 2004. An analysis of methods for determination of onsets of asphaltene phase separations. J. Petrol. Sci. Eng. 42 (2004), 145–156.
Nellensteyn, F.J., 1924. The constitution of asphalt. J. Inst. Petrol. Technol. 10, 311.
Nellensteyn, F.J., 1938. The colloidal structure of bitumens. In: Brooks, B.T. (Ed.), The Science of Petroleum, Volume 4, Oxford University Press, London, pp. 2760.
Nicksic, S.W., Jeffries-Harris, M.J., 1968. Acid precipitation of crude-oil asphaltenes-structural implications. J. Inst. Petrol. Technol. 54, 107–114.
Orr, W.L., 1977. Sulfur in heavy oils, oil sands, and oil shales Preprints. Div. Fuel Chem. Am. Chem. Soc. 22 (3), 86–99.
Otzisk, B., Kempen, H., 2009. Stabilization of asphaltenes. Petrol. Technol. Quart. Q1, 31–36.
Parkash, S., Moschopedis, S.E., Speight, J.G., 1980. The thermal decomposition of resins. Fuel 59, 64.
Petersen, J.C., Barbour, F.A., Dorrence, S.M., 1974. Oxygen identification in asphaltenes. Proc. Assoc. Asphalt Paving Technol. 43, 162–169.
Petersen, J.C., Branthaver, J.F., Robertson, R.E., Harnsberger, P.M., Duvall, J.J., Ensley, E.K., 1993. Effects of physicochemical factors on asphalt oxidation kinetics Transportation Research Record no. 1391. Transportation Research Board, National Research Council., pp. 1–10.
Pfeiffer, J.P., Saal, R.N., 1940a. Asphaltic bitumen as a colloidal system. Phys. Chem. 44, 139–149.
Pfeiffer, J.P., Saal, R.N., 1940b. The resin-peptized model of asphaltene dispersion. Phys. Chem. 44, 139.
Redelius, P.G., 2000. Solubility parameters and bitumen. Fuel 79, 27–35.
Redelius, P.G., 2004. Bitumen solubility model using hansen solubility parameter. Energy Fuels 18, 1087–1092.
Reynolds, J.G., 1998. Metals and heteroatoms in heavy crude oils Petroleum Chemistry and Refining. Taylor & Francis Publishers, Washington, DC, (Chapter 3).
Ritchie, R.G.S., Roche, R.S., Steedman, W., 1979. Phenols from asphaltene constituents. Fuel 58, 523.
Rose, K.D., Francisco, M.A., 1987. Characterization of acidic heteroatoms in heavy petroleum fractions by phase-transfer methylation and NMR spectroscopy. Energy Fuels 1, 233.
Rose, K.D., Francisco, M.A., 1988. A two-step chemistry for highlighting heteroatom species in petroleum material using ^{13}C NMR spectroscopy. J. Am. Chem. Soc. 110, 637.
Schucker, R.C., Keweshan, C.F., 1980. Reactivity of cold lake asphaltene constituents Preprints. Div. Fuel Chem. Am. Chem. Soc. 25, 155.
Sheu, E.Y., 2002. Petroleum asphaltenes—properties, characterization, and issues. Energy Fuel 16, 74–82.
Speight, J.G., 1970. Thermal cracking of athabasca bitumen, athabasca asphaltenes, and athabasca deasphalted heavy oil. Fuel 49, 134.

Speight, J.G., 1971. Structural analysis of athabasca asphaltene constituents by proton magnetic resonance spectroscopy. Fuel 50, 102.
Speight, J.G., 1972. The application of spectroscopic techniques to the structural analysis of coal and petroleum. Appl. Spectrosc. Rev. 5, 211.
Speight, J.G., 1979. Studies on bitumen fractionation: (a) fractionation by a cryoscopic method (b) effect of solvent type on asphaltene solubility Information Series No. 84. Alberta Research Council, Edmonton, AB.
Speight, J.G., 1984. The chemical nature of petroleum asphaltene constituents. In: Tissot, B. (Ed.), Characterization of Heavy Crude Oils and Petroleum Residues Editions Technip, Paris.
Speight, J.G., 1992a. Asphalt Kirk–Othmer Encyclopedia of Chemical Technology, fourth ed. John Wiley & Sons Inc., New York, NY, pp. 689–724.
Speight, J.G., 1992b. A chemical and physical explanation of incompatibility during refining operations Proceedings 4th International Conference on the Stability and Handling of Liquid Fuels. U.S. Department of Energy., (DOE/CONF-911102). p. 169.
Speight, J.G., 1994. Chemical and physical studies of petroleum asphaltene constituents Yen, T.F. Chilingarian, G.V. (Eds.), Asphaltene Constituents and Asphalts. I. Developments in Petroleum Science, vol. 40 Elsevier, Amsterdam, Netherlands. (Chapter 2).
Speight, J.G., 1998. Asphaltenes and the structure of petroleum Petroleum Chemistry and Refining. Taylor & Francis Publishers, Philadelphia, PA, (Chapter 4), pp. 103–120.
Speight, J.G., 2001. Handbook of Petroleum Analysis. John Wiley & Sons Inc., Hoboken, NJ.
Speight, J.G., Ozum, B., 2002. Petroleum Refining Processes. Marcel Dekker Inc., New York, NY.
Speight, J.G., 2014. The Chemistry and Technology of Petroleum, fifth ed. CRC Press, Taylor & Francis Group, Boca Raton, FL.
Speight, J.G., 2015a. Handbook of Petroleum Product Analysis, second ed. John Wiley & Sons Inc., Hoboken, NJ.
Speight, J.G., 2015b. Fouling in Refineries. Gulf Professional Publishing, Elsevier, Amsterdam, Netherlands.
Speight, J.G., Long, R.B., 1981. Spectroscopy and asphaltene structure. In: Filby, R.H. (Ed.), Atomic and Nuclear Methods in Fossil Energy Research Plenum Press, New York, NY, pp. 295.
Speight, J.G., Moschopedis, S.E., 1981. On the molecular nature of petroleum asphaltenes. In: Bunger, J.W., Li, N.C. (Eds.), Chemistry of Asphaltenes American Chemical Society, Washington, DC. Advances in Chemistry Series No. 195.
Speight, J.G., Pancirov, R.J., 1984. Structural types in asphaltene constituents as deduced from Pyrolysis-Gas Chromatography-Mass spectrometry. Liquid Fuels Technol. 2, 287.
Speight, J.G., Long, R.B., Trowbridge, T.D., 1982. On the definition of asphaltene constituents Preprints. Div. Fuel Chem. Am. Chem. Soc. 27 (3/4), 268.
Speight, J.G., Long, R.B., Trowbridge, T.D., 1984. Factors influencing the separation of asphaltene constituents from heavy petroleum feedstocks. Fuel 63, 616.
Speight, J.G., Wernick, D.L., Gould, K.A., Overfield, R.E., Rao, B.M.L., Savage, D.W., 1985. Molecular weights and association of asphaltenes: a critical review. Rev. Inst. Franç. Pétrole 40, 27.
Swanson, J.M., 1942. A contribution to the physical chemistry of asphalts. J. Phys. Chem. 46, 141.
Taft, R.W., Berthelot, M., Laurence, C., Leo, A.J., 1996. Hydrogen bonds and molecular structure. Chemtech, 20. July.
Udourioh, G.A., Ibezim-Ezeani, M.I., Ofodile, S.E., 2014. Comparative investigation of heavy organics precipitation from crude oil using binary mixtures and single *n*-alkane organic solvents. J. Petrol. Gas Explor. Res. 4 (4), 53–59.

Weiss, V., Edwards, J.M., 1980. The Biosynthesis of Aromatic Compounds. John Wiley & Sons Inc., New York, NY.
Yen, T.F., 1974. The structure of asphaltene and its significance. Energy Sources 1, 447.
Yen, T.F., 1975. The role of trace metals in petroleum. Ann Arbor Science Publishers Inc., Ann Arbor, MI.
Yen, T.F., 1994. Multiple structural orders of asphaltenes" Asphaltenes and Asphalt. 1Yen, T.F. Chilingarian, G.V. (Eds.), Developments in Petroleum Science, vol. 40 Elsevier, Amsterdam, Netherlands.

PART II

Asphalt Technology

CHAPTER 8

Uses of Asphalt

8.1 INTRODUCTION

Asphalt has been known and used for approximately 6000 years as a caulking agent, as a waterproofing agent, and as roadway cement (see Chapter 2). The demographic of ancient asphalt use (i.e., the relative amounts used for the various tasks) is not known, although it can be speculated that asphalt use as a building mastic took priority, particularly in ancient Mesopotamia (Speight, 1978). In the modern world, the primary use of asphalt is in road construction, where it is used as the glue or binder mixed with aggregate particles to create asphalt concrete. Other main uses are for surfaces for parking areas (from heavy vehicle parks to private domestic driveways), waterproofing products, including production of roofing felt and for sealing flat roofs against moisture incursion (Abraham, 1945; Traxler, 1961; Hoiberg, 1964). There are, however, other specialized applications of asphalt and it is the purpose of this chapter to present descriptions of the other uses of asphalt.

The largest nonhighway use of asphalt in some countries is for surfacing work on airfields. Most runways on both civil and military airfields are surfaced with these materials. In the case of runways on airports, special porous surfaces are used (over high-strength asphalt bases) to ensure rapid dispersal of surface water to reduce the risk of aircraft aquaplaning. At the other end of the scale, the runways of small privately owned airfields are often surfaced with asphalt to provide a bump-free all-weather surfacing.

8.2 TYPES

Asphalt is not asphalt is not asphalt because crude oil is *not crude oil is not crude oil.* Each batch of asphalt produced by a refinery is different depending upon (i) the crude oil source from which the asphalt was produced, (ii) the process by which the asphalt was produced, and (iii) the types of additives that are blended into the asphalt to ensure that the product meets

Asphalt Materials Science and Technology.
DOI: http://dx.doi.org/10.1016/B978-0-12-800273-5.00008-8

the required specifications. In fact, the increasing commercial importance of asphalt has been manifested by past and continued efforts to perfect and standardize a variety of test methods for the asphalt binders (see Chapter 4) and for the various types of aggregate (see Chapter 5) through agencies such as the American Association of State Highway & Transportation Officials and ASTM International.

This is especially important since each type of asphalt can be subcategorized into several subcategories that are named according to use, but it must always be remembered that each subcategory of asphalt is also different according to the two criteria given above. Thus, the subcategories (presented alphabetically rather than by importance or by tonnage use) are: (i) asphalt emulsion, (ii) asphalt paint, (iii) cutback asphalt, (iv) hot-mix asphalt (HMA), warm mix asphalt, and cold-mix asphalt (CMA), (v) polymer-modified asphalt, (vi) porous asphalt, (vii) road asphalt, (viii) road oil, (ix) roofing asphalt, and (x) rubber-modified asphalt. A note on the miscellaneous uses of asphalt is also included.

8.2.1 Asphalt Emulsion

Asphalt may also be *emulsified* to produce a liquid product that can be pumped and transported by pipeline, mixed with aggregate, or sprayed through nozzles. Emulsified asphalts consist of minute globules of asphalt (generally in the colloidal size range) suspended in chemically treated water. Asphalt–water emulsions are usually prepared in a high-speed mixing apparatus, which emulsifies the asphalt into the water. The chemical used in treating the water is usually some type of caustic soap, clay, or other such agent, which assists in promoting the emulsification process and which also controls some of the properties of the emulsion.

When the emulsion is deposited upon the surfaces of soil or stone, the asphalt particles coalesce through chemical action on the surface of the stone or by evaporation of the water. In the jargon of the trade, the emulsion "breaks" and in so doing leaves a thin film of asphalt on the aggregate. There are three standard types of emulsified asphalt and many specialty products made for specific applications: (i) rapid setting, (ii) medium setting, and (iii) slow setting. The names given to these types indicate the relative rates at which coalescence of the asphalt particles can be expected. However, it is more accurate to think of the rapid-setting grades as breaking on contact with stone, medium-setting grades as breaking on mixing with fine aggregates or prolonged mixing with coarse aggregates, and slow-setting grades as being capable of prolonged mixing even with fine aggregates and breaking only on final evaporation of the water from the mix. These properties are controlled principally by the amount and type of emulsifying agent used in the manufacture of the product.

Each type of emulsified asphalt is available in one or more standard grades, varying either in viscosity of the emulsified asphalt or in consistency (penetration) of the asphalt cement used in preparing the emulsified asphalt. A specific grade of a given type is usually designated by a number following the letters identifying the type, for example, RS-I or RS-2. In general, the higher grade numbers indicate materials of higher viscosity or lower consistency of asphalt cement or a combination of these two factors. Some agencies, however, use other systems of identification for emulsified asphalts. Several test procedures are employed to identify the various types and grades of liquid asphalts and to prescribe characteristics as required by their use.

Asphalt emulsions are fine dispersions of heated asphalt (base asphalt used in HMA applications) in water with an emulsifying agent. The emulsions are classified as cationic emulsion (electropositively charged micelles containing asphalt molecules) or anionic emulsion (electronegatively charged micelles containing asphalt molecules) depending on the emulsifying agent, and are graded according to chemical setting time. The emulsions can be applied as sprays or in cold-mix applications for seal coating, maintenance and repair.

To emulsify, the asphalt cement is ground into globules of 5–10 μm and smaller (1 μm is equal to one millionth of a meter), which are then mixed with water. An emulsifying agent is added, which reduces the tendency of the asphalt and water to separate. The emulsifying agent may be colloidal clay, soluble or insoluble silicate minerals, soap, or sulfonated vegetable oil.

These emulsions are normally of the oil-in-water type, which reverse or break on application to a stone or earth surface, so that the oil clings to the stone and the water disappears. In addition to their usefulness in road and soil stabilization, they are useful for paper impregnation and waterproofing. The emulsions are chiefly (i) the soap or alkaline type and (ii) the neutral or clay type. The former break readily on contact, but the latter are more stable and probably lose water mainly by evaporation. Good emulsions must be stable during storage or freezing, suitably fluid, and amenable to control for speed of breaking.

In the emulsification process, hot binder is mechanically (in a colloid mill) separated into minute globules and dispersed in water treated with a small quantity of emulsifying agent. The water is called the continuous phase and the globules of binder are called the discontinuous phase. The binder globules are extremely small, mostly in the colloidal size range, and by proper selection of an emulsifying agent and other manufacturing controls, emulsified asphalts are produced in several types and grades. By choice of emulsifying agent, the emulsified asphalt may be: (i) anionic—the binder globules are electronegatively charged or (ii) cationic—the binder globules are electropositively charged.

Because particles having a like electrostatic charge repel each other, asphalt globules are kept apart until the emulsion is deposited on the surface of the soil or aggregate particles. At this time, the asphalt globules coalesce through neutralization of the electrostatic charges or water evaporation. Coalescence of asphalt globules occurs in rapid- and medium-setting grades, resulting in a phase separation between asphalt and water. When this coalescence occurs, it is usually referred to as the break or set.

Emulsified asphalts can be used with cold as well as heated aggregates, and with aggregates that are dry, damp, or wet.

8.2.2 Asphalt Paint

Although widely used for a time in paint during the nineteenth century, asphalt (or naturally occurring bitumen) ultimately proved unstable for use in oil painting, especially when mixed with the most common diluents, such as linseed oil, varnish, or turpentine. Unless thoroughly diluted, asphalt never fully solidifies and in time corrupted the other pigments with which it came into contact. The use of asphalt as a glaze to set in shadow or mixed with other colors to render a darker tone resulted in the eventual deterioration of a good many paintings. The use of asphalt caused the brilliant colors to degenerate into dark greens and blacks and the paint and canvas to buckle. In more modern times, asphalt is used as an anticorrosive agent and is sold under the name asphalt coating or bituminous coating.

Asphalt coating can be used both externally and internally on carbon steel pipes. It is usually black and, when applied properly, this coating offers outstanding cathodic protection needed for almost all structural steel pipes situated underground. It is regarded as one of the least expensive ways to provide cathodic protection to pipes. It is also the least complex coating process, making it suitable for casing pipes such as those used in road bore. Almost all asphalt coatings are ready to use and free from solvent. These are chemical- and UV-resistant, making them reliable protective coatings against wear and corrosion. The coating can be applied through various methods using a roller, brush, or sprayer and is effective in waterproofing surfaces, especially on surfaces that are in direct contact with seepage water, moisture, and nonpressurized liquid or water.

Asphalt coating is used to build a vapor-proof and flexible protective coat in accordance with its formulation and polymerization grade. Its flexibility and protection against vapor and water can be influenced by the polymer grade as well as reinforcement of fiber. The most common applications of asphalt coatings include areas that are beneath screed wet. It is an excellent protective coating and waterproofing agent, especially on surfaces such as concrete foundations.

8.2.3 Cutback Asphalt

Asphalt may next be blended or *cut* (*diluted*) with a volatile solvent resulting in a product that is soft and workable at a lower temperature than pure uncut asphalt. Thus, cutback asphalts are a mixture of asphalt with volatile petroleum diluents such as white spirits, kerosene, or gas oil to render them more fluid for ease of handling and application. When the *cutback* asphalt is used for paving or construction, the volatile element evaporates when exposed to air or heat, leaving the hard asphalt. These products are used in spray applications as surface treatments and are handled at temperatures ranging from ambient to 150°C (300°F).

Petroleum solvents used for dissolving binder are sometimes called distillate, diluent, or cutter stock. If the solvent used in making the cutback asphalt is highly volatile (low-boiling) it will quickly escape by evaporation. Solvents of lower volatility evaporate more slowly. On the basis of relative speed of evaporation, cutback asphalts are divided into three types: (i) rapid-curing asphalt—a mixture of asphalt and a volatile solvent or light distillate, generally in the gasoline or naphtha boiling range, (ii) medium-curing asphalt—a mixture of asphalt and a solvent of intermediate volatility or medium distillate, generally in the kerosene boiling range, and (iii) slow-curing asphalt—a mixture of asphalt and an oily diluent of low volatility; slow-curing asphalt is often called *road asphalt* or *road oil* and this terminology originated in the early decades of the twentieth century when asphaltic residual oil was used to give roads a low-cost, all-weather surface.

In the process for preparing cutback asphalt, heated asphalt is mixed with a gas-oil-type solvent from the distillation process or to produce slow-curing asphalt, the asphalt is mixed with kerosene for medium-curing, and with gasoline or naphtha for the rapid-curing asphalt. The degree of fluidity obtained in each case depends on the grade of asphalt cement, volatility of the solvent, and proportion of solvent to binder. The degree of fluidity results in several grades of cutback asphalt. Some cutback asphalts are fluid at ordinary atmospheric temperatures and others are somewhat more viscous and may require heating to melt them enough for construction operations. Cutback asphalts can be used with cold aggregates, with a minimum of heat. Rapid-curing and medium-curing types of cutback asphalts are used in a variety of highway construction methods. Among the more important uses are road mixing operations, stockpiling mixes, and spray applications.

8.2.4 HMA, Warm Mix Asphalt, and CMA

The construction of asphaltic roadways is not merely covering the base (such as soil) with asphalt—because the roadway is much more flexible than a roadway constructed of Portland cement concrete, asphaltic roadways are often referred to as flexible pavements. Asphaltic roadways are actually engineered structures that are composed of several different layers. The visible part of an asphalt concrete pavement, the part that directly supports truck and passenger vehicles, is called the surface course or wearing course and is typically 1.5–3 in. (40 and 75 mm) thick and consists of crushed aggregate and asphalt binder. Surface course mixtures tend to have a relatively high asphalt content, which helps these mixtures stand up better to traffic and the effects of sunlight, air, and water. However to construct such a roadway, there is the need to design the road courses and the type of asphalt mix that will be employed. This involved the use of one of three type of mix: (i) HMA, (ii) warm mix asphalt, and (iii) CMA.

HMA is a blend of asphalt paving cement and mineral aggregate. Asphalt paving cement is the straight reduced or vacuum processed asphalt used mainly as a binder (4–10%, w/w) of hot mixed asphalts to hold the aggregate together. HMA materials comprise a majority of all paving products and are the most important commercially and in terms of numbers of workers exposed.

Warm mix asphalt is an asphalt mix produced at lower temperatures as compared to those typically associated with rolled asphalt pavement; produced and placed at temperatures typically 10–40°C (50–100°F) lower than conventional rolled asphalt; the differentiation between warm mix asphalt and HMA is not often obvious.

Cold-mix asphalt (*cold mix asphalt concrete*, *cold placed mixture*, CMA) is generally a mix made with emulsified or cutback asphalt. Aggregate material may be anything from a dense-graded crushed aggregate to a granular soil having a relatively high percentage of dust. At the time of mixing, the aggregate may either be damp, air-dried, or artificially heated and dried. CMA may be used for surface, base, or sub-base courses if the pavement is properly designed. Cold-mix surface courses are suitable for light and medium traffic; but such surface courses normally require a seal coat or hot asphalt overlay as surface protection. When used in the base or sub-base, they may be suitable for all types of traffic.

8.2.5 Polymer-Modified Asphalt

Conventional binders, such as binder with 100 penetration grade (100 pen.) and 60 penetration grade (60 pen.) have been used in road pavement construction. However, increased axle loading and braking power of vehicles in recent years required the durability and strength of the binder to resist the (i) rutting and (ii) cracking tendencies of road pavements.

Rutting (often referred to as permanent deformation, longitudinal deformation) is a common problem in flexible pavements. When a heavy vehicle moves across an asphalt concrete pavement, the pavement deflects to a small extent and the amount of the deflection can increase in warm weather. After the stress (weight) passes, the pavement tends to move back to its original position (pavement recovery) but the move may not be complete and there will be a small amount of permanent deformation in the wheel path. The deformation increases with increased traffic loads leading to a rutted pavement, which becomes a serious problem when the ruts are sufficiently large to contribute to a rough riding surface and collect water during rainy or snowy weather causing vehicles to hydroplane and lose control.

Other related forms of permanent deformation include: (i) *shoving* and (ii) *wash boarding*. Shoving occurs at intersections when vehicles stop, which exerts a lateral force on the asphaltic surface and causes deformation across the pavement (transverse deformation) rather than in the form of wheel ruts (longitudinal deformation). Wash boarding is a similar phenomenon to shoving but the deformation takes the form of a series of ripples transversely across the pavement surface. In addition, rutting, shoving, and wash boarding are not specific to deformation in the surface course of the roadway but can result from permanent deformation in any part of the pavement, such as in the subgrade and the sub-base.

Excessive permanent deformation in one or more of the bound layers is the result of an asphalt concrete mixture that lacks strength and stiffness at high temperatures. This can arise from poor mix design, such as selecting an asphalt binder that is too soft for the given climate and traffic level, and can make it prone to rutting and other forms of permanent deformation.

Fatigue cracking (like rutting) results from the large number of loads applied over time to a pavement subject to traffic but fatigue cracking tends to occur when the pavement is at moderate temperatures, rather than at the high temperatures that cause rutting. The cracks that first appear are difficult to detect (they are said to be *microscopic cracks* and are not continuous). However, under repeated traffic loads the cracks grow in size and number evolving into severe fatigue cracking (alligator cracking and crocodile cracking) (see Chapter 9). These large cracks will significantly affect pavement performance, by weakening the pavement, contributing to a rough riding surface, and allowing air and water into the pavement, which will cause additional damage to the pavement structure. Eventually fatigue cracking can lead to extensive areas of cracking, large potholes, and total pavement failure.

One means of achieving a more resilient asphalt mix (to reduce the amount of deformation and cracking) is to modify the asphalt with polymers (Zhang et al., 2013a; Haggam et al., 2014). The addition of polymers, chains of repeated small molecules, to asphalt has been shown to improve performance. Pavement with polymer modification exhibits greater resistance to rutting and thermal cracking, and decreases fatigue damage, stripping, and temperature susceptibility. Polymers that are used for asphalt modification can be grouped into three main categories: (i) thermoplastic elastomers, (ii) plastomers, and (iii) reactive polymers.

Thermoplastic elastomers are able to confer good elastic properties on thermomodified binder; while *plastomers* and *reactive polymers* are added to improve rigidity and reduce deformations under load. Belonging to the first category, styrene–butadiene–styrene (SBS) block copolymers are probably the most frequently used asphalt modifiers for paving applications (Zhang and Yu, 2010; Al-Hadidy et al., 2011; Martin et al., 2013; Peiliang et al., 2013). Examples of the plastomeric types of polymers that were investigated for asphalt modification are polyethylene (PE) and ethylene–butyl acrylate (EBA) random copolymers (Esmaeil et al., 2011; Moatasim et al., 2011; Karim et al., 2012; Zhang et al., 2013b). Due to its low compatibility with asphalt, PE is not widely used for paving applications, and thus ethylene copolymers are preferred.

Recently, reactive polymers have been introduced as asphalt modifiers. Their "reactivity" is due to the presence of functional groups supposedly able to bond with asphalt molecules. Polarity of the polymer can enhance its solubility and compatibility with base asphalt. Polar groups present in the polymer molecules can react with the polar constituents of asphalt. Subsequently, phase separation is prevented, which in turn enhances the material's consistency and decreases oxidative aging (Polacco et al., 2005; Edwards et al., 2007; Kim et al., 2011; Merusi and Giuliani, 2011). Among polar polymers, a very limited number of investigations have focused on the fundamental properties of modified asphalt with acrylate polymers. Most frequently used acrylates as asphalt-modifying agents in road applications are chemicals such as ethylene vinyl acetate, glycidyl methacrylate terpolymer, and EBA copolymer (Fawcett and McNally, 2001; Airey, 2002; Iqbal et al., 2006).

8.2.6 Porous Asphalt

Porous asphalt pavements offer developers a new option for managing storm water. Porous asphalt pavements, used mostly for parking lots, allow water to drain through the pavement surface into a stone recharge bed and infiltrate into the soils below the pavement. Such pavements have been proving their worth since the mid-1970s, and recent changes in storm-water regulations have prompted many consulting engineers and public works officials to seek information about them.

With proper design and installation, porous asphalt can provide pavements with a lifespan of more than 20 years and provide storm-water management systems that promote infiltration, improve water quality, and often eliminate the need for a detention basin. The performance of porous asphalt pavements is similar to that of other asphalt pavements. And, like other asphalt pavements, they can be designed for many situations.

Porous asphalt pavements are constructed in the form of an underlying, open-graded stone bed that provides drainage for the water. As the water drains through the porous asphalt and into the stone bed, it slowly infiltrates into the soil. However, the stone bed size and depth must be designed so that the water level never rises into the asphalt. This stone bed, often 18–36 in. in depth, provides a sub-base for the asphalt paving.

While slightly coarser than standard asphalt, porous asphalt pavements are acceptable. Moreover, the porous asphalt pavement does not require proprietary ingredients and does not require special paving equipment. Because of the open structure of the pavement, porous asphalt offers a means to replenish water tables and aquifers rather than forcing rainfall into storm sewers, porous asphalt also helps to reduce demands on storm sewer systems.

8.2.7 Road Asphalt

Road asphalt is, as the name implies, asphalt used for roadway construction which, in contrast to road oil, may also be pulverized to produce powdered asphalt. In the process, the asphalt is crushed and passed through a series of fine mesh sieves to ensure uniform size of the granules. Powdered asphalt can be mixed with road oil and aggregate for pavement construction. Heat and pressure in the road slowly amalgamate the powder with the aggregate and binding oil, and the substance hardens to a consistency similar to regular asphalt cement.

Briefly, aggregates (mineral aggregates) are hard, inert materials such as sand, gravel, crushed stone, slag, or rock dust. Properly selected and graded aggregates are mixed with the cementing medium asphalt to form pavements. Aggregates are the principal load-supporting components of an asphalt pavement and constitute 90–95% (w/w) or 75–85% (v/v) of the mixture.

Selection of an aggregate material for use in an asphalt pavement depends on the availability, cost, and quality of the material, as well as the type of construction for which it is intended. To determine whether an aggregate material is suitable for use in asphalt construction, it should be evaluated in terms of the following properties: (i) size and grading, (ii) cleanliness, (iii) toughness, (iv) soundness, (v) particle shape, (vi) surface texture, (vii) adsorption, and (viii) stripping.

In terms of *size and grading*, the maximum size of an aggregate is the smallest sieve through which 100% of the material will pass. The end use of the asphalt–aggregate mix determines not only the maximum aggregate size but also the desired gradation (distribution of sizes smaller than the maximum). Asphalt must also be *clean* since foreign or deleterious substances can render asphalt–aggregate mixtures unsuitable for paving. In addition, *toughness* or *hardness* is the ability of the aggregate to resist crushing or disintegration during mixing, placing, and compacting; or under traffic loading. *Soundness*, although similar to toughness, is the ability of the aggregate to resist deterioration caused by natural elements such as the weather.

Weather climate patterns throughout most countries can differ significantly, which can result in significant differences in the temperature profiles of pavements that are the main factors that influence the behavior of asphalt binder and pavement performance. For different pavement temperature profiles, asphalt binders in pavements experience unequal amounts of binder oxidation over time. Understanding these differences is a necessary aspect of pavement design to mitigate the binder hardening that occurs during service life. Moreover, different pavement temperature profiles can change the rate of binder hardening and, therefore, the asphalt binders with appropriate rheological properties must be selected to compensate for binder hardening due to the diversity of climate zones. Furthermore, integration of the asphalt binder oxidation process with pavement temperature is critical to binder oxidation in pavements under various climate conditions and thus to pavement performance modeling and maintenance planning.

In addition, the *shapes of aggregate particles* influence the strength of the asphalt–aggregate mixture as well as workability and density achieved during compaction—when compacted, irregular particles such as crushed stone tend to lock together and resist displacement. Workability and pavement strength are influenced by *surface texture* of the aggregate—a rough, sandpapery texture results in a higher strength than a smooth texture. Although smooth-faced aggregates are easy to coat with an asphalt film, they are generally not as good as rough surfaces—it is more difficult for the asphalt to adsorb onto a smooth surface.

The porosity of an aggregate permits the aggregate to *adsorb* asphalt and form a bond between the particle and the asphalt. A degree of porosity is desired, but aggregates that are highly absorbent are generally not used. *Stripping* occurs when the asphalt film separates from the aggregate because of the action of water. Aggregates coated with too much dust also can cause poor bonding, which results in stripping. The problems of aggregates readily susceptible to stripping paving mixes can be mitigated by use of an antistripping agent.

For paving purposes, asphalt and aggregate (the mineral matrix) are combined in a mixing facility where they are heated, proportioned, and mixed to produce the desired paving mixture. Hot-mix facilities may be permanently located (*stationary facilities*) or may be portable and moved from job to job. Hot-mix facilities may be classified as either a *batch facility* or a *drum-mix facility* and both can be either stationary or portable units.

Batch-type hot-mixing facilities use different size fractions of hot aggregate which are drawn in proportional amounts from storage bins to make up one batch for mixing. The combination of aggregates is dumped into a mixing chamber (pug mill) in which the asphalt, which has also been weighed, is then thoroughly mixed with the aggregate. After mixing, the material is then emptied from the pug mill into trucks, storage silos, or surge bins. The drum-mixing process heats and blends the aggregate with asphalt all at the same time in the drum mixer.

When mixing is complete, the hot-mix is then transported to the paving site and spread in a partially compacted layer to a uniform, even surface with a paving machine. While still hot, the paving mixture is further compacted by heavy rolling machines to produce a smooth pavement surface.

The quality of an asphalt product is affected by the inherent properties of the petroleum crude oil from which it was produced. Different oil fields and areas produce crude oils with very different characteristics and the refining method also affects the quality of the asphalt cement. For engineering and construction purposes, there are three important factors to consider: (i) consistency, also called the viscosity or the degree of fluidity of asphalt at a particular temperature, purity, and safety, (ii) purity, and (iii) ductility.

The *consistency* or viscosity of asphalt varies with temperature, and asphalt is graded based on ranges of consistency at a standard temperature. Careless temperature and mixing control can cause more hardening damage to asphalt cement than many years of service on a roadway. A standardized viscosity or penetration test is commonly specified to measure paving asphalt consistency—air-blown asphalts typically use a softening point test.

The *purity* of asphalt can be easily tested since (by definition) it is composed almost entirely of material which is soluble in carbon disulfide. Refined asphalts are usually more than 99.5% soluble in carbon disulfide and any impurities that remain are inert. Because of the hazardous flammable nature of carbon disulfide, trichloroethylene, which is also an excellent solvent for asphalt, is also used in the test for purity by solubility.

Asphalt must be free of water or moisture as it leaves the refinery. However, transports loading the asphalt may have moisture present in their tanks. This can cause the asphalt to foam when it is heated above 100°C (212°F), which is a safety hazard. Specifications typically require that asphalt does not foam at temperatures up to 175°C (345°F). Asphalt, if heated to a sufficiently high temperature, will release fumes which will flash in the presence of a spark or open flame (the *flash point*) and should be well above temperatures normally used in paving operations. Because of the possibility of asphalt foaming and to ensure an adequate margin of safety, the flash point of the asphalt is measured and controlled.

Ductility, which is a measure of the ability of the asphalt to be pulled, drawn, or deformed, is another important engineering property of asphalt.

For asphalt, the presence or absence of ductility is usually more important than the actual degree of ductility because asphalt with a high degree of ductility is also more temperature-sensitive. Ductility is measured by an "extension" test, whereby a standard asphalt cement briquette molded under standard conditions and dimensions is pulled at a standard temperature (normally 25°C, 77°F) until it breaks under tension. The elongation at which the asphalt cement sample breaks is a measure of the ductility of the sample.

Finally, in keeping with environmental protection issues (see Chapter 12) and the stringent codes limiting water flow as well as particulate and smoke emissions from oil refineries, asphalt processing plants are also subject to these laws. The products (or by-products) formed during asphalt production, if unchecked, create odoriferous fumes and pollutants which will stain and darken the air. Pollutants emitted from asphalt production are controlled by enclosures which capture the exhaust and then recirculate it through the heating process. This not only eliminates the pollution but also increases the heating efficiency of the process.

Higher costs of asphalt cement, stone, and sand have forced the industry to increase efficiencies and recycle old asphalt pavements. In asphalt pavement recycling, materials reclaimed from old pavements are reprocessed along with new materials. The three major categories of asphalt recycling are (i) hot-mix recycling, where reclaimed materials are combined with new materials in a central plant to produce hot-mix paving mixtures, (ii) cold-mix recycling, where reclaimed materials are combined with new materials either onsite or at a central plant to produce cold-mix base materials, and (iii) surface recycling, a process in which the old asphalt surface pavement is heated in place, scraped down (*scarified*), remixed, re-laid, and rolled. Organic asphalt recycling agents may also be added to help restore the aged asphalt to desired specifications.

Because of solvent evaporation and volatility, use of cutback asphalts, especially rapid-cure cutback asphalt which uses gasoline or naphtha, is becoming more restricted or prohibited while emulsified asphalts (in which only the water evaporates) are becoming more popular because of various environmental regulations.

Mastic asphalt is a mixture of asphalt and fine mineral material in proportions so that it may be poured hot in place then compacted by hand troweling to a smooth surface for flooring, roofing, and paving. While mastic is sometimes used for specific paving applications in Europe, it is not used in the United States and should not be confused with the predominant and typical asphalt paving applications in both the United States and Europe. Application methods, equipment, and job tasks for mastic asphalt vary from those of conventional asphalt paving. Mastic, as used in road paving, can be spread by hand or with a special paving machine. Harder asphalt grades are used in mastic asphalt, resulting in mixing and placement at 180–250°C (356–482°F), which is higher than the temperatures for typical asphalt pavements.

8.2.8 Road Oil

Road oil is, as the name implies, a liquid asphalt that is intended for easy application to earth roads; it does not provide a strong base or a hard surface but maintains a satisfactory passage for light traffic. Both straight-run and cracked residua have been employed successfully. Binding quality and adhesive character are important in governing the quality of the road produced; resistance to removal by emulsification has some influence on its permanence. Liquid road oils, cutbacks, and emulsions are recently discovered products, but the use of asphaltic solids for paving goes back to a European practice of about 1835. The asphaltic constituents employed may have softening points up to approximately 110°C (230°F). Put simply, road oil is an asphalt product used in sealing and protecting roadways and consists of an almost-black viscous fluid that is typically applied (using a sprayer) to the surface of the road. Manufacturers of asphalt products usually produce road oil and can formulate the product by customer request.

Road oil can be used to reseal a road periodically without having to repave the road. Potholes can be filled with a solution of asphalt pavement crushed into the roadway and covered with road oil to seal it, although this measure is generally only temporary, as the paving will flake away at the edges of the pothole. Eventually, wear and tear on the road will make it necessary to tear up the asphalt paving and replace it with a fresh layer of pavement.

8.2.9 Roofing Asphalt

As a roofing material, asphalt far exceeds the popularity of all other types of roofing material combined. Asphalt shingles are prepared by impregnating suitable roofing felt with an asphalt saturant, after which both sides of the felt are coated with a harder, tougher asphalt. Mineral granules are embedded into the surface, and the material is then cut into shingles or strips. Two types of asphalt roll roofing are in common use: (i) smooth surfaced and (ii) mineral surfaced. These, too, are asphalt-saturated felts, coated with a harder grade of asphalt. Asphalt built-up roofing is composed of several layers of asphalt-saturated felt applied to a flat or sloping roof deck. Between the felt layers, asphalt is applied for adhesion. Over the top layer, a flood coat of asphalt is applied and usually covered with gravel or slag.

Roofing asphalts are graded into four types (Types I, II, III, or IV, in increasing order of hardness) and are liquefied by heating and applied directly during construction. Mopping grade roofing asphalts are used as an interply adhesive or top coating for asphalt-saturated felts on built-up roofs. To insure proper performance and longevity, and to avoid product degradation due to overheating, roofing asphalts are typically heated to temperatures on the order of 235–275°C (450–525°F) on the job site and applied at lower temperatures of approximately 165–230°C (330–445°F) at the point of mopping. The slope of the roof dictates the grade of asphalt used; as the slope increases so does the hardness and grade of asphalt.

The asphalt for use in asphalt shingles (saturated felts, coated fabrics, coated glass fibers) is produced by blowing air through heated residuum, which is mixed with mineral filler at the roofing plant and applied to an organic or inorganic matting to produce granule-surfaced shingles, smooth-surfaced shingles, smooth roll ply sheets, or granule-surfaced roll sheets. For roll roofing materials, nonblown viscosity-graded asphalt cement is heated and mixed with fillers and a polymer or copolymer, which is then impregnated onto inorganic reinforcing matting and formed as a granule-surfaced or smooth-surfaced roll. This material is normally installed on a roof as a multilayer membrane system.

In addition, recycled roofing shingles have been used in pavement applications as a blend with HMA. The use of roofing shingles can result in a reduction in optimum binder content and the use of roofing shingles enhances the ability of the mixtures to densify under compaction. However, the effect of the shingles is dependent upon the source as well as on the source and properties of the asphalt binder (Newcomb et al., 1993).

8.2.10 Rubber-Modified Asphalt

One of the most successful methods of disposing of scrap tires and improving the properties of asphalt is the use of crumb rubber or rubber latex in asphalt pavements (Bahia and Davis, 1994; Zanzotto and Kennepohl, 1996; Kim et al., 2001; Airey et al., 2003; Tuntiworawit et al., 2005; Kim et al., 2008; Xiao et al., 2009). The time required to react the rubber with an asphalt binder is dependent on many factors including the chemistry of the asphalt binder and the rubber, as well as the particle size and texture of the rubber and the reaction temperature. In this investigation, rubber-modified binders were produced using seven reaction times (5, 30, 60, 90, 120, 240, and 480 min) at the reaction temperature of 177°C (351°F).

More recent work has shown that the combination of the crumb rubber and the warm mix asphalt additive in asphalt binder is beneficial for improving the rheological properties of both the original and aged binders (Xiao et al., 2009). There is an increase in the mixing and compaction temperatures due to the addition of crumb rubber but which can be offset by adding the warm asphalt additives, which lowers the mixing and compaction temperatures of rubberized mixtures comparable to conventional HMA.

However promising the results may be, the increasing usage of crumb rubber in flexible pavements requires a better understanding of its effects on the physical, chemical, and rheological properties of rubber-modified binders (Wahhab et al., 1999; AASHTO T315; AASHTO T316). In general, the application of crumb rubber in an asphalt binder is intended to improve the binder properties by reducing the binder's inherent temperature susceptibility. The improvement of the properties of rubber-modified binders depends on the interaction between crumb rubber and binder where the rubber particles swell in the binder to form a viscous gel with an increase in viscosity of the rubber-modified binder (Green and Tolonen, 1977; Heitzman 1992; Bahia and Davis, 1994; Zanzotto and Kennepohl, 1996; Kim et al., 2001; Airey et al., 2003).

The time required to react the crumb rubber with an asphalt binder is dependent on many factors including the chemistry of the asphalt binder and the crumb rubber as well as the particle size and texture of the crumb rubber and the reaction temperature.

8.2.11 Miscellaneous Uses

Modifications (by fluxing or emulsification) of paving-grade asphalt have specific secondary roles within the asphalt paving industry. Fluxed asphalt involves the mixing of a specific asphalt with lower-viscosity diluents to produce a cutback asphalt which allows application at lower temperatures. It should be noted that cutback asphalt has been largely replaced with the more environmentally friendly asphalt emulsion. Emulsification involves the fine dispersion of asphalt in a solution of water and surfactant. Like cutback asphalt, emulsified asphalt can be applied at lower temperatures.

These products are commonly used to provide a waterproof layer under new pavement surfaces and sometimes to improve bonding between various layers of asphalt pavement; in these cases they are known as *tack coats* or *bond coats*. They are also used in some surface sealing applications, such as surface dressing and slurry sealing, and to produce a cold-mix patching material that can be stored for longer periods. These special asphalts are typically applied at ambient temperature.

In addition, asphalt has also been used in many parts of the world, particularly in the United States and Italy, as a base under new rail track, in place of the traditional rail-track ballast. Asphalt is well worth considering when a new rail-track is being laid as, although it will initially be more expensive than the traditional form of construction, it is likely to require far less maintenance.

REFERENCES

AASHTO T315, 2004. Determining the Rheological Properties of Asphalt Binder Using a Dynamic Shear Rheometer (DSR). AASHTO Standards, American Association of State Highway and Transportation Officials, Washington, DC.

AASHTO T316, 2004. Viscosity Determination of Asphalt Binder Using Rotational Viscometer. AASHTO Standards, American Association of State Highway and Transportation Officials, Washington, DC.

Abraham, H., 1945. Asphalts and Allied Substances: Their Occurrence, Modes of Production, Uses in the Arts and Methods of Testing, 5th Edition, Van Nostrand, New York.

Airey, G.D., 2002. Rheological evaluation of ethylene vinyl acetate polymer modified bitumens. Constr. Build. Mater. 16, 473–487.

Airey, G.D., Rahman, M.M., Collop, A.C., 2003. Absorption of bitumen into crumb rubber using the basket drainage method. Int. J. Pavement Eng. 4 (2), 105–119.

Al-Hadidy, A.I., Tan, Y.Q., Ayman, T.H., 2011. Starch as a modifier for asphalt paving materials. Constr. Build. Mater. 25, 14–20.

Bahia, H.U., Davis, R., 1994. Effect of crumb rubber modifiers (CRM) on performance related properties of asphalt binders. J. Assoc. Asphalt Paving Technol. 63, 414–449.

Edwards, Y., Tasdemir, Y., Isacsson, U., 2007. Rheological effects of commercial waxes and polyphosphoric acid in bitumen 160/220—high and medium temperature performance. Constr. Build. Mater. 21, 1899–1908.

Esmaeil, A., Majid, Z., Mohamed, R.K., Mahrez, A., Payam, S., 2011. Using waste plastic bottles as additive for stone mastic asphalt. Mater. Des. 32, 4844–4849.

Fawcett, A.H., McNally, T., 2001. Studies on blends of acetate and acrylic functional polymers with bitumen. Macromol. Mater. Eng. 286, 126–137.

Green, E.L., Tolonen, W.J., 1977. The Chemical and Physical Properties of Asphalt-Rubber Mixtures Report No. ADOT-RS-14 (162). Arizona Department of Transportation, Phoenix, AZ.

Haggam, R.A., Ibrahim, I.M., El-Shafie, A.M., El Latief, A., Elrhman, M.A., El-Kholy, S.A., 2014. Improvement of asphalt properties using polymethyl methacrylate. Open J. Org. Polymer Mater. 2014 (4), 43–54.

Heitzman, M., 1992. Design and construction of asphalt paving materials with crumb rubber modifier Transportation Research Record: Journal of the Transportation Research Board, No. 1339. Transportation Research Board, National Research Council., pp. 1–8.
Hoiberg, A.J., 1964. Bituminous Materials: Asphalts, Tars, and Pitches. John Wiley and Sons, New York, NY.
Iqbal, M.H., Hussein, I.A., Wahhab, H., Amin, M.B., 2006. Rheological investigation of the influence of acrylate polymers on the modification of asphalt. J. Appl. Polym. Sci. 102, 3446–3456.
Karim, G., Azam, J.A., Mahsa, N., 2012. Statistical investigation on physical-mechanical properties of base and polymer modified bitumen using artificial neural network. Constr. Build. Mater. 37, 822–831.
Kim, S., Loh, S.W., Zhai, H., Bahia, H., 2001. Advanced characterization of crumb rubber-modified asphalts, using protocols developed for complex binder Transportation Research Record: Journal of the Transportation Research Board, No. 1767. Transportation Research Board, National Research Council, Washington, DC, pp. 15–24.
Kim, H.S., Geiger, A., Amirkhanian, S.N., Parl, T.S., Kim, K.W., 2008. Effects of asphalt ratios on the properties of crumb rubber modified asphalts. In: Proceedings. 6th International Conference on Road and Airfield Pavement Technology (ICPT), Sapporo, Japan. July 20–23.
Kim, H., Lee, S.J., Amirkhanian, S.N., 2011. Rheology of warm mix asphalt binders with aged binders. Constr. Build. Mater. 25, 183–189.
Martin, J., Dusan, B., Daryl, M., Ludo, Z., 2013. Preparation and properties of conventional asphalt modified by physical mixtures of linear SBS and montmorillonite clay. Constr. Build.Mater. 38, 759–765.
Merusi, F., Giuliani, F., 2011. Rheological characterization of wax-modified asphalt binders at high service temperatures. Mater. Struct. 44, 1809–1820.
Moatasim, A., Cheng, P.F., Al-Hadidy, A.I., 2011. Laboratory evaluation of HMA with high density-polyethylene as a modifier. Constr. Build Mater. 25, 2764–2770.
Newcomb, D., Stroup-Gardiner, M., Weikle, B., Drescher, A., 1993. Influence of Roofing Shingles on Asphalt Concrete Mixture Properties Report No. MN/RC-93/09. University of Minnesota, Minneapolis, Minnesota and Office of Research Administration, Minnesota Department of Transportation, St. Paul, MN.
Peiliang, C., Peijun, X., Mingliang, X., Shuanfa, C., 2013. Investigation of asphalt binder containing various crumb rubbers and asphalts. Constr. Build. Mater. 40, 632–641.
Polacco, G., Berlincioni, S., Biondi, D., Stastna, J., Zanzotto, L., 2005. Asphalt modification with different polyethylene-based polymers. Eur. Polym. J. 41, 2831–2844.
Speight, J.G., 1978. The Use of Bitumen as an Adhesive and Building Material at the Ancient Cities of Babylon, Calah, Nineveh, and Ur. Report of Investigations. College of Science, University of Mosul, Iraq.
Traxler, R.N., 1961. Asphalt: Its Composition, Properties, and Uses. Reinhold Publishing Corp., New York, NY.
Tuntiworawit, N., Lavansiri, D., Phromsorn, C., 2005. The modification of asphalt with natural rubber latex. Proc. Eastern Asia Soc. Transp. Stud. 5, 679–694.
Wahhab, H.I., Asi, I.M., Ali, F.M., Dubabi, I.A., 1999. Prediction of asphalt rheological properties using HP-GPC. J. Mater. Civil Eng., ASCE 11 (1), 6–14.
Xiao, F., Zhao, W., Amirkhanian, S.N., 2009. Fatigue behavior of rubberized asphalt concrete mixtures containing warm asphalt additives. Constr. Build. Mater.. 23 (10), 3144–3151.
Zanzotto, L., Kennepohl, G., 1996. Development of rubber and asphalt binders by depolymerization and devulcanization of scrap tires in asphalt Transportation Research Record: Journal of the Transportation Research Board, No. 1530. Transportation Research Board, National Research Council, Washington, DC, pp. 51–59.

Zhang, F., Yu, J.Y., 2010. Research for high-performance SBR compound modified asphalt. Constr. Build. Mater. 24, 410–418.
Zhang, H.L., Jia, X.J., Yu, J.Y., Xue, L.H., 2013a. Effect of expanded vermiculite on microstructures and aging properties of styrene–butadiene–styrene polymer modified bitumen. Constr. Build. Mater. 40, 224–230.
Zhang, H.Y., Wu, X.W., Cao, D.W., Zhang, Y.J., He, M., 2013b. Effect of linear low density-polyethylene grafted with maleic anhydride (LLDPE-g-MAH) on properties of high density-polyethylene/styrene–butadiene–styrene (HDPE/SBS) modified asphalt. Constr. Build. Mater. 47, 192–198.

CHAPTER 9
Asphalt Technology

9.1 INTRODUCTION

The history of the use of asphalt-type materials as construction materials began as a result of the occurrence of bitumen (natural asphalt) in seepages to the surface of the Earth, in asphalt lakes and in rock asphalt (a mixture of sand, limestone, and bitumen) (see Chapter 1). In fact, the first recorded (documented) use of bitumen as a road building material was in Babylon in 625 BC during the reign of King Nabopolassar (Gillespie, 1992). Following on from this, the ancient Greeks were familiar with asphalt and its properties—the word *asphalt* comes from the Greek *asphaltos*, meaning *secure* after which the ever-practical Romans, with their massive building programs (especially the aqueducts that carried water into major cities to reservoirs and baths, all of which required water-proofing to prevent leaks), used bitumen and the word became *asphaltus*. Many centuries later, in 1595 Sir Walter Raleigh described a *plain* (or lake) of bitumen on the island of Trinidad, off the coast of Venezuela and he used the material to supplement and strengthen the caulking of his ships. However, despite these early uses of bitumen, it was not until the eighteenth century that European or American builders tried it as a paving material as a means of improving road conditions.

The production of manufactured asphalt was used in 1869 in New Jersey and the use was extended to New York in 1872–1873. After this the construction of asphalt roadways was practiced in Pennsylvania and in Washington, DC, main roads were paved using bitumen from the Trinidad Lake and the quality (durability) of this pavement showed that the bitumen available in the Americas was as good as the asphalt or bitumen imported from Europe. By 1910 it was the most commonly used paving material in US cities. Early construction was by hand, and mechanical spreading of these asphalt mix did not occur until 1928. Since that time, vehicle speeds and load weight have increased with concomitant increases in vehicle weight and tire pressures, and traffic frequency, which have been accompanied by changes in asphalt design and construction. Asphalt roadways are now built using plant-based construction techniques for reasons

Asphalt Materials Science and Technology.
DOI: http://dx.doi.org/10.1016/B978-0-12-800273-5.00009-X

of speed, quality, economy, worker safety, and preference. The technique for asphalt surfacing by hand has been substantially lost (SABITA, 1992, 1993a,b).

The advantages of the use of asphalt for roadway construction were quickly recognized and asphalt roadways became an important industry. With the surge in the construction of asphalt roadways, the need for warranties arose and focused on the quality (through specifications) of the components of the roadways. By the early 1900s, with the growth in popularity of the automobile, the need for quality roadways led to changes and innovations in the production of asphalt and laying of asphalt roadways (Mallick et al., 2009).

9.2 EARLY ASPHALT TECHNOLOGY

The earliest plants to produce asphalt mix for road construction consisted of shallow iron trays heated over open coal fires. The aggregate was dried on the tray and hot asphalt was placed on top of the hot aggregate and the mixture was stirred manually, which dictated the quality of the mix. In 1854, mechanical mixers were introduced and by the end of the nineteenth century, mixers and dryers were available in a variety of forms. Some roadway construction companies developed portable mixing machines and mixing units that were mounted on railroad cars. By 1901, the continued evolution of asphalt technology had resulted in the first production facility to contain the different facets of a modern facility followed by the installation of drum mixers and dryer-mixers in 1910. Improved cold feed units for portable and semi-portable production systems occurred in the 1920s and the next 10 years saw the introduction of vibrating screens and pressure injection systems. The earliest methods for laying asphalt roadways involved the extremely labor-intensive *hands-on* job and involved the use of brooms, lutes, squeegees, and tampers. In fact, it was only after the asphalt was tipped at the site, spread, and smoothed by hand that the horse-drawn roller and, later, the steam roller were used to complete the road construction.

With the evolution of, and improvements to, road construction in the 1930s, the demand for asphalt mix tended to outstrip availability. However, by the late 1930s and the beginning of World War II the tonnage of asphalt mix production was increasing and matched demand. Electronic leveling controls were in use in the 1950s and automated screed controls came into use in the early 1960s—the screed is a leveling device that is designed to flatten the material and is attached to the rear of the paving machine. At the same time, the asphalt plants of the early 1950s, which typically included an aggregate dryer and an asphalt–aggregate mixer, were expanded to include air pollution controls such as a wet scrubber and/or a baghouse that are typical of modern gas processing plants (Mokhatab et al., 2006; Speight, 2007, 2014). Then extra-wide finishers that were capable of paving two lanes simultaneously were in regular use by the late 1960s.

Recycling (or reclaiming for recycle) used asphalt pavement was common in the early twentieth century, but fell out of use for a variety of reasons—the prime reason appears to have been the plentiful supply of petroleum that could be used to manufacture high-grade asphalt. However, during the 1970s, recognition of the depletion of petroleum resources—especially petroleum that could be used for the manufacture of high-quality asphalt—and the energy crises of 1973–1974 and 1979 emphasized the need for reclaiming programs. Indeed, the move towards the conservation of natural resources emphasized the necessity (or wisdom) of asphalt reclaiming and recycling and since that time both base and surface courses of asphalt roadways have incorporated an increasing amount of recycled asphalt in the mixes (Al-Qadi et al., 2012). In addition, quality improvement programs have been adopted and specifications for asphalt mix have been, and continue to be, under continual scrutiny (Kennedy et al., 1994, 1998a,b).

Over the last 30 years, the versatility of asphalt has led to its increased use in other applications, such as airport runways, solutions to water storage, flood control, erosion, lining drinking water reservoirs and fish hatcheries, and conservation-related issues (see Chapter 12). Moreover, asphalt (with the approval of the US Environmental Protection Agency) is used successfully as a primary liner for both sanitary and hazardous-waste landfills. With continuous improvements in production and the necessary equipment, and with the continued emphasis on quality, there is increased durability and in-service life of asphalt roadways and in other uses where asphalt is a key component.

9.3 MODERN ASPHALT TECHNOLOGY

To use more modern parlance, an *asphalt concrete pavement* consists of a combination of layers, which include an asphalt concrete surface constructed over a granular or asphalt concrete base and a sub-base. The production of the asphalt mix—a mixture of fine and coarse aggregate with asphalt cement binder that is mixed, placed, and compacted in a heated condition—involves heating and mixing the components (asphalt binder, aggregate, and any additives) at a central plant and placing on the road using an asphalt spreader. The entire pavement structure, which is constructed over the sub-grade, is designed to support the traffic load and distribute the load over the roadbed. The pavements can be constructed using hot mix asphalt (HMA) or cold mix asphalt (CMA) but surface treatments may be necessary during pavement construction. In such cases, surface treatment can act as a means of waterproofing the existing pavement surface and will also provide resistance to abrasion by traffic.

Asphalt concrete mixtures can be classified in many different ways. Perhaps the most general type of classification is by whether or not the asphalt mix must be heated prior to transport, placement, and compaction. For example, the mix commonly referred to as HMA concrete must be thoroughly heated during mixing, transport, placement, and compaction. The asphalt binder used in asphalt mix is not fluid at room temperatures, so that once this type of asphalt concrete cools it becomes semi-solid or solid and strong enough to support heavy traffic. CMA, on the other hand, is normally handled, placed, and compacted without heating. This material can be handled cold because it uses liquid asphalt in the form of an asphalt emulsion and/or cutback asphalt, both of which are fluid at room temperature.

Asphalt emulsions are mixtures of asphalt, water, and specific chemical additives (surfactants) that allow the other two materials to be blended into a stable liquid—the same principle is used in the recovery of heavy oil or tar sand bitumen by the use of surfactants (Speight, 2008, 2009, 2014). When blended with aggregate, the asphalt separates from the water (the emulsion breaks) and coats the aggregate. Cutback asphalt—a blend of asphalt binder and a petroleum-based solvent, usually a type of naphtha—is mixed with the aggregate and the mix (typically a cold mix since temperature influences the behavior and effectiveness of the cutback asphalt) and the mix cures as the petroleum-based solvent evaporates from the asphalt concrete. However, this procedure gives rise to various environmental concerns as the solvent evaporates into the atmosphere, as well as worker health and safety (see Chapter 12).

Thus, CMA is a mixture of emulsified asphalt and aggregate, produced, placed, and compacted at ambient air temperature. The production of the mix asphalt is economical because the production of the mix does not require large amounts of energy to heat the mix during the production and placement processes and the cooling of the mix during transportation to the work site is not an issue, as it can be with warm mix asphalt (WMA) and HMA. However, on the technical side, the cold mix is difficult to compact thoroughly and in general is not as durable as the temperature-produced mixes. However, CMA may be used for base course construction and is also commonly used for patching and repairing pavement (see Chapter 11). The use of CMA is usually limited to relatively low-volume rural roads. For higher traffic applications, CMA pavement usually requires an overlay of asphalt mix or surface treatment to resist traffic action. The components of CMA can be mixed at a central plant or *in situ* with a traveling mixer.

A third type of mix—WMA—has recently been developed for use in asphaltic roadway construction (Button et al., 2007). In this type of asphalt–aggregate mixture, various methods are used to significantly reduce mix production temperature by 17°C to approximately 56°C (30°F to approximately 100°F). In addition, other methods that are used to significantly reduce mix production temperature and include techniques such as: (i) the use of chemical additives to lower the high-temperature viscosity of the asphalt binder, (ii) addition of water to the binder, causing it to become *foamy asphalt* or *asphalt foam*, and (iii) use of a two-stage process which involves the addition of hard and soft asphalt at different points during mix production. In respect of the actual temperature of the mix, that is so variable that it is difficult if not impossible to define. Furthermore, the rate of cooling of the asphalt mix during transportation to the work site also needs to be taken into account. The rate of cooling of the hot mix is higher than the rate of cooling of the warm mix and it is conceivable that both are applied to the pavement construction at approximately the same temperature. The WMA technology has several benefits, including (i) lower cost—significantly less fuel is required to bring the mix to the desired temperature, (ii) the emissions are lower and so there is a decreased environmental impact, and (iii) there is the potential for improved performance because of decreased time for hardening (age hardening) (Sargand et al., 2009; You et al., 2011).

The selection of the asphalt emulsion for CMA is based on (i) compatibility of the emulsion with the aggregate, (ii) the ability of the emulsion to coat the aggregate at the design rate, and (iii) the conditions for use. The key determinants are the type of aggregate, the gradation of the mixture, the amount of material passing the #200 sieve, and the use of the mix.

Surface treatments consist of an application (or sometimes multiple applications) of emulsified or liquid asphalt and select aggregate, placed over a prepared granular base or existing surface. Following placement of the aggregate, the mixture is rolled and compacted to provide a drivable, dust-free surface. This type of pavement is common on light- to medium-volume roads that may or may not already have an existing bituminous surface.

9.3.1 Materials

The components of asphalt concrete include (i) the asphalt binder and (ii) the aggregate. In some cases, mineral filler may be added to the aggregate to improve the quality and performance of the asphalt concrete, such as reducing the propensity of the asphalt concrete to fatigue damage, low-temperature cracking, and moisture damage (NCHRP, 2011).

9.3.1.1 Aggregate

Approximately 85% (v/v) (approximately 95% w/w) of an asphalt mix consists of the aggregate and, consequently, pavement performance is greatly influenced by the characteristics of the aggregate—contrary to some opinions, the binder is not the only influential aspect of the performance of asphalt concrete. In fact, the aggregate material(s) used in asphalt concrete has a profound influence on the properties and performance of the mixture (Table 9.1). Furthermore, to produce quality asphalt pavements, quality aggregates must be used and should be clean, hard, sound, and durable, with a majority of crushed particles. Proper aggregate grading, strength, toughness, and shape are needed for mixture stability. Generally, the aggregate used in asphalt–aggregate mixtures may be either crushed stone or crushed gravel and must be of good quality to ensure the resulting pavement will perform as expected. The aggregate material must be thoroughly crushed, and the resulting particles should be cubical rather than flat or elongated. Aggregates should be free of dust, dirt, clay, and other deleterious materials and, because aggregate particles carry most of the load in asphalt mix pavements, the aggregate should be tough and abrasion-resistant.

Aggregates can be divided into three types according to the size of the particles: (i) coarse aggregate, (ii) fine aggregate, and (iii) mineral filler. Coarse aggregate is defined as having particles that are retained on the 2.36-mm sieve while fine aggregate is defined as having particles that pass through a 2.36-mm sieve but are retained on the 0.075-mm sieve. On

Table 9.1 Factors That Influence Pavement Performance

Materials	Mixing	Construction	Mix Design	Pavement Structure	Traffic Load	Environment
Asphalt type	Cold or hot	Transportation	Dense-graded	Thickness	Types	Temperature
Asphalt properties	Mixing plant	Compaction	Open-graded	Base strength	Load repetitions	Water
Aggregates		Quality control	Asphalt content	Sub-base strength	Tire properties	Oxidation
Asphalt–aggregate affinity			Film thickness	Sub-grade strength		Spring thaw
Additives			Air voids			

N.B.: The *Time Factor* refers to the duration of the pavement to any of the other effects.
The *Maintenance Factor* refers to the maintenance techniques employed and the frequency of maintenance.

the other hand, mineral filler is defined as that portion of the aggregate passing a 0.075-mm sieve (ASTM D242) and, consequently, is a very fine material and is also referred to as *mineral dust* or *rock dust*.

Gravel refers to a coarse aggregate made up mostly of rounded particles. Gravels are often dredged from rivers and are sometimes mined from deposits. Because of the rounded particle size, gravels are not suitable for use in asphalt mixtures unless they are well crushed. Poorly crushed gravels will not interlock when used in asphalt mixes, and the resulting mixture will have poor strength and rut resistance. Crushed stone is coarse aggregate that is mined and processed by mechanical crushing. It tends to be a very angular material and, depending on its other properties, can be well suited for use in asphalt pavements. One potential problem with crushed stone is that the particles sometimes will tend to be flat, elongated, or both, which can cause problems in asphalt mixtures. Ideally, the particles in crushed stone aggregate should be cubical and highly angular.

The fine aggregate, or sand, used in asphalt mixes can be natural sand, manufactured sand, or a mixture of both types. Natural sand is dredged from rivers or mined from deposits and is then processed by sieving to produce a fine aggregate having the desired particle size distribution. On the other hand, manufactured sand is produced by crushing quarried stone and, like natural sand, followed by sieving to produce the desired gradation. The particles in manufactured sand tend to be more angular than those in natural sand and have the potential to produce an asphalt mix (subject to binder suitability) with a higher strength and rut resistance compared to an asphalt mix composed of binder and natural sand. However, just as caution is needed when choosing the binder, care is also needed when selecting fine aggregate for use in an asphalt mix. The fine aggregate angularity test (AASHTO T304; ASTM C1252) can help to evaluate the angularity of both natural and manufactured sand aggregates (see Chapter 5).

Aggregate toughness and abrasion resistance have also been shown to be related to pavement performance. Aggregate particles that are tough and resistant to abrasion will not break down during the construction process, which helps ensure that an asphalt mix can be properly constructed, placed, and compacted. Tough, abrasion-resistant aggregates also tend to produce a mix that is resistant to pop-outs and raveling. Because aggregate pop-outs and broken aggregate particles near the pavement surface make it easier for water to flow into a pavement, tough and abrasion-resistant aggregates help improve the moisture resistance of asphalt pavements.

Aggregates with poor abrasion resistance can also polish under the action of traffic. This can cause the pavement surface to lose skid resistance, especially when wet. An aggregate property that is closely related to toughness and abrasion resistance is durability and soundness. Freeze–thaw cycles and alternate periods of wetting and drying in a pavement can weaken poor-quality aggregates, causing pop-outs (typically, small, shallow pits or depressions in the surface of either concrete or hot-mix asphalt pavements that result mostly from the weathering of individual aggregate particles) and raveling (the progressive disintegration of an asphalt mix layer from the surface downward as a result of the dislodgement of aggregate particles). Aggregates that possess good durability and soundness will resist the actions of wet–dry and freeze–thaw cycles during the life of the pavement.

Another contribution by the aggregate is that asphalt mixtures contain small amounts of air voids, usually falling into the range of 3–5% (v/v), depending on the type of mixture being designed and the design procedure being used. Properly constructed asphalt mix pavements will usually contain about 6–8% (v/v) air voids immediately after placement and compaction. After construction, as traffic passes over a pavement, the asphalt mix in the wheel paths will normally gradually compact to air-void levels approaching the design value of 3–5% (v/v). However, if the pavement is not compacted adequately during construction, compaction under traffic will fail to reduce the air-void content to the design value and, as a result, the pavement will be permeable to air and water, potentially leading to moisture damage and excessive age hardening.

Furthermore, the empirical nature of mix design method development has led to significant differences in the volumetric composition for an acceptable asphalt mix. A key to the successful mix design is the balance between the volumetric composition and the used raw materials (binder, aggregate, filler, and additives) at specific climatic and traffic conditions. Thus, a balanced design can be achieved only when the climatic considerations are taken into account when selecting the volumetric criteria. A delicate balance between mixture stiffness and shear strength appears to provide the key to a successful design (Kim and Lee, 1995; Pellinen, 2003). Some of the properties for aggregates that are used in asphalt paving mixes include: (i) gradation, (ii) particle shape, (iii) particle strength, (iv) durability, (v) specific gravity, (vi) unit weight, (vii) absorption, (viii) volume stability, and (ix) the presence of deleterious components (Table 9.2).

Gradation refers to the size distribution of the aggregate particles, which should be a combination of sizes that results in the optimum

Table 9.2 Asphalt Paving Aggregate Test Procedures

Property	Test Method	References
General Specifications	Coarse Aggregate for Bituminous Paving Mixtures	ASTM D692
	Fine Aggregates for Bituminous Paving Mixtures	ASTM D1073
		AASHTO M29
	Steel Slag Aggregates for Bituminous Paving Mixtures	ASTM D5106
	Aggregate for Single or Multiple Surface Treatments	ASTM D1139
	Crushed Aggregate For Macadam Pavements	ASTM D693
Gradation	Sieve Analysis of Fine and Coarse Aggregates	ASTM C136
		AASHTO T27
	Sizes of Aggregate for Road and Bridge Construction	ASTM D448
		AASHTO M43
Particle Shape	Index of Aggregate Particle Shape and Texture	ASTM D3398
	Flat and Elongated Particles in Coarse Aggregate	ASTM D4791
	Uncompacted Void Content of Fine Aggregate (As Influenced by Particle Shape, Surface Texture, and Grading)	ASTM C1252
		AASHTO TP33
Particle Texture	Accelerated Polishing of Aggregates Using the British Wheel (Not Widely Recognized in North America)	ASTM D3319/T279
	Insoluble Residue in Carbonate Aggregates (Indirect Measure of Resistance of Aggregate to Wear, by Determining Amount of Carbonate Rock Present)	ASTM D3042
	Centrifuge Kerosine Equivalent (Only Used as Part of the Hveem Mix Design Procedure)	ASTM D5148

Particle Strength	Resistance to Degradation of Large-Size Coarse Aggregate by Abrasion and Impact in the Los Angeles Machine	ASTM C535
	Resistance to Degradation of Small-Size Coarse Aggregate by Abrasion and Impact in the Los Angeles Machine	ASTM C131 AASHTO T96
	Degradation of Fine Aggregate Due to Attrition	ASTM C1137
Durability	Aggregate Durability Index	ASTM D3744 AASHTO T210
	Soundness of Aggregates by Use of Sodium Sulfate or Magnesium Sulfate	ASTM C88 AASHTO T104
	Soundness of Aggregates by Freezing and Thawing	AASHTO T103
Specific Gravity and Absorption	Specific Gravity and Absorption of Coarse Aggregate	ASTM C127 AASHTO T85
	Specific Gravity and Absorption of Fine Aggregate	ASTM C128 AASHTO T84
Unit Weight	Unit Weight and Voids in Aggregate	ASTM C29 AASHTO T19
Volume Stability	Potential Expansion of Aggregates from Hydration Reactions (Developed to Measure Expansion Potential of Steel Slag Aggregates)	ASTM D4792
Deleterious Components	Sand Equivalent Value of Soils and Fine Aggregate (Indirect Measure of Clay Content of Aggregate Mixes)	ASTM D2419
	Clay Lumps and Friable Particles in Aggregates	ASTM C142

balance of voids (density) and pavement strength. *Particle shape* refers to the shape of the aggregate particles, which should be angular and nearly equidimensional or cubical in shape to minimize surface area. Angular and rough-textured aggregates are desirable within asphalt mixes to resist permanent deformation and fatigue cracking. Very angular and rough-textured aggregates provide better interlock between the aggregate particles which helps prevent plastic deformation (rutting; see Chapter 8) within asphalt mix layers. Angular and rough-textured aggregates also help improve the strength of asphalt mixtures, which can help prevent fatigue cracking. Angular aggregates with good surface texture also improve the frictional properties of pavement layers, an important safety consideration in the design of asphalt mixtures for pavements.

The presence of flat or elongated particles within the asphalt–aggregate mix is undesirable because these particles tend to break down during production and construction. Aggregates that break during production and construction will reduce the durability of the asphalt mix layer, leading to raveling, pop-outs, and potholes. On the other hand, in terms of particle texture, the particles should have a rough, rather than smooth, texture to minimize the stripping of asphalt cement and the *particle strength* should be sufficient to resist degradation or breakdown under compaction or traffic and the particles must be of sufficiently high *durability* to remain intact under variable climatic conditions and/or exposure to a variety of chemicals. Thus, durability is the resistance of an asphalt–aggregate mixture to disintegration due to exposure to the combined effects of weathering and traffic. The surface and wearing courses have the most severe exposure, because they are subjected directly to damage by both traffic loading and the environment.

Environment is a direct consideration in each of the design procedures presented in this manual. The environment in which the pavement will be constructed determines the performance grade of binder that will be used for all mixture types. When considering an open-graded friction course (OGFC) as a wearing course in freezing climates, it is important to recognize that these surfaces may require different maintenance practices. The structure of OGFCs causes these mixtures to freeze more quickly than dense-graded and gap-graded asphalt mixtures (see Chapter 10), resulting in the need for earlier and more frequent application of deicing chemicals. Additionally, sand should not be used with the deicing chemicals because the sand will plug the pores of the OGFC, decreasing their effectiveness.

The exposure for intermediate and base courses is less, except during some construction schedules when the intermediate or base layer may be temporarily subjected to traffic loads for extended time periods. Mixtures subjected to more severe exposure conditions must have greater durability and should be efficiently compacted during construction. In general, both the binder content and the amount of fines in the aggregate blend will increase with decreasing nominal maximum aggregate size (NMAS), which is a prime reason for the use of smaller NMAS mixtures in surface courses. The effective binder content (EBC) of gap-graded asphalt mixes is higher than typical due to the gap-graded structure of these mixes. OGFC asphalt mixes typically incorporate modified asphalt binders (as well as fibrous material) to increase the binder content of these mixtures and improve their durability.

The *specific gravity* of an aggregate is an important property that is needed in order to properly design and proportion an asphalt mix. In relation to this, the *unit weight* of an aggregate is an indicator of the compacted density of an asphalt paving mix containing this aggregate and the pavement yield (the volume of pavement that will be required for a given pavement mass).

The property of an aggregate known as *absorption* refers to the amount of void spaces within a particle that may be filled with asphalt binder (or air or water), and is a measure of the tendency of an aggregate to absorb asphalt—the higher the absorption capacity, the more asphalt cement will be needed. In addition, there must be *volume stability* since certain aggregates may undergo volumetric expansion following prolonged exposure to moisture and other effects such as exposure to de-icing salts, which may contribute to pop-outs, raveling, and random cracking in an asphalt pavement.

Finally, the aggregate must be free of *deleterious components* which can be harmful insofar as there are amounts of potentially reactive components (such as shale, chert, sulfates, alkalis, and expansive silicates) which can also contribute to pop-outs, raveling, and pavement cracking. Cleanliness is a term used to characterize the coatings on some aggregate particles. These coatings are often very fine clay-like materials and can affect the adhesion between the asphalt binder and aggregate particles leading to increased potential for moisture damage. Deleterious materials are particles in an aggregate stockpile that are weak, prone to freeze–thaw damage or damage through repeated wetting and drying, or that otherwise can cause a pavement to deteriorate. Some examples of deleterious materials are clay lumps, friable particles, shale, coal, free mica, and vegetation. These types of materials are not as strong as mineral aggregates and break down during the life of a pavement layer. When this happens, pop-outs and potholes can occur.

9.3.1.2 Mineral Filler

Mineral fillers consist of finely divided mineral matter such as rock dust, slag dust, hydrated lime, hydraulic cement, fly ash, loess (fragments of pre-existing minerals and rock forming a predominantly silt-sized sediment that is formed by the accumulation of wind-blown dust), or other suitable mineral matter. The mineral filler consists of very fine, inert mineral matter that is added to the asphalt mix to improve the density and strength of the mixture. Mineral fillers make up less than 6% (w/w) of the asphalt mix concrete by mass, and generally less than about 3% (w/w). A typical mineral filler completely passes a 0.060 mm (No. 30) sieve, with at least 65% of the particles passing the 0.075 mm (No. 200) sieve.

The portion of the mineral filler that is finer than the thickness of the asphalt film and the asphalt cement binder form a mortar or mastic that contributes to improved stiffening of the mix. The particles larger than the thickness of the asphalt film behave as mineral aggregate and hence contribute to the contact points between individual aggregate particles. The gradation, shape, and texture of the mineral filler significantly influence the performance of asphalt mix. Some of the more important properties of mineral filler used in asphalt concrete applications are as follows: gradation, plasticity, and the presence of deleterious components (Table 9.3).

Finally as a point of clarification, in terms of *gradation*, mineral fillers should have 100% of the particles passing 0.60 mm (No. 30 sieve), 95–100% passing 0.30 mm (No. 40 sieve), and 70% passing 0.075 mm (No. 200 sieve). In addition, the mineral filler should be *non-plastic* so the particles do not bind together and the percentage of *deleterious materials* such as clay and shale in the mineral filler must be minimized to prevent particle breakdown.

Table 9.3 Mineral Filler Test Procedures

Property	Test Method	Reference
General Specifications	Mineral Filler for Bituminous Paving Mixtures	ASTM D242 AASHTO M17
Gradation	Sieve Analysis of Mineral Filler for Road and Paving Materials	ASTM D546
Plasticity	Liquid Limit, Plastic Limit, and Plasticity Index of Soils	ASTM D4315
Deleterious Materials	Sand Equivalent Value of Soils and Fine Aggregate (Indirect Measure of Clay Content of Aggregate Mixes)	ASTM D2419

9.3.1.3 Aggregate Storage

One aspect of asphalt technology that must also be given consideration is the issue of aggregate storage. Provision should be made for adequate storage and stockpiling facilities for all materials that are components of the asphalt mix. Moreover, sufficient material should be on hand prior to the start of daily operations to insure continued daily availability of the materials for the project.

Aggregates should be handled, hauled, and stored in a manner that will minimize segregation and degradation, and avoid contamination. The aggregate should be stockpiled in the vicinity of the plant on ground that is denuded of vegetation, hard, well-drained, or otherwise prepared to protect the aggregate from contamination. In addition, the stockpiles should be separated to prevent intermingling, which can be accomplished by positive separation of stockpiles by using bins or adequate bulkheads. If bulkheads are used, each bulkhead should extend to the full depth of the adjacent stockpiles and should be strong enough to withstand the pressure exerted as the stockpile increases. Furthermore, the stockpiles should be constructed in layers (analogous to strata) rather than in cone formations—the individual truckloads should be tipped close together and spread to cover as much of the stockpile surface as possible. Similarly, if a crane is used for stockpile construction, each bucket load should be deposited adjacent to another over the entire area so that the thickness of the layers is uniform. In all cases, care should be exercised in loading and unloading to prevent segregation material and care should also be excised to prevent aggregate degradation by the hauling equipment when withdrawing material from a stockpile.

Finally, since mineral filler is subject to caking or hardening from moisture, it should be handled differently than other aggregate materials, and separate storage is provided to keep it protected from dampness.

9.3.1.4 Binder

Although the asphalt binder (also called *asphalt cement*) component typically comprises a minority of the asphalt–aggregate mix—approximately 5–6% (w/w) of the asphalt paving mixture—it does have a strong and meaningful effect insofar as the binder coats and binds the aggregate particles together. In fact, the selection of the proper grade of asphalt (asphalt cement or emulsion) for the traffic and climatic conditions to which the paving mixture is to be exposed is essential to the performance of the mix. Some of the more important properties of asphalt cement that are used to distinguish between different cements and to evaluate their quality include: (i) the needle penetration test, (ii) viscosity, (iii) ductility, and (iv) incompatibility (Table 9.4). In fact, over the years it has been recognized that compatibility/incompatibility and the contribution of stability/instability of the asphalt mix are important aspects of binder quality. Either phenomenon (stability/instability) can be manifested when binders for different sources are blended with the result being phase separation of incompatible (typically asphaltene) material (see Chapter 7).

It is the asphalt binder that holds the aggregate in asphalt mix together—if the binder was ineffective, the asphalt mix would merely be crushed stone or gravel with some sticky bituminous-like material as a non-participating part of the mixture. However, it is essential to remember that not all binders are the same and the physical properties of asphalt binder vary tremendously with the source from which they were derived as well as with temperature. In fact, at high temperatures, the binder is a fluid with a consistency similar to that of lubricating oil whereas at ambient temperature the most binders have the consistency of a highly viscous non-mobile liquid or semi-solid. At sub-zero temperatures, the binder is brittle and will shatter if dropped on to a hard surface, as will coal or coke (ASTM D40; ASTM D3038).

Table 9.4 Asphalt Binder Test Procedures

Property	Test Method	Reference
General Specifications	Recovery of Asphalt from Solution by the Abson Method	ASTM D1856
	Graded Asphalt Cement for Use in Pavement Construction	ASTM D946
	Graded Asphalt Cement for Use in Pavement Construction	ASTM D3381
	Emulsified Asphalt	ASTM D977
Rheology	Penetration of Bituminous Materials	ASTM D5
	Preparation of Viscosity Blends for Recycled Bituminous Materials	ASTM D4887
	Kinematic Viscosity of Asphalts	ASTM D2170
	Ductility of Bituminous Materials	ASTM D113
	Effect of Heat/Air on Asphaltic Materials by Thin-Film Oven Test	ASTM D1754
	Binder Testing	SHRP Mix Design Manual A-407
Incompatibility	Storage Stability Test	Shell Bitumen Industrial Handbook, 1995

Additives (usually on the order of <3%, w/w, of the total mix) and admixtures are added to many asphalt mixes to enhance performance or workability. These additives include fibers, crumb rubber, and anti-strip additives. Liquid asphalt, which is asphalt cement dispersed in water with the aid of an emulsifying agent or solvent, is used as the binder in surface treatments and CMA pavements. The properties of binders are often improved or enhanced by using additives or modifiers to improve adhesion (stripping resistance), flow, oxidation characteristics, and elasticity. Modifiers include oil, filler, powders, fibers, wax, solvents, emulsifiers, wetting agents, as well as other proprietary additives and are used in order for the asphalt cement to meet specifications.

Penetration is a measure of the relative softness or hardness of an asphalt cement (or emulsion) at a given temperature, while viscosity is a measure of the resistance of an asphalt cement to flow at a given temperature. On the other hand, ductility is a measure of the ability of asphalt cement to undergo elongation under tensile stress at a given temperature. Finally, incompatibility is a measure of phase separation of the components of polymer-modified asphalt binders during storage and use. Such a separation is undesirable since it results in significant variation in the properties of the binder and the asphalt in which it is used.

Asphalt cement modification has been practiced for over 50 years but has received added attention in the past decade or so. There are numerous binder additives available on the market today. The benefits of modified asphalt cement can only be realized by a judicious selection of the modifier(s); not all modifiers are appropriate for all applications. In general, asphalt cement should be modified to achieve the following types of improvements (Roberts et al., 1996): (i) lower stiffness (or viscosity) at the high temperatures associated with construction which facilitates pumping of the liquid asphalt binder as well as mixing and compaction of asphalt mix; (ii) higher stiffness at high service temperatures, which reduces rutting and shoving; (iii) lower stiffness and faster relaxation properties at low service temperatures, which reduces thermal cracking; and (iv) increased adhesion between the asphalt binder and the aggregate in the presence of moisture, which reduces the potential for stripping of the asphalt binder from the aggregate.

9.3.2 Asphalt Concrete

The current uses of asphalt for road and street paving began in the late nineteenth and early twentieth centuries (see Chapter 1). With the evolution of materials and traffic patterns into the twenty-first century, asphalt technology has made considerable advances and modern roads and the equipment used for road-laying are much more advanced. It has been realized over the decades that an asphalt roadway is only as good (in terms of performance and longevity) as the materials and effort that are used to lay the roadway. High-quality materials cannot make up for poor road-laying techniques and modern innovative equipment cannot make up for poor-quality materials.

Asphalt concrete is composed primarily of binder (see Chapter 4) and aggregate (see Chapter 5) and the aggregate typically makes up approximately 94–95% (w/w) of an asphalt mix mixture, whereas asphalt binder makes up the remaining 5–6% (w/w). By volume, a typical asphalt mixture is about 85% (v/v) aggregate, 10% (v/v) asphalt binder, and 5% (v/v) air voids and the voids are an important aspect of the performance of asphalt concrete.

Asphalt concrete is composed of two components: (i) the asphalt binder and (ii) the aggregate (see Chapters 1 and 2). For general purposes of the definition, the asphalt binder is (or should be) a *performance-graded binder* or some variation of a *performance-graded binder* and constitutes 5–10% (w/w) of the asphalt–aggregate mixture. The aggregate (see Chapter 5), which will be chosen according to the requirement of the roadway is generally a fine aggregate or a coarse aggregate and typically amounts to 90–95% (w/w) of the asphalt–aggregate mixture.

The asphalt binder is petroleum-based (see Chapter 3) and is characterized by the properties at different temperatures and stages of evolution simulated by laboratory aging (see Chapters 3, 4, and 6) (Daniel et al., 1998). The consistency of the binder (the degree of fluidity or plasticity measured as the viscosity) varies with temperature and is typically graded based on ranges of consistency at a standard temperature. When the binder is exposed to air in thin films and is subjected to prolonged heating (i.e., during mixing with the aggregate), hardening of the binder occurs, thereby increasing the consistency (viscosity) of the binder within specified limits. However, lack of careful temperature and mixing control can cause more damage to the binder, through hardening, than many years of service on the finished roadway.

By definition, the binder should be completely *soluble in carbon disulfide* (or a similar solvent approved by a reliable testing authority). Any insoluble material, such as phase-separated asphaltene constituents (see Chapters 6 and 7) should be inert and if the insoluble material is arbitrarily less than 0.5% (w/w) of the binder may be tolerable without causing an adverse effect on the binder properties and on the binder–aggregate mix. Higher amounts of insoluble material in excess of the arbitrary 0.5% (w/w) may be deleterious to the performance of the asphalt–aggregate mix, remembering that the properties of the asphalt binder vary with the source of the asphalt and the method of production (see Chapters 2 and 3).

Typically, the binder should have negligible (if any) *water or moisture content* as it leaves the refinery. However, tank transports loading binder may have some moisture present in their tanks. If any water is inadvertently present in the binder, it may cause the binder to foam when it is heated above 100°C (212°F). Binder (asphalt) foaming can be a safety hazard and binder specifications usually require that asphalt does not foam at temperatures up to 177°C (350°F). If the binder is heated to a sufficiently high temperature that is above the temperature typically used in paving operations (such as temperatures on the order of 350°C, 650°F), the binder will *flash and ignite* in the presence of a spark or open flame. The operative event is the thermal decomposition of the binder at the high temperature to produce flammable hydrocarbon products that cause the flash and ensuing fire. To combat such a potential (and disastrous) event, the flash point of the binder (see Chapter 4) should be known.

The *durability* of the binder is the measure of how well a binder retains its original characteristics when exposed to normal weathering and aging processes. The performance grading of current binders includes laboratory tests that simulate the weathering and aging processes and establishes pass/fail limits on the test results. Even if the binder passes the weathering and aging tests, pavement performance (i.e., asphalt–aggregate performance) is still affected by variables such as mix design, aggregate characteristics, and pavement lay-down. Other properties related to durability include (i) *adhesion*, which is the ability of the binder to adhere to the aggregate in the paving mixture, and (ii) *cohesion*, which is the ability of the binder to hold the aggregate particles in place in the finished pavement and during in-service life.

As already mentioned (see Chapters 3 and 4) all binders are *thermoplastic* and the properties are susceptible to temperature changes at which a binder will soften (become less viscous) as the temperature increases and harden (become more viscous) as the temperature decreases. The *temperature susceptibility* is a valuable asset as an understanding of the temperature susceptibility of the binder indicates (i) the correct temperature at which to mix the binder with the aggregate and (ii) the correct temperature at which to compact the mixture on the roadbed. It should be understood that it is essential that a binder exhibit temperature susceptibility. Thus, the binder should be sufficiently fluid at elevated temperatures to permit coating of the aggregate particles during mixing and to allow these particles freedom of movement during compaction. The binder must then become sufficiently viscous at ambient air (and in-service roadway) temperatures to hold the aggregate particles in place in the pavement. Again, the temperature susceptibility of the binder varies with the petroleum sources and according to the method of production, even if the two binders (from different sources and production methods) have identical grades.

Asphalt *hardens* in the paving mixture during construction of the roadway and during the service life of the binder. The hardening is caused primarily by *oxidation* (binder combining with oxygen), a process that occurs most readily at higher temperatures (such as construction temperature) and in thin binder films (such as the film coating aggregate particles) (see Chapter 4). During mixing, the binder is both at a high temperature and in a thin film as it coats the aggregate particles. Thus, mixing is the most likely stage at which the most severe oxidation (and hardening) occurs. However, not all binders harden at the same rate when heated in thin films and it is essential that each binder (recalling again that binder properties vary with the source and precaution process) should be tested to determine the aging characteristics so that construction techniques can be adjusted to minimize hardening. Anti-aging (or anti-hardening protocols) usually involve mixing the binder with the aggregate at the lowest possible temperature for the shortest possible (practical) time.

However, binder hardening continues in the pavement after construction—oxidation is the main cause—and polymerization has also been suggested to be a cause of asphalt hardening. Whether or not it is a true polymerization reaction or a coupling reaction is not really known. It is known that asphaltene deposition due to incompatibility of oxidized asphaltene constituents (see Chapters 6 and 7) does occur and (due to molecular association due to the incorporation of additional oxygen functional groups) can cause an increase in molecular weight and give the impression of polymerization. In any case, the processes can be retarded by maintaining a low number of connected voids (air spaces) in the final pavement and the binder coating on the aggregate particles thick.

9.3.2.1 Standard Mix

The mix proportions for a properly compacted asphalt concrete paving mixture are determined in the laboratory during mix design testing. The ability of a properly proportioned asphalt paving mix to resist the potentially damaging effects of the asphalt binder stripping from the aggregate particles is also routinely evaluated in the laboratory. To perform properly in the field, a well-designed asphalt paving mixture must be placed within the proper temperature range and must be adequately compacted. Asphalt concrete paving mixtures should be evaluated for the following properties: stability, flow, air voids, stripping resistance, resilient modulus, compacted density, and unit weight (Table 9.5).

The stability refers to the load that a well-compacted paving mixture can accept before failure—sufficient mix stability is required to satisfy the demands of traffic without distortion or displacement. The flow is the maximum diametric compressive strain measured at the instance of failure—the ratio of Marshall stability to flow approximates the load-deformation characteristics of the mix and therefore indicates the material's resistance to permanent deformation in service. Air voids are the percentage of void spaces within the aggregate–binder matrix that are not filled with binder—sufficient voids should be provided to allow for a slight amount of additional compaction under traffic and a slight amount

Table 9.5 Asphalt Paving Material Test Procedures

Property	Test Method	Reference
Stability and Flow Characteristics (also Air Voids)	Marshall Method	AASHTO T245
	Hveem Method	AASHTO T246, T247
	Asphalt Institute Recommended Cold Mix Method	Asphalt Institute Cold Mix Manual
	Resistance to Plastic Flow of Bituminous Mixtures Using Marshall Apparatus	ASTM D1559
Stripping Resistance	Immersion—Marshall Method	ASTM D4867
	Immersion—Marshall Method	AASHTO T283 (Modified Lottman Method)
Resilient Modulus	Superpave Mix Design	Asphalt Institute Superpave Series No. 1 (SP-1)
		Asphalt Institute Superpave Series No. 2 (SP-2)
Unit Weight	Theoretical Maximum Specific Gravity and Density of Bituminous Paving Mixtures	ASTM D2041
Compacted Density	In-Place Density of Compacted Bituminous Paving Mixtures	ASTM D2950

of asphalt expansion due to temperature increases, without flushing, bleeding, or loss of stability.

The stripping resistance is the ability of a paving mixture to resist the loss of tensile strength due to stripping of the asphalt cement from the aggregate—low resistance to stripping could result in mix disintegration. The resilient modulus is a measure of the stiffness of a well-compacted paving mixture under prescribed conditions of load application. In fact, an asphalt–aggregate mix having a low resilient modulus would be susceptible to deformation whereas an asphalt–aggregate mix having a high resilient modulus would be brittle and subject to cracking. The compacted density is the maximum unit weight or density of a properly designed paving mixture compacted under prescribed laboratory compaction procedures. Finally the unit weight is a measure of the density of a paving mixture compacted in the field in accordance with project specifications.

Aggregate gradation influences such key asphalt mix parameters as stability, durability, permeability, workability, fatigue resistance, frictional resistance, and resistance to moisture damage (Roberts et al., 1996; Castelo Branco et al., 2008). Additionally, the maximum aggregate size can be influential in compaction and lift thickness determination. The Superpave mix design also specifies aggregate gradation control points, through which aggregate gradations must pass and the control points are a starting point for a mix formula. Furthermore, it is rare to obtain a desired aggregate gradation from a single aggregate stockpile. Therefore, Superpave mix designs usually draw upon several different aggregate stockpiles and blend them together in a ratio that will produce an acceptable final blended gradation. It is quite common to find a Superpave mix design that uses three or four different aggregate stockpiles. Typically, several aggregate blends are evaluated prior to performing a complete mix design. Evaluations are done by preparing an asphalt mix sample of each blend at the estimated optimum asphalt binder content then compaction of the blend. Results from this evaluation can show whether or not a particular blend will meet minimum voids requirements.

In order to ensure the proper amount of material passing the 0.075 mm (No. 200) sieve (called *silt-clay* by the soil classification system (Table 9.6) and *dust* by the Superpave definition) in the mix, Superpave specifies a range of dust-to-binder ratio by mass which is given by the relationship:

$$P_{0.0075} / P_{\mathrm{be}}$$

Table 9.6 Soil Classification

Major Divisions	Group Name		
Coarse-grained soils; more than 50% retained on or above No. 200 (0.075 mm) sieve	Gravel; >50% of coarse fraction retained on No. 4 (4.75 mm) sieve	Clean gravel; <5% smaller than #200 sieve	Well-graded gravel, fine to coarse gravel
			Poorly graded gravel
		Gravel, >12% fines	Silty gravel
			Clayey gravel
	Sand; ≥50% of coarse fraction passes No. 4 sieve	Clean sand	Well-graded sand, fine to coarse sand
			Poorly graded sand
		Sand with >12% fines	Silty sand
			Clayey sand
Fine-grained soils; 50% or more passing the No. 200 sieve	Silt and clay; liquid limit <50	Inorganic	Silt
			Clay of low plasticity, lean clay
		Organic	Organic silt, organic clay
	Silt and clay; liquid limit ≥50	Inorganic	Silt of high plasticity, elastic silt
			Clay of high plasticity, fat clay
		Organic	Organic clay, Organic silt
Highly organic soils			Peat

In this equation, $P_{0.0075}$ is the mass (weight) of particles passing the 0.075 (No. 200 sieve) and P_{be} is the EBC:

$$\text{EBC} = (\text{total asphalt binder content of the mix}) - (\text{asphalt absorbed into the aggregate})$$

The dust-to-binder ratio specifications are typically in the range 0.6–1.2, but a ratio of up to 1.6 may be used at an agency's discretion.

The requirements (frequently referred to as *consensus requirements*) arose because of the need to specifically address aggregate properties and also the need for guidance associated with the Superpave mix design method. Therefore, a consensus on several aggregate property requirements was reached and it was recommended that (i) minimum angularity, (ii) flat or elongated particle, and (iii) clay content should be given consideration. Moreover, the desired aggregate properties are different depending upon the amount of traffic loading which is based on the anticipated traffic level on the design lane over a 20-year period regardless of actual roadway design. In addition, the desired aggregate properties vary depending upon their intended use as it relates to depth below the pavement surface. These requirements are imposed on the final aggregate blend and not the individual aggregate sources.

Coarse aggregate angularity is important to mix design because smooth, rounded aggregate particles do not interlock with one another nearly as well as angular particles. This lack of interlock makes the resultant asphalt mix more susceptible to rutting. Fine aggregate angularity is important to mix design for the same reasons as coarse aggregate angularity—rut prevention.

Rutting (or permanent deformation) results from the accumulation of small amounts of unrecoverable strain as a result of repeated loads applied to the pavement. Rutting can occur as a result of problems with the subgrade, unbound base course, or permanent deformation caused by asphalt mix problems. Permanent deformation in an asphalt mix is caused by consolidation and/or lateral movement of the asphalt mix under traffic. Shear failure (lateral movement) of the asphalt mix courses generally occurs in the top 100 mm of the pavement surface, however, it can occur deeper if satisfactory materials are not used. Rutting in pavement usually develops gradually with increasing numbers of load applications, usually appearing as longitudinal depressions in the wheel paths sometimes accompanied by small upheavals to the sides. It is typically caused by a combination of densification (decrease in volume and, hence, increase in density) and shear deformation and can occur in any one or more of the asphalt mix layers as well as in the unbound materials underneath the asphalt mix (Eisenmann and Hilmer, 1987).

An excessive amount of flat or elongated aggregate particles can be detrimental to asphalt mix. Flat/elongated particles tend to break down during compaction (giving a different gradation than determined in mix design), decrease workability, and lie flat after compaction (resulting in a mixture with low voids content) (ASTM D4791) (Roberts et al., 1996).

The required rut resistance of a mixture depends on the traffic level and the location of the mixture in the pavement structure. Pavements with higher traffic levels require greater rut resistance than pavements with low traffic volumes. Surface and intermediate layers require greater rut resistance than base layers. Rut resistance is a consideration in each of the design procedures presented in this manual. For dense-graded mixtures, aggregate angularity, binder grade, compaction energy, and some volumetric properties vary with traffic level and layer depth to provide adequate rut resistance.

9.3.2.2 Warm mix

Asphalt mixtures are normally divided into three categories (i) asphalt mix, (ii) WMA, and (iii) CMA. In general asphalt mix properties are much better than CMA so asphalt mix is used for higher-volume traffic. The type of mix and grade of any binder largely governs the temperature of the mix. Although the CMAs are more environmentally friendly, unfortunately cold mix properties are poor due to the poor coating of the aggregate and the presence of water in the mix reduces the effect of the compaction. CMA also needs a curing time to open up to traffic, therefore, the CMAs are not used for higher traffic volume roads.

To overcome some of these problems, an asphalt mix at somewhat lower temperatures (warm asphalt) than conventional asphalt mix has been used. The testing on this warm asphalt was successfully conducted in both laboratory and in the field in Norway, Britain, and the Netherlands before they concluded that the test results were satisfactory. The production temperature of WMA is in the range of 100–140°C (212–285°F), whereas for asphalt mix the production temperature is on the order of 150–170°C (300–340°F). After WMA was introduced in 2000, it attracted considerable attention of the highway engineering community because of its advantages over both HMA and CMA. The main advantage with the warm asphalt is to reduce temperatures of production and placement, producing less fumes, less emissions at the plant, less energy consumption, less wear and tear on the plant, and less aging of the binder (Hurley and Prowell, 2005a,b, 2006; Gandhi and Amirkhanian, 2007).

The utilization of WMA is simple and does not require any major plant modifications to the existing HMA plant system. The WMA technologies are patented and the manufacturing processes are different. Aspha-min® is a fine powder which releases its hydration bound water and creates foaming to asphalt binder, the lubricating action keeps the mix workable at a temperature range between 130°C and 140°C (265°F and 285°F) (Hurley and Prowell, 2005a; Button et al., 2007).

Sasobit® (Button et al., 2007) is a synthetic wax that is produced in the coal gasification process that is manufactured by Sasolwax, a part of the Sasol Company that gasifies coal in South Africa (Speight, 2013). It is usually blended with asphalt binders at temperatures above 115°C (240°F). This wax has a melting point in the range between 85°C and 115°C (185°F and 239°F). Moreover, Sasobit® forms a crystalline structure within the asphalt binder at ambient temperatures. Therefore, Sasobit® reduces the viscosity of the binder at temperatures above 115°C (240°F) and helps improve the stability at ambient temperatures. The optimum amount of Sasobit® to be used is on the order of 3–4% (w/w) of the asphalt binder which in turn would result in reducing the production temperatures by approximately 8–30°C (15–54°F). In addition, the manufacturer does not recommend blending the solid Sasobit® with the asphalt binder during mixing because it will result in a non-homogeneous distribution of Sasobit® within the mixture, resulting in instability of the mixture and phase separation (see Chapter 4). Therefore, Sasobit® is blended with the hot asphalt binder stream to ensure homogeneous distribution. When blended correctly, the Sasobit®, which melts at approximately 100°C (212°F), chemically changes the temperature-viscosity behavior of the asphalt binder (Hurley and Prowell, 2005b). Thus, the mixture remains workable at a low temperature of 90°C (194°F). Warm asphalt technology seems quite promising since it reduces carbon dioxide emissions by 30%, it consumes 30% less energy and reduces dust emissions by 50–60% compared to conventional HMA production. In addition, the various evolutionary and innovative paths of WMA technologies have also led to Evotherm (Hurley and Prowell, 2006; Button et al., 2007). In the original process, an emulsion is mixed with hot aggregates to produce a mix temperature between 85°C and 115°C (185°F and 240°F). In the new process, the emulsion is produced using a chemical package which is designed to enhance coating, adhesion, and workability.

During production, the Evotherm emulsion is simply used in place of the traditional asphalt binder. The emulsion is mixed with the aggregate in the asphalt plant. Water in the emulsion is liberated in the form of steam when it is mixed with the hot aggregate. The resulting WMA has the appearance of HMA.

There are many ways to classify WMA technologies; one of them is by the degree of temperature reduction. Generally, the WMA production and lay down temperatures are 20–30°C (36–54°F) below HMA at temperatures slightly above 100°C (212°F). If the temperature of the mix at the plant is less than 100°C (212°F), the mix is considered as half warm mix, which is between cold mix and warm mix. The other ways of classifying WMA are by using water to produce foam in the asphalt and using some inorganic additives or wax to reduce temperature. When a small amount of water is added to the hot asphalt, the water turns to steam at atmospheric pressure and the asphalt binder expands (Yunus and Boles, 1994). The volume expansion of the binder phase corresponds to reducing the viscosity of the mixture. The amount of expansion varies depending on a number of factors including the amount of water added and the temperature of the binder. The organic additive processes like Fischer–Tropsch wax, Montana wax, and fatty amides have shown considerable reduction in viscosity of the asphalt binder. Care should be taken to select the wax, such that the melting point of the wax should be considered for service temperature otherwise permanent deformation may occur.

The overall conclusion is that the WMA should provide equal or better performance than HMA. The moisture sensitivity for WMA samples was lower compared to HMA mixtures. The fatigue tests indicated similar results. In general, however, the paving was conducted at cooler temperatures, the density of mix was within the specifications. In addition, they concluded that other advantages included the ability to haul the mix longer distances and still have workability to place and compact, and the ability to run high percentages of reclaimed asphalt pavement (recycled asphalt pavement, RAP).

Another important consideration related to traffic loading is the resistance of the asphalt–aggregate mixture to fatigue cracking. Since two types of fatigue cracks have been identified in asphalt pavements: (i) top-down and (ii) bottom-up, fatigue resistance is an important consideration for both surface and base course mixtures. Pavements with higher traffic levels require surface and base courses with greater resistance to fatigue cracking. One of the most important mixture design factors affecting fatigue resistance is the EBC of the asphalt–aggregate mixture. Fatigue resistance increases with increasing EBC; therefore, to resist top-down cracking, dense-graded mixtures of smaller NMAS and gap-graded asphalt mixtures should be considered for high traffic levels. The dense-graded mixture design procedure provides the flexibility to increase the design requirements of voids in the mineral aggregate (VMA) up to 1.0% to produce mixtures with improved fatigue resistance and durability. Increasing the VMA requirement increases the EBC of these mixtures over that for normal dense-graded mixtures. The use of dense-graded mixtures with higher EBC should be considered for base courses in perpetual pavements. One of the structural design considerations for a perpetual pavement is that bottom-up fatigue cracking never occurs in the pavement.

When reclaimed asphalt pavement is included in a mix design there is cause for concern about the inherent asphalt binder that the mix receives from the reclaimed asphalt pavement. The asphalt has been significantly aged through its initial production (short-term) and then throughout its life (long-term) as a pavement structure. The asphalt is referred to as age-hardened asphalt due to its deteriorated rheological properties from extensive oxidation. There are two issues that need to be addressed by designers when including reclaimed asphalt pavement in a mix design, the first of which is to make a decision regarding the availability of binder in reclaimed asphalt pavement material and the second of which is the issue of stiffness.

The first issue, binder availability, tends to be addressed through one of three accepted concepts. The three concepts are: (i) *black rock*—all age-hardened asphalt acts as aggregate, (ii) *capable of use as a blend stock*—all age-hardened asphalt becomes fluid and totally blends with virgin asphalt binder, (iii) *partially reusable*—some age-hardened asphalt is reusable in the new mixture with the extent being dependent on several factors including aged binder properties, temperature, aging time, and additives (Nourendin and Wood, 1987; Doyle and Howard, 2010).

The second issue relates to the stiffness of the age-hardened asphalt in the reclaimed asphalt pavement—the age-hardened asphalt will exhibit a loss in ductility as it hardens, resulting in cracking and raveling of a pavement structure containing a high amount of reclaimed asphalt pavement where stiffness of the mix was not properly addressed (Karl et al., 1979). Particularly when the Superpave method was adopted, reclaimed asphalt pavement use became very conservative due to the difficult and limiting procedures associated with the incorporation of reclaimed asphalt pavement in a Superpave mix design. However, it has been demonstrated that mixes with a high reclaimed asphalt pavement content can exhibit good performance and can be produced with existing quality control procedures.

However, it cannot be overemphasized that the quality of asphalt cement is affected by the inherent properties of the petroleum from which it is produced. Different oil fields and areas produce crude oils with very different characteristics (Speight, 2012, 2014). The refining method also affects the quality of asphalt and for construction purposes, there are three important factors to consider: (i) consistency, also called the viscosity or the degree of fluidity of asphalt at a particular temperature, (ii) purity, and (iii) safety. In addition, the consistency or viscosity of asphalt varies with temperature, and further variability of the properties is introduced if the properties (for comparison between asphalt samples) are not measured at a standard (or quoted) temperature. Generally, the *purity* of asphalt can be tested by a solubility test (ASTM D2042) (see Chapter 4).

Even though some investigations have been conducted on the compatibility of the warm asphalt additives on the asphalt binders (Gandhi and Amirkhanian, 2007), and on HMA mixtures (Hurley and Prowell, 2005a,b, 2006), the level of investigation of the compatibility of these warm asphalt additives with the crumb rubber modified binders and their effects on rubberized mixtures has been a much lower research effort (Button et al., 1987).

WMA modified mixes were not effective in resisting rutting, due to the fact that the mixture will undergo less aging temperature compared to conventional HMA. The research performed on the high-temperature properties of the WMA modified crumb rubber modified binders has concluded that these additives will improve the rutting resistance of the crumb rubber modified binders compared to the conventional binders by improving the failure temperatures of WMA (Bahia and Davis, 1994). The results indicated that the rubberized binders containing Sasobit® showed the highest stiffness and lowest *m*-value properties irrespective of the binder source, indicating that the addition of Sasobit® may result in the binder being less resistant to low-temperature cracking. Furthermore, asphalt binders with Sasobit® showed the higher stiffness and lower *m*-value compared to conventional binders due to the crystallization of the high-molecular-weight paraffin derivatives (as wax), which usually increases the resistance of plastic deformation of asphalt binders. In addition, the effect of wax on binder performance will depend on the source and properties of the binder itself and the amount of the additive, which are also related to the crystallization of the wax as well as the melting point (or softening point) of the wax in the binder. It has also been observed that the viscosity of rubberized binders with Aspha-min® increased due to the filling effect of the additive, and Sasobit® was found to decrease the high-temperature viscosity of the binders. Furthermore, the addition of Sasobit® increased the viscosity of binders modified with the crumb rubber at mid-range temperatures (60°C, 140°F). This means that the addition of Sasobit® reduces the viscosity of the binder at high temperatures and increases the viscosity at mid-range temperatures, which makes it more workable at higher temperatures, and stiff, and therefore, more resistant to penetration and rutting at mid-range temperatures (Gandhi and Amirkhanian, 2007). Sasobit®-modified CRM binders have shown more recovery from deformation in the creep recovery tests.

The benefits of WMA technology are dependent upon which technology is utilized. There are varying degrees of benefits for each method. One of the driving forces of WMA technology research is the potential for it to reduce energy and fuel consumption and therefore reduce emissions. The typical WMA technology production temperature is in the range of 30–100°F lower than typical HMA. Oftentimes only a slight reduction in temperature is achieved (10–15°F, 18–59°F) but the reduction can lead to energy savings and significantly reduce emissions. The WMA technology is available for potentially greater temperature reductions.

The WMA technology reduces the temperature of the asphalt at the time of paving and there are several resulting benefits. These include an improved and cooler working environment and decreased exposure to asphalt fumes. In unison with reduction of fumes, is the reduction of odors. As the asphalt production temperatures are reduced through operation of the WMA technologies, this reduction would commonly be associated with plant and paving operations.

There are also other benefits from WMA use, including: less compaction effort, longer haul distances, and a better workability with high reclaimed asphalt pavement material (Hurley and Prowell, 2006; Tao and Mallick, 2009). Another potential benefit of using WMA is that asphalt can be transported over longer distances—the smaller differential between the mix temperature and the ambient temperature results in a slower rate of cooling as well as better compactability at a lower temperature. An additional benefit is increased smoothness when crack sealant is on the underlying layer.

9.3.2.3 Open-Graded Mix

An open-graded mix consists primarily of coarse aggregate and asphalt, with a minimum amount of sand or fine aggregate. The mix has a high air-void content after compaction, typically 15–25% (v/v). Open-graded mixes are very permeable, allowing for the passage of a large quantity of water through the mix—hence they can act as a drainage course.

Open-graded mixes require fractured aggregate to ensure good interlocking of the aggregate for stability of the placed mix—noncrushed aggregates should not be used. A properly produced open-graded mix will have the asphalt uniformly distributed throughout and all aggregate particles uniformly coated. Coating is a function of the amount and type of asphalt used. If an excess of asphalt must be used to get acceptable coatings, or if drain down or runoff occurs, the emulsion supplier should be notified to rectify the problem. Generally, an emulsion modification must be made at the asphalt emulsion plant to correct these problems.

Open-graded cold mixes are generally used for base and binder courses. They are free-draining, have high stone-on-stone contact for stability, are resistant to fatigue and reflective cracking, and are self-healing when placed over bases that move or flex during frost periods in the road. They are used extensively on town and county roads in many states for base and binder courses, then topped with dense mix or sealed in some way, such as surface treating. Roads over which open-graded mixes are placed must have good drainage and ditching to allow water to get out of the pavement and to the ditch. When used as a base course, a positive moisture seal should be provided under the open-graded mix to prevent water from entering and weakening the subgrade materials.

The *OGFC* is a gap-graded mixture with a high air-void content. The high air-void content and open structure of the mixture provides macrotexture and high permeability to drain water from the tire–pavement interface. This minimizes the potential for aquaplaning, improves wet weather skid resistance, and reduces splash and spray. Other benefits of the OGFC include reduced noise levels, improved wet weather visibility of pavement markings, and reduced glare. OGFCs are made with durable, polish-resistant aggregates and usually contain modified binders and fibers to increase the binder content and improve their durability.

9.3.2.4 Dense-Graded Mix

A dense-graded mix has a balance of aggregate particle sizes distributed from coarse particles to fine particles, resulting in a mixture that has relatively low void space when compacted. Typically, after compaction the air-void content is 3–10% (v/v) of the mix, depending on the mix type.

Dense-graded mixtures are the most common HMA mix type. The term dense-graded refers to the dense aggregate gradation used in these types of mixtures, which means that there is relatively little space between the aggregate particles in such mixtures. Historically, dense-graded mixtures were popular because they required relatively low asphalt binder contents, which kept their cost down. However, experience has shown that HMA with binder contents that are too low can be difficult to place and compact and may be prone to surface cracking and other durability problems. Therefore, many dense-graded HMA mixtures do not use a true maximum density gradation, but use somewhat open gradations that deviate slightly from maximum density. Such mixtures have more space between the aggregate particles and can be designed to contain more asphalt binder.

Dense-graded asphalt–aggregate mixtures can be used in any layer of the pavement structure for any traffic level and are the most commonly used mixtures in the United States. Traffic level is a direct consideration in the design of dense-graded mixtures. Aggregate angularity, clay content, binder grade, degree of compaction, and some volumetric properties vary with traffic level in the dense-graded mixture design procedure.

Dense-graded mixtures also provide the mixture designer with the greatest flexibility to tailor the mixture for the specific application and can also be designed as fine or coarse mixtures (AASHTO M323). For each NMAS, a primary control sieve has been identified. If the percent passing the primary control sieve is equal to or greater than a specified value, the mixture is classified as a fine mixture; otherwise it is classified as a coarse mixture. Fine mixtures have smoother surface texture, lower permeability for the same in-place density, and can be placed in thinner lifts than coarse mixtures.

Dense-graded mixtures can vary in size from large stones in base courses to sand size in surface courses. These mixes consist of blends of various coarse aggregate depending on the depth and top size required and 30–50% sand size material to provide dense gradation. Dense-graded mixes are relatively impervious to water. Dense-graded mixes are best if mixed using high-quality, fractured coarse and fine aggregate to ensure good stability of the produced mix.

A properly produced dense-graded cold mix will have the bituminous material uniformly distributed throughout without any unevenness of the asphalt or fines distribution. The coarse aggregate is generally not totally coated. If it is totally coated it is usually an indication of an excessive amount of asphalt in the mixture. Normal mixes are usually 85–95% coated. Dense-graded mixes are used for base, binder, and surface courses. The coarser mixes (1"–1½" top size) are used for base and binder, and the finer mixes (¼"–½" top size) are used for shim and top courses. When used as surface courses, dense-graded cold mixes can be left unsealed. However, often they are sealed in some manner, such as fog sealing, surface treating, slurry seal, or hot mix overlay to enhance the durability.

9.3.2.5 Coarse-Graded Mix

Mixtures that are somewhat coarser than the maximum density gradation are called coarse-graded mixtures, while mixtures somewhat finer than the maximum density gradation are called fine-graded mixtures. This terminology can be somewhat confusing, since both coarse- and fine-graded mixtures should be considered variations of dense-graded HMA. A more appropriate terminology is to refer to the three types of dense-graded HMA as dense/dense-graded, dense/coarse-graded, and dense/fine-graded mixtures. When engineers and technicians first began developing mix designs using the Superpave system in the 1990s, there was a clear trend toward dense/coarse-graded mixtures, in order to increase the rut resistance of HMA pavements. However, in the past few years, many agencies have shifted back toward finer mixtures (dense/dense or dense/fine), to help improve the durability of surface course mixtures. Also, recent research has suggested that dense/fine HMA mixtures can, in most cases, be designed to have just as much rut resistance as dense/coarse mixtures.

The procedure for designing dense-graded HMA mixtures suggests that a range of gradations be evaluated during the mix design process and that the gradation most effective in meeting the given mixture specifications should be selected. The suggested volumetric requirements do include a slight increase in the allowable range for dust-to-binder ratio and an optional table for high-durability mixtures that includes an even higher dust-to-binder ratio and an increase in minimum VMA. Both of these changes will probably reduce the number of dense/coarse-graded HMA mixtures being designed under this system.

During the past 20 years, stone-matrix asphalt has become increasingly common in the United States and Europe. Stone-matrix asphalt is a special type of HMA designed specifically to hold up under very heavy traffic. Stone-matrix asphalt is composed of high-quality coarse aggregate, combined with a large amount of mastic composed of a high-performance asphalt binder, mineral filler, and a small amount of fibers. The aggregate used in stone-matrix asphalt contains a large amount of coarse aggregate and a large amount of very fine material (called mineral filler), but not much sand-sized material. For this reason, such aggregates are called gap-graded, and stone-matrix asphalt and similar HMA types are referred to as gap-graded mixtures, or gap-graded HMA. A well-developed coarse aggregate structure in combination with a relatively large volume of high-performance binder helps ensure that a properly designed stone-matrix asphalt mixture will exhibit high-quality performance. Stone-matrix asphalt is usually only used on very heavily trafficked roadways, where high performance is necessary.

9.3.2.6 Superpave Mix

Superpave mix (superior performing pavement mix) is not specifically HMA and refers to a complete paving system. The mix is designed to accommodate the traffic loading expected as well as the historical climatic conditions of the location for the pavement. The components for the HMA mixture are carefully selected, each having to meet specific requirements. The binder grade is established by looking at both the historical high and low temperatures of the pavement at the site. In addition, the aggregate must meet specific consensus properties including coarse and fine aggregate angularity, flat and elongated particles, and clay content. Because of the design criteria, Superpave mixes must have a strong aggregate structure which, in general, results in a coarser aggregate blend and lower asphalt contents than standard mixes (Witczak et al., 2002). The Superpave mixes are designed to match the expected traffic loads and the high and low expected temperatures of the pavements.

The primary issue that led to the development of the Superpave paving system was pavement deformation in the form of rutting. The resultant mixes have a much greater stone-on-stone contact than previously designed and are more rut-resistant but are also more difficult to compact. Heavier rollers and a greater compaction effort are required to achieve the proper mix density. To ensure that the proper density is being achieved, contractors are often required to monitor the densities with devices such as a nuclear density or a non-nuclear gauge. Higher proportions of coarse aggregate are needed to achieve the greater stone-on-stone contact and also produce a mix that is much easier to segregate whenever it is moved. Paving crews need to be aware of this and take appropriate steps to minimize it.

Superpave mix design and analysis is performed at one of three increasingly rigorous levels of performance. Superpave Level 1 is an improved materials selection and volumetric mix design procedure; Level 2 uses the same volumetric mix design procedure as Level 1 as a starting point, in conjunction with a battery of tests to predict the mix performance; and Level 3 involves a more comprehensive array of tests to achieve a more reliable level of performance prediction.

The Superpave mix design consists of seven stages: (i) aggregate selection, (ii) asphalt binder selection, (iii) sample preparation, including compaction, (iv) performance tests, (v) density and voids calculations, (vi) optimum asphalt binder content selection, and (vii) moisture susceptibility evaluation.

Aggregate selection is achieved in two ways: (i) restrictions are placed on aggregate gradation by means of broad control points, (ii) requirements are placed on coarse and fine aggregate angularity, flat and elongated particles, and clay content. Other aggregate criteria such as properties that are considered to be source-specific such as abrasion, soundness, and water absorption are also used in Superpave.

The Superpave method effectively limits the content of reclaimed asphalt pavement material in the binder to 40% (w/w). The adoption of the Superpave method discredited many of the practices that used reclaimed asphalt pavement material and which were developed through the 1970s. It is recommended that no more than 15–30% reclaimed asphalt pavement material content should be included in a mix design without additional specialized testing. Common practice includes the use of binder bumping and blending charts to achieve desired binder properties.

9.3.2.7 Perpetual Pavements

Perpetual pavements are intended to provide an exceptionally long service life—approximately 20 years for the surface course and approximately 50 years (or more) for the underlying pavement layers. In this type of pavement, the base material should be flexible and fatigue-resistant, meaning it should be designed as a NMAS mixture. Improved fatigue resistance will usually be obtained through the use of fine aggregate gradations and increased asphalt binder content. The high-temperature asphalt binder grade for the base material should be high enough to prevent any rutting, but not excessively high, otherwise the fatigue resistance of the material might be compromised. The low-temperature binder grade should, in general, be one grade higher than that required at the surface.

The intermediate layer should be a strong, rut-resistant mixture. Early indications were that relatively coarse-graded mixtures with large NMAS mixtures provided optimum rut resistance, more recently it has become evident that equal or even better rut resistance can be obtained using fine-graded mixtures with large NMAS gradations. The high-temperature binder grade for this layer should be the same as that required for the surface mixture and to ensure that the intermediate layer has a high modulus, the low-temperature binder grade should be one grade higher than that used for the surface mixture.

Selection of mix type for the surface coarse mixture will depend on the traffic level. For very heavy traffic levels, gap-graded asphalt–aggregate mixtures will provide the best performance and greatest assurance of a long pavement life. At intermediate to high traffic levels, carefully designed dense-graded asphalt–aggregate mixtures should perform well. Normal procedures for binder grade selection should be followed in designing the asphalt–aggregate mix for the surface course of a perpetual pavement.

9.4 PAVEMENT DESIGN AND STRUCTURE

Ensuring good performance in pavement design and construction is accomplished through: (i) proper structural (thickness) design for the existing soil, anticipated loads, and existing pavement condition for overlays, (ii) the optimum number of layers to facilitate stability, and smoothness, (iii) selection of the appropriate mix types for each of the layers to achieve stability and smoothness, and (iv) construction that complies with or exceeds the specifications for uniformity, smoothness, and, compaction. Although not specifically mentioned in the preceding categories, economy is also an important aspect of pavement design and construction. The goal is to provide a pavement structure that has adequate strength to distribute the wheel loads to the soil without undue deflection, compaction, or consolidation and to provide a surface that is adequately stable so as to not deform under traffic load, is weather-resistant, has adequate skid resistance, is adequately smooth and is sufficiently wear-resistant.

In order to satisfy these categories, selection of an appropriate asphalt mix for a specific paving application is important in designing new pavements and for rehabilitation strategies for existing pavements. The type of mixture selected for the various layers of a pavement has a major effect on the construction and long-term performance of the pavement as well as the cost. A mixture with a low binder content and low-quality aggregate is less expensive. To facilitate placement and compaction, thinner layers should be made with smaller NMAS mixtures, while thick base layers should be made with larger NMASs. Mixtures at the surface of a pavement should have relatively high binder content to make them more resistant to the damaging effects of traffic and the environment. Lower binder contents can be used in mixtures for intermediate and base courses because they are protected by the layers above them. Careful consideration of mix type is an important factor when staged construction is used, because the base or intermediate courses must serve temporarily as the surface during the first stages of construction.

Pavement design is an essential part of asphalt technology and the correct design serves to give the asphalt pavement prolonged in-serve life. Pavements are made up of several layers of different materials. In some cases, when the soils in the sub-grade are poor (in terms of composition and structure) construction of the sub-base layer may be necessary to serve as a construction platform that will (i) prevent the intrusion of fines into the base, (ii) improve drainage, or (iii) reduce damage from frost action. The sub-grade fill (or embankment fill) actually bears the traffic loads and the upper pavement layers support the wheel loads and distribute the loads across a wider area of the sub-grade.

An asphalt concrete pavement is not merely a thin covering of asphalt concrete over indigent soil. Each pavement involves the placement of several courses (layers) and must be designed as an engineering structure and each course in the pavement must be constructed carefully if each of the several different layers is to operate efficiently and in concert with the other layers. Furthermore, asphalt concrete is much more flexible than Portland cement concrete and, hence, asphalt concrete pavement is often referred to as a *flexible pavement*. Thus, the various aspects pf pavement design that require careful attention are: (i) the surface course, (ii) the base course, (iii) the sub-base, and (iv) the base. Each of these criteria contributes to the durability and longevity of the pavement.

9.4.1 Surface Course

The surface course is the uppermost structural layer in an asphalt pavement and is typically the top layer of the pavement, which also serves as the wearing course. This course is directly exposed to traffic and environmental forces and it must be produced with the highest-quality materials to provide the following characteristics: (i) adequate wet weather friction for safety, (ii) high resistance to load-induced rutting, shoving, and surface cracking, (iii) high resistance to thermally induced cracking, (iv) low permeability to minimize surface-water infiltration, (v) high durability to resist disintegration due to the combined effects of aging, traffic loading, and freeze–thaw effects, (vi) appropriate surface texture for noise control, safety, and (vii) smoothness.

Surface courses contain highly angular aggregates and an appropriate performance-graded binder to resist traffic and environmental forces. If the surface course is also the top layer in the pavement, then the aggregates must be resistant to polishing under traffic loading to provide appropriate skid resistance over the service life of the pavement. Also, the surface course should efficiently and effectively withstand the adverse effects of environmental factors such as air, water, and temperature changes without showing a high propensity for failure. In addition, the surface course should be impermeable and prevent water from infiltrating into the underlying layers as well as providing sufficient smoothness and skid resistance from various types of traffic.

The visible part of the asphalt concrete pavement—the part that directly supports truck and passenger vehicles—is called the *surface course* or *wearing course* and is typically between approximately 1.5 and 3 inches (40 and 75 mm) thick and consists of crushed aggregate and asphalt binder. Furthermore, the mixture used for a surface course tends to have a relatively high asphalt content, which helps the mixture have a higher resistance to traffic loads as well as the effects of sunlight, air, and water. Surface course mixtures also are usually made using maximum aggregate sizes less than 0.75 inch (9 mm), which helps to provide for a smoother less wearing ride on the vehicles and the human occupants of the traffic.

If the aggregate size is larger than 0.75 inch (larger than 19 mm) it can be more difficult to obtain mixtures with sufficient asphalt binder contents to provide adequate durability for surface course mixtures, since the lower aggregate surface area of these aggregates results in a lower demand for asphalt binder. On the other hand, the lower binder content needed for these mixtures can make them more economical than mixtures made using smaller aggregates but this may be detrimental to the properties and durability of the finished roadway.

9.4.2 Base Course

The *base course*, which lies below the surface course of the pavement, may consist of crushed stone, crushed slag is often regarded as the most important structural layer of a pavement. This course ensures efficient distribution of the traffic load and enables adequate sub-surface drainage. The base course is usually constructed from good-quality gravel or crusher run aggregates. When heavy traffic is expected, the base layer is usually HMA. The base course helps to provide the overall thickness to the pavement that is necessary to ensure that the pavement can withstand the projected traffic loads over the in-service life of the roadway. Base courses may be approximately 4–12 inches (100–300 mm) thick with a higher thickness being necessary to accommodate heavier traffic loads and reduce the strain within the pavement and make it less prone to fatigue cracking.

The base course is the primary load-carrying element in deep-strength flexible pavements, full-depth flexible pavements, and rubblized Portland cement concrete pavements. Because base course is deep in the pavement structure, the course does not have to be highly rut-resistant. The mixture for the base course should be relatively easy to compact to ensure that the base course is durable and resistant to bottom-up fatigue cracking. Traditionally, base course aggregate mixtures have been designed using larger aggregate size than surface course aggregate mixtures—the maximum aggregate size had tended to be in the range 0.75–1.5 inches (19–37.5 mm) which helps to produce a lean mixture with low asphalt binder content thereby reducing the cost of such mixtures but this must not be at the expense of efficiency. More recently, base course mixtures have been designed to be more like surface course mixtures with smaller aggregate size and a higher content of asphalt binder. This has helped to improve both fatigue resistance as well as resistance to moisture damage—generally, increased asphalt binder content in HMA tends to improve fatigue resistance and also reduce permeability to water.

In some cases, an *intermediate course* (sometimes called a *binder course*) is placed between the surface course and the base course of the pavement system. The purpose of the intermediate course is to add thickness to the pavement when additional structural capacity is required in new flexible pavements, rehabilitated asphalt pavements, and rubblized Portland cement concrete pavements. An intermediate course may also be used in overlays of intact Portland cement concrete pavement to provide additional thickness to delay reflective cracking or to provide an additional layer to improve pavement smoothness. Since intermediate courses are close to the surface of the pavement, they must be resistant to rutting. However, they can be constructed with mixtures having lower binder contents than surface courses because the intermediate course is not directly subjected to traffic loading or the damaging effects caused by water and oxidative hardening of asphalt binder. Thus, the intermediate course is typically on the order of 2–4 inches (50–100 mm) thick and consists of a mixture with intermediate aggregate size and intermediate asphalt binder content.

In some cases, there may be a *combined surface and base course*, which is a single-layer bituminous paving that functions as both a base course and wearing course. Thus type of course is typically used for less heavily trafficked roads and asphalt pavements with a small total thickness as well as a low demand on the deformation resistance. The mix usually consists of particle-graded mineral mix and road construction asphalt. Since round grain material (natural sand, gravel) is frequently used, the material can also be highly sensitive to scuffing during compaction work. The layer thickness is on the order of 2–4 inches (50–100 mm).

9.4.3 Sub-Base Course

The surface course, intermediate course (if present), and base course together are collectively referred to as bound material or bound layers, because of the use of the asphalt binder. In a typical asphalt concrete pavement, the bound layers are supported by a granular sub-base that in turn lies over the sub-grade.

The *sub-base course* (as the name implies) lies beneath the base course and provides structural support to the sub-grade layer. This course reduces intrusion of fines (typically micron-size mineral mater) from the sub-grade and makes a substantial contribution to the foundation of the roadway structure. In cases where the pavement is to be constructed over a resilient and high-quality sub-grade, a sub-base may not be required.

Granular sub-base material is typically crushed stone or gravel, usually 4–12 inches (100–300 mm) thick—the nominal maximum size of the aggregate varies but should always be well compacted prior to placement of the base course.

On occasion, the granular sub-base course may be from a pavement and a relatively thick base course is placed directly on the sub-grade soil. Such a pavement structure (called a *full-depth asphalt pavement*) has the advantage that this type of construction can be thinner because of the increased strength and stiffness of the supporting pavement. However, it should be remembered that such a base course mixture is typically more expensive than granular sub-base course, since it contains an asphalt binder.

9.4.4 Sub-Grade Course

The *sub-grade course* is typically the natural soil as well as chosen particle size aggregate which is compacted to specific levels to relative stress from the weight of the above course. Thus, the indigent soil serves as the foundation that supports the road and if the roadway is constructed on an embankment, the imported fill materials (embankment) become the subgrade. This course has (or, should have) the capability of absorbing any weight or load stress that is transmitted from the courses lying above.

After removal of topsoil and other organic materials and assuming that the soil is strong, the sub-grade may be stabilized by compaction alone. However, many soils contain clay minerals and are classed as *unstable* because of the significant shrinkage and expansion that occur with moisture expulsion and moisture intrusion, respectively. When the moisture intrusion is high, such soils are weak and unreliable. In order to combat such effects, such a subgrade should be stabilized (before pavement construction) by blending in lime, Portland cement, or other additives, or by treating the soil with asphalt emulsion, followed by thorough compaction.

9.4.5 Soil

Soil is the mixture of minerals, organic matter, gases, liquids, and organisms that together support plant life. Thus, soil consists of a solid phase (minerals and organic matter) as well as a porous phase that holds gases and water. Most soils have a density between 1 and 2 g/cm^3 (Canarache et al., 2006; Chesworth, 2008).

Asphalt pavement design projects commence with a simple but fundamental consideration of the underlying soil strength (ASTM D1241). The properties and characteristics of the subgrade soil will determine the pavement thickness needed to carry the expected traffic loads. The properties and characteristics of the soil that are important are (i) gradation, (ii) permeability, (iii) capillarity, (iv) plasticity, (v) susceptibility to frost, and (vi) spring thaw.

9.4.5.1 Gradation

Soil *gradation* refers to the size and distribution of the mineral grains and, typically, soils are divided into two broad categories: (i) coarse-grained soils and (ii) fine-grained soils. Coarse-grained soils are subdivided into three types: boulder, gravel, or sand fractions, crushed stone, and slag depending on their grain size. On the other hand, fine-grained soils are either silt or clay.

Briefly, *gravel* consists of rounded or angular particles of rock that will pass through a three-inch square sieve (76 mm) and be retained on a No. 10 US sieve (2 mm). Gravel is generally a glacial deposit. Different types of gravel are pea gravel, river gravel, and bank gravel. Gravels are usually crushed and graded into various sizes. *Sand* is the product of the breakdown of rock or sandstone—the result is a fine granular material. Different types of sand are blow sand, lake sand, gravel pit sand, bank run sand, manufactured sand, and river sand. Sand consists of particles that will pass the No. 10 sieve and be retained on the No. 200 sieve (0.075 mm). *Silt* is the material that will pass through the No. 200 sieve and is greater than 0.005 mm, and exhibits little or no strength when dried, while *clay* is the portion of soil that is finer than 0.005 mm and exhibits a plastic (putty-like) behavior within a wide range of water content and which also exhibits considerable dry strength. The *fines* fraction is the portion of the soil that passes through the No. 200 sieve and, therefore, this category can include both silt and clay. *Crushed stone* is produced from crushing ledge rock or granites and producing 100% crushed aggregate of various sizes. *Slag* is a non-metallic product containing silicates, derived from the production of steel.

Most naturally occurring roadway sub-grades involve mixtures of gravel, sand, silt, and clay. Materials that have very little silt and clay (typically less than 5% w/w) are termed *clean* and they usually are strong even when wet. Materials having 12% (w/w) or more fines are termed *dirty* and they tend to be susceptible to frost penetration and have low strength when they are wet.

9.4.5.2 Permeability

Permeability refers to the ease with which a fluid (especially water) flows through the pores of the soil. Since too much water in a soil reduces its strength, the ability to drain water is a beneficial effect that enables the soil to acts as a sub-grade course for the roadway. Particle texture, gradation, and degree of compaction have a strong influence on the ability of the soil to transmit water. As expected, coarse-grained soils are usually more permeable than fine–grained soils, but small amounts of fines (silt or clay) can have a detrimental effect on the permeability—silt and clay have a very low permeability and will retain moisture for long periods of time, thus influencing the permeability, even of course-grained soil.

9.4.5.3 Capillarity

Capillarity (*capillary action*, *capillary attraction*) refers to the upward movement of water by capillary action through a fine soil. It is, in fact, a manifestation of surface tension by which the portion of the surface of a liquid coming in contact with a solid is elevated or depressed, depending on the adhesive or cohesive properties of the liquid.

Some soils not only retain water because of their low permeability, but also draw in water by capillary action and this is a property that influences the frost susceptibility of soil.

9.4.5.4 Plasticity

Plasticity is the ability or propensity of a material (in this case, soil) to undergo permanent deformation under load when compressed. In the current context, plasticity also refers to the ability of the material to deform under a load without cracking, and to maintain the deformed shape after the load is released. At the right moisture content, soil exhibiting plasticity can be rolled into a thin thread that will not crumble.

9.4.5.5 Frost Susceptibility

In the United States, Canada, and other northern countries, *frost susceptibility* (leading to frost heaves, i.e., road deformation) is a serious problem whenever freezing temperatures occur for prolonged periods. The road deformation is the result of ice lenses forming in the soil under the pavement and these ice lenses grow and expand in a downward direction and are propagated by water drawn up from below by capillary action.

For ice lenses to form and grow in a soil there must be (i) soil temperatures at or below the freezing temperature, (ii) a water table close to the frost line, and (iii) a soil favorable to the rapid movement of capillary water. When the freezing temperatures are prolonged (as in winter in many northern states and northern countries) the frost line moves downward, ice lenses grow, and the ground and pavement deform (heave). In order to combat such an effect, the action involves eliminating the water source or using solids that are not susceptible to frost. However, eliminating the water source and lowering the water table are of limited effectiveness due to the high level water can be raised from capillary action in soil that is susceptible to frost—such soils tend to exhibit both high capillarity and permeability. Clay-soil is not as susceptible to frost as silt-soil but both are susceptible to frost heaving. Gravel and/or sand have little, if any, capillarity and are the preferred materials to use to eliminate frost heave and deformation of the roadway.

9.4.5.6 Spring Thaw

The effects that arise from the spring thaw (common to northern climates), which lead to weakening of the subgrade during the thawing period, are considered to be the worst aspects of frost action. In fact, the soil and moisture conditions that cause frost heaving are just as severe for damage to the roadway during thawing. Moreover, the greatest loss of support from the sub-grade occurs when a wet autumn is followed by a winter with several freeze–thaw cycles and a rapid period of thawing in the spring. Potholes formed in the spring in northern cities are a testament to these effects.

Thawing of the roadbed occurs from the top down and usually starts under the center of the roadway. The shoulders stay frozen and trap water in the base and subgrade, because snow is left mounded there. The saturated material is very weak and will probably fail prematurely.

9.4.6 Drainage

An extremely important aspect of roadway design and construction relates to the drainage of the roadway which affects the ability of a pavement to withstand weather and traffic phenomena and can contribute to long in-service performance. Problems related to drainage are either surface-related problems or subsurface-related problems and each must be dealt with separately.

When developing a roadway drainage system, the principal purposes of the system are: (i) collection and draining of surface and subsurface water, (ii) prevention or retardation of erosion, (iii) interception of water from surrounding areas and removing the water from the roadway area, and (iv) lowering the groundwater table.

When water intrudes into the roadway pavement, significant weakening of the roadway structure can occur which will eventually lead to premature failure of the roadway (i.e., failure before the projected in-service life of the roadway has been attained). Some soils (such as clay-laden soils) expand significantly with the addition of water, causing differential heaving (deformation) of the pavement in addition to the weakening of the pavement structure. In colder climates, such as in northern parts of the United States, Canada, and other northern countries, frost heaving (deformation) results from water freezing in and under the pavement and is the primary cause of volume changes that lead to pavement breakup and potholes. Open-graded and/or poorly compacted asphalt mixes will allow water to penetrate through the course which can cause destruction of the course(s) due to freeze–thaw action or stripping of the asphalt from the aggregate(s). The end result of these actions is premature pavement failure and, thus, removal of water from above and below the pavement is critical to assuring prolonged service life of the pavement.

Surface drainage of water requires disposal of all water on the pavement and on the nearby ground surface. Efficient drainage of the pavement and shoulder surfaces is necessary to prevent saturation of the pavement structure and the sub-grade, and to minimize hazardous driving conditions (auaplaning) due to rainwater on the driving lanes. This can be achieved by means of full-width (travel lanes and shoulder) chip sealing or paving with HMA or CMA.

Briefly, chip seal (also chipseal) is an asphalt roadway surface treatment that combines one or more layer(s) of asphalt with one or more layer(s) of fine aggregate. In the United States, chip sealing (also called *tar and chip*, *sprayed seal*, *surface dressing*) is typically used on rural roads carrying lower traffic volumes, and the process is often referred to as *asphaltic surface treatment*. Chip sealing is accomplished by evenly distributing a thin base of hot asphalt on to an existing pavement and then embedding finely graded aggregate into the asphalt. The aggregate is evenly distributed over the seal spray, and then rolled into a smooth pavement surface. A chip-seal-surfaced pavement can optionally be sealed with a top layer, which is referred to as a *fog seal* or *enrichment*.

To assist (or encourage) efficient draining, the roadway pavement should be built with a crown or cross-slope to enable water to run off the roadway. Some water can enter the pavement structure but the amount can be minimized by maintaining the pavement to be free of small holes, potholes, and cracks.

REFERENCES

AASHTO M323, 2014. Standard Specification for Superpave Volumetric Mix Design. American Association of State Highway and Transportation Officials, Washington, DC.

AASHTO T283, 2014. Resistance of Compacted Hot-Mix Asphalt (HMA) to Moisture-Induced Damage. American Association of State Highway and Transportation Officials, Washington, DC.

AASHTO T304, 2014. Uncompacted Void Content of Fine Aggregate. American Association of State Highway and Transportation Officials, Washington, DC.

Ali, N., Chan, J.S., Potion, A., Bushman, R., Bergen, A., 1995. Mechanistic evaluation of asphalt concrete mixtures containing reclaimed roofing materials Proceedings. 74th Annual Meeting of Transportation Research Board. Technical University of Nova Scotia and University of Saskatchewan, Canada.

Al-Qadi, I.M., Aurangzeb, Q., Carpenter, S.H., Pine, W.J., Trepanier, J., 2012. Impact of High RAP Contents on Structural and Performance Properties of Asphalt Mixtures Research Report FHWA-ICT-12-002. University of Illinois at Urbana-Champaign/Bureau of Material and Physical Research, Illinois Department of Transportation, Urbana, IL/Springfield, IL.

ASTM C1252, 2014. Standard Test Methods for Uncompacted Void Content of Fine Aggregate (as Influenced by Particle Shape, Surface Texture, and Grading) Annual Book of Standards. ASTM International, West Conshohocken, PA.

ASTM D40, 2014. Standard Test Method of Drop Shatter Test for Coal Annual Book of Standards. ASTM International, West Conshohocken, PA.
ASTM D242, 2014. Standard Specification for Mineral Filler for Bituminous Paving Mixtures Annual Book of Standards. ASTM International, West Conshohocken, PA.
ASTM D1241, 2014. Standard Specification for Materials for Soil-Aggregate Subbase, Base, and Surface Courses Annual Book of Standards. ASTM International, West Conshohocken, PA.
ASTM D2042, 2014. Standard Test Method for Solubility of Asphalt Materials in Trichloroethylene Annual Book of Standards. ASTM International, West Conshohocken, PA.
ASTM D3038, Standard Test Method for Drop Shatter Test for Coke. Annual Book of Standards, ASTM International, West Conshohocken, PA.
ASTM D4791, 2014. Flat or elongated particles in coarse aggregate Annual Book of Standards. ASTM International, West Conshohocken, PA.
Bahia, H.U., Davis, R., 1994. Effect of crumb rubber modifiers (CRM) on performance related properties of asphalt binders. J. Assoc. Asphalt Paving Technol. 63, 414–449.
Button, J.W., Little, D.N., Kim, Y., Ahmed, J., 1987. Mechanistic evaluation of selected asphalt additives. Proc. Assoc. Asphalt Paving Technol. 56, 62–90.
Button, J.W., Estakhri, C., Wimsatt, A., 2007. A Synthesis of Warm-Mix Asphalt Report No. FHWA/TX-07/0-5597-1. Research and Technology Implementation Office, Texas Department of Transportation, Austin, TX.
Canarache, A., Vintila, I., Munteanu, I., 2006. Dictionary of Soil Science. Elsevier Science, Amsterdam, Netherlands.
Castelo Branco, V., Masad, E., Bhasin, A., Little, D.N., 2008. Fatigue analysis of asphalt mixtures independent of mode of loading Transportation Research Record No. 2057. Transportation Research Board, National Research Council., pp. 149–156.
Chesworth, W. (Ed.), 2008. Encyclopedia of Soil Science Springer Science, Dordrecht, Netherlands.
Daniel, J.S., Kim, Y.R., Lee, H.J., 1998. Effects of Aging on Viscoelastic Properties of Asphalt-Aggregate Mixtures. Transportation Research Record, TRB No. 1630, pp. 21–27.
Doyle, J.D., Howard, I.L., 2010. Compactability and Bitumen Utilization Of 100% Warm Mixed RAP. Transportation Research Board, National Research council, Washington, DC, p. 22.
Eisenmann, J., Hilmer, A., 1987. Influence of Wheel Load and Inflation Pressure on the Rutting Effect at Asphalt-Pavements-Experiments and Theoretical Investigations. In: Proceedings. 6th International Conference on the Structural Design of Asphalt Pavements, Vol. I, pp. 392–403.
Gandhi, T., Amirkhanian, S., 2007. Laboratory Investigation of Warm Asphalt Binder Properties—A Preliminary Investigation. In: Proceedings. 5th International Conference on Maintenance and Rehabilitation of Pavements and Technological Control (MAIREPAV5), vol. 5, Park City, UT, pp. 475–480.
Gillespie, H., 1992. A Century of Progress: The History of Hot Mix Asphalt. National Asphalt Pavement Association, Lanham, MD.
Hurley, G., Prowell, B., 2005a. Evaluation of Aspha-Min® for use in Warm Mix Asphalt NCAT Report No. 05-04. National Center for Asphalt Technology, University of Auburn, Auburn, AL.
Hurley, G., Prowell, B., 2005b. Evaluation of Sasobit® for use in Warm Mix Asphalt NCAT Report No. 05-06. National Center for Asphalt Technology, University of Auburn, Auburn, AL.
Hurley, G., Prowell, B., 2006. Evaluation of Evotherm® for use in Warm Mix Asphalt NCAT Report No. 06-02. National Center for Asphalt Technology, University of Auburn, Auburn, AL.

Karl, W.J., Santucci, L.E., Coyne, L.D., 1979. Hot mix recycling of asphalt pavements. Proc. Assoc. Asphalt Paving Technol., 192–220.
Kennedy, T.W., Huber, G.A., Harrigan, E.T., Cominsky, R.J., Hughes, C.S., Von Quintus, H., et al., 1994. Superior performing asphalt pavements (Superpave): the product of the Shrp asphalt research program Report No. SHRP-A-410. Strategic Highway Research Program. National Research Council, Washington, DC.
Kennedy, T.W., Tam, W.O., Solaimanian, M., 1998a. Optimizing use of reclaimed asphalt pavement with the superpave system. Proc. Assoc. Asphalt Paving Technol., 311–333.
Kennedy, T.W., Tam, W.O., Solaimanian, M., 1998b. Effect of Reclaimed Asphalt Pavement on Binder Properties Using the Superpave System Research Report No. 1250-1. Center for Transportation Research, Bureau of Engineering Research, University Of Texas at Austin, Austin, TX, September.
Kim, Y.R., Lee, Y.C., 1995. Interrelationships among stiffness of asphalt-aggregate mixtures. J. Assoc. Asphalt Paving Technol. 64, 575–606.
Mallick, R.B., Tao, M., O'Sullivan, K.A., Frank, R., 2009. Use of 100% Reclaimed Asphalt Pavement (RAP) Material in Asphalt Pavement Construction. In: Proceedings. 89th Conference, International Society of Asphalt Pavement, Nagoya, Japan.
Mokhatab, S., Poe, W.A., Speight, J.G., 2006. Handbook of Natural Gas Transmission and Processing. Elsevier, Amsterdam, Netherlands.
NCHRP, 2011. Test methods and specification criteria for mineral filler used in hot mix asphalt Research Results Digest 357. National Cooperative Highway Research Program. Transportation Research Board, Washington, DC.
Nourendin, A., Wood, L., 1987. Rejuvenator Diffusion in Binder Film for Hot-Mix Recycled Asphalt Pavement Transportation Research Record No. 1115. Transportation Research Board, National Research Council, Washington, DC.
Pellinen, T.K., 2003. The Effect of Volumetric Properties on Mechanical Behavior of Asphalt Mixtures Proceedings. Annual Meeting. Transportation Research Board, National Research Council, Washington, DC.
Roberts, F.L., Kandhal, P.S., Brown, E.R., Lee, D.Y., Kennedy, T.W., 1996. Hot Mix Asphalt Materials, Mixture Design, and Construction. Education, Foundation. National Asphalt Pavement Association, Lanham, MD.
SABITA, 1992. Guidelines for the Manufacture and Construction of Asphalt Manual No. 7. Southern African Bitumen Association, Cape Town, South Africa.
SABITA, 1993a. Labor Enhanced Construction for Bituminous Surfacings Manual No. 11. Southern African Bitumen Association, Cape Town, South Africa.
SABITA, 1993b. Methods and Procedures—Labor Enhanced Construction for Bituminous Surfacings Manual No. 12. Southern African Bitumen Association, Cape Town, South Africa.
Sargand, S., Ludwig Figueroa, J., Edwards, W., Al-Rawashdeh, A.S., 2009. Performance Assessment of Warm Mix Asphalt (WMA) Pavements Report No. FHWA/OH-2009/08. Office of Research and Development, Ohio Department of Transportation, Columbus, OH.
Speight, J.G., 2007. Natural gas: a basic handbook GPC Books. Gulf Publishing Company, Houston, TX.
Speight, J.G., 2008. Synthetic Fuels Handbook: Properties, Processes, and Performance. McGraw-Hill, New York, NY.
Speight, J.G., 2009. Enhanced Recovery Methods for Heavy Oil and Tar Sands. Gulf Publishing Company, Houston, TX.
Speight, J.G., 2012. Crude Oil Assay Database. Elsevier-Knovel, New York, NY, http://www.knovel.com/web/portal/browse/display?_EXT_KNOVEL_DISPLAY_bookid=5485&VerticalID=0.
Speight, J.G., 2013. The Chemistry and Technology of Coal, third ed. CRC Press, Taylor & Francis Group, Boca Raton, FL.

Speight, J.G., 2014. The Chemistry and Technology of Petroleum, fifth ed. CRC Press, Taylor & Francis Group, Boca Raton, FL.

Tao, M., Mallick, R., 2009. An evaluation of the effects of warm mix asphalt additives on workability and mechanical properties of reclaimed asphalt pavement (RAP) material Proceedings. 88th Annual Meeting. Transportation Research Board, National Research Council, Washington, DC.

Witczak, M.W., Kaloush, K., Pellinen, T., El-Basyouny, M., Von Quintus, H., 2002. Simple Performance Test for Superpave Mix Design NCHRP Report No. 465. Transportation Research Board, National Research Council, Washington, DC.

You, Z., Goh, S.W., Dai, Q., 2011. Laboratory Evaluation of Warm Mix Asphalt RC-1556. Michigan Department of Transportation, Lansing, MI.

Yunus, C., Boles, M., 1994. Thermodynamics: an Engineering Approach, second ed. McGraw-Hill, Inc., New York, NY.

CHAPTER 10

Asphalt Paving

10.1 INTRODUCTION

The term *hot mix asphalt* is used generically to include many different types of asphalt–aggregate mixtures that are produced in an asphalt plant. Most commonly hot mix asphalt is divided into the various components (Figure 10.1) and thereafter into three different types of mix depending upon the aggregate: (i) *dense-graded mix*, (ii) *open-graded mix*, and (iii) *gap-graded mix* according to the gradation of the aggregate used in the mix. Furthermore, the dense-graded mix is subdivided into *continuously graded mix* asphalt, *conventional hot mix asphalt*, *large-stone mix asphalt*, and *sand mix asphalt*. To add further clarification to this multi-classification system, the open-graded asphalt includes the subtypes: open-graded friction course and asphalt-treated permeable base. The gap-graded type encompasses both gap-graded asphalt concrete mixes and stone-matrix asphalt mixes. The different mix types are employed to satisfy the different demands of pavement performance and also to accommodate the varying nature of the available aggregates and asphalt cement supplies.

Hot mix asphalt has two primary ingredients: (i) asphalt binder, also called asphalt cement (see Chapters 1 and 4) and (ii) mineral aggregate (see Chapters 1 and 5). Furthermore, in the United States asphalt binders have been commonly graded by one of three methods: (i) the penetration-grading system, (ii) the viscosity-grading system, and (iii) the performance-grading system. The performance-grading system has been recently adopted by many states and agencies since it was introduced in the early 1990s. The aggregate used is typically a combination of coarse and fine materials, with mineral filler added as needed. The mix design system determines the correct proportion of asphalt binder and mineral aggregate required to produce an asphalt mix with the properties and characteristics needed to assure in-service roadway performance.

In the United States, mix design generally uses one of three methods: (i) the Marshall method, (ii) the Hveem method, and (iii) the Superpave® method—the *su*perior *per*forming asphalt *pave*ment method. In the Superpave method, compacted samples are tested and analyzed in order to

Asphalt Materials Science and Technology.
DOI: http://dx.doi.org/10.1016/B978-0-12-800273-5.00010-6

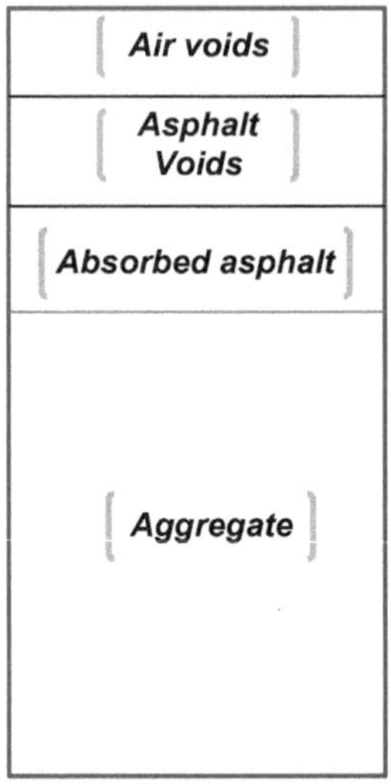

Figure 10.1 Components of hot-mix asphalt (not drawn to scale).

help predict the performance of the hot mix asphalt during roadway service by estimating fatigue failure, permanent deformation (rutting), moisture susceptibility, and thermal (low-temperature) cracking.

For any asphalt paving project, mix design is an essential part of the planning and is the starting point for producing the most appropriate asphalt mix for the project.

10.2 MIX DESIGN

Mix design is essential in order to enhance or mitigate the various roadway performance issues that are to be addressed and these issues include: (i) resistance to permanent deformation, (ii) resistance to fatigue and reflective cracking, (iii) resistance to low-temperature thermal cracking, (iv) durability, (v) resistance to moisture damage, (vi) workability, and (vii) skid resistance.

In order for an asphalt pavement to exhibit *resistance to permanent deformation*, the mix should not distort or displace under traffic loading, especially when the temperature is high (during summer) which causes the asphalt binder whereupon the traffic load is carried predominantly by the aggregate structure. Resistance to permanent deformation is controlled through improved aggregate properties (crushed faces), proper gradation, and proper asphalt grade and content.

Resistance to pavement fatigue and reflective cracking is inversely related to the stiffness of the roadway mix. Stiffer (harder) mixes are appropriate to deter the roadway from developing ruts, but focused design for rut resistance as the main property of the mix can be detrimental to the performance if the design allows pavement fatigue or reflective cracking. Stiff mixes perform more than adequately when used in thick hot mix asphalt pavements and when used as a thin overlay on a continuously reinforced concrete pavement. In addition, a thin hot mix asphalt mat placed on an unbound base or on a surface that is prone to reflective cracking (such as jointed rigid pavements and bound bases subject to shrinkage cracking) should use a mix that is designed for a balance between rut resistance and crack resistance. Proper selection of the asphalt binder can mitigate pavement fatigue and reflective cracking—another option for mitigating cracking is the inclusion of a specially designed crack-resistant interlayer in the roadway structure.

Low-temperature thermal cracking occurs in roadways that are subject to cooler temperatures and often occurs in areas where the differential between the daytime temperature and the night-time temperature is high. Thermal cracking can be mitigated by the selection of an asphalt binder with adequate low-temperature properties.

In order to exhibit in-service *durability*, the mix must contain sufficient asphalt binder to ensure an adequate film thickness of the blinder around the particles of the aggregate in order to minimize hardening and aging of the asphalt binder during production and also during in-service life. If the binder content is sufficient, this will also ensure adequate compaction in the field, keeping air voids within a range that minimizes permeability and aging.

The *resistance to moisture damage* (stripping) protects from the loss of adhesion between the aggregate surface and the asphalt binder, which has been assigned to deficiencies in the properties of the aggregate. However, the properties of the binder and the role of the binder properties in the asphalt–aggregate mix cannot be ignored (see Chapters 3 and 4). If it is assumed, during the design process, that moisture will eventually find a pathway into the pavement structure, the asphalt mix should be designed to resist stripping by (i) choice of a suitable binder, (ii) choice of a suitable aggregate, and (iii) selection, where necessary, of suitable anti-stripping agents.

Workability refers to the use of mixes that can be adequately compacted under laboratory conditions but which may not be easily compacted in the field. Too often laboratory data cannot be projected to performance of a product in the commercial world. In the case of an asphalt mix, adjustments may need to be made to the mix design to ensure the mix can be used in a roadway without sacrificing performance over the performance noted in the laboratory.

Finally, *skid resistance* relates to the need for an asphalt–aggregate mix to exhibit sufficient resistance to skidding, particularly under wet weather conditions. Aggregate properties such as texture, shape, size, and resistance to polish are all factors related to skid resistance. Many States have a wet accident reduction program (WARP) which is used to identify locations with high incidence of wet weather accidents and lower surface friction so the appropriate corrective measures can be taken. Under such a program, mineral aggregates are classified into four categories (such as a, b, c, or d) based on a combination of frictional and durability properties and the proper aggregate or blend may be selected to achieve the assessed skid-resistance rating.

10.3 MIX DESIGN METHODS

When hot mix asphalt pavements first began to be constructed, mixture composition was determined based on the judgment and experience of the contractor or used proprietary mix designs. Some of these pavements performed well, others did not. In the 1930s, a more rational system for designing hot mix asphalt mixtures was developed, which became common by the 1950s and continued to be widely used through the 1980s. The mix design procedure relies on compacting specimens using a standard drop hammer over a range of asphalt binder contents. The binder content was selected to produce proper air-void content and voids in the mineral aggregate (VMA). An essential part of the design method is the stability and flow test, which is an empirical procedure used to evaluate the strength and flexibility of the hot mix asphalt mixture.

At about this time, another hot mix asphalt mix design was developed which used a centrifuge kerosene equivalent test (CKE test) to determine an initial estimate of the asphalt binder content for a given aggregate. The kerosene equivalent test is no longer in use, although some laboratories may still use it as a guideline to estimate optimum asphalt content (Roberts et al., 1996). More recently, the performance-grading system is now used to specify binders in the United States and other countries—the Superpave system of mix design and analysis. In this method, laboratory specimens are prepared using a gyratory compactor. Superpave also includes a comprehensive set of requirements for aggregate gradations and property requirements. The Superpave method of mix design was meant to include mixture test methods and an associated computer program that would predict the performance of hot mix asphalt pavements, as an aid in the design and analysis of mixtures.

The Superpave grading system appears to present more realistic estimates of binder quality in ensuring that asphalt binders have adequate stiffness at high temperature while remaining flexible at low temperatures for a wide range of applications, helping to provide resistance to both rutting (see Chapter 8) and low-temperature cracking in asphalt mix pavements. To date, excessive rutting does not appear to occur in pavements constructed of mixes that have been designed using the Superpave system but there has been concern over surface cracking and high permeability in pavements using an asphalt mix designed using the Superpave system. This has required modification of the performance design system to mitigate such problems as well as problems associated with local materials and environmental conditions.

10.3.1 Asphalt Binders

The asphalt binder, sometimes referred to as the asphalt cement binder or the asphalt cement, is an essential component of asphalt concrete—it is as the name implies the cement that holds the aggregate together. The asphalt binder is a co-product of the petroleum-refining system that produces gasoline, diesel fuel, lubricating oil, as well as several other petroleum products. The asphalt binder is produced from the thick, heavy residue (residuum) that remains after distillation of petroleum to remove fuels and lubricants (see Chapter 2). The residuum is processed further by methods such as steam treatment (steam distillation) to remove any vestiges of gas oil or lubricating oil constituents and/or by oxidation until the treated residuum meets the desired specifications required of an asphalt binder. Once the binder has been produced, for demanding, high-performance applications, small amounts of additives may be blended into the binder to produce a modified binder—for example, polymer additives produce a polymer-modified binder.

The practice of mixing an asphalt binder with crushed aggregate to form paving material (asphalt concrete) has been in use for well over 100 years, which has proved to be a useful and valuable material for constructing flexible pavement on a worldwide basis. However, asphalt binders have very unusual engineering properties that must be carefully controlled in order to ensure good performance. One of the most important characteristics of asphalt binders that must be addressed in test methods and specifications is that their precise properties almost always depend on their temperature. Thus, asphalt binders may tend to become stiff and brittle at low temperatures, a thick viscous fluid at high temperatures, and a semi-solid at intermediate temperatures. Such extreme changes in properties with temperature (frequently referred to as *temperature susceptibility*) can cause performance problems in pavements. For example, at high temperatures that exist during the summer months in many southern states and regions, a pavement with a binder that is too soft will be prone to rutting and shoving. On the other hand, a pavement that contains a binder that is too stiff at low temperatures that exist during the winter months in many northern states and regions will be prone to low-temperature cracking.

Thus, the specifications for asphalt binders must control properties at high, low, and intermediate temperatures. Furthermore, the standard test methods that are used to specify the desired properties of an asphalt binder must be conducted with very careful temperature control; otherwise, the results will not be reliable (see Chapter 4) (Speight, 2001, 2014, 2015). Another characteristic of asphalt binders that complicates the pre-construction testing of these materials is that, for various chemical and physical reasons, such binders tend to harden with time (see Chapters 6 and 7). For example, when asphalt binders are heated to high temperatures, as happens when mixing and transporting asphalt mix, some of the remaining lighter volatile oil fractions of the asphalt vaporize, which can harden the remaining asphalt binder. At the same time, some of the chemical compounds making up asphalt binders can oxidize, which can also result in an increase in stiffness. While oxidation and the potentials for changes in the properties (albeit small) occur during mixing, transport, and placement of the asphalt mix, the slow long-term oxidation of the binder continues to occur over many in-service years, which results in a slow increase in stiffness of the binder. In fact, sometimes asphalt binder age-hardening can be so severe that it can lead to serious cracking of the pavement surface.

Other types of hardening occur in asphalt binders without any loss of volatiles or oxidation; these include steric hardening and physical hardening. These phenomena are not yet well understood but may be due to the orientation of the binder constituents in relation to the aggregate surface (see Chapters 6 and 7) which occurs slowly over time, resulting in a gradual increase in stiffness. Unlike other types of hardening, steric hardening and physical hardening can be reversed if the asphalt is heated until fluid and then cooled, all or most of the hardening will be removed, which is in keeping with a change in orientation of the higher-molecular-weight constituents of the binder relative to the aggregate surface. This is one of the reasons it is important to thoroughly heat and stir asphalt concrete samples prior to performing any laboratory tests and in order to assure accuracy as well as repeatability and reliability of the data (see Chapters 4 and 5) (Speight, 2001, 2014, 2015).

Asphalt binders are complex materials that are difficult to specify and test and it has been a major effort to develop standard test methods as well as effective and meaningful specifications for asphalt binders. One of the earliest tests for asphalt binders was the needle penetration test or variations thereof (ASTM D5; ASTM D139; ASTM D217; ASTM D312; ASTM D449; ASTM D937; ASTM D1403) in which a small lightly weighted needle was allowed to penetrate the asphalt for a set period of time (typically 5 or 60 s) (see Chapter 4). The distance the needle penetrates into the asphalt is measured and then used as an indication of binder stiffness.

Other such empirical tests were the ring and ball softening point temperature, and the ductility test and while these tests were useful to a degree (many are still used in specifications in Europe and other parts of the world), the test method had serious shortcomings that left serious doubt about their meaning in terms of the ability of the in-service performance of the asphalt binder and the asphalt concrete. For example, the test method did not measure any fundamental property of the asphalt binder such as the modulus or strength and the results were also sometimes highly variable and were not always in close agreement from laboratory to laboratory. In the 1960s, specifications based on viscosity measurements began to be adopted by many highway agencies. The viscosity test method and the data produced were superior to the data derived from earlier empirical tests and provided information on a fundamental characteristic of the asphalt binder and provided reasonably repeatable results among laboratories.

However, there are drawbacks to viscosity testing and the viscosity data were insufficient for predicting in-service performance and left much to be desired. For example, viscosity data were derived at higher temperatures than actual in-service temperatures and often involved extrapolation to in-service temperatures and the extrapolation was not always accurate. Furthermore, at high temperatures the behavior of the asphalt binder approaches the behavior of an ideal fluid. At low and intermediate temperatures (the range of in-service temperatures) viscosity tests become difficult to perform and the data are even more difficult to interpret. Furthermore, the data from viscosity test methods provide a limited amount of information on the flow properties of the binder. Two different asphalt binders (i.e., binders from different sources), which may be seemingly the same from bulk property data, can have identical viscosity values at a given temperature but might behave very differently because of differences in the degree of elasticity exhibited in their behavior. When loaded, the asphalt binders might deform the same amount, but when the load is removed, one binder might *recover* to nearly the initial shape while the other binder might hardly recover at all and remain in the deformed shape. Of the two sensibility similar binders, the binder that behaved in a more elastic fashion (i.e., that showed more recovery) would tend to provide better rut resistance in paving applications compared to the binder that exhibited poor elasticity (i.e., poor recovery). Viscosity tests provide no information about recovery or about the degree of elasticity exhibited by a material under loading. The shortcomings in both older empirical tests and in the newer viscosity tests eventually led to the necessary development of a more effective system of grading asphalt binders, the Superpave system (see Chapters 1 and 9).

10.3.2 Binder Evaluation

Superpave uses its own asphalt binder selection process, which is, of course, tied to the Superpave asphalt binder performance grading (PG) system and the (necessary) associated specifications. A Superpave performance-graded asphalt binder is selected based on the expected pavement temperature extremes in the area of intended use. Superpave software is used to calculate these extremes and select the appropriate performance-graded asphalt binder using one of the following three alternate methods (Roberts et al., 1996): (i) pavement temperature—the designer inputs the design pavement temperatures directly, (ii) air temperature—the designer inputs the local air temperatures, then the software converts them to pavement temperatures, and (iii) geographic area—the designer inputs the project location (i.e., state, county, and city) from which climate conditions from a weather database can be converted to air temperatures and pavement temperatures. Once the design pavement temperatures are determined they can be matched to an appropriate performance-graded asphalt binder.

The Superpave mix design method determines both a high and a low design pavement temperature. These temperatures are determined as follows: (i) high pavement temperature—based on the 7-day average high air temperature of the surrounding area, (ii) low pavement temperature—based on the 1-day low air temperature of the surrounding area. Using these temperatures as a starting point, Superpave then applies a reliability concept to determine the appropriate performance-graded asphalt binder. performance-graded asphalt binders are specified in increments of 6°C (11°F).

The original intent of the Superpave mix design method was to subject the mix design to a series of meaningful performance-related tests that are similar to the Hveem method that uses the stabilometer and cohesiometer, or the Marshall method with the stability and flow test.

Briefly, as a recap, the basic concepts are (Vallerga and Lovering, 1985): (i) the asphalt mix requires enough asphalt binder to coat each aggregate particle to an optimum film thickness, which allows for absorption of the binder into the aggregate, (ii) the asphalt mix requires sufficient stability to resist traffic loading, which is generated by internal friction between aggregate particles and cohesion or tensile strength created by the binder, and (iii) the durability of the asphalt mix increases with thicker asphalt binder film thicknesses. Based on these concepts, the design asphalt content is selected as that asphalt content resulting in the highest durability without dropping below a minimum allowable stability. In other words, as much asphalt binder as possible should be used while still meeting minimum stability requirements (Roberts et al., 1996).

In addition, the basic concepts of the Marshall mix design method move to selecting an asphalt binder content at a desired density that satisfies minimum stability and range of flow values (NAPA, 1982; Roberts et al., 1996). This mix design method consists of six basic steps: (i) aggregate selection, (ii) asphalt binder selection, (iii) sample preparation, including compaction, (iv) stability determination using the Hveem stabilometer, (v) density and void calculations, and (vi) optimum asphalt binder content selection.

Design pavement temperature calculations are based on hot mix asphalt pavements subjected to fast-moving traffic loads (Roberts et al., 1996). The standard test method (ASTM D7175) and the test temperature for this test are related to the temperature experienced by the pavement in the geographical area for which the asphalt binder is intended to be used. The complex shear modulus is an indicator of the stiffness or resistance of asphalt binder to deformation under load and the data define the resistance of the binder (and, hence, the asphalt concrete) to shear deformation of the asphalt binder in the linear viscoelastic region. The complex modulus and the phase angle are used to calculate performance-related criteria (AASHTO M320; ASTM D6373). This standard (ASTM D7175) is appropriate for un-aged materials and aged material produced under standard test methods (ASTM D2872; ASTM D6521).

The Superpave design method is also applicable to pavements that are subject to significantly slower (or stopped) traffic, such as traffic at intersections and toll booth lines, which, of necessity, need a relatively stiff asphalt binder compared to the binder that is used for fast-moving traffic. The Superpave design method allows the high-temperature grade to be increased by one grade for slow transient loads and by two grades for stationary loads. Additionally, the high-temperature grade should be increased by one grade for anticipated 20-year loading in excess of 30 million equivalent single axel loads (ESALs). Thus, the Superpave design method, like other mix design methods, creates several trial aggregate–asphalt binder blends, each with different asphalt binder content. By evaluating the performance of each blend, the optimum asphalt binder content can be selected. However, in order for the concept to be meaningful, the trial blends must contain a range of asphalt contents both above and below the optimum asphalt content—a standard analytical method for comparing optional changes in the sample to the control sample. Therefore, the first step in sample preparation is to estimate optimum asphalt content after which the range of asphalt binder in each trial blend can be determined.

Each sample is heated to the anticipated mixing temperature, aged for a short time (up to 4 h) and compacted with the gyratory compactor—a device that applies pressure to a sample through a hydraulically operated or mechanically operated load. Mixing and compaction temperatures are chosen according to asphalt binder properties so that compaction occurs at the same viscosity level for different mixes. The Superpave gyratory compactor establishes three different gyration numbers:

1. $\boldsymbol{N_{initial}}$: The number of gyrations used as a measure of mixture compactability during construction. Mixes that compact too quickly (air voids at $N_{initial}$ are too low) may be tender during construction and unstable when subjected to traffic. Often, this is an indication of aggregate quality—an asphalt mix with excess natural sand will frequently fail the $N_{initial}$ requirement. A mixture designed for greater than or equal to 3 million equivalent single axel loads (3×10^6 ESALs) with 4% air voids at N_{design} should have at least 11% (v/v) air voids at $N_{initial}$.

2. N_{design}: The number of gyrations required to produce a sample with the same density as that expected in the field after the indicated amount of traffic. A mix with 4% (v/v) air voids at N_{design} is desired in mix design.
3. N_{max}: The number of gyrations required to produce a laboratory density that should never be exceeded in the field. If the air voids at N_{max} are too low, the field mixture may compact too much under traffic, resulting in excessively low air voids and potential rutting. The air void content at N_{max} should never be below 2% (v/v) air voids.
   Typically for the test method, the samples are compacted to N_{design} to establish the optimum asphalt binder content and then additional samples are compacted to N_{max} as the control check.

Thus, all mix design methods use density and voids to determine basic characteristics of the asphalt mix (AASHTO TP4). Two different measures of densities are typically taken: (i) bulk specific gravity, G_{mb} and (ii) theoretical maximum specific gravity, TMD. G_{mm} and the test data are used to calculate the volumetric parameters of the asphalt mix. Measured void expressions are typically (i) air voids, V_a, sometimes expressed as voids in the total mix, VTM, (ii) voids in the mineral aggregate, VMA, and (iii) voids filled with asphalt, VFA, and these values should meet local or state or regional specification criteria. The optimum asphalt binder content is selected as that asphalt binder content that results in 4% (v/v) air voids at N_{design}. This asphalt content then must meet several other requirements that are dependent upon state or regional or national specifications, whichever is the operative agency. If any requirements are not met the mixture needs to be redesigned.

Moisture susceptibility testing (AASHTO T283, also known as the modified Lottman test) is the only performance test incorporated in the Superpave mix design procedure as of early 2002. The test method is used to compare the split tensile strength of unconditioned samples to samples partially saturated with water. The test subjects the conditioned group to partial vacuum saturation and an optional freeze–thaw cycle. Although it is expected that the water-conditioned samples will have a lower tensile strength, excessively low values indicate the potential for moisture damage.

10.4 TYPES OF MIX

The language of asphalt technology is variable (see Chapter 1) and is designed to describe in easy terms the type of mix that might be used in a roadway or for a particular project. For example, *dense-graded hot mix asphalt* is composed of an asphalt cement binder and a well-graded or continuously graded aggregate. *Conventional hot mix asphalt* consists of mixes with a nominal maximum aggregate size in the range of 0.5–0.75 inch. This type of material makes up the majority of the hot mix asphalt used in many regions of the United States. *Large-stone mixes* contain coarse aggregate with a nominal maximum size larger than 1 inch and this type of mix typically has a higher percentage of coarse aggregate than the conventional mix.

Sand asphalt mix (sometimes called *sheet asphalt*) is composed of aggregate that passes the 0.375-inch sieve. The binder content of the mix is higher than that of conventional hot mix asphalt because of the increased VMA. Use of manufactured sand or rough-textured natural sand in the mix maintains a reasonably high resistance to rutting. The sand mix asphalt can be produced in a batch plant or in a drum-mix plant with no significant changes in the plant operation. However, when compacted, sand mix asphalt may tend to shove and check under steel wheel rollers, especially when constructed in relatively thick layers (>2 inch). *Open-graded asphalt mix* consists of an aggregate with relatively uniform grading and an asphalt cement or modified asphalt binder. The primary function of the open-graded asphalt mix is to serve as a drainage layer, either at the pavement surface or within the structural pavement section.

There are two types of open-graded asphalts mixes: (i) a mix used as a surface course to provide a free-draining surface in order to prevent hydroplaning, reduce tire splash, and reduce tire noise—this type of mix is frequently termed an *open-graded friction course*, and (ii) an asphalt-treated permeable base which consists of a uniformly graded aggregate of larger nominal maximum size than that used for open-graded friction course and is used to drain water that enters the structural pavement section from either the surface or subsurface. Open-graded mixes are produced in a manner similar to that of dense-graded mixes except that a lower temperature is used for the open-graded materials to prevent drain-down (of the binder) during temporary storage. More recently, polymers and fibers have been used in open-graded friction courses to reduce drain-down and improve the durability of the mixes.

Gap-graded asphalt mix is similar in function to dense-graded asphalt mixes insofar as they provide dense impervious layers when properly compacted. This type of mix contains aggregate that ranges in size from coarse to fine, with some intermediate sizes missing or present in small amounts.

The *stone-matrix asphalt mix* is a type of gap-graded mix. Production of the stone matrix asphalt mix requires the addition of a significant amount of mineral filler to the normal aggregate in order to achieve the required 8–10% passing the 0.075 mm (No. 200) sieve. As with open-graded mixes, the discharge temperature of the mix needs to be carefully controlled at the plant to prevent drain-down of the binder during temporary mix storage in the silo and during transport to the job site. Fibers, polymer, or both are normally used with stone matrix asphalt mix to prevent drain-down of the binder.

10.5 GRADING ASPHALT CEMENT

Grading asphalt cement has evolved over the past three decades with the realization that penetration grading and viscosity grading are limited in their ability to fully characterize asphalt binder for use in asphalt mix pavements. As part of the Superpave system, new binder tests and specifications were developed to more accurately and fully characterize asphalt binders for use in asphalt mix pavements. As a result, test methods and specifications have been specifically designed to address asphalt mix pavement performance parameters such as rutting, fatigue cracking, and thermal cracking. In order to compare the various grading methods, this section presents each method in order to follow, assess, and differentiate between the methods.

10.5.1 Penetration and Viscosity Grading Systems

The penetration of asphalt cement (indentation measured by a standard needle in units of 0.1 mm or 1.0 dmm) is determined at 25°C (77°F). The grading is based on the concept that the stiffer the asphalt, the lower the penetration and the stiffer the asphalt–aggregate mix at a given temperature.

Grading of asphalt cement by viscosity is achieved by determining the viscosity at 60°C (140°F) of the material in its original condition (as received from the refinery; AC) or on a binder considered to be comparable to the binder after it has passed through the hot-mix process (AR). In this (AC) system, a mix containing AC-20 asphalt will be stiffer than a mix containing AC-10 asphalt at the same temperature. And, similarly, using the AR grading system, a mix containing AR-4000 asphalt will be stiffer than a mix containing AR-2000 asphalt at the same temperature.

10.5.2 Temperature-Viscosity Characteristics

Data for the viscosity-temperature characteristics of the asphalt binder are important in the production and placement of hot mix asphalt pavements. At high temperatures that are associated with mixing the binder and the aggregate in the mix facility, the flow characteristics of the binder (as measured by viscosity) must be known to provide assurance that the binder can be pumped and handled in the facility. Similarly, in mix placement, compaction of the hot mix is influenced by the stiffness of the binder. As the binder becomes stiffer or more viscous, a greater effort is required to achieve compaction to a given prescribed density.

A binder that is highly temperature-susceptible is a binder that exhibits a substantial change in viscosity for a small change in temperature. In addition, two samples of asphalt that have the same penetration at 25°C (77°F) may not necessarily have the same viscosity at 135°C (275°F) since their temperature-susceptibility characteristics may vary considerably. The temperature-susceptibility characteristics of the binder can also influence the compaction process. A mix containing a binder with high temperature susceptibility will stiffen more quickly with a drop in temperature than a mix containing a binder that has lower temperature susceptibility. Thus, if the temperature susceptibility characteristics of the binder in the mix change during production—such as when a different binder source is used for the same grade of binder—it may be necessary (if not essential) to change the compaction procedures in order to achieve the prescribed level of density.

10.5.3 Superpave Performance-Grading System

The grading system based on penetration and/or viscosity has been used for several decades and, for the most part has been satisfactory, but the results have not always been directly applicable to in-service roadway performance and longevity. In order to circumvent the anomalies (or inapplicability of the data), the performance-grading system involves use of a suite of test methods that allow measurement of physical properties that can be related directly to field performance (Table 10.1).

In the AC (AR) grading system, the principal grading (at 25°C, 77°F) may not accurately reflect low-temperature asphalt binder rheology and when using the grading system, thin-film oven test residue viscosities can vary greatly with the same grade (Roberts et al., 1996). Therefore, although asphalt binders are of the same grade they may behave differently after construction. Thus, the tests are performed at loading times,

Table 10.1 AC and AR grading systems compared to the superpave grading system

Limitations of Penetration, AC, and AR Grading Systems	Superpave Binder Testing and Specification Features that Address prior Limitations
Penetration and ductility tests are empirical and not directly related to pavement performance	The physical properties measured are directly related to field performance by engineering principles
Tests are conducted at one standard temperature without regard to the climate in which the asphalt binder will be used	Test criteria remain constant, however, the temperature at which the criteria must be met change in consideration of the binder grade selected for the prevalent climatic conditions
The range of pavement temperatures at any one site is not adequately covered—for example, there is no test method for asphalt binder stiffness at low temperature to control thermal cracking	The entire range of pavement temperatures experienced at a particular site is covered
Test methods only consider short-term asphalt binder aging (thin-film oven test) although long-term aging is a significant factor in fatigue cracking and low-temperature cracking	Three critical binder ages are simulated and tested: (i) original asphalt binder prior to mixing with aggregate, (ii) aged asphalt binder after hot mix asphalt production and construction, and (iii) long-term aged binder
Asphalt binders can have significantly different characteristics within the same grading category	Grading is more precise and there is less overlap between grades
Modified asphalt binders are not suited for these grading systems	Tests and specifications are intended for asphalt binders to include both modified and unmodified asphalt cement

See also ASTM D6373.
Viscosity grading can be performed on the original (as-supplied) asphalt binder samples (AC grading) or aged residue samples (AR grading). The AR viscosity test is based on the viscosity of aged residue from the rolling thin-film oven test. With AC grading, the asphalt binder is characterized by the properties it possesses before it undergoes the manufacturing process. The AR grading system is an attempt to simulate asphalt binder properties after it undergoes a typical manufacturing process and should be more representative of how asphalt binder behaves in pavements.

temperatures, and aging conditions that represent more realistically those encountered by in-service pavements. The derived data assist in the selection of a binder that will limit the propensity of the binder to low-temperature cracking, permanent deformation (rutting), and fatigue cracking of the asphalt pavement within the range of climate and traffic loading that are operative at the roadway site.

An important difference between the performance-graded specifications and those based on penetration or viscosity is the overall format of the requirements. For the performance-graded asphalt binders, the physical properties remain constant but the temperatures at which those properties must be achieved vary depending on the climate in which the binder is expected to serve.

10.6 MIX DESIGN PROCEDURES

In order to produce an asphalt mix design, asphalt binder and aggregate are blended together in different proportions in the laboratory and the resulting mixes are evaluated using a standard set of criteria to permit selection of appropriate binder content. The type and grading of the aggregate and the stiffness and amount of the asphalt binder influence the physical properties of the mix. The design (or optimum) binder content is selected to ensure a balance between the long-term durability of the mix and its resistance to rutting (stability).

10.6.1 Marshall Method

The Marshall method resulted from developments for a mix design procedure for airfield pavements during World War II and subsequent modifications. The procedure was adapted, in modified form, for the design of asphalt–aggregate mixes for highway pavements and during the 1990s was used by many highway organizations in the United States and in many other countries.

The design procedure includes a density–voids analysis of the compacted specimens to determine the percent air voids and percent voids filled with asphalt (see Chapter 5). After these determinations, the specimens are tested at 60°C (140°F), and the Marshall stability (maximum load observed in the test) and flow value (deformation corresponding to the maximum load) are obtained. Data resulting from these mix evaluations are plotted as a series of curves and include (i) density versus asphalt content, (ii) percent air voids versus asphalt content, (iii) percent voids filled with asphalt versus asphalt content, (iv) Marshall stability versus asphalt content, and (v) flow value versus asphalt content. The design asphalt content is determined as the average of the four contents selected corresponding to the peak density, 4% air voids, 75% voids filled with asphalt, and maximum Marshall stability. This asphalt content is then checked to ensure that the resulting air-void content and percent voids filled with asphalt fall within prescribed limits, that the Marshall stability exceeds a specified minimum level, and that the flow value does not exceed a prescribed maximum value.

10.6.2 Hveem Method

The Hveem method has been used since the early 1940s (Vallerga and Lovering, 1985). As is the case with the Marshall method, actual design criteria vary among organizations using this method, although the equipment for mix evaluation is essentially the same. The design philosophy embodied in this procedure is as follows: (i) stability is a function primarily of the surface texture of the aggregate, (ii) optimum asphalt content is dependent on the surface area, surface texture and porosity of the aggregate, and asphalt stiffness, and (iii) if required, the design asphalt content is adjusted to leave a minimum of 4% calculated air voids to avoid bleeding or possible loss of stability.

Kneading compaction (ASTM D1561) is used to prepare specimens for laboratory testing over a range of asphalt contents. The Hveem stabilometer, a closed-system triaxial compression test, provides the key performance measure in this method. Mix specimens are tested in this device at 60°C (140°F) over a range of binder contents, and a stability curve as a function of asphalt content is produced. By setting a minimum level of stability consistent with the applied traffic, the design asphalt content is selected.

10.6.3 Superpave Method

The Superpave mix design method was designed and developed to replace the Marshall mix design method and Hveem mix design method. The volumetric analysis common to the Marshall and Hveem methods provides the basis for the Superpave mix design method and the Superpave system ties asphalt binder and aggregate selection into the mix design process, and considers traffic and climate as well. The compaction devices from the Hveem and Marshall procedures have been replaced by a gyratory compactor and the compaction effort in mix design is tied to expected traffic (Roberts et al., 1996; Asphalt Institute, 2001).

The volumetric mix design is accomplished in four steps: (i) selection of component materials, (ii) selection of design aggregate structure, (iii) selection of design asphalt content, and (iv) evaluation of moisture susceptibility. Selection of the component materials includes selection of the appropriate binder performance grade and aggregate with requisite characteristics for the traffic applied. Both the high temperature and low temperature at the pavement site establish the binder grade to be used. Aggregate characteristics include coarse aggregate angularity, fine aggregate angularity, flat and elongated particles, and clay content.

10.7 LABORATORY- AND PLANT-PRODUCED MIXES

As with any operation that evolves from a laboratory, differences may exist between the properties of an asphalt mix designed in the laboratory and the same commercial formula produced in a batch-mix plant or in a drum-mix plant (see Chapter 11). It is important to examine those differences and understand how and why the test properties or characteristics of a mix produced in a plant may vary significantly from the results predicted by tests conducted on laboratory-produced material.

The objective of testing plant-produced asphalt mixes is to compare the test results with the laboratory formula with the objective that both sets of results (data) are the same (or similar). This may be difficult to accomplish because of all the variables that exist at the plant—from the type of plant used to the particular plant operating conditions. There are often major differences between laboratory and plant mixes, such as in the gradation of the aggregates, the rounding of the aggregates as they pass through the plant, the degree of hardening of the asphalt cement, and the segregation of fines through the operation of an emission-control system. The primary causes of these differences include mixing method, moisture content, and increased fine content. In addition, compaction conditions are considerably different between the laboratory and the actual mix compaction under various rollers in the field.

The job-mix formula produced in the laboratory, therefore, should serve as an *initial mix design* and as part of a quality control/quality assurance program the desired properties of the mix should be checked and verified on the plant-produced, laboratory-compacted asphalt mixture. A daily test program should be instituted to determine the characteristics of the manufactured mix (*mix verification*) and all of the mix values should be within the range required by the mix design process. If the test results on the plant-produced mix indicate compliance with the design requirements, the plant is operating within the limits of variability. If one or more of the mix properties are outside the desired range, it will be necessary to determine the cause and extent of the variability. In addition, if major differences in gradation exist between the aggregate used in the laboratory mix design process and the aggregate used in the plant, the job-mix formula should be adjusted or a new mix design developed.

10.7.1 Asphalt Cement Binder

In an asphalt cement storage tank, the binder is stored in bulk and usually is circulated continuously by a pump. During this time, the occurrence of aging and hardening should be minimal. However, in the laboratory, the asphalt cement can be heated in an oven for various periods of time and, as a result, laboratory samples may undergo more aging because they are usually handled in small quantities in open containers. In addition, binder modifiers are added in the field and are not evaluated during the mix design phase and, in such cases, the properties of the laboratory-produced mix and the properties of the plant-produced mix may vary. Therefore, it is recommended that all the materials used in the field also be used in the laboratory mix design and that quality assurance and quality control procedures should be followed as dictated by the relevant monitoring and control protocols that are relevant to the petroleum products industry (Speight, 2001, 2014, 2015). For example, the laboratory mixing of asphalt and aggregate is accomplished either by hand or by means of a mechanical mixer, and mixing efficiency and mixing times may vary. After mixing, the loose mix is aged to allow for asphalt absorption and, presumably, some additional stiffening. As a control of sample variability, the Superpave system incorporates an aging time (termed short-term oven aging) to produce a mix stiffness comparable to that which will exist early in the pavement life, usually less than 1 year.

10.7.2 Aggregate

The Superpave method requires washed sieve analysis of all fractions, including filler. As the aggregate passes through a batch plant dryer or drum mixer, variability is introduced because the gradation of the aggregate can be changed (to a variable degree) and the amount of the change (such as an increase in the amount of fines in the mix) is a function of many variables but is related primarily to the hardness of the aggregate. As the abrasion resistance of the aggregate decreases, the amount of fines generated inside the dryer or the drum normally increases. For example, in the case of a hard, durable aggregate, the amount passing the 0.075-mm (No. 200) sieve may increase no more than 0.2% when processed but when a soft aggregate is used, the amount of the aggregate passing the 0.075-mm (No. 200) sieve may increase markedly and be on the order of 1–2%.

Another source of variability between laboratory and plant data sets is aggregate dryness. For example, the aggregate used to make laboratory samples is (typically or always) completely dry and there is no moisture in the material. However, when the aggregate is heated in a batch plant dryer or in the dryer on a counter-flow drum-mix plant, the moisture content will undoubtedly be reduced but there may be residual moisture on the order of about 0.1% (w/w) of the aggregate but in some cases the moisture content in the aggregate could be as high as 0.5 to even 1.0% (w/w), depending on the amount of moisture in the incoming aggregate, the production rate of the dryer, and the aggregate discharge temperature. The amount of moisture retained in the plant-produced mix can have a significant effect on the tenderness of the mix and the ability to densify the hot mix asphalt under the compaction equipment.

Furthermore, aggregate variability can also be introduced in the emissions control system. For example, if a wet scrubber is used (as is used in an emission control and gas cleaning system) (Mokhatab et al., 2006; Speight, 2007, 2014) on either a batch or drum-mix plant, any fines (fine mineral matter) captured are carried out of the dryer or drum mixer and are not recycled (i.e., the fines are said to be *wasted*)—these fines are no longer part of the aggregate gradation. If a baghouse is used (as is used in an emission control and gas cleaning system) (Mokhatab et al., 2006; Speight, 2007, 2014) as the emission-control device on either type of plant, some or all of the collected fines can be returned to the mix but if the fines from the baghouse are wasted, a slightly different aggregate gradation will exist in the mix, similar to that which occurs when the plant is equipped with a wet scrubber system. If all of the baghouse fines are returned to the asphalt–aggregate mix, the gradation of the aggregate still may be different from that tested in the laboratory because of aggregate breakdown in the plant.

Thus, the type of emission-control equipment used on the batch or drum plant can significantly affect the properties of the asphalt mixture. The amount of fines can change the dust-to-asphalt ratio, and thus the stiffness of the resulting asphalt mix, which is not usually taken into account in the laboratory mix design procedure. However, some mix plants add baghouse fines to the mix during the mix design process to simulate the mix gradation after breakdown of material in the plant.

Baghouses operate at different efficiencies, depending on the pressure drop between the dirty and clean sides of the filter bags (Mokhatab et al., 2006; Speight, 2007, 2014). If the bags are clean and the pressure drop is small, the fines-laden exhaust gases pass through the fabric filter, and some of the very fine particles pass through the plant stack. As the bags become more heavily coated with material and the pressure drop increases, more of the fines are captured on the coating already on the bags. Thus, as the loading on the bags is increased, the baghouse actually becomes (some would say, fortuitously) more efficient, and a greater volume of fines as well as a finer gradation of material is returned to the mix in either a batch or drum-mix plant. The change in the amount of fines captured and sent back to the plant can be substantial.

If the plant is equipped with only a dry collector (knockout box or cyclone), most of the fines returned to the mix will be larger than the 0.300-mm (No. 50) sieve. When a fabric filter is employed (Mokhatab et al., 2006; Speight, 2007, 2014), particles as small as 5 microns (smaller than the asphalt cement film thickness on the aggregate) can be reincorporated into the mix—for reference purposes, 1 micron is 10^{-6} meter or 10^{-3} millimeter (25×10^{-3} inch). These ultra-fine particles can have an influence on mix response during the construction process. Further, the baghouse fines must be returned consistently and uniformly to the plant for incorporation into the mix.

10.7.3 Mixing Process

If reclaimed asphalt pavement is incorporated into the mix, it is usually mixed in the laboratory until thoroughly heated and blended with the new aggregate. However, the mix plant, the degree of mixing and the transfer of heat from the new aggregate to the reclaimed material are functions of many variables, such as: (i) the amount of reclaimed asphalt pavement in the mix, (ii) the point of introduction of the reclaimed asphalt pavement, (iii) the temperature of the new aggregate, and (iv) the mixing time. Blending of reclaimed asphalt pavement with the new aggregate differs for a batch plant, a parallel-flow drum-mix plant, and a counter-flow drum-mix plant. However, blending of the reclaimed asphalt pavement with the new aggregate in the laboratory is not subject to the same variability and is subject to the type of plant that will be used to manufacture the hot mix asphalt mix.

In the laboratory, the reclaimed asphalt pavement used in the mix design process may be a representative sample of the materials to be recovered from the paving project. In most instances, however, the aggregate gradation and asphalt content of the reclaimed asphalt pavement actually incorporated into the mix may vary from the values obtained from the representative sample. The milling and processing to reclaim material may add a significant amount of fines [percent passing the 0.075-mm (No. 200) sieve] to the mix. The extent of this expected variability needs to be considered during the laboratory mix design process.

The laboratory mixing process is accomplished by hand or by machine, with the time necessary to blend the asphalt cement and aggregate depending on the efficiency of the mixing process. Usually several minutes are required to obtain complete coating of the aggregate. During this period, the asphalt cement is exposed to the air, and some hardening takes place. The degree of hardening is a function of the aggregate temperature and the mixing time. The change in asphalt cement properties will differ from that which will occur during mix production in a batch or drum-mix plant.

Thus, the amount of hardening that occurs in the binder material depends on the time between manufacture and testing, as well as on the storage conditions (temperature and availability of oxygen). The process of reheating the sample, including the time and temperature of heating and any remixing of the sample, also can have a significant effect on the measured properties of the mix. Thus the laboratory handling process can affect the differences found between plant-prepared samples and laboratory-prepared samples.

10.7.4 Compaction

The purpose of any laboratory compaction process is to approximate, as closely as possible, the particle orientation produced in the field by the rollers and some amount of traffic loading. Several methods including impact compaction, kneading compaction, and Superpave gyratory compaction, are used to compact hot mix asphalt specimens in the laboratory. Thus, the degree of compaction obtained in the laboratory can be highly variable, depending on the method used.

The compaction process in the laboratory is very quick, and is usually completed in a comparatively short time (less than 5 min). This is in contrast to the roller operations in the field, which use many different roller combinations, roller passes, and roller patterns, and in which final density levels may not be attained until 30 min or longer after the mix has been placed by the paver. Also, during the laboratory compaction process, the temperature of the mixture is relatively constant whereas in the pavement, the temperature of the material continually decreases with time. In the laboratory, the compaction effort is usually applied before the mix temperature drops to 115°C (240°F) for the Marshall and Superpave methods (depending on the viscosity characteristics of the binder material) or 105°C (220°F) for the Hveem method. In the field, the mix may cool to 80°C (175°F) before the compaction process has been completed.

In the laboratory, the asphalt mix is compacted against a solid foundation, whereas in the field a wide variety of base types and stiffness is encountered. An asphalt mix can be placed as part of a newly constructed pavement, as the first layer on top of a soft subgrade soil, or as the surface course on a full-depth asphalt pavement structure. The material can be used as an overlay on distressed asphalt pavement. The ability to obtain a particular level of density in an asphalt mixture depends in part on the rigidity of the base being overlaid and on the type of compaction equipment used. Thus, the differences between some pavement and laboratory base conditions can be significant.

10.8 METHOD SPECIFICATIONS

Method specifications were probably the most widely used type of specification in highway construction until the mid-1980s. This type of specification allows a highway agency to direct the contractor to use specific methods, including materials, proportions, and equipment. The placement process is also explicitly defined, with each step being either controlled or directed, and in some cases actually performed by a representative of the agency.

Relative to asphalt mix production, method specifications require that the component materials (asphalt cement, aggregates, and additives) are pretested and pre-approved before asphalt laydown. The mixture must also meet other specific requirements related, for example, to air voids, stability, and flow. As an example, for hot mix asphalt compaction, the agency might tell the contractor what equipment to use, when to roll, and how many passes to make with each roller.

Method specifications have evolved with experience and reflect a lack of quick acceptance tests for assessing the quality of materials and construction. In most instances, the quality control and acceptance decisions are based on individual test results. Terminology such as *substantial compliance* and *reasonably close conformity* is associated with method specifications and the contractor (in terms of materials and equipment) should be within reasonably close conformity with these requirements.

10.8.1 Advantages and Disadvantages

Method specifications offer an advantage when a measure of quality is particularly difficult to define, such as asphalt–mix segregation. Segregation is an undesirable feature, but the allowable degree of segregation is difficult to measure or to specify. Thus, method specifications can be used to specify what a contractor must do to prevent segregation.

Also, if the quality of the product is measured and found to be less than desirable, the contractor has no legal responsibility to improve it. On the other hand, the quality attained is difficult to relate to the performance of the finished product. The major weakness of this type of specification is that there is no assurance it will produce the desired quality of construction. Most important, by explicitly specifying the material and procedures, the owner or agency obligates itself to a large degree to accept the end product. Such a specification is also very difficult to enforce uniformly. The terms *reasonably close conformity* and *substantial compliance* cannot be precisely defined and in the absence of a clearly defined quality level as well as an established means of measuring compliance, decisions become arbitrary, and acceptance procedures become inconsistent in their application. Limits that are based on subjective judgment or experience are often difficult to meet because of the lack of definition of the capabilities of the production process and the properties or quality of the desired product.

10.8.2 Quality Control

Many paving projects have been carried out successfully with method specifications. Every successful hot mix asphalt contractor controls the quality of the asphalt mix throughout the production process. Testing to ensure quality begins with raw aggregate (see Chapter 5) as well as raw binder (see Chapter 4), and ends with finished pavement. Each test has a place in the overall control system, from designing the job mix through proportioning, mixing, and placing the hot mix asphalt.

Design qualities are the main consideration when selecting aggregates for a job-mix formula. The decision concerning which aggregate to use is based solely on data from standard test methods (see Chapter 5) originating at the source pits or quarries, long before the material reaches the mix production plant. The general characteristics and physical properties of aggregates for hot mix asphalt surface and base courses are defined in detail (AASHTO PP6) and, in addition, many local, state, or regional agencies typically have their own standards.

Table 10.2 Basic elements pertaining to mixtures that require quality control testing

Mix Design	Plant Control Tests
Selection of asphalt cement	Voids (VMA, VFA)
Selection of aggregates	Density
Development of the job-mix formula	Flow
Selection of a mixing temperature	Stability
Day-to-day plant control and tests	Asphalt content of mix
Adjustments of mixing time and temperature	Gradation of aggregate

Asphalt is generally purchased from a source tested and accepted by the agency or accepted on the basis of the supplier's certification. Strict quality control procedures may also require that the hauler supplying material to the plant furnish a *prior load certificate*, which protects the supplier of the load from disputes resulting from contamination during transport. These requirements should be specified when executing a purchase agreement. In addition, it is good practice to randomly sample incoming loads of asphalt cement for future testing if necessary. The agency may also sample asphalt at the plant and run tests in the agency laboratory—in this case, samples stored on site are useful should any question arise about the quality of the asphalt.

Variations in the properties of the asphalt binder are often missed because these properties are not frequently tested. However, the properties of the binder can change minimally or substantially from lot to lot and, coupled with the variability of the aggregate on a lot-to-lot basis, the mix properties and laydown characteristics of the asphalt mix will also change. These variations can be monitored if the plant quality control program is active, part of which is the maintenance record of the certificates of tests submitted by the asphalt supplier. In addition, the temperature of the incoming asphalt must be closely monitored. Specifications set limits on the allowable temperature in the asphalt storage tanks and overheating by the supplier or hauler is cause for rejection of the asphalt cement.

Finally, plant control of the quality of the asphalt mix includes a series of elements that should be closely monitored and interpreted. In fact, any particular test method may include a variety of functions and satisfy a number of these quality control needs (Table 10.2).

REFERENCES

AASHTO M320, 2014. Standard Specification for Performance-Graded Asphalt Binder. American Association of State Highway & Transportation Officials, Washington, DC.
AASHTO PP6. Practice for Grading or Verifying the Performance Grade of an Asphalt Binder. American Association of State Highway & Transportation Officials, Washington, DC.
AASHTO TP4. Preparing and Determining the Density of Hot-Mix Asphalt (HMA) Specimens by Means of the Superpave Gyratory Compactor. American Association of State Highway & Transportation Officials, Washington, DC.
AASHTO T283. Resistance of Compacted Bituminous Mixture to Moisture-Induced Damage. American Association of State Highway & Transportation Officials, Washington, DC.
Asphalt Institute, 2001. Superpave Mix Design. Asphalt Institute, Lexington, KY, Superpave Series No. 2 (SP-02).
ASTM D5, 2014. Standard Test Method for Penetration of Bituminous Materials Annual Book of Standards. ASTM International, West Conshohocken, PA.
ASTM D139, 2014. Standard Test Method for Float Test for Bituminous Materials Annual Book of Standards. ASTM International, West Conshohocken, PA.
ASTM D217, 2014. Standard Test Methods for Cone Penetration of Lubricating Grease Annual Book of Standards. ASTM International, West Conshohocken, PA.
ASTM D312, 2014. Standard Specification for Asphalt Used in Roofing Annual Book of Standards. ASTM International, West Conshohocken, PA.
ASTM D449, 2014. Standard Specification for Asphalt Used in Damp-proofing and Waterproofing Annual Book of Standards. ASTM International, West Conshohocken, PA.
ASTM D937, 2014. Standard Test Method for Cone Penetration of Petrolatum Annual Book of Standards. ASTM International, West Conshohocken, PA.
ASTM D1403, 2014. Standard Test Methods for Cone Penetration of Lubricating Grease Using One-quarter and One-half Scale Cone Equipment Annual Book of Standards. ASTM International, West Conshohocken, PA.
ASTM D1561, 2014. Standard Practice for Preparation of Bituminous Mixture Test Specimens by Means of California Kneading Compactor Annual Book of Standards. ASTM International, West Conshohocken, PA.
ASTM D2872, 2014. Standard Test Method for Effect of Heat and Air on a Moving Film of Asphalt (Rolling Thin-Film Oven Test) Annual Book of Standards. ASTM International, West Conshohocken, PA.
ASTM D6373, 2014. Standard Specification for Performance Graded Asphalt Binder Annual Book of Standards. ASTM International, West Conshohocken, PA.
ASTM D6521, 2014. Standard Practice for Accelerated Aging of Asphalt Binder Using a Pressurized Aging Vessel (PAV) Annual Book of Standards. ASTM International, West Conshohocken, PA.
ASTM D7175, 2014. Standard Test Method for Determining the Rheological Properties of Asphalt Binder Using a Dynamic Shear Rheometer Annual Book of Standards. ASTM International, West Conshohocken, PA.
Mokhatab, S., Poe, W.A., Speight, J.G., 2006. Handbook of Natural Gas Transmission and Processing. Elsevier, Amsterdam, Netherlands.
NAPA, 1982. Development of Marshall Procedures for Designing Asphalt Paving Mixtures Information Series No. 84. National Asphalt Pavement Association, Lanham, MD.
Roberts, F.L., Kandhal, P.S., Brown, E.R., Lee, D.Y., Kennedy, T.W., 1996. Hot Mix Asphalt Materials, Mixture Design, and Construction. National Asphalt Pavement Association Education Foundation, Lanham, MD.
Speight, J.G., 2001. Handbook of Petroleum Analysis. John Wiley & Sons Inc., Hoboken, NJ.

Speight, J.G., 2007. Natural Gas: A Basic Handbook. GPC Books, Gulf Publishing Company, Houston, TX.
Speight, J.G., 2014. The Chemistry and Technology of Petroleum, fifth ed. CRC Press, Taylor & Francis Group, Boca Raton, FL.
Speight, J.G., 2015. Handbook of Petroleum Product Analysis, second ed. John Wiley & Sons Inc., Hoboken, NJ.
Vallerga, B.A., Lovering, W.R., 1985. Evolution of the Hveem stabilometer method of designing asphalt paving mixtures. In: Proceedings, Association of Asphalt Paving Technologists, vol. 54, pp. 243–264.

CHAPTER 11

Asphalt Laydown

11.1 INTRODUCTION

Asphalt laydown is the general term used to describe the construction of an asphalt roadway. However, asphaltic roadway construction is not merely a matter of laying asphalt on to a surface but a sequence of carefully planned stages that result in the finished project (see Chapter 9).

Historically, the oldest known paved road was laid in Egypt at some time between 2600 and 2200 BC. Paved streets—using stone with asphalt as the fixative—have been found in the city of Ur and date to 4000 BC—the asphalt mastic was believed to have been introduced to hold the stones in place approximately one millennium later (Speight, 1978). In 500 BC, the Persians (by the command of Darius the Great) started an extensive road system, including the Royal Road—which ran from Susa (in southwest Persia) through Babylon (a short distance from Baghdad) to Sardis (in Asia Minor to the south of the Dardanelles)—which was one of the finest highways of its time and there is evidence for the use of asphalt is some segments of the road (Speight, 1978). In the later part of the fourth century BC (approximately 315–312 BC) the Romans started the construction of long and straight roads (some have fragmentary evidence for the use of asphalt as a water-proofing course under the stones; asphalt was used as a sealant in Roman aqueducts) throughout Europe and in support of its military campaigns—the US Interstate Highway system is modeled after such roads and these asphaltic highways were built during the early years of the Cold War.

In the eighth century AD, many roads were built throughout the Arab Empire—the most prominent were those in Baghdad which were constructed using bitumen which was most likely collected from the deposits at Hit (Speight, 1978). However, Arabian scientists were knowledgeable in the technique known as destructive distillation which would produce volatile products and an asphaltic residuum (albeit a cracked residuum) from the high-boiling constituents of petroleum. From that time on, road construction has been through many stops and starts resulting in the modern use of asphalt for roadway construction. The medieval period was

Asphalt Materials Science and Technology.
DOI: http://dx.doi.org/10.1016/B978-0-12-800273-5.00011-8

particularly bleak in terms of road construction in the countries of western Europe and it was not until the mid-to-late eighteenth century that many western European countries undertook roadway construction projects, with asphalt (usually coal-tar asphalt) becoming a major contributor to the roadway.

Thus, in the early nineteenth century there was a surge of interest, and asphalt became widely used for pavements, flat roofs, and the lining of cisterns (see Chapters 2 and 8)—at that time coal tar still played a prominent role in the production of asphalt. The discovery of petroleum and subsequent commercialization of petroleum in Pennsylvania in 1859 led to further innovations and developments in petroleum refining (Speight, 2014).

Because of the predominant use in roadway construction, it is the purpose of this chapter to present to the reader a general description of the means by which asphalt is delivered from the plant to the work site and laid down in a manner that will be conducive to the production of an asphalt roadway. The description of the process will also present to the reader the need for knowledge of the various properties of the asphalt and asphalt mix and how these properties can affect the roadway construction process—the same logic can also apply when asphalt is used for other purposes. Although the types of equipment used will also be presented, it is not the intent here to present a full description of each piece of equipment—such descriptions are more in keeping with a mechanical engineering document or a roadway engineering document.

11.2 ASPHALT MANUFACTURE

The use of asphalt for roadway surfacing is common (one might say, essential) on high-traffic roads and/or high-strength pavements. The historic laying of asphalt by hand has been superseded by the use of technologically advanced equipment as quality standards for asphalt surfaces have changed with increasing vehicle loading and speed. On site, labor-intensive construction is organized for site preparation, priming/tacking of the base course, stockpiling of the asphalt, spreading and laying the asphalt, and screeding (leveling the asphalt mix on the roadway). Emphasis is placed on obtaining the quality and finish expected of the asphalt-wearing course.

There are two industries that fall under the category of asphalt manufacture: (i) plants for the manufacture of road asphalt, which produce asphalt paving materials and (ii) plants for the manufacture of roofing asphalt, which produce asphalt-saturated felt rolls, fiberglass, and organic (felt-base) shingles, and surfaced and smooth roll roofing for use mainly in roof construction.

11.2.1 Mixing Process

Hot mix asphalt is created by mixing and heating size-graded, high-quality aggregate (which can include reclaimed asphalt pavement) with liquid asphalt cement and can be manufactured by batch mix equipment, continuous mix equipment, parallel flow drum mix equipment, or counter-flow drum mix equipment (see Chapter 10). In the batch mix process, the aggregate is sent to a rotary dryer and after exit from the dryer, the aggregate is sorted into size grades (*graded*) and weighed, then dropped into the pug mill (*mixer*), where it is combined with liquid asphalt. In the parallel flow drum mix process, sized aggregate is introduced to the dryer drum at the burner end and, as the drum rotates, the aggregate and combustion products move toward the opposite end of the drum. The liquid asphalt cement is added midway down the drum, therefore facilitating heating and mixing in the same unit. In the counter-flow drum mix process, the flow of aggregate is opposite to the flow (*counter flow*) of exhaust gases, and the liquid asphalt cement is added beyond the mid-point of the drum (after the burner flame) to reduce contact of the hot asphalt cement with the hot exhaust gases and to mitigate any potential thermal changes to the asphalt.

In a batch-equipment plant, the aggregate is stored in hot bins prior to mixing with asphalt in discrete batches before being stored or loaded into trucks. In drum-equipment plants, mixing of the aggregate and the asphalt takes place in the same drum, after which the mix is stored in a silo before being loaded into trucks for delivery. Because of the variability of pavement type, various asphalt mix formulas are used and the mix formulas are designed to meet the requirements (in-service use) of the pavement. In the case of major roads, highways, and airport runways, the owners are typically governmental entities and governmental design formulas will be followed. In the case of parking areas, low-volume roads, and other facilities, many owners are from the commercial market but they, for the most part, also follow (by choice or by regulation) specifications from government agencies.

The asphalt binder is usually stored in heated tanks on site between 150°C (302°F) and 180°C (356°F), which enables the viscous liquid to be pumped through insulated pipes to the mixing plant. The mineral aggregate—stone, sand, or gravel—is stored in stockpiles at ambient temperature. In addition to virgin aggregates, most asphalt mix facilities may also have stockpiles of reclaimed asphalt pavement (recycled asphalt pavement, RAP) which are typically segregated stockpiles in which the aggregate has been carefully sorted by type and size. At the initiation of the process, the aggregate and any reclaimed material are taken from various stockpiles and loaded into specific bins—care must be taken to ensure that the material is representative of the whole stockpile. Each size of aggregate and reclaimed asphalt material is fed onto conveyor belts in proportions specified by the design mix formula (which will vary according to the job) and transported to be dried in a drum before mixing with the binder.

For the actual mixing process at a batch plant, the aggregate is heated in a rotating drum prior to mixing by tumbling the aggregate through a stream of hot air after which the aggregate and any filler material are mixed in batches with the exact proportions of asphalt and possibly RAP in a pug mill. In contrast, at a drum mix plant, the asphalt is added to the dried aggregates and continuously mixed in the same drum used for drying. However, as already stated, the asphalt binder (and any reclaimed asphalt pavement) is added to aggregate far downstream from the source of heat to prevent thermal decomposition and changes to the asphalt binder, which can be extremely sensitive to heat thereby causing loss of desirable (and necessary) properties (see Chapters 6 and 7) (Moschopedis et al., 1978; Parkash et al., 1980; Speight, 2014). Although such temperature susceptibility is unlikely to manifest itself during roadway service (with the exception of the ignition of spilled oil on the roadway), the exposure of the asphalt to excessive heat in the asphalt mix plant (by injection of the asphalt on to the aggregate in the heated mixer) may be sufficient to cause irreversible chemical and physical changes to the binder and the loss of necessary and desirable properties.

Every part of the plant has enclosures and/or control technologies. For example, most asphalt mix plants are fuelled by natural gas or fuel oil, and state-of-the-art scrubbers and emission control equipment (Mokhatab et al., 2006; Speight, 2007, 2014) keep combustion-related emissions very low. Dust is controlled in the baghouse, where fines and dust are collected on the outside of filter bags, while clean air passes through the center of the bags. The fines are periodically subjected to bursts of air which force them to the floor of the baghouse, where they are collected for metering back into the paving mix. While most asphalt mix plants are on permanent sites and have emission control equipment built into the plant, portable mixing plants must also be conscious of emissions and install similar advanced environmental controls that are included in plants on permanent sites.

Approximately 85% of all plants use the counter-flow drum mix design, while 10% use batch mix, and 5% use parallel flow drum mix. Continuous mix plants are not readily used and make up less than 0.5% of the existing hot mix asphalt plants. Once mixed, the hot mix asphalt is conveyed to hot storage silos, or loaded onto trucks to be hauled to the job site.

Asphalt roofing materials are produced through a series of steps, including asphalt preparation, followed by felt saturation, coating, mineral surfacing, cooling and drying, product finishing, and packaging. Preparation of the asphalt is accomplished through a blowing process which involves bubbling air through liquid asphalt flux at 260°C (500°F) for 1–10 h in a blowing still. The amount of blowing time depends on the desired characteristics (specifications) of the roofing asphalt and inorganic salts (such as ferric chloride, $FeCl_3$) can also be used as catalysts to achieve desired properties and increase the rate of reaction, which decreases the time required for each blowing operation. After asphalt preparation, the various asphalt roofing products can be made through the production steps. Saturated felt is produced by passing organic felt through the spray section of the saturator where asphalt at 205–250°C (400–480°F) is sprayed on one side of the felt. The saturated felt is then run over a series of rollers in the saturator dip section, where the bottom rollers are submerged in hot asphalt.

In the next step, heated drying drums and the wet looper (sometimes called the *hot looper* and where the asphalt is drawn into the felt as it cools) provide heat and time, respectively, for the asphalt to penetrate the felt. As the last step, the saturated felt is passed through water-cooled rolls and onto the finish floating looper, rolled, and cut to product size. Production of asphalt roofing shingles, mineral-surfaced rolls, and smooth rolls is similar to the production line of the saturated felt rolls, including saturation, coating, application of a granule and sand or banking surface, pressing, cooling, finishing through floating looper, and finally rolling, processing through a seal-down applicator and shingle cutter for shingles, or processing through a laminating applicator and laminating operation for laminated shingles.

The dryer operation is the main source of pollution at hot mix asphalt manufacturing plants. The capacity of the dryer burners is usually less than

100×10^6 Btu/h, but may be as large as 200×10^6 Btu/h. Natural gas is the preferred source of heat used by the industry, although oil, electricity, and combinations of fuel and electricity are used. The reaction of nitrogen and oxygen in the dryer creates nitrogen oxide (NOx) emissions in the combustion zone, while sulfur dioxide (SO_2) emissions are the product of oxidation of sulfur-containing compounds in fuels. Particulate matter emissions result from the volatilization of materials that later form condensates and also from material handling operations. Volatile organic compound (VOC) emissions are the by-product of incomplete combustion.

As for asphalt roofing manufacturing plants, emissions of particulate matter and VOCs are largely from asphalt storage tanks, blowing stills, saturators, coater-mixer tanks, and coaters. The particulate matter from these units consists primarily of recondensed asphalt fumes, while sealant strip and laminant applicators are sources of small amounts of particulate matter and VOCs. Asphalt heaters are also sources of the combustion products, nitrogen oxides and sulfur oxides, due to combustion of natural gas and/or fuel oil.

11.2.2 Truck Loading and Transport

After the aggregates have been dried and thoroughly mixed with the asphalt at the required temperature, the asphalt pavement material may either be temporarily stored in silos on the plant site or discharged directly into a truck for transport to the paving site. The asphalt mix is transported from the plant site to the paving site in trucks. Transport distances vary, but are normally on the order of up to 50 miles, during which time the rate of cooling can affect the temperature of the asphalt mix to be laid on the roadway. Thus, the distance for transporting the heated mix to the paving site is (must be) limited since the asphalt mix must be delivered to the paving site while it is still warm enough to be placed and compacted on the road.

Thus, after preparation of the asphalt according to the specification of the area or region in which it is to be used, the asphalt is transported to the mix plant. After the mix (the asphalt–aggregate mixture) is produced according to design specifications, the first exposure of the asphalt–aggregate combination is transportation to the site where paving operations are occurring.

Three haulage vehicles, which vary in type depending upon how the mix is unloaded, are predominantly used: (i) end-dumpers, which unload the mix from the end of the truck bed, (ii) belly-dumpers, which unload the mix from the mid-point of the truck bed, or (iii) live-bottom dumpers, which unload the mix by discharging the mix using a conveyor belt without the need to raise the bed. The live-bottom truck moves the asphalt mix *en masse*, which minimizes segregation as compared with end-dump trucks.

However, during the unloading procedure the hot asphalt is susceptible to aerial oxidation. This is the time when the asphalt spreads as it unloads creating a relatively large heated surface area leaving the asphalt susceptible to aerial oxidation and cooling which can affect compaction. Therefore, the unloading operation must be accomplished without delay to minimize any change in the characteristics of the mix during the delivery process, and without segregation of the asphalt from the aggregate or segregation of the mix by size of the aggregate particles. After transfer of the mix to the paving machine (the *paver*), the uniformity (or non-uniformity) of the volume and character of the mix being fed to the paving machine dictates the ability of the paver to create a quality (or sub-quality) asphalt roadway.

The transportation of the asphalt mix is, seemingly to many observers, a matter of taking along the truck and delivering the asphalt. However, without due care and attention, the asphalt may be delivered in an (unknowingly) unsatisfactory form (dies to binder oxidation or mix segregation to mention only two issues) that can cause problems with the roadway during service life and even shorten service life. Thus there are several important factors that must be given consideration when monitoring truck loading, transportation, and unloading operations.

The truck bed should be free of all contaminants (deleterious materials) and any debris in the bed from previous use of the truck should be removed. If not, this can lead to the presence of deleterious materials in the mix that will have a serious effect on mix performance in the roadway. In addition, the truck bed should be lightly and uniformly coated with a non-petroleum release agent to prevent the bed being contaminated with older and cooler asphalt (from pervious transportation journeys) that can lead to sub-par performance of newer asphalt mixes. There has been a tendency in the past to use diesel fuel—trucks are fueled by diesel fuel, which is close at hand—but diesel fuel can have an adverse effect on the mix either by weakening the asphalt–aggregate bonding or by causing incompatibility within the asphalt (see Chapters 6 and 7).

If necessary due to environmental conditions, the sides and bottom of the truck bed should be insulated and the insulation should be close to the bed (tight against the body of the bed) and there should be no gaps between the side of the truck and the insulation through which wind could enter. In some cases, a coarse-graded mix, such as friction courses, stone-matrix asphalt, and coarse-graded Superpave, will cool more quickly than a fine-graded mix, which causes changes in the properties of the mix and influences the compaction of the mix.

Since the hot asphalt in the mix is susceptible to oxidation and other chemical changes during transportation, it is advisable that the truck should be equipped with a tarpaulin that is in good condition, with minimal (preferably without) tears and holes. The tarpaulin should be large enough to cover the bed and wrap over its sides and end and have a sufficient number of fasteners so that it can be tied down completely and will not flap in the wind. If truck bed side rails are used to hold the tarpaulin in place, the tarpaulin should be stretched tightly over the load of mix. This prevents excessive exposure of the hot asphalt mix to environmental conditions, such as a hot day in the sun or (in some countries) exposure to the ever-present rain. The former can initiate the process known as oxidative aging while the latter can lead to initiation of the process known as moisture damage (see Chapter 6).

When at the site, the mix should be deposited uniformly to enable the paver operator to maintain a uniform head of material in front of the paver screed (a flat board or a purpose-made aluminum tool is used to smooth and true materials, such as asphalt mix, after they have been placed on the surface or to assist in flattening).

The distance for transportation of the mix to the site has not been regulated but the shortest possible distance has minimal effect (through minimal cooling of the mix) on the workability of the mix while it is passing through the paver and the ability to compact the mix once it has been set in place by the paver. The rate of cooling of the mix depends on such variables as (i) the temperature at the time of production, (ii) the ambient air temperature, and (iii) the efficiency of any insulation used on the sides and bottom of the truck. Furthermore, when transported over (undefined) long distances without a tarpaulin cover, crusting will occur on the top of the mix as it cools. While the formation of a top crust may appear to be beneficial insofar as the crust insulates the remainder of the mix thereby reducing the rate of cooling for the remainder of the material, there is also an adverse effect of the crust. The crust must be destroyed and be compatible with the remainder of the mix before being used in the paver and unevenness will occur in the finished asphalt surface leading to problems with the roadway and shortening of in-service life.

During periods of rainfall, there is always the potential for the risk of water-damage. Even though asphalt mix placed during rainfall can perform adequately over time, it is considered to be better practice to refrain from placing the mix if the rain is heavy and the pavement surface is very wet or standing water exists on the surface. Furthermore, any mix that is placed during rain will cool very quickly as a result of the rapid heat transfer from the mix to the wet underlying pavement surface and the cooling of the mix by the rain itself. It then becomes difficult to obtain the desirable mix density because of the low temperature which leads to poor compaction and high air-void content in the mix with the resulting poor pavement performance.

11.3 SITE PREPARATION

The performance of an asphalt mix under traffic loads is directly related to the condition of the surface on which the pavement layers are placed. If the asphalt mix layers are placed on top of a new, untreated granular base course, then (i) the base material should be stable, (ii) the surface should be dry, and (iii) the base should not be distorted by the trucks that carry the mix to the paver.

On the other hand, for an asphalt mix laid on top of existing asphalt layers, the surface should be properly prepared, which includes: (i) potholes should be filled, (ii) cracks sealed, and (iii) the surface should be cleaned. In addition, a tack coat should also be used to ensure a bond between the existing pavement surface and the new asphalt overlay.

11.3.1 Base Preparation—New Asphalt Mix Pavement

If the asphalt pavement is to be placed directly on the subgrade soil, that subgrade material should meet all applicable requirements for (i) moisture content, (ii) density, (iii) structural support, (iv) smoothness and lack of distortion, as well as (v) the ability to support the traffic loads.

In some instances, a prime coat of asphalt emulsion or cutback asphalt is placed on the subgrade soil, especially when the soil is a silt material or clay material or a mixture of silt and clay. The prime coat acts as a temporary waterproofing layer that protects the base course and prevents it from absorbing excess moisture during times of rain before the paving is complete. The prime coat also allows (i) the base course to be used for light traffic, (ii) binds together any dust on the surface of the granular base layer, (iii) promotes the bond between the base-course material and the new asphalt mix overlay, and (iv) prevents slippage of thin overlying pavement layers. In many instances, cutback asphalt has been used successfully as the prime coat material and the rate of application of the prime coat typically varies with the porosity of the base-course material. However, a prime coat should not be used as a substitute for proper preparation of the subgrade soil and the decision to use a prime coat is typically site-specific. If the subgrade consists of sandy material which may displace under the wheels of heavy vehicles, stabilization of the subgrade with a binder material should be done before paving to achieve the required load-bearing properties. In such cases, the use of a prime coat is not a guarantee that sandy soil will remain in place during paving operations.

If the asphalt layer is to be laid immediately on top of a new or existing untreated granular base layer, that base material should meet all the requirements for (i) moisture content, (ii) density, (iii) structural strength, and (iv) smoothness. If the base material is stable and dry and does not deflect and indent significantly under the wheels of a loaded truck, placement of the prime coat or the new asphalt mix is in order but if the condition of the granular material is unsatisfactory, the base course should be stabilized until it is in the correct condition for overlay.

11.3.2 Existing Asphalt Surfaces

The degree of preparation needed to overlay a fresh asphalt mix on an existing asphalt mix pavement depends on the condition of the pavement. As a preliminary operation: (i) potholes should be properly patched, (ii) cracks cleaned out and sealed, and (iii) ruts filled in or, preferably, removed by cold milling—a process of removing at least part of the surface of the paved area to level and smooth the surface to a full-depth removal.

Generally, covering failed areas with new overlay material is not recommended unless a very thick asphalt mix overlay is to be employed. In fact, asphalt mix and granular base materials that have failed should be excavated or cold milled and then either recycled or sent for disposal. In addition, any subgrade distortion should be repaired and the means for subsurface drainage should be installed as needed. Furthermore, new granular base course material, stabilized base course layers, or asphalt mix should be put in place in order to bring the strength of the pavement structure in each failed area to the same level of predicted performance as the surrounding pavement layers.

Cracked pavement sections, especially those with pattern cracking (e.g., map cracking—also called *large pattern cracking*, *alligator cracking*, or *crocodile cracking* which is characterized by interconnected or interlaced cracks in the wheel path, forming a series of small polygons), must be patched or replaced. If the cracks are narrow, it is doubtful that the crack-sealing material will actually enter the crack and may (preferably) pool on the pavement surface—such cracks should be widened before sealing is attempted.

Depending on the cause of the pavement cracking, the amount of *reflective cracking* that occurs in an overlay can often be reduced by the use of a surface treatment (seal coat) on the existing pavement. Briefly, reflective cracking is the occurrence of cracks in a flexible overlay over an existing crack or joint. The cracks occur directly over the underlying cracks or joints. Joint reflection cracking specifically refers to reflection cracks arising from underlying Portland cement concrete (PCC) pavement joint movement.

If the pavement contains a large (undefined) number of cracks, consideration should be given to applying a surface treatment instead of filling individual cracks. After cleaning the cracks, the surface treatment should consist of a single application of asphalt binder material (asphalt cement, cutback asphalt, or asphalt emulsion) and cover aggregate.

11.3.3 Portland Cement Concrete

When asphalt mix is placed over a PCC pavement, any severely distressed areas in the concrete slabs should be cut out, removed, and replaced with either PCC or asphalt mix. If necessary, corrective work should also be completed on the underlying sub-base or subgrade material. In addition, and depending on the condition of the PCC pavement, procedures such as re-seating or rubblizing of the existing pavement can be used before the overlay is placed, particularly if the slabs are showing signs of movement as a result of high traffic load.

The purpose of a tack coat is to ensure a bond between the existing pavement surface and the new asphalt overlay. The tack coat material is typically an asphalt emulsion but can also be asphalt cement or cut-back asphalt. Prior to application, the tack coat material should be heated to the proper temperature so that it is sufficiently fluid to be sprayed uniformly from the nozzles instead of coming out in strings. The amount of tack coat that is needed will depend on the condition of the pavement surface—an open-textured surface requires more tack coat than a surface that is tight or dense. Furthermore, it is essential to differentiate between the residual tack coat rate (the amount of asphalt cement remaining on the pavement surface after the water has evaporated) and the application rate (the amount of emulsion sprayed from the distributor).

When the tack coat is an asphalt emulsion it is brown in color because it contains both asphalt cement and water but when the emulsion breaks there is a color change from brown to black—indications that the water is evaporating. The rate of evaporation will depend on the type and grade of the emulsion used, the application rate, the temperature of the existing pavement surface, and ambient environmental conditions. Once all the water is gone, the emulsion is *set* and the rate of set depends on the same conditions that control the rate of break of the emulsion—type and grade of the emulsion used, the application rate, the temperature of the existing pavement surface, and the ambient environmental conditions. If asphalt cement is used as the tack coat material, it will cool to ambient temperature very quickly. Further, because there is no carrier material (water) to evaporate, paving may immediately follow the asphalt cement tack coat application.

11.4 MIX PLACEMENT

The primary goal in any road-paving project is to place the asphalt mix to the desired width, grade, cross slope, and thickness, and to produce a uniform mat texture for the roadway.

In the beginning of the twentieth century, hot asphalt mixes were spread manually into the roadway channel by hand (shovel and brush technique). As the century evolved, asphalt paving machines (mechanical spreaders) were introduced and, beginning in the late 1930s, these paving machines were provided with floating screeds for better leveling and pre-compaction of the asphalt paving mixture. The earliest paving machines used mechanical leveling methods and they were superseded by hydraulic devices and later by electronic leveling controls and vibratory screeds.

In modern paving machines, which incorporate the latest technology, trucks discharge the hot asphalt mix into a hopper on the paving machine. The material then is conveyed through the paving machine where it is spread across the width of the machine by an auger at the rear of the machine. As the auger distributes the material along the screed (an asphalt leveling device attached to the end of the paver), the paver continues to move forward, so that the screed keeps the paving mat level and smooth. The asphalt mix cools throughout this process (the rate of cooling is for the most part unknown so that actual temperature at which the mix is unloaded and then compacted is often subject to estimation or guesswork) and must be quickly compacted by a roller to the required pavement density and smoothness by one or more rollers following the paving machine. A paving crew typically consists of one or two paver operators, one or two screed operators, and two or three laborers with rakes and lutes but each roller has its own individual operator.

Briefly, the road roller is a piece of equipment primarily used in the construction to uniformly compact the asphalt mix into the bed. Although the road roller may seem like a modern piece of equipment, it has been used to asphalt laydown for centuries. Although it was first used under horse power, the horses were eventually replaced with steam power in 1860 and by the 1900s, the steam-powered roller was replaced by rollers that were powered by other sources, such as diesel.

Typically, a modern road roller is comprised of a roller body and at least one metal drum, but, sometimes, a road roller may have two or even three drums, depending on the project. Generally, a road roller works when the weight of the drum that is attached to the roller body causes the asphalt mix to compress and become compact. The weight of the drum may vary significantly, depending upon the project and the means of propulsion of the roller (human or mechanical). The drum of a manual road roller (powered by human effort) may only weigh 100 pounds whereas a mechanical roller used to construct roads and highways can weigh more than 25 tons (50,000 lbs) or more.

The structure of an asphalt roadway should be such that the roadway is sufficiently reliable to carry the traffic loads and to discharge these to the substructure or sub-base, in order to prevent harmful deformation—this applies to (i) deformation from load burden resulting from weather and (ii) deformation as a result of temperature stress. Moreover, while the binder–aggregate mix composition and the paving are important, the compaction of the mix is of utmost importance with respect to the quality of the road, which dictates the in-service performance and life of the road.

The simplified structure of an asphalt roadway involves (i) base course, (ii) asphalt binder course, and (iii) asphalt wearing course. Asphalt concrete, stone mastic asphalt, and porous asphalts are used for asphalt wearing courses and the laid mix must be compacted to such a degree that an increase in density or a reduction of porosity in the asphalt layer is achieved. In addition, the degree of compaction must be such that the required void content should be as designed (planned). This results in an enhanced stability of the various layers which, in turn, results in good compaction and higher resistance against deformation with a positive effect on the wear resistance of the wearing course. At the same time the compaction machine should produce a level asphalt layer as desired for driving comfort, and a wearing course with closed and smooth surface structure, but with maximum grip for vehicle tires. However, since the composition of asphalt mixes varies widely depending on the source of the binder and the aggregate and the temperature of the asphalt–aggregate mix (see Chapters 9 and 10), all of which confer different compaction properties on the mix, there are also the expected occurrences in the expected loads caused by traffic and weather.

On a molecular scale, the ability of the mix to undergo compaction also depends on the composition of minerals as well as the quality and viscosity of the asphalt and thus also on the temperature of the mix. Thus, asphalt–aggregate mixes for a roadway with a high traffic load have to be designed to exhibit high resistance to deformation. Such mixes are characterized by a bulky mineral skeleton structure due to high stone content, coarse grain, a high proportion of crushed particles in the sand range (crushed sand) and as well as relatively stiff binder—such mixes require extra effort to compact. On the other hand, an asphalt mix for a roadway with low traffic volume would typically contain a lower proportion of stones, a considerably high amount of natural sand, and soft bituminous mortar. Such mixes are relatively easy to compact but often when hot after laying or the use of heavy compaction equipment or an early start of compaction work which can cause early deformation of the laid mix can have adverse effects on the performance of such asphalt–aggregate mixes.

The temperature of the mix during compaction is of high significance for the compaction effort required. If the temperature of the mix is too low, compaction by the roller is supported by the low viscosity of the asphalt. The asphalt acts as a lubricant and reduces the internal friction in the mineral mix and due to the progressing stiffening of the asphalt caused by cooling, the compaction effort increases considerably under low temperatures. In one of the better mix laydown scenarios, compaction temperatures between 100°C and 140°C (212°F and 285°F) have been found to be suitable for conventional asphalt mixes. However, the temperature is affected by the rate of cooling and the distance of the transportation from the mix plant to the laydown site.

With a low pre-compaction by the finisher, the use of a light roller to initially press down the mix is an option—a heavy roller may adversely affect the evenness of the layer and, depending on the stability of the hot mix, may cause undesired displacements and misplacements in the material. High initial compaction has a favorably effect on the evenness of the layer and, at the same time, enables an early start of compaction, while the mix temperatures are sufficiently high. Because of the flexing effect of their wheels, pneumatic-tired rollers achieve homogeneous distribution of the mix and close the pores on the surface. The main areas of application are the pre-profiling and processing of easy-to-compact mixes and the finish of the surfaces of wearing and base courses or of light-traffic roads. Vibratory rollers are versatile and require considerably less passes than static rollers. The vibration reduces the internal friction in the mineral mix, so that the interaction between deadweight and dynamic load increases the density.

For new road construction, the asphalt base course (single or multilayer) is laid on the unbonded frost blanket layer (gravel or crushed rock layer) of the pavement. This course consists predominantly of a mixture of coarse particle aggregate with additives and asphalt. The base course has the function of distributing (or ability to distribute) traffic loads and should therefore be particularly stable as well as paved true to the design profile. Moreover, the base course should protect the substructure against moisture (rain and melted snow).

On roads with higher traffic loads the binder course is applied to the tack-coated base course as a single layer and serves not only to bond the wearing course to the base course but also serves as a compensating layer to minimize any evenness. Apart from this, the binder course also absorbs a significant portion of the shearing strain caused by traffic.

The wearing course is the top course of the asphalt pavement and, therefore, needs to be properly sealed and resistant to deformation. In addition, the surface of the wearing course must keep its form when subjected to traffic and should also be wear-resistant, weather-proof, and flexible. Generally, this course consists of a well-graded mineral mix with asphalt acting as binder and typical applications are roadways and motorways, airports, layers on bridges, agricultural roads, and cycle paths. Depending on the category of road and the expected traffic loads, different wearing courses can be used—for example, coarse mix with a higher layer thickness can be used for roads with higher traffic loads.

An aggregate consisting of a fine-particle mix with low air-void content and lower layer thickness is particularly suitable for a roadway with a low traffic load. Typical mix types used for asphalt binder courses must have very high and durable deformation resistance properties. Multiple crushed stones with durable edges are used for this purpose—the mix has a high proportion of coarse particles with crushed sand and shows a tendency for size segregation during haulage to the construction site.

Once the asphalt–aggregate is delivered, it is up to the paving machine to be operated and controlled so as to lay the mix according to the planned strategy. The paver consists of two primary parts: (i) the tractor unit and (ii) the screed unit. The *tractor unit*—typically equipped with either rubber tires or tracks—fulfills all of the functions necessary to receive the asphalt mix directly from the transportation trucks and then distribute it across the width of the *screed*. The tractor unit has the following components: (i) the truck push rollers, (ii) a material feed system consisting of a mix-receiving hopper, slat conveyors, material flow gates (usually), a pair of augers, and (iii) a screed that is pulled behind the tractor.

The push rollers, which are located on the front of the paver hopper, are used to maintain contact with the tires of the haul truck and to push that truck ahead of the paver. The rollers must be clean and free to rotate in order to allow smooth forward travel of the paver. If the push rollers are not cleaned on a regular basis and do not rotate freely, adverse effects will occur, such as uneven laying of the asphalt mix which will lead to poor performance and severe in-service problems.

11.4.1 Effect of Temperature

Temperature—including ambient (air) temperature, and mix temperature (aggregate and existing asphalt temperature), and hot mixed asphalt temperatures—are critical to obtaining compaction and longevity of newly paved surfaces and patches. Hot mixed asphalt is manufactured at temperatures between 135°C and 165°C (270°F and 325°F) but this temperature cannot always be maintained—depending on the environmental conditions and on the distance from the hot mix plant to the project, asphalt mix can lose between 3°C and 14°C (5°F and 25°F) and even more depending upon the rate of cooling, which is dependent upon the distance from the plant to the work site. At best, the asphalt mix arrives at the work site (at best estimates) at temperatures between 120°C and 150°C (250°F and 300°F) but there are no guarantees. If the air and base temperatures are colder than required or specified, then the asphalt pavement will cool much faster than the best estimates, which causes the mix to set making it very difficult to obtain the required or specified compacted density.

In order to assure a successful paving project, there are three basic steps involving temperature: (i) monitoring the ambient temperature, (ii) monitoring the base temperature, and (iii) monitoring the temperature of the asphalt mix. In addition, fore-knowledge of the expected high and low temperatures that are expected on the day of paving, as well as monitoring the *ambient temperature* during the work hours, is an important aspect of the project. The usual requirement is that the ambient temperature should be at least 10°C (50°F). If there is wind or precipitation, the asphalt mix will cool, thereby depriving the asphalt mix of the desired compaction.

The final step is checking the *temperature of the asphalt mix* prior to installing it on the base. This should be accomplished by taking the temperature of the pavement in the haul truck, at the front of the laydown machine, and behind the screed (after the laydown machine has passed). Also, wind will cool the asphalt mix very rapidly after it has been placed on the base so caution should be taken when paving on windy days and breakdown rolling will need to be adjusted for the effects of wind velocity.

11.4.2 Thickness of the Asphalt Layer

The screed unit, which is towed by the tractor unit, establishes the thickness of the asphalt layer and provides the initial texture to the new surface. In addition, through its weight and vibratory action, the screed imparts some level of density to the material being placed. The screed unit is attached to the tractor unit at only one point (tow point, pull point) on each side of the paver.

There are two primary effects which constantly act on the paver screed as the paver places the mix. The first effect is the towing force of the tractor, which varies as the speed of the paver increases and decreases. The second effect is the amount of force of the mix pushing against the screed. As the amount of asphalt–aggregate mix in the auger chamber increases and pushes against the screed, this changes the net force acting on the screed. As the forces acting on the screed change, the screed must come to a new angle of attack to compensate for the change in force acting on it. In addition, the line of pull—the angle at which the tractor pulls the screed forward—will influence the angle of the screed and, hence the thickness of the asphalt mix layer.

To ensure a good laying pattern of mix thickness, the forces on the screed must be in equilibrium, i.e., the sum of all the forces on the screed must equal zero and not preference one side or the other. If this equilibrium is achieved, the screed will remain at a constant angle as it is towed by the tractor unit. A change in any one force will cause the screed to rise or fall and the thickness of the asphalt mix layer will change accordingly. The screed will react to the change in the force against it until it reaches a new equilibrium elevation where the forces are in balance and there will be a constant thickness of the asphalt mat. Furthermore, in order to achieve the smoothest possible mat behind the paver screed, it is essential to keep the paver moving at a constant speed at all times (without stopping for the next truck-load of asphalt mix)—the trucks must be ready when required in order to maintain the operation at a steady pace in order to produce the equal-thickness asphalt mat. Another factor that affects the uniformity of the head of material in front of the screed is the temperature of the asphalt mix. If cold material is deposited in the paver hopper and carried back to the screed, the colder, stiffer mix increases the force acting on the screed and causes the screed to rise, increasing the thickness of the layer placed.

If, on the other hand, a hot load of mix is delivered to the paver, the decrease in viscosity of the binder material reduces the stiffness of the asphalt mix and reduces the force of the mix on the screed when the mix is deposited in front of it, which causes the screed to fall and reduces the layer thickness. With these forces in control and force-equivalency being the name of the game, operation of the paving machine at a constant speed without stopping will lay down a smooth uniform mix as dictated by the mix design for the project.

11.4.3 Effect of Grade

Grade sensors are used to monitor the elevation of the existing pavement surface in a longitudinal direction and there are three basic types of grade references that can be used to maintain the elevation of the screed tow points: (i) erected stringline, (ii) mobile reference or ski, and (iii) joint matching shoe. Each project may use different grade references.

The grade sensor is used to monitor the elevation of the tow points (in order to assure a uniform thickness of the asphalt mat) which will vary depending on the roughness of the existing pavement surface, so that the angle of attack of the screed will remain constant as the tractor unit follows the underlying pavement grade. The change in elevation of the tow points, however, should be smooth and should not move up and down rapidly or constantly as the paver travels forward.

11.5 JOINT CONSTRUCTION

During the construction of asphalt mix pavements, two main types of joint are encountered: (i) a transverse joint, which is constructed whenever the paving operation is interrupted for a period of time, and (ii) a longitudinal joint, which is built when a lane of asphalt mix is constructed adjacent to a previously placed lane of mix. There may be various versions of the two joints as well as combinations of the joints but for the purposes of this book, two types of joint are sufficient.

The method of construction of the *transverse joint* depends on whether or not traffic will be traveling over the asphalt mix before the paving is restarted. If traffic will not be passing over the end of the pavement, a vertical butt joint can be constructed—a butt joint is a joint in which the two parts of the asphalt roadway are pushed together. For a butt joint, a vertical face is constructed by hand across the width being paved. This operation consists of raking, shoveling, and then removing the mix that is located downstream of the selected joint location. The mix thus removed is discarded or returned to the plant to be recycled. The mix that is in place up-stream of the joint is not touched in any manner.

On the other hand, a *tapered joint* (also called a *ramp*) will be used. For this type of joint, it is essential that the paver operator keeps the head of material in front of the paver screed as consistent as possible up to the point at which the joint is to be built to ensure that the thickness of the mix being placed will be uniform up to the joint. This can be accomplished more easily with a tapered joint than with a butt joint because a portion of the asphalt mix left in the paver hopper and in front of the screed can be used to form the taper.

An advantage of the tapered joint is that the compaction equipment can run over the edge of the joint and down the taper without rounding the joint. Because the rollers can pass over the end of the mat easily, the compaction of the mix upstream of the joint is usually superior to that of the mix adjacent to a butt joint. A second advantage is that there generally is less mix to shovel from the joint because some of the extra mix is used to make the taper. However, there is a disadvantage to the taper joint insofar as the mix must eventually be removed before paving restarts on the downstream side of the transverse joint.

Thus, if a transverse butt joint is to be built, the paver operator should maintain a constant head of material in front of the screed to a point downstream of the location where the joint is to be constructed.

The *longitudinal joint* is typically the joint between two traffic lanes. Two key factors that affect the long-term durability of a longitudinal joint are built into the pavement during construction of the first lane: (i) the importance of running the paver in a straight line so the joint can be matched on the next pass of the paver, and (ii) the need to properly compact the unconfined edge of the first lane.

For the longitudinal joint, compaction of the longitudinal joint should be accomplished by rolling from the hot side of the layer with the roller wheels lapping approximately 6 inches over on to the cold mat. It is much more efficient to have the first pass of the roller extend over the joint for a short distance onto the cold side of the joint. One disadvantage of the longitudinal joint construction relates to an excessive amount of overlap of the paver screed over the previously placed mat. This problem may arise when the end of the screed is allowed to hang too far over the first lane. Alternatively, it may also arise because the edge of the joint was too uneven (rather than a nearly straight longitudinal edge) on the first paver pass. Since the additional asphalt mix cannot be pushed into the already compacted mat of the first lane, it is normally raked on to the new mat. If the longitudinal edge of the first lane is straight and if the correct amount of overlap is used, the joint can be successfully closed and the amount of raking required will be minimal.

11.6 PAVEMENT FAILURE AND REPAIR

Different types of asphalt concrete have different performance characteristics in terms of surface durability, tire wear, braking efficiency, and roadway noise. Roadway noise has only recently become an issue when environmental constraints against roadway noise were legislated and the constructions of high buffer walls to separate housing areas from highway lanes became the norm. Moreover, the determination of appropriate asphalt performance characteristics must take into account the volume of traffic in each vehicle category, and the performance requirements of the friction course—an asphalt concrete roadway typically generates less roadway noise than a PCC surface.

Factors that cause asphalt concrete to deteriorate over time mostly fall into one of three categories: (i) construction quality, (ii) environmental considerations, and (iii) traffic loads, but quite often the damage results from a combination of factors from all three categories. Asphalt deterioration can include crocodile cracking potholes, upheaval, raveling, bleeding, rutting, shoving, stripping, and grade depressions. In cold climates, frost heaves can crack asphalt even in one winter. Filling the cracks with asphalt is a temporary fix, but only proper compaction and drainage can slow this process. Thus, construction quality is critical to pavement performance and includes the construction of utility trenches and appurtenances that are placed in the pavement after construction. Lack of compaction in the surface of the asphalt, especially on the longitudinal joint, can considerably reduce the life of a pavement. Service trenches in pavements after construction have been said to reduce the life of the pavement by even more due to the lack of compaction in the trench, and also because of water intrusion through improperly sealed joints.

Environmental factors include (i) large climatic temperature variations, (ii) the presence of water in the sub-base or subgrade soil underlying the pavement, and (iii) pavement heaving due to frost. High temperatures soften the asphalt binder, allowing heavy tire loads to deform the pavement into ruts although, in many cases, high temperatures and strong sunlight also cause the asphalt to oxidize and/or increase the rate of oxidation, becoming stiffer and less resilient, leading to crack formation. Cold temperatures can cause cracks as the asphalt contracts and cold asphalt is also less resilient and more vulnerable to cracking.

Water trapped under the pavement softens the sub-base and subgrade, making the road more vulnerable to traffic loads. Any water under a roadway will freeze and expand in cold weather, causing and enlarging cracks in the pavement. During the spring thaw, ground typically thaws from the top down due to increased air temperature, so water is trapped between the pavement above and in the still-frozen soil underneath. This layer of saturated soil provides little support for the road above, leading to the formation of potholes, which is more of a problem for soil containing silt and clay. Some roadway jurisdictions have passed so-called *frost laws* to reduce the allowable weight of trucks during the spring thaw season in an effort to protect the roads.

Traffic damage mostly results from trucks and buses and the damage caused by a vehicle is proportional to the axle load raised to the fourth power—doubling the weight an axle carries actually causes 16 times as much damage to the roadway. Wheels cause the road to flex slightly, resulting in fatigue cracking, which often leads to crocodile cracking (Huang et al., 2004). Vehicle speed also plays a significant role in roadway damage and slow-moving vehicles stress the road over a longer period of time, increasing ruts, cracking, and corrugations in the asphalt roadway. Other causes of damage include heat damage from vehicle fires, or solvent action from chemical spills.

11.6.1 Pavement Failure

Pavement failure (also called *pavement distress*) can be the result of traffic loadings, environmental factors, poor drainage and/or soils, deficient materials, poor construction practices, or external causes such as a utility cut. However, rather than one cause, most pavement failures can be ascribed to a combination of these factors. For example, poor soil may not be a problem unless there is also heavy traffic and moisture.

For general purposes, asphalt pavement failure is typically divided into four classes: (i) cracking, (ii) distortion, (iii) disintegration, and (iv) surface defects. Furthermore, within each class there are several subcategories, each with different causes.

11.6.1.1 Cracking

There are several types of cracks and the selection for roadway maintenance and/or rehabilitation treatment is dependent upon the cause and extent of the pavement damage (often referred to as *pavement distress*) and for effective maintenance and repair, is necessary to be able to recognize the type of crack and consequently the cause of the crack. Upon repeated loading and stress, longitudinal cracks will connect forming patterns similar to those on the back of alligator cracks.

Generally, all cracks greater than 0.25 inch (6.3 mm) in width should be sealed to prevent the infiltration of water and incompressible materials (such as rocks and sand) into the crack and the underlying pavement structure. Water entering the base soil through a crack can cause significant weakening of the base and, if left untreated, will lead to the formation of additional cracks and eventually serious pavement failure. The different types of cracks are: (i) reflective cracks, (ii) fatigue cracks, also called alligator cracks, (iii) thermal cracks, (iv) block cracks, (v) linear cracks, (vi) edge cracks, and (vii) slipping cracks.

Reflective cracks form in an asphalt overlay directly over cracks in the underlying pavement. The pattern reflects the original crack pattern and can be longitudinal, transverse, diagonal or block. Horizontal expansion or contraction movement due to temperature or moisture changes in the original pavement causes reflective cracking. They also can be the result of vertical movements due to traffic loads. They are especially prominent in asphalt overlays on PCC pavements. Reflective cracks can appear very quickly. The rule of thumb is 1–2 years per inch of overlay.

Fatigue cracks (alligator cracks, also called *crocodile cracks*) are a series of interconnected cracks that look like the patterns on the skin of an alligator or the pattern projected by chicken wire. The cracks are caused by repeated large deflections (or deformations) of the asphalt pavement under heavy traffic. The bending and unbending of the pavement under these heavy loads cause the pavement to crack due to fatigue. Generally, alligator cracked areas, due to a poor base, are not large. When associated with repeated heavy loads, however, they can cover entire pavement sections. With thin asphalt pavements (less than 4 inches (100 mm) thick), areas that have suffered alligator cracks can quickly graduate to potholes. Depending on the dimension of the pavement, the cracks initiate where the tensile stress is highest and extends to form longitudinal cracks. Hence, in thin pavements cracks begin from the bottom and extend to the surface giving rise to bottom-up cracking, while top-down cracking is predominant in thick pavements.

This type of crack pattern indicates one or more of the following: (i) the asphalt pavement was too thin, (ii) the pavement base or subgrade materials were of poor quality or saturated with water, (iii) the traffic loads were too much for the pavement design, (iv) the asphalt in the mix has become oxidized and brittle, and (v) there is high moisture content and poor drainage beneath the pavement. Careful selection of the asphalt mix with an appropriate mix design can significantly reduce fatigue cracking during in-service life.

Like rutting, fatigue cracking results from the large number of vehicle loads applied over time to a pavement subject to heavy traffic flow. However, fatigue cracking tends to occur when the pavement is at moderate temperatures, rather than at the high temperatures that cause rutting. Because the hot mix asphalt at moderate temperatures is stiffer and more brittle than at high temperatures, it tends to crack under repeated loading rather than deform. When cracks first form in a hot mix asphalt pavement, they are so small that they cannot be seen without a microscope. The cracks at this point will also not be continuous. Under the action of traffic loading, these microscopic cracks will slowly grow in size and number, until they grow together into much larger cracks that can be clearly seen with the naked eye. These large cracks will significantly affect pavement performance, by weakening the pavement, contributing to a rough riding surface, and allowing air and water into the pavement, which will cause additional damage to the pavement structure. Eventually fatigue cracking can lead to extensive areas of cracking, large potholes, and then total pavement failure.

Traditionally, it was believed that fatigue cracks first formed on the underside of the hot mix asphalt layers, and gradually grew toward the pavement surface. It has become clear during the past two decades that pavements are also subject to top-down fatigue cracking in which the cracks begin at or near the pavement surface and grow downward, typically along the edges of the wheel paths. It is likely that most asphalt mix pavements undergo both bottom-up and top-down fatigue cracking. However, as asphalt mix pavements have become thicker and as asphalt mix overlays on top of PCC pavements have become more common, top-down cracking has become more commonly observed than bottom-up cracking.

Although fatigue cracking in hot mix asphalt pavements is still not completely understood, it is generally agreed that there are several ways that the composition of the mix can affect fatigue resistance in asphalt mix pavements. One of the most important factors affecting fatigue resistance is asphalt binder content. For example, an asphalt mix with a low asphalt content tends to be less fatigue-resistant than an asphalt mix with richer asphalt content. In addition, poor mix compaction also contributes significantly to surface cracking by reducing the strength of the pavement surface. A high air-void content in the mix (often the result of poor compaction) will also increase pavement permeability, which will then allow air and water into the pavement—both of these effects can cause damage to the asphaltic pavement and increase the rate of fatigue cracking.

Thermal cracks or *low-temperature cracks* are transverse cracks perpendicular to the road centerline, and are generally equally spaced along the roadway. These types of cracks result from low-temperature contraction of the pavement, especially where pavement contraction results in a build-up of tensile stress, and are usually initiated at the surface, and grow downward into the pavement. As the pavement contracts with decreasing temperature, the asphalt binder is unable to relieve the associated stresses and the pavement cracks. Temperature cycling will, more than likely, accelerate this problem.

Temperature has an extreme effect on the properties and behavior of asphalt binders (see Chapter 4). At temperatures on the order of 150°C (300°F), an asphalt binder is typically a fluid (unless it is an extremely hard binder) that can be conveniently pumped through pipes and mixed with hot aggregate. At temperatures of approximately 25°C (77°F), an asphalt binder typically has the consistency of stiff putty or a soft rubber. On the other hand, at temperatures of approximately −20°C (−4°F) and lower, an asphalt binder can become very brittle. As a result, hot mix asphalt pavements in many northern regions of the United States and most of Canada will be stiff and brittle during the winter. When cold fronts move through an area causing rapid drops in temperature, the rate of cooling increases significantly and asphalt mix pavements can suffer accordingly. Like most materials, the asphalt mix tends to contract as it cools and unlike PCC pavements, flexible asphalt pavements have no contraction joints and the entire pavement surface will develop tensile stresses during rapid drops in temperature during cold weather. The resulting tensile stresses in the asphalt pavement can cause cracks to appear and these low-temperature cracks will stretch transversely across part or the entire pavement, their spacing can range for approximately 10 feet to 40 feet (3–10 meters). Although low-temperature cracking may not at first cause a significant problem in a pavement, the cracks tend to become more numerous and wider with time, thereby causing a significant performance problem after several years.

Low-temperature cracking of an asphalt pavement is attributed to tensile strain induced in the asphalt mix as the temperature drops to some critically low level. As the name indicates, low-temperature cracking is a distress type of cracking that is caused by low pavement temperatures rather than by applied traffic loads, although it is entirely likely that traffic loads do play a role. Low-temperature cracks form when an asphalt pavement layer shrinks in cold weather and, as the pavement shrinks, tensile stresses build within the asphalt mix layer. At some point along the pavement, the tensile stress exceeds the tensile strength and the asphalt layer cracks. Thus, low-temperature cracks often occur from a single stress event cause by the low temperature of the pavement.

Low-temperature cracking can also be a fatigue phenomenon resulting from the cumulative effect of many cycles of cold weather—the magnitude and frequency of low temperatures and stiffness of the asphalt mixture on the surface are major factors in the occurrence and intensity of low-temperature transverse cracking. The crack starts at the surface and works its way downward and the stiffness of the asphalt mix, which is primarily related to the properties of the asphalt binder, is probably the greatest contributor to low-temperature cracking. Low-temperature cracking in an asphalt mix pavement can be minimized or even eliminated by careful selection of the asphalt binder. In fact, one of the main premises behind the development of the current system for grading asphalt binders was to assist in the prevention of the low-temperature cracking phenomenon.

While aggregate is the major contributor to pavement strength, the resistance to thermal cracking is almost wholly determined by the properties of the asphalt and is more common when hard asphalt is used. Also, extreme aging (oxidation), as a result of high air-void content, can contribute to low-temperature cracking. Thermal cracking is predominantly a cold-region problem, but it also occurs in southern climates where harder asphalt (that will crack at higher temperatures) is typically used to combat hot-weather pavement problems such as rutting. The change to asphalt selection by the performance-graded (PG) system should help to minimize this mode of pavement failure.

In *block cracks*, the pavement divides into rectangular pieces. The pieces range in size from approximately one-foot square to ten-feet square. It is more common on low-volume roads and on large paved areas such as parking lots. Because these pavements typically are not as well compacted as major roads, a higher number of air voids are a suspected cause. Typically these cracks are caused by the same factors as thermal cracking. Block cracking is not load-related—on some occasions, longitudinal and transverse cracks merge to appear as block cracks, which can be load-related.

Linear cracks are caused by many factors. The longitudinal paving joint crack is the most common and can be found on all classes of roads and is caused by the inability to fully compact the unconfined edge of a paving pass. The resulting low-density mix has insufficient strength to prevent cracking and will allow the asphalt to prematurely oxidize, harden, and eventually ravel. Another longitudinal crack frequently occurs between the wheel-paths and is attributed to a certain popular brand of paver. Segregation occurs where the two paver slat conveyors drop the hot mix asphalt in front of the paver augers. The segregation usually does not show up in the mat surface behind the screed but the crack will eventually appear in time.

Longitudinal cracks in the wheel-path are usually load-related and can lead to alligator cracking. Another source of longitudinal cracks is frost heaves in the base or subgrade. Localized heaving pushes the pavement up from below, and spring thaw provides a weakened area. These cracks often appear in the wheel-paths, or immediately adjacent to them.

Edge cracks are longitudinal cracks away from the pavement edge with or without transverse cracks branching out to the shoulder. Lack of lateral or shoulder support is the most common cause. Settlement of the base soils, poor drainage, and frost heaves are other causes. Edge cracks combined with a buildup of winter sand deposits at the pavement edge can make the problem much worse. The sand buildup prevents water from running off the shoulder so that it must drain through the cracks and further weakens the underlying soils.

Slippage cracks are crescent-shaped cracks resulting from traffic-induced horizontal forces and they are caused by a lack of a bond between the surface layer and the course below. This lack of bond may be due to dust, dirt, oil, or the absence of a tack coat prior to placing the surface course.

Thus, fatigue cracking, thermal cracking, rutting, and moisture-induced damage are the most prevalent forms of distress in asphalt pavements. To circumvent such problems, it is necessary to: (i) understand the mechanisms of the aforementioned distresses, (ii) accurately quantify and predict the resistance of materials to these distresses from a design standpoint, and (iii) engineer material modification and production techniques that will result in an overall longer serviceable life of asphalt pavements.

In practice, however, test specimens used to evaluate the fatigue-cracking resistance of asphalt mixtures or mortars have only been through a short-term aging program prior to compaction. As a result, it is preferable that asphalt–aggregate mixes that have been through a long-term aging program should be examined. The extent and uniformity of aging in compacted samples when subjected to such limited aging times can be questioned. Therefore, a relevant question that arises related to whether or not the relative fatigue cracking resistance of different mixtures (as determined only after short-term aging) changes significantly after long-term aging and gives meaningful data that can be equated to mix performance.

11.6.1.2 Distortion

Pavement *distortion* is the result of asphalt mix instability, movement, or weakness in the granular base or subgrade and, furthermore, pavement cracking may accompany some types of distortion. Pavement distortion can take many forms, but rutting and shoving are the most common.

Rutting (the formation of channelized depressions, *ruts*) occurs in the wheel tracks of a pavement surface. Rutting is a serious problem because the ruts contribute to a rough riding surface and can fill with water during rain or snow events, which can then cause vehicles traveling on the road to aquaplane and lose control.

Rutting (often referred to as permanent deformation) is a common form of distress in flexible pavements and is caused when (loaded or heavy) truck tires move across an asphalt concrete pavement, the pavement deflects a very small amount. These deflections range from much less than a tenth of a millimeter in cold weather—when the pavement and sub-grade are very stiff—to a millimeter or more in warm weather—when the pavement surface is hot and very soft. If the elastic modulus of the mix is sufficient, the pavement tends to spring back to its original position after the truck tire passes over a given spot in the pavement,. Often, however, the pavement surface will not completely recover leaving a very small (but not insignificant) amount of permanent deformation of the pavement in the wheel path. Accordingly, after many wheel loads have passed over the pavement—the number of wheel loads will vary with the quality of the pavement—the rutting can become significant leading to severely rutted pavements with 0.75 inch or more (20mm or more) in depth. Rut depths of about 10mm or more are usually considered excessive and a significant safety hazard.

Rutting of the pavement subgrade is caused by consolidation or displacement of materials due to repeated traffic loading. In severe cases, uplift on the outside of the rut may occur. This failure mode occurs in the base or sub-base as a result of excessive stresses, moisture infiltration, or a basic design failure. Hence, it could appear as a depression in the wheel path or as uplift along the sides of the rut. Rutting normally occurs when a pavement is newly constructed and becomes minimal as the asphalt binder hardens and becomes aged. Rutting is also caused by trucks carrying heavy loads on tires with greatly increased inflation pressures on pavements not designed to withstand that kind of stress. Those mixes had too high an asphalt content, too high fines content, round and smooth textured aggregates, and too soft asphalt. Improvements in mix design procedures, aggregate specifications and testing, and PG binders have significantly reduced rutting problems.

Finally, rutting is more common in summer when elevated temperatures soften the asphalt cement. On cold mix asphalt (CMA) pavements, lack of aeration or curing of the emulsion can result in a tender mix that will rut. This is especially a problem when cold mix paving is constructed late in the year. Other related forms of permanent deformation include shoving and wash boarding.

Shoving (*corrugations*, *wash-boarding*) is a form of plastic deformation that results in ripples across the pavement surface. They typically occur when there is severe horizontal stress where traffic starts and stops, on downgrades where vehicles brake, at intersections, and on sharp horizontal curves. Corrugations are usually caused by too much asphalt or using a very soft asphalt in the mix.

Shoving typically occurs at intersections when vehicles stop, exerting a lateral force on the surface of the hot mix causing it to deform excessively across the pavement, rather than within the wheel ruts. Wash boarding is a similar phenomenon but, in this case, the deformation takes the form of a series of large ripples across the pavement surface. Rutting, shoving, and wash boarding can be the result of permanent deformation in any part of the pavement—the subgrade, the granular sub-base, or any of the bound layers. Excessive permanent deformation in one or more of the bound layers is the result of an asphalt concrete mixture that lacks strength and stiffness at high temperatures. Several problems with a mix design, such as selecting an asphalt binder that is too soft for the given climate and traffic level, can make it prone to rutting and other forms of permanent deformation.

Disintegration is the breakup of a pavement that begins with the loss of the fine aggregate particles from the pavement surface and progresses to the formation of potholes. It is the progressive downward distress into the lower layers of pavement resulting in large chunks of unbound fragments. This failure can be a result of fatigue loading as huge potholes are formed on pavement as the fragments are dislodged by traffic. The various types of pavement disintegration are raveling/weathering, delamination, stripping, and potholes.

Raveling (*weathering*) is the progressive loss of aggregate from the pavement surface. Raveling is one of the complications that result from stripping. It can be described as the progressive loss of surface material by weathering or surface abrasion. It starts when fine aggregates become detached from the asphalt cement leaving small rough patches on the pavement surface, which increase as larger aggregate particles are dislodged from the pavement surface. As it progresses, larger aggregate particles come loose because of the lack of support from surrounding fines. Raveling is accelerated in the wheel-paths by traffic. Weathering occurs over the entire pavement surface including the non-traffic areas. Both water and traffic are usually needed to cause extensive raveling. Raveling is caused by high air voids in hot mix asphalt due to poor compaction or late season paving. In addition, a lean asphalt mix (too little asphalt) or overheating of the asphalt in the mix asphalt plant (leading to disruption of an aggregate-related binder structuring) can also result in raveling. Raveling also occurs when tires dislodge aggregate particles from the surface of a hot mix asphalt pavement. Many of the same factors that contribute to poor fatigue resistance will also contribute to raveling, including low asphalt binder contents and poor field compaction. Because the pavement surface is exposed to water from rain and snow, poor moisture resistance can also accelerate raveling in hot mix pavements.

Stripping is the gradual loss of adhesion between the asphalt film and the aggregate surface, resulting in loss of integrity of the hot asphalt mix. Therefore, the pavement becomes susceptible to various forms of failure by losing its structural stiffness. Factors that contribute to stripping include; water on pavement with high traffic load, high temperature, nature of aggregates and binder, and poor compaction. Stripping can lead to other forms of distress such as cracking and rutting.

Water does not flow easily through asphalt mix pavements that have been constructed carefully and efficiently but it will flow very slowly even through well-compacted material. Water can work its way between the aggregate surfaces and asphalt binder in a mixture, weakening or even totally destroying the bond between these two materials (stripping). Moisture damage can occur quickly when water is present underneath a pavement, as when pavements are built over poorly drained areas and are not properly designed or constructed to remove water from the pavement structure. In fact, occasional exposure to water can cause moisture damage in asphalt mixes because of faulty design or faulty construction or use of poor materials.

The physicochemical processes that control moisture damage are complex insofar as different combinations of asphalt binder and aggregate will exhibit widely varying degrees of resistance to moisture damage. In fact, it is difficult to predict the moisture resistance of a particular combination of asphalt and aggregate, although asphalt mix produced with aggregates containing a high proportion of silica (such as sandstone, quartzite, chert, and some types of granite) tend to be more susceptible to moisture damage. Proper construction, especially thorough compaction, can help reduce the permeability of hot mix asphalt pavements and, thus, significantly reduce (even mitigate) the likelihood of moisture damage. Anti-stripping additives can be added to asphalt mixes to improve the moisture resistance—hydrated lime [$CaO \cdot H_2O$, $Ca(OH)_2$] is one of the most common and most effective of such additives.

The moisture resistance of hot mix asphalt mixtures can be evaluated (often referred to as the Lottman procedure). In this test, six cylindrical hot mix asphalt specimens are compacted in the laboratory. Three of these are subjected to conditioning—vacuum saturation, freezing, and thawing—while the other three are not conditioned. Both sets of specimens are then tested using the indirect tension test. The percentage of strength retained after conditioning is called the tensile strength ratio and is an indication of the moisture resistance of that particular mixture. Many highway agencies require a minimum tensile strength ratio of 70–80% for hot mix asphalt mixtures but it must be remembered that the data from this test method may not always be 100% reliable and may only provide an approximate indication of the resistance of the mix to moisture incursion and damage.

Delamination is the localized loss of the entire thickness of an overlay and is caused by the lack of a bond between the overlay and the original pavement. Again, water is the causative agent, specifically when the water incursion occurs between two layers of the pavement. However, delamination is usually confined to the wheel-path area and takes several years after the overlay to become a serious problem, but once it occurs the roadway is difficult to patch. Cleaning the old surface and applying an asphalt emulsion as the tack coat (a thin layer of asphalt used in the construction or refacing of roads and highways) will help to alleviate the problem and is especially helpful when the overlay thickness is two inches (50 mm) or less.

Potholes are bowl-shaped holes of various sizes in a pavement resulting from localized disintegration under traffic. They cause faulty wheel alignment and can start with a small crack that lets water in and weakens the road's base, or a small area of raveling that goes full depth, or a whole bunch of potholes can occur overnight in an alligator-cracked area of a thin pavement. Poor soils, poor drainage, too thin an asphalt surface, poor compaction, and poor pavement maintenance can all lead to the formation of potholes.

11.6.1.3 Surface Defects

Bleeding asphalt or *flushed asphalt* is caused by the upward movement of asphalt to the pavement surface. This results in a pavement with a smooth, black, shiny appearance that is sticky in hot, sunny weather. The most common causes of bleeding are the loss of stone cover in a chip seal and the over-compaction of a tender hot mix asphalt mix by heavy traffic. Overly rich asphalt mixes and application of a very heavy tack coat or excessive crack seal materials are other possible causes. A bleeding surface is very smooth and almost as slippery as ice when wet. Since most bleeding occurs in the wheel-paths, it is a serious safety hazard that must be dealt with immediately.

Polished aggregate is the result of use of a soft aggregate (such as limestone) included in an asphalt surface mix, which becomes polished (smoothed) by traffic and the aggregate loses its ability to squeeze out the water, and the pavement can become slippery. Pavement skid resistance is dependent on contact between the vehicle tires and the aggregate in the pavement surface. Clean, dry pavements have high friction values but a small amount of rainwater can dramatically change the vehicle stopping distance and ability to negotiate a curve. To develop adequate friction, a tire must squeeze the water out from between its tread and the stone in the pavement surface. Many specifications require incorporation of non-polishing high-friction aggregates, such as granite and trap rock (also known as *trap*, which is any dark-colored, fine-grained, non-granitic intrusive or extrusive igneous rock) in all surface course mixes.

11.6.2 Pavement Repair

The life of a road can be prolonged through good design, construction, and maintenance practices. During design, engineers measure the traffic on a road, paying special attention to the number and types of trucks. They also evaluate the subsoil to determine how much load it can withstand. The pavement and sub-base thicknesses are designed to withstand the wheel loads. Sometimes, geogrids are used to reinforce the sub-base and further strengthen the roads. Drainage, including ditches, storm drains, and underdrains are used to remove water from the roadbed, preventing it from weakening the sub-base and subsoil.

Good maintenance practices center on keeping water out of the pavement, sub-base, and subsoil. Maintaining and cleaning ditches and storm drains will extend the life of the road at low cost. Sealing small cracks with bituminous crack sealer prevents water from enlarging cracks through frost weathering, or percolating down to the sub-base and softening it. For somewhat more distressed roads, a chip seal, or similar surface treatment, may be applied. As the number, width and length of cracks increase, more intensive repairs are needed. In order of generally increasing expense, these include thin asphalt overlays, multicourse overlays, grinding off the top course and overlaying, in-place recycling, or full-depth reconstruction of the roadway.

Generally, there are several treatments that can be applied to asphalt pavements that are showing signs of distress. Treatments can be divided into four general categories: (i) preventive maintenance, (ii) corrective maintenance, (iii) pavement rehabilitation, and (iv) pavement reconstruction.

Preventive maintenance may correct a minor defect, but is primarily intended to prevent further deterioration and extend the life of a pavement. Microsurfacing, surface treatments, and crack sealing are examples of preventative maintenance treatment which may improve riding comfort but do not add structural strength to the road.

Corrective maintenance treatments correct existing defects and deficiencies until more permanent and extensive treatments are necessary. Single course 25–37.5 mm (1 to 1½″) overlays, hot in–place recycling 25 mm (<1″), and "mill and fill" overlays are examples.

Pavement rehabilitation treatments are intended to address serious pavement failures and restore the roadway to a condition that will ensure a service life of 10–20 years with normal maintenance. Pavement rehabilitation treatments include two-course overlay, hot in-place recycling with an overlay, cold in-place recycling with an overlay, and full–depth reclamation.

Pavement reconstruction treatment essentially invokes: (i) digging up the pavement, (ii) disposal of the excavated material, and (iii) commencement of pavement rebuilding, assuming that the existing pavement is in such a condition that it can no longer serve a useful purpose, and a life-cycle cost analysis shows that other treatments are not cost-effective.

On the other hand, even if preventative maintenance measures have been applied, road repair typical requires more drastic measures, such as the use of recycled asphalt in various forms (Nourendin and Wood, 1989; Ali et al., 1995; Button et al., 1995; Kennedy et al., 1998a,b; Watson et al., 1998; Sengoz and Topal, 2004; Shen et al., 2007).

11.6.2.1 Recycling

Asphalt concrete is often touted as being 100% reclaimable or recyclable and, as a result, very little asphalt concrete is actually sent for disposal in landfills. Asphalt recycling on a large scale (in-place asphalt recycling or asphalt recycling performed at a hot mix plant) and asphalt recycling on a smaller scale are used. For small-scale asphalt recycling, the user separates asphalt material into three different categories: (i) blacktop cookies, (ii) reclaimed asphalt pavement, and (iii) asphalt millings.

11.6.2.2 Blacktop Cookies

In this technology, pieces of uncompacted virgin hot mix asphalt are used for pothole repair. The use of blacktop cookies has been investigated as a less-expensive, less-labor-intensive, more durable alternative to repairing potholes with cold patch. New hot mix asphalt wafers can be spread on the ground and, once cooled, the wafers could be stored until reheated in a hotbox to make minor road repairs. Blacktop cookies may also be produced from leftover material from paving jobs.

11.6.2.3 Reclaimed Asphalt Pavement

Reclaiming (recycling) asphalt pavement became popular in the 1970s due to the high cost of crude oil during the Arab oil embargo. Even with the volatile fluctuation in oil process, this concept has continued. While several factors influence the use of reclaimed asphalt pavement in asphalt pavement, the two primary factors are economic savings and environmental benefits. Reclaimed asphalt pavement is a useful alternative to virgin materials because it reduces the use of virgin aggregate and the amount of virgin asphalt binder required in the production of hot mix asphalt. The use of reclaimed asphalt pavement also conserves energy, lowers transportation costs required to obtain quality virgin aggregate, and preserves resources. Additionally, using reclaimed asphalt pavement decreases the amount of construction debris placed into landfills and does not deplete non-renewable natural resources such as virgin aggregate and asphalt binder. Ultimately, recycling asphalt creates a cycle that optimizes the use of natural resources and sustains the asphalt pavement industry. However, in order for the concept to be successful, RAP must perform well and be environmentally sound. To ensure that these requirements are met, it is important that (i) the use of recycled material in the construction of highways is to the maximum economical and practical extent possible with equal or improved performance, and (ii) the use of reclaimed asphalt pavement in hot mix asphalt is advantageous because reclaimed asphalt pavement can have a significant environmental and engineering impact in pavement recycling (Mohammad, 2006; Copeland, 2011).

In this technology, pieces of asphalt that have been removed from a road, parking lot, or driveway are considered reclaimed asphalt pavement material. Typically, these pieces of asphalt are ripped up when making a routine asphalt repair, manhole repair, catch basin repair, or sewer main repair. Because the asphalt has been compacted, reclaimed asphalt pavement material is a denser asphalt material and typically takes longer to recycle than blacktop cookies.

11.6.2.4 Asphalt Millings

Small pieces of asphalt produced by mechanically grinding asphalt surfaces are referred to as asphalt millings. Large millings that have a rich, black tint, indicating a high asphalt cement content, are best for asphalt recycling purposes. Surface millings are recommended over full-depth millings when choosing asphalt millings to recycle. Full-depth millings usually contain sub-base contaminants such as gravel, mud, and sand. These sub-base contaminants will leach oil away from original asphalt and dry out the material in the recycling process. Asphalt milled from asphalt is better than asphalt milled from concrete. When milling asphalt from concrete the dust that is created is not compatible with asphalt products because it is not asphalt.

Small-scale asphalt recycling will usually involve high-speed on-site asphalt recycling equipment or overnight soft heat asphalt recycling. Small-scale asphalt recycling is used when wanting to make smaller road repairs versus large-scale asphalt recycling which is done for making new asphalt or for tearing up old asphalt and simultaneously recycling/replacing existing asphalt. Recycled asphalt is very effective for pothole and utility cut repairs. The recycled asphalt will generally last as long or even longer than the road around since new asphalt cement has been added back to the material.

For larger-scale asphalt recycling, several in-place recycling techniques have been developed to rejuvenate oxidized binders and remove cracking, although the recycled material is generally not very water-tight or smooth and should be overlaid with a new layer of asphalt concrete. Cold in-place recycling mills off the top layers of asphalt concrete and mixes the resulting loose millings with asphalt emulsion. The mixture is then placed back down on the roadway and compacted. The water in the emulsion is allowed to evaporate for a week or so, and new hot mix asphalt is laid on top.

Asphalt concrete that is removed from a pavement is usually stockpiled for later use as aggregate for new hot mix asphalt at an asphalt plant. This reclaimed material (reclaimed asphalt pavement) is crushed to a consistent gradation and added to the hot mix asphalt production process. Sometimes waste materials, such as asphalt roofing shingles, crushed glass, or rubber from old tires, are added to asphalt concrete as is the case with rubberized asphalt but there is a concern that the hybrid material may not be readily recyclable.

The use of recycled rubber particles in asphalt binder applications is an environmentally efficient valorization of the reclaimed material. Unlike using tire rubber as fuel for cement kilns, asphalt rubber binder takes advantage of the exceptional elastic properties of rubber while allowing further recycling of the asphalt mixture at the end of the lifespan. Asphalt rubber production is assumed to be essentially a physical interaction between the asphalt and the rubber particles. The main change in the asphalt–rubber composite system is the swelling of the rubber due to the migration and interaction of constituents of the asphalt resulting in the formation of a cross-linked asphalt–rubber molecular network.

The swelling of rubber goes hand in hand with an increase in viscosity which is a major concern for satisfactory pumping and mixing conditions. Heating of asphalt rubber during the curing process induces some breaks in rubber crosslinks referenced by some authors as devulcanization. This phenomenon induces a progressive decrease in viscosity which is considered in some papers as the initial degradation of the polymer chains. Both swelling and devulcanization are present during the asphalt rubber curing, even for the so-called dry process (rubber is introduced as a filler in the asphalt mixing plant). The predominance of one of the phenomena relies on the processing parameters and essentially on blending temperature and blending duration.

The development of an effective asphalt rubber binder is mainly an experimental iterative process. Assuming a predetermined choice of components, a large number of parameters (contents, temperatures, blending duration) are investigated to select the optimum combination of these factors. Any modification of the previous parameters alters the final product properties and imposes an experimental correction. The main limitation in investigations of the composite asphalt–rubber medium is the complexity of a direct investigation of the components or phases at the end of the curing process. Rheological measurements are carried out on the base components as well as on the final blend but rarely on the effective final constituting phases.

11.6.2.5 Methods to Mitigate Stripping

Although include in this section on pavement repair, the following is a presentation of the preventative actions that can be taken in order to limit or mitigate stripping of the asphalt from the aggregate.

Anti-stripping additives have been developed to address the issues of poor pavement performance and high maintenance costs experienced by moisture-susceptible pavements. These additives, whether solid or liquid, are used to promote adhesion of asphalt cement onto the aggregate surface. For example, water-sensitive asphalt mixes, or mixes to be used in very wet locales, can greatly benefit from the use of anti-stripping agents. The most commonly used anti-stripping agents are lime additives and liquid additives.

Lime (and the associated derivative) has been shown to be very effective at preventing stripping in asphalt mixes due to the chemical reaction that occurs when lime interacts with silicate aggregates to form a layer of calcium around the aggregate. This calcium layer bonds to the aggregate well, while still maintaining a sufficient level of porosity to allow asphalt cement to penetrate the voids in the aggregate. Both hydrated lime [$Ca(OH)_2$] and quick lime (CaO) are effective, although the former is most commonly used. Dolomitic lime has also been used as an anti-stripping additive but the carbonate ($CaCO_3$) lime is not as effective. In the process, hydrated lime (a dry powder) is added to the aggregate before mixing with the asphalt cement. Hydrated lime can also be added to undried aggregate with water content from 3–5% (w/w).

Most *liquid anti-stripping agents* are of the surface active type and reduce surface tension of any existing water on the aggregate and promote better adhesion of the asphalt cement to the aggregate. The simplest method of applying the liquid anti-stripping additive is to mix it directly with the asphalt cement immediately before its application onto the aggregate. Another method is to treat the aggregate with the liquid anti-stripping additive before the addition of the asphalt cement.

However, taking the correct steps in *aggregate preparation* improves the ability of the aggregate to absorb and bond with asphalt cement. The asphalt binder is hydrophilic with weaker bonds existing between the aggregate and binder, thus when moisture is present and the asphalt mix is loaded repeatedly, the asphalt binder strips from the aggregate, resulting in a loss of adhesion. Siliceous aggregate sources are prone to stripping due to a high silica dioxide component. Stripping is the weakening or eventual loss of the adhesive bond between the aggregate surface and asphalt binder in the asphalt mix due to the presence of moisture.

The most important of the preparation processes involves heating of the aggregate to remove any excess moisture. Other commonly taken steps are to wash the aggregate to remove very fine dust, or to process the aggregate in a crusher. In addition, proper aggregate selection can play an important role in the ability of the binder to adhere to the aggregate.

REFERENCES

Ali, N., Chan, J.S., Potion, A., Bushman, R., Bergen, A, 1995. Mechanistic evaluation of asphalt concrete mixtures containing reclaimed roofing materials Proceedings. 74th Annual Meeting ofTransportation Research Board. Technical University of Nova Scotia and University of Saskatchewan, Canada.

Button, J.W., Williams, D., Scherocman, J.A., 1995. Shingles and Toner in Asphalt Pavements. FHWA Report No. FHWA/TX-96/1344-2F. Texas Transportation Institute, College Station, TX.

Copeland, A., 2011. Reclaimed Asphalt Pavement in Asphalt Mixtures: State of the Practice. Report No. FHWA-HRT-11-021. Office of Infrastructure Research and Development Federal Highway Administration, McLean, VA, April.

Huang, B., Egan, B.K., Kingery, W.R., Zhang, Z., Zuo, Z., 2004. Laboratory Study of Fatigue Characteristics of HMA Surface Mixtures Containing RAP. Transportation Research Board, National Research Council, Washington, DC.

Kennedy, T.W., Tam, W.O., Solaimanian, M., 1998b. Effect of Reclaimed Asphalt Pavement on Binder Properties Using the Superpave System Research Report No. 1250-1. Center for Transportation Research, Bureau of Engineering Research, University Of Texas at Austin, Austin, TX, September.

Mohammad, L., 2006. Performance Tests for HMA Including Fundamental and Empirical Procedures Publication No. STP 1469. ASTM International, West Conshohocken, PA.

Mokhatab, S., Poe, W.A., Speight, J.G., 2006. Handbook of Natural Gas Transmission and Processing. Elsevier, Amsterdam, Netherlands.

Moschopedis, S.E., Parkash, S., Speight, J.G., 1978. The thermal decomposition of asphaltenes. Fuel 57, 431.

Nourendin, A.S., Wood, L.E., 1989. Variation in Molecular Size Distribution Virgin and Recycled Binders Associated with Aging Transportation Research Record: Journal of the Transportation Research Board, No. 1228. Transportation Research Board, National Research Council, Washington, DC, pp. 191–197.

Parkash, S., Moschopedis, S.E., Speight, J.G., 1980. The thermal decomposition of resins. Fuel 59, 64.

Sengoz, B., Topal, A., 2004. Use of asphalt roofing shingle waste in HMA. Constr. Build. Mater. 19, 337–346.

Shen, J., Amirkhanian, S., Tang, B., 2007. Effects of rejuvenator on performance-based properties of rejuvenated asphalt binder and mixtures. Constr. Build. Mater. 21, 958–964.

Speight, J.G., 1978. The Use of Bitumen as an Adhesive and Building Material at the Ancient Cities of Babylon, Calah, Nineveh, and Ur. Report of Investigations. College of Science, University of Mosul, Iraq.

Speight, J.G., 2007. Natural Gas: A Basic Handbook. GPC Books, Gulf Publishing Company, Houston, TX.

Speight, J.G., 2014. The Chemistry and Technology of Petroleum, fifth ed. CRC Press, Taylor & Francis Group, Boca Raton, FL.

Watson, D.E., Johnson, A., Sharma, H.R., 1998. Georgia's Experience with Recycled Roofing Shingles in Asphaltic Concrete Transportation Research Record No. 1638. Transportation Research Board, National Research Council, Washington, DC, pp. 129–133.

CHAPTER 12

Environmental Aspects of Asphalt Use

12.1 INTRODUCTION

Asphalt (sometimes referred to as bitumen in some countries of the world; see Chapter 1) is produced from the distillation residuum. In addition to road asphalt, a variety of asphalt grades for roofing and waterproofing are also produced. Asphalt has complex chemical and physical compositions that usually vary with the source of the crude oil and it is produced to certain standards of hardness or softness in controlled vacuum distillation processes (Speight and Ozum, 2002; Parkash, 2003; Hsu and Robinson, 2006; Gary et al., 2007; Speight, 2014).

There are wide variations in refinery operations and in the types of crude oils so different asphalts will be produced that have different environmental effects (US EPA, 1996; Speight, 2005). Blending with higher and lower softening point asphalts may make asphalts of intermediate softening points. If lubricating oils are not required, the reduced crude may be distilled in a flash drum that is similar to a distillation tower but has few, if any, trays. Asphalt descends to the base of the flasher as the volatile components pass out of the top. Asphalt is also produced by propane deasphalting and can be made softer by blending the hard asphalt with the extract obtained in the solvent treatment of lubricating oils. Furthermore, a softer asphalt can be converted into a harder asphalt by oxidation (air blowing).

Road oil is liquid asphalt materials intended for easy application to earth roads and provides a strong base or a hard surface and will maintain a satisfactory passage for light traffic. *Cutback asphalt* is a mixture in which hard asphalt has been diluted with a lower boiling oil to permit application as a liquid without drastic heating. *Asphalt emulsions* are usually the oil-in-water-type emulsion that breaks on application to a stone or earth surface, so that the oil clings to the stone and the water disappears. In addition to their usefulness in road and soil stabilization, asphalt emulsions are also used for paper impregnation and waterproofing.

Asphalt Materials Science and Technology.
DOI: http://dx.doi.org/10.1016/B978-0-12-800273-5.00012-X

Asphalt is an important part of modern life, especially when used for road paving and in the past four decades it has been recognized that the environmental evaluation of road infrastructure is as important as the technical and structural design of the road way. In fact, it is now common for environmental criteria to be included in roadway planning. In fact, the introduction of environmental criteria is a relatively current practice because it requires special policies by various levels of government and the respective road construction agencies. Furthermore, in order to combat any threat to the environment, it is necessary to understand the nature and magnitude of the problems involved (Ray and Guzzo, 1990). It is in such situations that environmental technology has a major role to play. Environmental issues even arise when outdated laws are taken to task. Thus the concept of what seemed to be a good idea at the time the action occurred no longer holds when the law influences the environment.

Considering only technical aspects, and often only roadway construction issues and costs, to evaluate a road project exposes the project to environmental risks without due consideration of the scientific and engineering aspects of the project. For example, design and construction decisions affect the environmental impact of the project in current and future contexts. The development of asphalt roadways needs to meet the needs of the present without endangering the roadway design and construction of environmental aspects of the future, particularly from the point of environmental sustainability.

As with any industrial product, it is helpful to understand the effect of the product on the environment and, in this context, any environmental risks that are caused by the chemical nature of the asphalt, which is dependent upon the source of the asphalt. For example, if coal tar or visbroken tar is employed, enhanced levels of polynuclear aromatic hydrocarbons must be anticipated in such products because the low-pressure hydrogen-deficient conditions favor aromatization of naphthene constituents and condensation of aromatics to form larger ring systems. To the extent that more compounds like benzo(*a*)pyrene are produced, the liquids from thermal processes will be more carcinogenic than the components of asphalt, steps must be taken to ensure that such compounds are not released into the environment. Another risk, going in the other direction, is that the formulation of a decision that the construction of asphalt roadways is to be discontinued for environmental considerations, some of which may be relevant and some of which may be based on emotion and incorrect information. There is the need for roadway construction agencies and environmental groups to work together to resolve such issues. When decisions are made without consideration of *all* of the facts, the roadway public are the losers.

In road construction and maintenance, it is necessary to distinguish, for environmental purposes, the operation phase from the construction/maintenance phase. The environmental consequences of the operation phase are poorly mitigated by constraints imposed by functional choices, needs for safety, and traffic flow and constraints. Also, in line with binder composition, asphalt (binder plus aggregate) composition, as well as the types of vehicles using the roadway, can significantly reduce or increase the environmental impact of the road during its service life (WHO, 2004). Indeed, the environmental harm caused by the construction and maintenance may be much lower than the harm resulting from the traffic. Thus it is at the design stage that all such factors must be taken into account in order to assess the minimum environmental criteria that are necessary for construction and maintenance, such as: as aggregate, binder, and additives.

Thus, the environmental impact and toxicological profile of asphalt is high because of the high content of polynuclear aromatic hydrocarbon derivatives (typically in the asphaltene fraction), some of which contain heteroaromatic ring systems. Soil contamination has been a growing concern, because it can be a source of groundwater (drinking water) contamination (Irwin et al., 1997). Contaminated soils can reduce the usability of land for development; and weathered petroleum residua are particularly recalcitrant and may stay bound to soil for years, if not decades.

12.2 IDENTITY AND ORIGIN OF ASPHALT

The term *asphalt* is used to describe the refinery product and the term *bitumen* is used to describe the naturally occurring material (see Chapter 1). Other terminology (often used incorrectly) that the investigator should be made aware of is the terms *tar* and *pitch*. These terms are often applied indiscriminately to any material that is black and viscous, no matter what the source. Confusion or misinterpretation of these terms with the terms *resid* and *asphalt* can lead any environmental investigator to a situation where inappropriate methods are applied to remediation of spills of such materials.

Tar is a product of the destructive distillation of many bituminous or other organic materials and is a brown to black, oily, viscous liquid to semi-solid material. *Tar* is most commonly produced from *bituminous coal* and is generally understood to refer to the product from coal, although it is advisable to specify *coal tar* if there is the possibility of ambiguity. The most important factor in determining the yield and character of the coal tar is the carbonizing temperature. Three general temperature ranges are recognized, and the products have acquired the designations: *low-temperature tar* (approximately 450–700°C; 540–1290°F); *mid-temperature tar* (approximately 700–900°C; 1290–1650°F); and *high-temperature tar* (approximately 900–1200°C; 1650–2190°F). Tar released during the early stages of the decomposition of the organic material is called *primary tar* since it represents a product that has been recovered without the secondary alteration that results from prolonged residence of the vapor in the heated zone. Treatment of the distillate (boiling up to 250°C, 480°F) of the tar with caustic soda causes separation of a fraction known as *tar acids*; acid treatment of the distillate produces a variety of organic nitrogen compounds known as *tar bases*. The residue left following removal of the heavy oil, or distillate, is *pitch*, a black, hard, and highly ductile material.

In the chemical-process industries, pitch is the black or dark-brown residue obtained by distilling coal tar, wood tar, fats, fatty acids, or fatty oils. *Coal tar pitch* is a soft to hard and brittle substance containing chiefly aromatic resinous compounds along with aromatic and other hydrocarbons and their derivatives; it is used chiefly as road tar, in waterproofing roofs and other structures, and to make electrodes. *Wood tar pitch* is a bright, lustrous substance containing resin acids; it is used chiefly in the manufacture of plastics and insulating materials and in caulking seams. *Pitches* derived from fats, fatty acids, or fatty oils by distillation are usually soft substances containing polymers and decomposition products; they are used chiefly in varnishes and paints and in floor coverings.

12.2.1 Asphalt

Asphalt is a major product of many petroleum refineries. Asphalt may be residual (straight-run) asphalt, which is made up of the non-volatile hydrocarbons in the feedstock, along with similar materials produced by thermal alteration during the distillation sequences, or they may be produced by air-blowing residua. Alternatively, asphalt may be the residuum from a vacuum distillation unit. In either case, the properties of the asphalt are, essentially, the properties of the residuum (Speight, 2014). If the properties are not suitable for the asphalt product to meet specifications, changing the properties by, for example, blowing is necessary.

Asphalt manufacture is, in essence, a matter of distilling everything possible from crude petroleum until a residue with the desired properties is obtained. This is usually accomplished in stages (Figure 12.1); crude distillation at atmospheric pressure removes the lower-boiling fractions and yields reduced crude that may contain higher-boiling (lubricating) oils, asphalt, and even wax. Distillation of the reduced crude under vacuum removes the oils (and wax) as overhead products and the asphalt remain as a bottom (or residual) product. The majority of the polar functionality in the original crude oil end to be non-volatile and concentrate in the vacuum residuum (Figure 12.2). It is this concentration effect that confers upon asphalt some of its unique properties. At this stage the asphalt is frequently and incorrectly referred to as pitch and has a softening point related to the amount of oil removed: the more oil distilled from the residue, the higher the softening point (Corbett and Petrossi, 1978).

Thus, asphalt is produced from residua (see Chapters 2 and 3), which are the residues obtained from petroleum after non-destructive distillation has removed all the volatile materials. The temperature of the distillation is usually maintained below 350°C (660°F) since the rate of thermal decomposition of petroleum constituents is minimal below this temperature but the rate of thermal decomposition of petroleum constituents is substantial above 350°C (660°F). *Residua* are black, viscous materials that are obtained by distillation of a crude oil under atmospheric pressure (atmospheric

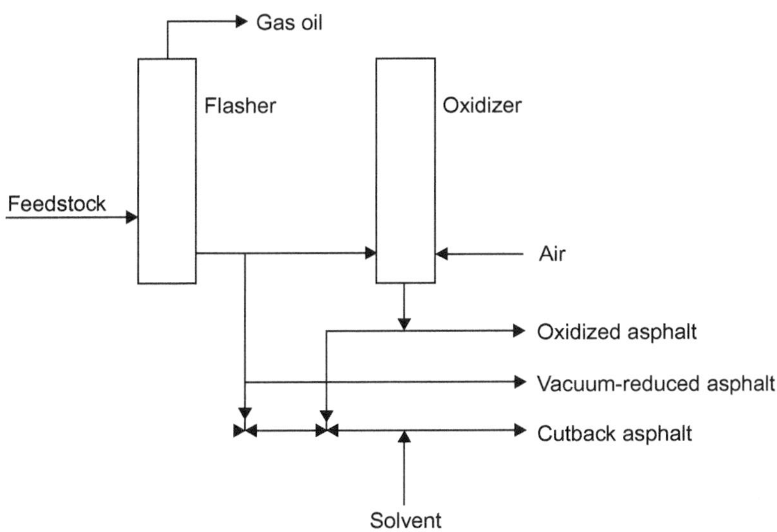

Figure 12.1 General representation of asphalt manufacture.

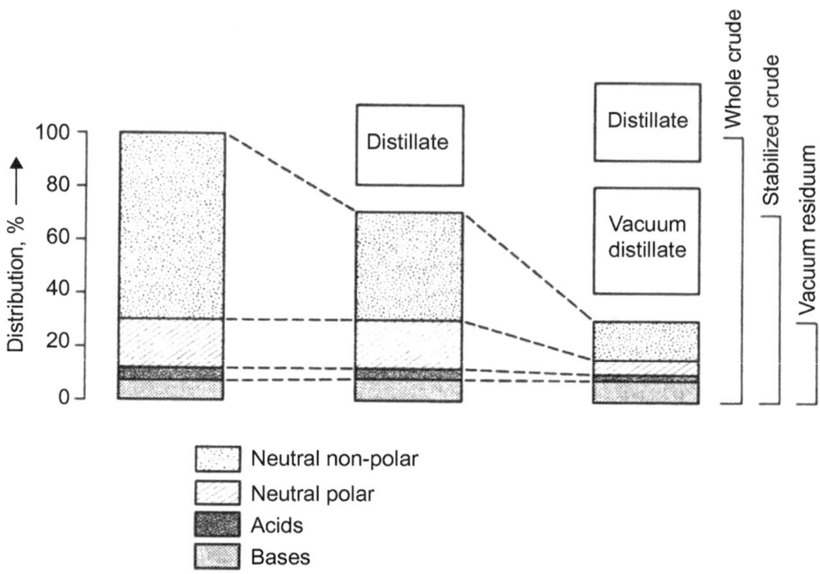

Figure 12.2 Accumulation of polar functions in residua during distillation.

residuum) or under reduced pressure (vacuum residuum) (see Chapters 2 and 3). The properties of residua are variable and crude-oil-dependent (Table 12.1)—they may be liquid at room temperature (generally atmospheric residua) or almost solid (generally vacuum residua) depending upon the nature of the crude oil (Speight, 2014). When a residuum is obtained from a crude oil and thermal decomposition has commenced, it is more usual to refer to this product as *pitch*. The differences between the parent petroleum and the residua are due to the relative amounts of various constituents present, which are removed or remain by virtue of their relative volatility.

Asphalt is not coal tar but has been confused with coal with coal tar and coal tar pitch because of the variable and inconsistent nomenclature that has been used over the past 100 years (see Chapter 1). Coal tar and coal tar pitch were used in road building before and during World War II due to a shortage of asphalt cement but coal tar materials have apparently not been used in asphalt paving formulations since the 1970s (Kriech et al., 1997; Blackburn et al., 1999). Coal tar contains relatively high levels of condensed-ring aromatic compounds with a greater proportion of unsubstituted polynuclear aromatic compounds (PNAs) in the toxicologically active 3–7 ring size range. In contrast, asphalts contain much larger

Table 12.1 Properties of various residua

Residuum	Gravity	Sulfur	Nitrogen	Nickel	Vanadium	Asphaltene Fraction	Carbon Residue
	API	% w/w	% w/w	ppm	ppm	(Heptane)	(Conradson, % w/w)
Alaska, North Slope, >650F	15.2	1.6	0.4	18.0	30.0	2.0	8.5
Alaska, North Slope, >1050F	8.2	2.2	0.6	47.0	82.0	4.0	18.0
Arabian Heavy, >650F	11.9	4.4	0.3	27.0	103.0	8.0	14.0
Arabian heavy, >1050F	7.3	5.1	0.3	40.0	174.0	10.0	19.0
Arabian light, >650F	17.7	3.0	0.2	10.0	26.0	1.8	7.5
Arabian light, >1050F	8.5	4.4	0.5	24.0	66.0	4.3	14.2
Kuwait, >650F	13.9	4.4	0.3	14.0	50.0	2.4	12.2
Kuwait, >1050F	5.5	5.5	0.4	32.0	102.0	7.1	23.1
Lloydminster (Canada), >650F	10.3	4.1	0.3	65.0	141.0	14.0	12.1
Lloydminster (Canada), >1050F	8.5	4.4	0.6	115.0	252.0	18.0	21.4
Taching, >650F	27.3	0.2	0.2	5.0	1.0	4.4	3.8
Taching, >1050F	21.5	0.3	0.4	9.0	2.0	7.6	7.9
Tia Juana, >650F	17.3	1.8	0.3	25.0	185.0		9.3
Tia Juana, >1050F	7.1	2.6	0.6	64.0	450.0		21.6

proportions of high-molecular-weight paraffinic and naphthenic hydrocarbons and their derivatives that, because of their size, viscosity, and limited solubility, are not readily bioavailable and have minimal toxicological activity (API, 2009).

The chemical composition of a residuum from an asphaltic crude oil is complex but physical methods of fractionation usually indicate high proportions of asphaltene constituents and resins, even in amounts up to 50% (or higher) of the residuum. In addition, the presence of ash-forming metallic constituents, including such organometallic compounds, such as those of vanadium and nickel, is also a distinguishing feature of residua. Furthermore, the deeper the *cut* into the crude oil, the greater is the concentration of sulfur and metals in the residuum and the greater the deterioration in physical properties (Speight, 2014).

Although the manufacturing process may change the physical properties of asphalt dramatically, the chemical nature of the asphalt does not change unless thermal cracking occurs. Raising the temperature increases the likelihood of cracking and leads to higher yields of volatile products to be released from the residuum (Speight, 2014). Solvent precipitation (usually using propane or butane) removes high-boiling components from a vacuum-processed asphalt and results in the production of a harder asphalt that is less resistant to temperature changes and often blended with straight-reduced or vacuum-processed asphalts. The air-blowing process can be a continuous or batch operation and since the continuous operation is faster and results in a softer asphalt, a continuous operation is preferred for processing paving asphalt (Speight, 1992). Air blowing is an exothermic process and has the potential to cause a series of chemical reactions, such as oxidation, condensation, dehydration, and dehydrogenation—polymerization is often cited as a possible reaction but has not been defined fully. These reactions cause the amount of asphaltene constituents (heptane-insoluble constituents) to increase and, at the same time, the oxygen content of the asphalt increases (Corbett, 1975; Moschopedis and Speight, 1973, 1975, 1977, 1978; Speight, 1992).

Although no two asphalts are chemically identical and chemical analysis cannot be used to define the exact chemical structure or chemical composition of asphalt, elemental analyses indicate that most asphalts contain 79–88% (w/w) carbon, 7–13% (w/w) hydrogen, traces to 3% (w/w) nitrogen, traces to 8% (w/w) sulfur, and traces to 8% (w/w) oxygen (Speight, 1992). While heteroatoms (i.e., nitrogen, oxygen, and sulfur) make up only a minor component of most asphalts, the heteroatoms profoundly influence the differences in physical properties of asphalts from different crude sources (Speight, 1992).

Asphalt is used for paving, roofing, industrial, and special purposes. Oxidized asphalt is used in roofing operations, pipe coating, undersealing for concrete pavements, hydraulic applications, membrane envelopes, some paving-grade mixes, and the manufacture of paints (Speight, 1992). From a scientific point of view, asphalts probably should be classified as to whether or not they have been oxidized. However, most publications have classified asphalts according to the performance specifications for which they were manufactured (e.g., paving asphalts and roofing asphalts). This greatly complicates an understanding of the chemistry and behavior of asphalt (i) in the manufacturing process, (ii) during use, and (iii) after a spill. The environmental issues related to asphalt use and (spillage) situation are further complicated by the addition of additives and modifiers.

However, since there are wide variations in refinery operations and crude petroleum, asphalt with softening points ranging from 25°C to 55°C (80°F to 130°F) may be produced. Blending with higher and lower softening point asphalt may make asphalt of intermediate softening points. If lubricating oils are not required, the reduced crude may be distilled in a flash drum, which is similar to a bubble tower but has few, if any, trays. Asphalt descends to the base of the drum as the oil components pass out of the top of the drum. If the asphalt has a relatively low softening point, it can be hardened by further distillation with steam or by oxidation. Asphalt is also produced by propane deasphalting (see Chapter 2) and the asphalt so produced may have a softening point of about 95°C (200°F). Softer grades are made by blending *hard asphalt* with the extract obtained in the solvent treatment of lubricating oils.

Soft asphalt can be converted into harder asphalt by oxidation, which promotes the formation of resins and asphaltene constituents from the lower-molecular-weight constituents:

$$\text{Oil constituents} \rightarrow \text{resin constituents} \rightarrow \text{asphaltene constituents}$$

The increase in the proportion of semisolid (resins) and solid (asphaltene constituents) constituents as a result of air blowing is accompanied by an increase in softening point with only a small loss in volume. A similar increase in softening point by removing oily constituents would cause a considerable decrease in volume. Oxidation is carried out by blowing air through asphalt heated to about 260°C (500°F) and is usually done in a tower (an oxidizer) equipped with a perforated pipe at the bottom through which the air is blown. The asphalt, in the batch mode or continuous mode, is heated until the oxidation reaction starts, but the reaction is exothermic, and the temperature is controlled by regulating the amount of air and by circulating oil or water through cooling coils within the oxidizer. Asphalt with softening points as high as 180°C (350°F) may be produced.

Asphalt, normally a liquid when applied at higher temperature, may be referred to as *non-volatile liquid asphalt*, but semisolid or solid asphalt may be made liquid for easier handling by dissolving them in a solvent and is referred to as *cutback asphalt*. The asphalt and solvent (naphtha, kerosene, or gas oil) are heated to about 105°C (225°F) and passed together through a mechanical mixer, and the effluent then enters a horizontal tank, which is used for further mixing. Liquid asphalt is pumped from the top of the tank into a perforated pipe lying at the bottom of the tank, and circulation is continued until mixing is complete.

Asphalt paving materials typically are composed of aggregate and/or sand (90–95% by weight) and asphalt (5–10%, w/w). The aggregate and/or sand is responsible for the primary load-bearing properties, while asphalt serves as the binder and as a protective coating. The asphalt binder functions best when the aggregate/sand particles are *wetted* by the asphalt. To ensure good adhesion between the asphalt and the aggregate or sand particles, moisture and clays are to be avoided.

Similar to petroleum residua, the spillage of asphalt (other than for its intended purpose of road use as part of an asphalt–aggregate mixture) on to soil and waterways is disadvantageous. The environmental impact and toxicological profile of petroleum residua is high because of the high content of polynuclear aromatic hydrocarbon derivatives (typically in the asphaltene fraction), some of which contain heteroaromatic ring systems. In addition, soil contamination has been a growing concern, because it can be a source of groundwater (drinking water) contamination; contaminated soils can reduce the usability of land for development; and weathered petroleum residua are particularly recalcitrant and may stay bound to soil for years.

The main hazard associated with asphalt is from the polynuclear aromatic hydrocarbon derivatives that can move into the ecosystem from the breakdown of asphalt. Since asphalt contains so many toxic and carcinogenic compounds and since leaching of harmful polynuclear aromatic hydrocarbon derivatives has been documented even in water pipe use, asphalt should be kept out of rivers, streams, and other natural waters to the extent possible (Irwin et al., 1997).

However, at ambient temperatures the semi-solid to solid nature of substances in the asphalt category and negligible vapor pressure and water solubility limit their distribution to different environmental compartments. Asphalt constituents tend to remain within the terrestrial or aquatic compartment to which they were released (Speight and Arjoon, 2012; Moretti et al., 2013). During the application of asphalts, fumes may be generated when the material is heated, and these fumes may condense onto local surfaces as the fumes cool and also have the potential to carry and deposit particulate matter as well as polynuclear aromatic hydrocarbons (Claydon et al., 1984; Roy et al., 2007). However, fumes generated from asphalt are primarily aliphatic with a high proportion of saturates and demonstrate much less toxicological activity than coal tar fumes which are comprised almost entirely of aromatic compounds (>99%) (API, 2009).

The importance of residua and asphalt to the environmental analyst arises from spillage or leakage in the refinery or on the road (see Chapter 12) (Speight, 2005). In either case, the properties of these materials are detrimental to the ecosystem in which the release occurred. As with other petroleum products, knowledge of the properties of residua and asphalt can help determine the potential cleanup methods and may even allow regulators to trace the product to the refinery where it was produced. In addition, the character of residua and asphalt render the usual test methods for *total petroleum hydrocarbons* ineffective since high proportions of asphalt and residua are insoluble in the usual solvents employed for the test (Speight and Arjoon, 2012). Application of the test methods for *total petroleum hydrocarbons* to residual fuel oil (see Chapter 10) is also subject to similar limitations.

Residua are the dark-colored near-solid or solid products of petroleum refining that are produced by atmospheric and vacuum distillation (see Chapters 2 and 3) (Speight and Ozum, 2002; Speight, 2014). Asphalt is usually produced from a residuum and is a dark-brown to black cementitious material obtained from petroleum processing and which contains very high-molecular-weight molecular polar species called asphaltenes that are soluble in carbon disulfide, pyridine, aromatic hydrocarbons, and chlorinated hydrocarbons (Speight, 2014, 2015).

Asphalt derives its characteristics from the nature of its crude oil precursor, the distillation process being a concentration process in which most of the heteroatoms and polynuclear aromatic constituents of the feedstock are concentrated in the residuum (see Chapter 2) (Speight, 1992,

2014). Asphalt may be similar to its parent residuum but with some variation possible by choice of manufacturing process. In general terms, asphalt is a hydrocarbonaceous material that consists of constituents (containing carbon, hydrogen, nitrogen, oxygen, and sulfur) that are completely soluble in carbon disulfide (ASTM D4). Trichloroethylene or 1,1,1-trichloroethane have been used in recent years as solvents for the determination of asphalt (and residua) solubility (ASTM D2042).

Asphalt manufacture involves distilling everything possible from crude petroleum until a residuum with the desired properties is obtained (see Chapter 2). This is usually done by stages in which distillation at atmospheric pressure removes the lower-boiling fractions and yields an atmospheric residuum (*reduced crude*) that may contain higher-boiling (lubricating) oils, wax, and asphalt. Distillation of the reduced crude under vacuum removes the oils (and wax) as overhead products and the residuum remains as a bottom (or residual) product. The majority of the polar functionalities and high-molecular-weight species in the original crude oil, which tend to be non-volatile, concentrate in the vacuum residuum thereby conferring desirable or undesirable properties on the residuum. At this stage the residuum is frequently, but incorrectly, referred to as pitch and has a softening point (ASTM D36; ASTM D61; ASTM D2319; ASTM D3104; ASTM D3461) related to the amount of oil removed and increases with increasing overhead removal. In character with the elevation of the softening point, the pour point is also elevated; the more oil distilled from the residue, the higher the softening point.

Propane deasphalting of a residuum also produces asphalt and there are differences in the properties of asphalts prepared by propane deasphalting and those prepared by vacuum distillation from the same feedstock. Propane deasphalting also has the ability to reduce a residuum even further and produce an asphalt product having a lower viscosity, higher ductility, and higher temperature susceptibility than other asphalts. However, such properties might be anticipated to be very much crude-oil-dependent. Propane deasphalting is conventionally applied to low-asphalt-content crude oils, which are generally different in type and sourced from those processed by distillation of higher-yield crude oils. In addition, the properties of asphalt can be modified by air blowing in batch and continuous processes (Speight, 1992, 2014). On the other hand, the preparation of asphalts in liquid form by blending (cutting back) asphalt with a petroleum distillate fraction is customary and is generally accomplished in tanks equipped with coils for air agitation or with a mechanical stirrer or a vortex mixer.

An *asphalt emulsion* is a mixture of asphalt and an anionic agent such as the sodium or potassium salt of a fatty acid. The fatty acid is usually a mixture and may contain palmitic, stearic, linoleic, and abietic acids and/or high-molecular-weight phenols. Sodium lignate is often added to alkaline emulsions to effect better emulsion stability. Non-ionic cellulose derivatives are also used to increase the viscosity of the emulsion if needed. The acid number is an indicator of its asphalt emulsification properties and reflects the presence of high-molecular-weight asphaltic or naphthenic acids. Diamines, frequently used as cationic agents, are made from the reaction of tallow acid amines with acrylonitrile, followed by hydrogenation. The properties of asphalt emulsions (ASTM D977; ASTM D2397) allow a variety of uses. As with other petroleum products, sampling is an important precursor to asphalt analysis and a standard method (ASTM D140) is available that provides guidance for the sampling of asphalts, liquid and semisolid, at the point of manufacture, storage, or delivery.

The properties of asphalt are defined by a variety of standard tests that can be used to define quality (see Chapter 4) and remembering that the properties of residua vary with cut-point, i.e. the volume % of the crude oil helps the refiner produce asphalt of a specific type or property. Roofing and industrial asphalts are also generally specified in various grades of hardness, usually with a combination of softening point (ASTM D61; ASTM D2319; ASTM D3104; ASTM D3461) and penetration (ASTM D312; ASTM D449).

The significance of a particular test is not always apparent by reading the procedure, and sometimes can only be gained through working familiarity with the test. The following tests are commonly used to characterize asphalts but these are not the only tests used for determining the property and behavior of an asphaltic binder. As in the petroleum industry, a variety of tests are employed having evolved through local, or company, use.

12.2.2 Other Constituents

Briefly, *asphalt fume* is a visible airborne condensation product of lower-boiling volatile components of petroleum asphalt that may be inhaled or deposited on skin and clothing. When asphalts are heated to facilitate paving or roofing applications, the lighter, more volatile components are distilled into the atmosphere. As these components cool, they condense forming small droplets of liquid (fume), some of which have an effective diameter of less than 12.5 microns and are considered respirable. The concentration of the lower-molecular-weight components of petroleum asphalts that include the tumorigenic 3–7-ring polynuclear aromatic compounds in fume condensate is likely to be higher than in the parent asphalts and hence the tumorigenic potential may be increased. The temperature of fume generation affects both the relative proportions of individual polynuclear aromatic compounds in the fume and the amount of fume generated. Furthermore, asphalt fumes and vapors may be absorbed following inhalation and dermal exposure but because of the complex chemical nature of asphalt, the pharmacokinetic behavior will vary depending upon the properties of the individual constituents. Therefore, it is inappropriate to make generalizations regarding the extent of absorption, distribution, and metabolism of asphalt (WHO, 2004).

Asphalt fume exposure potential, including the quantity and nature of organic compounds, is directly dependent upon the specific application process conditions including temperature. Emissions of polynuclear aromatic compounds from asphalt are highly temperature-dependent. It has also been reported that only simple aromatics and very low amounts of 2-ring and 3-ring polynuclear aromatic compounds are emitted at temperatures typically employed for asphalt paving applications (140–160°C; 284–320°F) (Brandt and de Groot, 1985, 1999; Brandt et al., 1985; Kitto et al. 1997; Gamble et al., 1999; Burstyn et al., 2000, 2002a,b, 2003a,b; Butler et al., 2000; Ekstrom et al., 2001; Heikkilä et al., 2002; Kriech et al., 2002; Lange and Stroup-Gardiner, 2007).

Lowering the asphalt temperature by 11–12.5°C (20–22.5°F) reduces benzene-soluble matter fume emissions by a factor of 2 (Asphalt Institute and Eurobitume, 2008). As was noted above, production of conventional asphalt pavement material is typically accomplished in the range of 140–160°C (280–320°F). In the past decade, warm-mix asphalt technologies have been developed and deployed to allow the production and placement of asphalt pavements at between 100°C and 140°C (212°F and 280°F). Exposure sampling and analytic methods for asphalt fumes generally fall into three main categories that measure the following: (i) particulate matter, (ii) solvent soluble particulate matter, and (iii) organic matter.

Total particulate matter includes aerosol matter from the asphalt and inorganic material such as dust, rock fines, and filler. Because methods for collection of total particulate matter include material from non-asphalt sources, the resulting values can suggest artificially high exposure values, especially in dusty environments.

The *solvent soluble fractions of particulate matter (BSM/BSF, benzene soluble matter/fraction) or the cyclohexane soluble matter/fraction (CSM/CSF)* methods rely on collection of the particulate fraction as described above. However, in order to reduce the confounding exposure to inorganic particulate matter a solvent is used to extract only the organic fraction of the particulates. Such methods more accurately define the exposure to the agent of interest (asphalt fume). A subset of such methods uses a special monitoring cassette to collect only a specific fraction of the particulate matter, e.g., the respirable, thoracic fractions or inhalable fraction.

Organic matter (total organic matter/total hydrocarbon, TOM/THC) is the sum of the organic part of the particulate fraction plus organic vapor phase collected using a back-up absorbent. At present, international standards for the assessment of exposure to asphalt fumes are variable and, as a result, reported values of exposures over time, between investigations within the same country and between investigations in various countries can be significantly different and must be given careful consideration as to the intended use.

Any vapor that remains suspended has the potential to photodegrade within a matter of hours to a few days. Although hydrocarbons can be utilized as an energy source by microorganisms, asphalts would not dissolve or disperse in a manner to augment microbial attack. Due to the bulk properties of asphalt, a release to the environment would not result in measureable biodegradation. In addition, the constituents in vacuum residues and asphalt are of such high molecular weight and low solubility that such materials would not be expected to cause acute or chronic toxicity in aquatic organisms (API, 2009).

In the early days of the industry, another type of binder, coal tar, was used in the paving industry, especially in Europe, southern Africa, Australia, and, to some extent, in the United States. Because of their similar appearance to petroleum-based asphalt, little distinction was made between asphalt and tar as a roadway construction material. However, the source of asphalt and tar dictates vast differences in chemical composition (Speight, 2013, 2014). Coal tar, because of its formation by the destructive distillation of coal, contains polynuclear aromatic hydrocarbons (PNAs, also called polycyclic aromatic hydrocarbons, PAHs). Furthermore, it is well established that coal tar has a high content of carcinogenic polynuclear aromatic hydrocarbons.

As a result, coal tar has not been used much in asphalt pavement applications in the United States since World War II, especially since the establishment of environmental issues relating to coal tar use in the 1970s. More recent applications for coal tar have been limited to a few non-road applications, such as airfields and emulsion application as a pavement sealer for parking lots, driveways, and bridges. State specifications typically prohibit the use of reclaimed asphalt pavement material known to contain coal tar (Mundt et al., 2009).

12.3 ASPHALT–AGGREGATE INTERACTIONS

The approach to design of roadways to protect the environment needs to focus, to some extent, on fatigue and deterioration of the roadway that is due (to a major extent) on the durability of the asphalt–aggregate interactions (see Chapter 4). Thus, maximum permissible load and the visible early evolution of a road towards collapse of the asphalt–aggregate interaction and/or progressive deterioration of the structural integrity of the roadway must be part of the planning and design process.

12.3.1 Stability

The stability of the asphalt–aggregate mix is essential for roadway performance and longevity. However, stabilization is where chemical bonding techniques render a binder less toxic and less harmful to the environment. Examples of such stabilization include (i) ion exchange of heavy metals in an alumina silicate matrix of a cementitious agent or (ii) sorption of heavy metals on fly ash in an aqueous system (D'Andrea et al., 2004). Stabilization of the asphalt–aggregate mix can be functionally described in terms of the final product, where the binder and the aggregate are transformed into a stable and durable matrix that is more compatible for roadway use. The chemical stability and environmental performance of asphalt is dependent on the nature and extent of the asphalt–aggregate bonds and much of the information concerning these bonds comes from the investigation of asphalt sensitivity to moisture-induced damage (Petersen et al., 1982).

Because of the polar/non-polar and hydrophobic/hydrophilic interactions among the various phases in the system, the functional groups in asphalt (Figure 12.3) will be concentrated at the interfaces between phases, such as the asphalt–aggregate interface, the polar micelle–asphalt oil interface, and the asphalt–water interface. Thus, the stability of the asphalt–aggregate mix is addressed on the basis of the relative tendency of the functional groups to be concentrated at the asphalt–aggregate interface, sorption affinity, and ability to be displaced by water.

In fact, the behavior of functional groups at the asphalt–aggregate interface determines the chemical stability of the asphalt to a large degree.

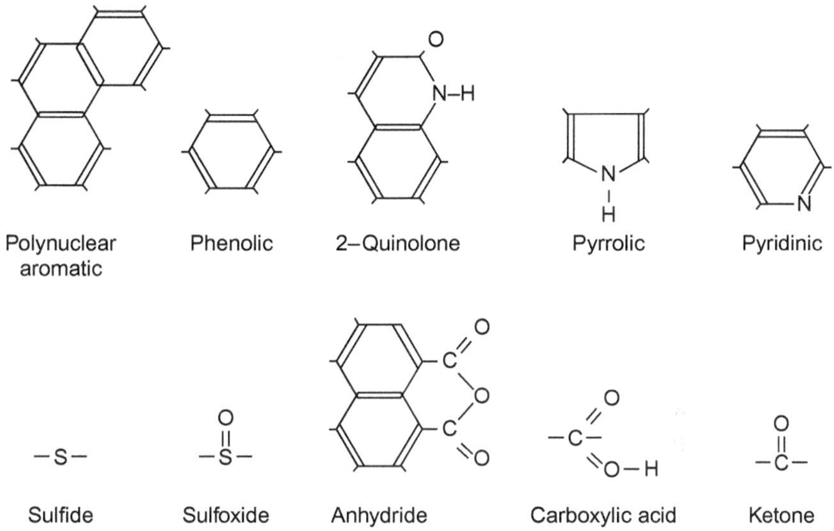

Figure 12.3 Representative polar and non-polar functional groups present in asphalt.

At the asphalt–aggregate interface, these groups are susceptible to interactions with the aqueous phases and any contaminants present in the affected soil. The relative tendency of the asphalt functional groups to be concentrated at the asphalt–aggregate interface has been determined (Petersen, 1986), in order of decreasing tendency, as:

Carboxylic acid derivatives > anhydride derivatives
Quinoline derivatives ≈ phenol derivatives
Quinoline derivatives, phenol derivatives > sulfoxide derivatives
Sulfoxide derivatives ≈ ketone derivatives
Ketone > pyrrole derivatives

The sorption affinity of the functional groups with the aggregate surface has also been investigated (Petersen et al., 1982) and is given, in order of decreasing tendency, as:

Pyridine derivatives ≈ carboxylic acid derivatives
Pyridine derivatives, carboxylic acid derivatives ≫ anhydride derivatives
Anhydride derivatives > quinoline derivatives
Quinoline derivatives ≈ phenol derivatives
Quinoline derivatives, phenol derivatives > sulfoxide derivatives
Sulfoxide derivatives > ketone derivatives
Ketone derivatives ≫ pyrrole derivatives
Pyrrole derivatives ≈ polynuclear aromatic derivatives

The third important tendency of the functional groups affecting stability is the ability of the functional group–aggregate bond to be displaced by water. This tendency, listed in order of decreasing tendency, is given as:

Anhydride derivatives > quinoline derivatives

Quinoline derivatives > carboxylic acid derivatives

Carboxylic acid derivatives > sulfoxide derivatives

Sulfoxide derivatives > ketone derivatives

Ketone derivatives > pyrrole derivatives, phenol derivatives

Combining these tendencies provides some indications of performance—for the best chemical performance, the asphalt should have high contents of pyridine, phenol, and ketone groups, which can be achieved by selectively choosing the source material or using additives, such as shale oil that is rich in pyridine-type basic nitrogen. Also, the presence of inorganic sulfur, monovalent salts, and high-ionic-strength solutions in the asphalt decreases the chemical stability of the asphalt cement by disruption of the functional group–aggregate bonds, and increases the system's permeability (Petersen et al., 1982). Addition of lime to the aggregate can thus be used, and often is, to counter this effect.

12.3.2 Contaminant Mobility

Contaminant mobility, especially of metals (Legret et al., 2005), in the asphalt cement will be affected by many factors, such as diffusivity, permeability, solubility, specification, complexation, redox reactions, sorption, and precipitation. Many of these properties are not known to the degree that specific contaminant release can be predicted and leaching tests are the primary method of evaluating contaminant mobility in these systems (Irwin et al., 1997; Brandt de Groot, 2001; Speight and Arjoon, 2012). However, the low diffusivity and permeability of asphalt is obviously the greatest factor in the retention of contaminants in asphalt cements. Conditions that adversely affect the diffusivity and permeability will have the greatest adverse effect on contaminant mobility and release. The asphalt acts primarily as a physical containment (i.e., via fixation or stabilization, solidification, and encapsulation) to the contaminants and the aggregate, reflecting on the asphalt cement overall composition and structure.

Contaminant mobility and leaching of contaminants from asphalt pavements was of little concern until the 1960s—asphalt products had been used since biblical times (see Chapter 1) not only to build roads but also to seal water reservoirs, fountains, aquariums, etc. without any noticeable health effects. This does not necessarily mean that leaching did not take place, but rather that the overall impact at that time was small and unnoticed, based on the existing road-network size and leachate detection limits (Kennepohl, 2008). With more advanced detection and analytical techniques now available, the situation must still be monitored (Irwin et al., 1997; Burstyn et al., 2000, 2003a,b; Boffetta et al., 2001; Brandt and de Groot, 2001; Legret et al., 2005; Winter, 2006; Speight and Arjoon, 2012).

Even though metals such as vanadium and nickel occur in asphalts at the hundreds of parts per million levels (Wolfe et al., 1986), as well as many toxic organic components, asphalt leachates and products have never produced toxic or contaminated solutions that are considered to be hazardous by the Environmental Protection Agency. Asphalt's non-volatile, viscoelastic properties result in the general observation that asphalt leachates do not contain reported contaminant concentrations and even in asphalts incorporating metal slags as an aggregate, metals do not become solubilized and do not leach from these asphalts in detectable concentrations when used with strong acidic and alkaline solutions (Testa, 1995). Leachates from asphalt that have contained detectable concentrations of contaminants have been obtained from investigations of asphalted nuclear wastes (Daugherty et al., 1988; Fuhrmann et al., 1989).

In summary, the issues related to the environmental aspects of the design, construction, and use of asphalt roadways must involve careful scientific, technological, and unemotional frameworks. It is essential that there are factual means for pursuing environmental policy objectives during the life and in-service aspect of asphalt roadways. This involves knowing the environmental consequences of each material or work related to road construction, maintenance, or rejuvenation of the roadway. Methodology that is not based on these factors will lead to inferior roadways that cause environmental damage. Methodology that is based on emotional issues without consideration of the scientific and technical facts will not lead to the identification and quantification of environmental impact of road materials and construction. The environmental impact of asphalt roadways must be analyzed in a logical and unbiased manner to provide a decision-making structure towards sustainability.

Finally, there is a report that coal-tar-sealant runoff causes toxicity and DNA damage (Mahler and LaVista, 2015). Runoff from pavement with coal-tar-based sealant is toxic to aquatic life, damages DNA, and impairs DNA repair. Pavement sealant is a black liquid sprayed or painted on the asphalt pavement of parking lots, driveways and playgrounds to improve the appearance and to protect the underlying asphalt. Pavement sealants that contain coal tar have extremely high levels of polynuclear aromatic hydrocarbons (PNAs), several of which are probable human carcinogens and some are toxic to fish and other aquatic life. Coal-tar sealants have significantly higher levels of polynuclear aromatic hydrocarbons and related compounds compared to asphalt-based pavement sealants and other urban sources, including vehicle emissions, used motor oil, and tire particles.

12.4 ASPHALT WASTE

Environmental protection legislation has led to the development of strict enforcement codes that have stated clearly that water flow, particulate emissions, and other forms of emissions from oil refineries and asphalt processing plants are to be limited (Speight, 2005; Speight and Arjoon, 2012). Not only dust but oxides of sulfur and many other emissions must be rigorously controlled. Electrostatic precipitators, primary dust collectors using single or multiple cone cyclones, and secondary collection units consisting of fabric filter collectors (baghouses) are all required equipment to control emissions. Hydrocarbons formed during asphalt production, if unchecked, create odoriferous fumes and pollutants which will stain and darken the air. Pollutants emitted from asphalt production are controlled by enclosures which capture the exhaust and then recirculate it through the heating process. This not only eliminates the pollution but also increases the heating efficiency of the process.

Higher costs of asphalt cement, stone, and sand have forced the industry to increase efficiencies and recycle old asphalt pavements. In asphalt pavement recycling, materials reclaimed from old pavements are reprocessed along with new materials. The three major categories of asphalt recycling are: (i) hot-mix recycling, where reclaimed materials are combined with new materials in a central plant to produce hot-mix paving mixtures, (ii) cold-mix recycling, where reclaimed materials are combined with new materials either onsite or at a central plant to produce cold-mix base materials, and (iii) surface recycling, a process in which the old asphalt surface pavement is heated in place, scraped down, remixed, re-laid, and rolled. Organic asphalt recycling agents may also be added to help restore the aged asphalt to desired specifications. In addition, because of solvent evaporation and volatility, use of cutback asphalts, especially rapid-cure cutback asphalts which use gasoline or naphtha, is becoming more restricted or prohibited while emulsified asphalts (in which only the water evaporates) are becoming more popular because of cost and environmental regulations.

Increasing economic and environmental needs will bring many new technical refinements to recycling old asphalt pavements, such as using microwaves to completely breakdown the pavement. Microwaves heat the crushed rock in asphalt pavement faster than the surrounding cement, which is then warmed by the radiant heat from the rock. This method prevents the asphalt cement from burning.

Alternative sources of raw material are being researched, such as the production of synthetic asphalt from the liquefaction of sewage sludge. To ensure consistent product quality, new methods are being developed for manufacturing modified asphalts and emulsions. Many new tests are being developed to help characterize asphalts, such as high-performance gel-permeation chromatography (HP-GPC), which allows many properties to be investigated and the results compiled in a relatively short timeframe. New processes, more efficient mixing and milling units, in-line liquid mass flow meters, on-line monitoring systems, and new safety equipment are some other areas being investigated for improvement.

Polymer-modified asphalt crack sealers are gaining in popularity, and many other asphalt modifiers are being developed. Modifiers are added to control pavement rutting, cracking, asphalt oxidation, and water damage. Some commercially available asphalt modifiers are polymers, including elastomers, metal complexes, elemental sulfur, fibers, hydrated lime, Portland cement, silicones, various fillers, and organic anti-strip agents. Many of these modifiers have not been extensively used and are being researched for further development. It might even be possible one day to have *smart asphalt cements* by mixing in certain asphalt fraction modifiers which would allow it to change characteristics depending on whether moisture was present.

12.5 BIODEGRADATION OF ASPHALT

The residuum or asphaltic fraction of petroleum is the fraction that is either (i) the residue remaining after completion of atmospheric distillation or (ii) the residue remaining after the completion of vacuum distillation. On the other hand, asphalt is the product produced from asphalt that is used for, amongst other uses, road and highway surfacing (see Chapter 1) (Abraham, 1945; Barth, 1962; Hoiberg, 1964; Luey and Li, 1993; Speight, 2014).

The chemical composition of a residuum from an asphaltic crude oil is complex. Physical methods of fractionation usually indicate high proportions of asphaltene constituents and resin constituents, even in amounts up to 50% (or higher) of the residuum. In addition, the presence of ash-forming metallic constituents, including such organometallic compounds as those of vanadium and nickel, is also a distinguishing feature of asphalt. Furthermore, the deeper the *cut* into the crude oil to produce the residuum from which asphalt is manufactured, the greater is the concentration of sulfur and metals in the residuum and the greater the deterioration in physical properties (Speight, 2014).

On the other hand, *asphalt* is manufactured from petroleum and is a black or brown material that has a consistency varying from a viscous liquid to a glassy solid. To a point, asphalt can resemble bitumen, hence the tendency to refer to bitumen (incorrectly) as *native asphalt*. It is recommended that there be differentiation between asphalt (manufactured) and bitumen (naturally occurring) other than by use of the qualifying terms *petroleum* and *native* since the origins of the materials may be reflected in the resulting physicochemical properties of the two types of materials. It is also necessary to distinguish between the asphalt which originates from petroleum by refining and the product in which the source of the asphalt is a material other than petroleum, e.g., *Wurtzilite asphalt*. In the absence of a qualifying word, it is assumed that the term *asphalt* refers to the product manufactured from petroleum.

When the asphalt is produced simply by distillation of an asphaltic crude oil, the product can be referred to as *residual asphalt* or *straight run asphalt*. If the asphalt is prepared by solvent extraction of residua or by light hydrocarbon (propane) precipitation, or if blown or otherwise treated, the term should be modified accordingly to qualify the product (e.g., *propane asphalt*, *blown asphalt*).

Asphalt softens when heated and is elastic under certain conditions. The mechanical properties of asphalt are of particular significance when it is used as a binder or adhesive. The principal application of asphalt is in road surfacing, which may be done in a variety of ways. Light oil *dust layer* treatments may be built up by repetition to form a hard surface, or a granular aggregate may be added to an asphalt coat, or earth materials from the road surface itself may be mixed with the asphalt.

Other important applications of asphalt include canal and reservoir linings, dam facings, and sea works. The asphalt so used may be a thin, sprayed membrane, covered with earth for protection against weathering and mechanical damage, or thicker surfaces, often including riprap (crushed rock). Asphalt is also used for roofs, coatings, floor tiles, soundproofing, waterproofing, and other building-construction elements and in a number of industrial products, such as batteries. For certain applications an asphaltic emulsion is prepared, in which fine globules of asphalt are suspended in water.

By definition, bioremediation is the use of living organisms, primarily microorganisms, to degrade the environmental contaminants into less toxic forms. It uses naturally occurring bacteria and fungi or plants to degrade or detoxify substances hazardous to human health and/or the environment. The microorganisms may be indigenous to a contaminated area or they may be isolated from elsewhere and brought to the contaminated site. Contaminant compounds are transformed by living organisms through reactions that take place as a part of their metabolic processes. Biodegradation of a compound is often a result of the actions of multiple organisms. When microorganisms are imported to a contaminated site to enhance degradation we have a process known as bio-augmentation. For bioremediation to be effective, microorganisms must enzymatically attack the pollutants and convert them to harmless products. As bioremediation can be effective only where environmental conditions permit microbial growth and activity, its application often involves the manipulation of environmental parameters to allow microbial growth and degradation to proceed at a faster rate. Bioremediation techniques are typically more economical than traditional methods such as incineration, and some pollutants can be treated on site, thus reducing exposure risks for cleanup personnel, or potentially wider exposure as a result of transportation accidents.

Bioremediation of asphalt has becomes increasingly important due to the damage caused by the spills of such materials. To lay a solid foundation of bioremediation processes, it is necessary (as with all petroleum products) that there is an understanding of the composition and properties of asphalt. Both the atmospheric residuum (which also contains a fraction known as vacuum gas oil) and the vacuum residuum residue (which is the atmospheric residuum without the vacuum gas oil) contain compounds that either are (i) non-biodegradable or (ii) degraded very slowly. One of the reasons that residuum, which is high in asphaltene constituents, makes an excellent road-paving material is because it is slow to degrade. Tar balls, like mousse, are difficult to degrade because their low surface area restricts the availability of oxygen and other nutrients.

Resin constituents include petroleum compounds containing nitrogen, sulfur, and/or oxygen as constituents. If not highly condensed, they may be subject to limited microbial degradation (Traxler et al., 1965; Speight and Arjoon, 2012). Asphaltene constituents and resin constituents are difficult to analyze and, to date, little information is available on the biodegradability of most compounds in these groups. Light oils may contain about 1–5% of both asphaltene constituents and resin constituents, whereas heavy oil or weathered oil may have up to 25% (w/w) asphaltene constituents and 20% (w/w) resin constituents.

12.5.1 Asphalt

In asphalt, which is characterized by high viscosity and high density or low API gravity (Speight, 2014), asphaltene constituents, resin constituents, and the matrix compose a dynamic stable system, in which the non-asphaltene and non-resin fraction act as the medium for dispersion in which the asphaltene constituents are stabilized by the resin constituents (Speight, 2014).

The inherent biodegradability of these individual high-boiling constituents is a reflection of their chemical structure, but is also strongly influenced by the physical state and toxicity of the compounds (Pendrys, 1989). Whilst *n*-alkanes as a structural group are the most biodegradable petroleum hydrocarbons, the higher-molecular-weight constituents become increasingly recalcitrant because not only the chemical structure but also the physical state of the constituents strongly influences biodegradation (Bartha and Atlas, 1977). In fact, asphaltene and resin constituents tend to increase during biodegradation in relative and sometimes absolute amounts (Okoh, 2006). This suggests that they not only tend to resist biodegradation but may also be formed by condensation reactions of biodegradation and photodegradation intermediates.

The biodegradation of asphalt requires a complex metabolic pathway which usually can be observed in a microbial community and many investigations have been carried out on the biodegradation of petroleum hydrocarbons using a consortium of microorganisms (Morais and de Tauk-Tornisielo, 2009; Speight and Arjoon, 2012). Moreover, it has been reported that in soils that are permeated with heavy hydrocarbons, bacteria including indigenous ones would survive and function after contamination (Whyte et al., 1997; Foght et al., 1990) seeped through the soil. Selection of bacterial communities for petroleum substances occurs rapidly after even short-term exposures of soil to petroleum hydrocarbons

(Van Der Meer et al., 1992; Van Der Meer, 1994). During adaptation of microbial communities to hydrocarbon components, particularly complex ones, genes for hydrocarbon-degrading enzymes that are carried on plasmids or transposons may be exchanged between species and new catabolic pathways eventually may be assembled and modified for efficient regulation (Rabus et al., 2005). Other cell adaptations leading to new ecotypes may include modifications of the cell envelope to tolerate solvents (Ramos et al., 2002) and development of community-level interactions that facilitate cooperation within consortia.

One of the limitation factors for biodegradation of petroleum products in oil is the bioavailability (Banat, 1995; Kim et al. 2001) of these components due to their chemical structure (Harvey, 1997; Cerniglia, 1992). One of the options to increase bioavailability of PAHs is the use of surfactants to increase desorption and apparent solubility in the aqueous phase and consequently enhance the oil mobility, improving the biodegradation rates (Barathi and Vasudevan, 2001; Burd and Ward, 1996a,b; Rosenberg et al., 1988; Sar and Rosenberg, 1983).

In the United States alone, several million tons of asphalt are used annually as cement in the construction of roads. Asphalt is the unrefined residue of the fractional distillation of petroleum crude oil. The chemical composition of asphalt depends not only on the source of the petroleum crude oil from which it is derived but also on the method of its production (Speight, 2014).

An early investigation indicated that organisms of the genus *Pseudomonas* vary to a greater or lesser extent in the mechanisms by which they oxidize an asphalt substrate (Phillips and Traxler, 1963). However, the use of a single asphalt for these investigations rules out the possibility that the findings were generally applicable to asphalts from various sources and manufactured by different processes. Other investigations have established that some bacteria can utilize asphalt but the complexity of asphalt has limited the number of investigations applicable to asphalt biodegradation (Atlas, 1981; Speight and Arjoon, 2012). Petroleum-degrading bacteria excrete biosurfactants which emulsify hydrocarbons of petroleum by reducing the interfacial tension between the hydrocarbons and water (Cooper and Zajic, 1980).

One petroleum-degrading bacterium, *Acinetobacter alcoaceticus*, excretes an emulsifier—*emulsan* (Goldman et al., 1982; Gutnick, 1987). Emulsan is an extracellular polysaccharide polymer with fatty acid side chains

(Shoham and Rosenberg, 1983; Shoham et al., 1983) (see Chapter 10). This extracellular molecule forms a complex with proteins. These proteins bind non-covalently to the polysaccharide backbone. This emulsifier loses most of its activity when these proteins are removed from the complex (Zosim et al., 1987; Foght et al., 1989).

Biodegradation of asphalt occurs mostly on the surface of the asphalt exposed to oxygen, where the primary nutrients (asphalt, oxygen, and other trace minerals) are all present in adequate levels. Surface degradation is limited by oxygen diffusion and diffusion or presence of needed minerals/ions (i.e., phosphorous and nitrogen) (Pendrys, 1989). As time progresses, degradation should diminish as all of these become limited. In addition, the mechanism for biodegradation may be such that a residue which is resistant to biodegradation remains on the surface, protecting the underlying asphalt from further biodegradation. As the saturate and naphthene aromatic hydrocarbons are emulsified, a residue remains on the surface which is resistant to biodegradation and protects the underlying asphalt from biodegradation. The most potent asphalt-degrading bacterium, *A. alcoaceticus* NAV2, excretes an emulsifier which is capable of emulsifying the saturate and naphthene aromatic fractions of asphalt cement-20 (Pendrys, 1989).

12.5.2 Asphaltene Constituents

The asphaltene constituents have drawn considerable attention due to problems caused by their detrimental effects in the extraction, transportation, and processing of residua because of their viscous and flocculating nature and their relative resistance to biodegradation following spills (Speight, 2014). The asphaltene constituents are the highest molecular weight and most polar fraction of crude oil. Despite that the structure of asphaltene constituents has not been fully elucidated, it is widely accepted that they are constituted by interacting systems of polyaromatic sheets bearing alkyl side-chains. Asphaltene molecules have a high content of O, N, and S heteroatoms as well as metals (V, Ni, and Fe) (Speight, 2014). The problems associated with asphaltene constituents have increased due to the need to extract heavier crude oils, as well as the trend to extract larger amounts of light fractions out of crude oil by cracking and visbreaking.

The asphaltene fraction is recognized as the most recalcitrant oil fraction. There is no clear evidence that asphaltene constituents can be degraded or transformed by microbial activity. Microorganisms have been found associated with bitumen (Wyndham and Costerton, 1981) and natural asphalt lakes (Naranjo et al., 2007), which contain high amounts of asphaltene constituents. Another investigation revealed that a wide range of phylogenetic groups existed within the Archaea and Bacteria domains in natural asphalt-rich tar pits and genes for encoding novel oxygenases were also detected in such samples (Kim and Crowley, 2007).

On the other hand, an extensive screening involving more than 750 strains of filamentous fungi was carried out to select strains able to modify untreated hard coal (Bublitz et al., 1994; Hofrichter et al., 1997). Only six of the 750 strains tested exhibited some activity, from which the most active fungi, *Panus tigrinus*, growing on wood shavings coated with coal asphaltene constituents, led to a decrease in the average molecular weight (Hofrichter et al., 1997), although the average molecular weight of any complex mixture is not a measure of the constituents that were actually biodegraded.

Furthermore, most investigations into the biodegradation of asphaltene constituents should be given careful consideration since the asphaltene content is typically determined using a gravimetric method after *n*-alkane precipitation, and thus the reported changes may be attributed to the disruption of the asphaltenic matrix by the production of surfactants during bacterial growth, liberating trapped hydrocarbons. Other investigations have shown that the asphaltene fraction does not support bacterial growth, and no changes in asphaltene content are found after bioconversion of heavy oil and asphaltene constituents (Lacotte et al., 1996; Thouand et al., 1999).

There have been claims of the biodegradation of asphaltene constituents by mixed bacteria (Bertrand et al., 1983; Rontani et al., 1985). However, none of these reports described the analytical results of extractable materials recovered from appropriate sterile controls. Therefore, most of the asphaltene losses during microbial activity could be considered abiotic losses (Lacotte et al., 1996).

It has been reported (Pineda et al., 2004) that a bacterial consortium is able to grow in the asphaltene fraction as the sole carbon source. Mineralization of the asphaltene constituents was estimated by measuring production of carbon dioxide and it was discovered as a result of two control experiments (inoculum without asphaltene constituents and non-inoculated asphaltene constituents), that the carbon dioxide production was equivalent to the carbon dioxide production found in the consortium growing in the asphaltene fraction. However, the microbial inoculum for consortium stabilization contained 1% crude oil, which could serve as carbon source. Thus, it is not possible to distinguish the origin of the carbon dioxide production. The first clear experimental evidence that enzymes are able to modify asphaltene molecules has been reported (Fedorak et al., 1993). Chloroperoxidase from the fungus *Caldariomyces fumago* and a chemically modified cytochrome c were able to transform petroporphyrin derivatives and asphaltene constituents in reaction mixtures containing organic solvents (Fedorak et al., 1993; Mogollon et al., 1998; Garcia-Arellano et al., 2004). Notable spectral changes in the petroporphyrin-rich fraction of asphaltene constituents were observed and the enzymatic oxidation of petroporphyrin derivatives led to the removal up to 74% of the nickel and 95% of the vanadium.

According to Fourier transform infrared spectroscopy (FTIR) spectra, the chemically modified *cytochrome c* catalyzed the oxidation of sulfur and carbon atoms in asphaltene molecules (Garcia-Arellano et al., 2004). The enzymatic treatment of asphaltene constituents is an interesting alternative for the removal of heavy metals. It would result in reduced catalyst poisoning during hydrotreatment and cracking processes. On the other hand, the introduction of polar groups in asphaltene molecules could positively affect their sedimentation properties and improve their behavior.

It has also been reported (Uribe-Alvarez et al., 2011) that a fungus isolated from a natural asphalt lake is able to grow using asphaltene constituents as the sole source of carbon and energy. Thus, a fungal strain isolated from a microbial consortium growing in a natural asphalt lake is able to grow in purified asphaltene constituents as the only source of carbon and energy. The asphaltene constituents were rigorously purified in order to avoid contamination from other petroleum fractions. In addition, most of the petroporphyrin derivatives were removed. The 18S rRNA and b-tubulin genomic sequences, as well as some morphologic characteristics, indicate that the isolate is *Neosartorya fischeri*. After 11 weeks of growth, the fungus is able to metabolize 15.5% of the asphaltenic carbon, including 13.2% transformed to CO_2. In a medium containing asphaltene constituents as the sole source of carbon and energy, the fungal isolate produces extracellular lactase activity, which is not detected when the fungus grows in a rich medium. The results obtained in this work clearly demonstrate that there are microorganisms able to metabolize and mineralize asphaltene constituents, which are considered the most recalcitrant petroleum fraction.

To overcome the shortcomings of conventional methods, microbial degradation of asphaltene has been accepted worldwide as the most promising environmentally sound technology for remediation of spills and discharges related to petroleum and petroleum products. Furthermore, bacterial metabolites (especially polysaccharides) are of great value as enhancers of oil recovery due to their surfactant activity and bio-emulsifying properties (Banat, 1995). Because the conditions in oil deposits are often saline, the use of salt-resistant metabolites may be advantageous to the recovery of oil. Furthermore, hypersaline water and soil are often contaminated with crude oils, heavy metals, or other toxic compounds from anthropogenic sources. However, conventional microbiological treatment processes do not function at high salt concentrations; therefore, the use of moderately halophilic bacteria should be considered (Hao and Lu, 2009).

In the past, biodegradation of asphaltene constituents through the use of a microbial consortium or mixed cultures isolated from soil samples, sediments contaminated with hydrocarbons and oil wells have taken place but in low proportions of 0.55–3.5% (Venkateswaran et al., 1995; Thouand et al., 1999). This is most likely due to the complex molecular structure of asphaltene constituents (Speight, 2014) which makes these molecular species resistant to biodegradation thereby causing their accumulation in the ecosystem where petroleum and its refining byproducts are spilled in either accidental or deliberate ways (Guiliano et al., 2000).

The focus of many investigations has generally been bioremediation of sites contaminated by total petroleum hydrocarbons (Iturbe et al., 2007; Machackova et al., 2008) and there is a general lack of detailed work that has been done on the biodegradation of asphaltene constituents. However, more recently, viscosity reduction by asphaltene degradation has been structurally characterized by FTIR. The work was focused on reduction of viscosity of heavy oil in order to improve enhanced recovery from the reservoir or deposit. The bacterium (*Garciaella petrolearia* TERIG02) also showed an additional preference to degrade toxic asphalt and aromatic compounds first, unlike the other known strains. Furthermore, these characteristics make the species *Garciaella petrolearia* TERIG02 a potential candidate for residua and asphalt biodegradation and a solution to degrading toxic aromatic compounds (Lavania et al., 2012).

In contrast to low-molecular-weight hydrocarbons, polycyclic aromatics and hydrocarbons included in the asphaltene fraction are usually considered as being only slightly biodegradable because of their insufficient susceptibility to microbial attack (Gibson and Subramanian, 1984; Cerniglia, 1992; Kanaly and Harayama, 2000). Among the pentacyclic triterpane derivatives, the hopane constituents are so stable that they are commonly used as ubiquitous biomarkers for the assessment of biodegradation levels of crude oil (Ourisson et al., 1979). They were shown to be only slightly biodegraded by specialized microflorae under laboratory conditions (Frontera-Suau et al., 2002).

The mechanism of the degradation is complex but is believed to be a sequential process in which *n*-alkane moieties are generally removed first, followed by the degradation of iso-alkane moieties, cycloalkane moieties, one-to-three ring aromatics, and finally polynuclear aromatic derivatives (Greenwood et al., 2008). However, the typical pattern of degradation varies with different bacteria as well as type and composition of oil (Greenwood et al., 2008; Zrafi-Nouira et al., 2009).

Microbes reduce the viscosity by degrading high-molecular-weight constituents into lower-molecular-weight constituents, such as biological surface active substances, acids, and gases. In addition, anaerobic fermentation leads to the production of acids, carbon dioxide, hydrogen, and alcohols. Anaerobic bacteria produce acetate and butyrate during the initial growth phase (acidogenic phase) of the fermentation process. As the culture moves to the second phase of fermentation, the stationary growth phase, there is a shift in the metabolism of the cells to solvent production (solvent-generation phase). These gaseous and liquid metabolites dissolve into the oil resulting in reduced viscosity (Bryant et al., 1998). Moreover, the reaction of asphalt degradation within an acidic background is preferable as the proton (H^+) effectively interacts with the polar functionalities in the asphaltene constituents and resin constituents thereby efficiently reducing the polar interactions which results in breaking the intermolecular associations that exist in the raw residuum or asphalt. Furthermore, when each of the fractions, namely aliphatic, aromatic, and asphaltic (asphaltene and resin), were treated with *Garciaella petrolearia* TERIG02 for 30 days it was found that maximum degradation was in the case of the asphalt followed by aromatic fraction (Lavania et al., 2012). Indications were that *Garciaella petrolearia* TERIG02 could tolerate the toxicity of these compounds and was capable of utilizing them as a carbon and energy source.

In addition, five asphaltene-degrading bacterial strains were obtained from crude oil and polluted soil samples of Dorood oilfield in the south of Iran. Maximum degradation abilities of 46% and 48% were observed by *B. lentus* and the mixed culture of five selected isolates at 28°C, respectively. Statistical optimization of asphaltene biodegradation was successfully carried out and the optimum values of pH, salinity, and asphaltene concentration for asphaltene biodegradation at 40°C (104°F) were obtained for pure cultures of *B. lentus* and the kinetic investigation allowed development of a model that showed a good fit for the biodegradation of asphaltene constituents.

REFERENCES

Abraham, H., 1945. Asphalts and Allied Substances: Their Occurrence, Modes of Production, Uses in the Arts and Methods of Testing, fifth ed. Van Nostrand, New York, NY.

API. 2009. Asphalt Category Analysis and Hazard Categorization. The American Petroleum Institute Petroleum HPV Testing Group. Report Submitted to the US Environmental Protection Agency, Washington, DC, July 14.

Asphalt Institute and Eurobitume, 2008. The Bitumen Industry—A Global Perspective: Production, Chemistry, Use, Specification and Occupational Exposure. The Asphalt Institute, Lexington, KY.

ASTM D4, 2014. Standard Test Method for Bitumen Content Annual Book of Standards. ASTM International, West Conshohocken, PA.

ASTM D36, 2014. Standard Test Method for Softening Point of Bitumen (Ring-and-Ball Apparatus) Annual Book of Standards. ASTM International, West Conshohocken, PA.

ASTM D61, 2014. Standard Test Method for Softening Point of Pitches (Cube-in-Water Method) Annual Book of Standards. ASTM International, West Conshohocken, PA.

ASTM D140, 2014. Standard Practice for Sampling Bituminous Materials Annual Book of Standards. ASTM International, West Conshohocken, PA.

ASTM D312, 2014. Standard Specification for Asphalt Used in Roofing Annual Book of Standards. ASTM International, West Conshohocken, PA.

ASTM D449, 2014. Standard Specification for Asphalt used in Dampproofing and Waterproofing Annual Book of Standards. ASTM International, West Conshohocken, PA.

ASTM D977, 2014. Standard Specification for Emulsified Asphalt Annual Book of Standards. ASTM International, West Conshohocken, PA.

ASTM D2042, 2014. Standard Test Method for Solubility of Asphalt Materials in Trichloroethylene Annual Book of Standards. ASTM International, West Conshohocken, PA.

ASTM D2319, 2014. Standard Test Method for Softening Point of Pitch (Cube-in-Air Method) Annual Book of Standards. ASTM International, West Conshohocken, PA.

ASTM D2397, 2014. Standard Specification for Cationic Emulsified Asphalt Annual Book of Standards. ASTM International, West Conshohocken, PA.

ASTM D3104, 2014. Standard Test Method for Softening Point of Pitches (Mettler Softening Point Method) Annual Book of Standards. ASTM International, West Conshohocken, PA.

ASTM D3461, 2014. Standard Test Method for Softening Point of Asphalt and Pitch (Mettler Cup-and-Ball Method) Annual Book of Standards. ASTM International, West Conshohocken, PA.

Atlas, R.M., 1981. Microbial degradation of petroleum hydrocarbons: an environmental perspective. Microbiol. Rev. 45, 180–209.

Banat, I.M., 1995. Biosurfactants production and possible uses in microbial enhanced oil recovery and oil pollution remediation: a review. Bioresour. Technol. 51, 1–12.

Barathi, S., Vasudevan, N., 2001. Utilization of petroleum hydrocarbons by *Pseudomonas fluorescens* Isolated from a petroleum-contaminated soil. Environ. Int. 26, 413–416.

Barth, E.J., 1962. Asphalt: Science and Technology. Gordon and Breach Publishers, Taylor & Francis Group, Abingdon, Oxfordshire.

Bartha, R., Atlas, R.M., 1977. The microbiology of aquatic oil spills. Adv. Appl. Microbiol. 22, 225–266.

Bertrand, J.C., Rambeloarisoa, E., Rontani, J.F., Giusti, G., Mattei, G., 1983. Microbial degradation of crude oil in sea water in continuous culture. Biotechnol. Lett. 5, 567–572.

Blackburn, G.R., Kriech, A.J., Kurek, J.T., Osborn, L.A., 1999. Detection of Coal Tar Materials in Asphalt Pavements Using Chemical and Biological Methods Transportation Research Record No. 1266. Transportation Research Board, Washington, DC.

Boffetta, P., Burstyn, I., Partanen, T., Kromhout, H., Svane, O., Langaard, S., et al., 2001. Epidemiological Study of Cancer Mortality Among European Asphalt Workers (IARC Internal Report No. 01/003). International Agency for Research on Cancer, Lyon, France, <ftp://ftp.iarch.fr/ece/asphaltl>(accessed 21.02.15.). <http://www.petroleumhpv.org/ProducCCategories/Asphalt/asphalt.htm>(accessed 21.02.15).

Brandt, H.C.A., de Groot, P.C., 1985. Sampling and analysis of bitumen fumes. Part 3: laboratory study of emissions under controlled conditions. Ann. Occup. Hyg. 29 (1), 59–74.

Brandt, H.C.A., de Groot, P.C., 1999. A laboratory rig for studying aspects of worker exposure to bitumen fumes. Am. Ind. Hyg. Assoc. J. 60, 182–190.

Brandt, H.C.A., de Groot, P.C., 2001. Aqueous leaching of polycyclic aromatic hydrocarbons from bitumen and asphalt. Water Res. 35 (17), 4200–4207.

Brandt, H.C.A., de Groot, P.C., Molyneux, M.K.B., Tindle, P.E., 1985. Sampling and analysis of bitumen fumes. Part 2: field exposure measurements. Ann. Occup. Hyg. 29 (1), 47–58.

Bryant, R.S., Bailey, S.A., Step, A.K., Evans, D.B., Parli, J.A., 1998. Biotechnology for heavy oil recovery. SPE Paper No. 36767. In: Proceedings. SPE/DOE Improved Oil Recovery Symposium, pp. 1–7.

Bublitz, F., Guenther, T., Fritsche, W., 1994. Screening of fungi for the biological modification of hard coal and coal derivatives. Fuel Proc. Technol. 40, 347–354.

Burd, G., Ward, O.P., 1996a. Physico-chemical properties of PM-factor, a surface active agent produced by *Pseudomonas marginalis*. Can. J. Microbiol. 42, 243–251.

Burd, G., Ward, O.P., 1996b. Bacterial degradation of polycyclic aromatic hydrocarbons on agar plates: the role of biosurfactants. Biotechnol. Tech. 10, 371–374.

Burstyn, I., Kromhout, H., Boffetta, P., 2000. Literature review of levels and determinants of exposure to potential carcinogens and other agents in the road construction industry. Am. Ind. Hyg. Assoc. J. 61, 715–726.

Burstyn, I., Ferrari, P., Wegh, H., Kromhout, H., 2002a. Characterizing worker exposure to bitumen during hot mix paving and asphalt mixing operations. Am. Ind. Hyg. Assoc. J. 63, 293–299.

Burstyn, I., Randem, B., Lien, J.E., Langård, S., Kromhout, H., 2002b. Bitumen, polycyclic aromatic hydrocarbons and vehicle exhaust: exposure levels and controls among norwegian asphalt workers. Ann. Occup. Hyg. 46, 79–87.

Burstyn, I., Boffetta, P., Kauppinen, T., Heikkila, P., Svane, O., Partanen, T., et al., 2003a. Estimating exposures in the asphalt industry for an International Epidemiological Cohort Study of Cancer Risk. Am. J. Ind. Med. 43, 3–17.

Burstyn, I., Ferrari, P., Wegh, H., Heederik, D., Kromhout, H., 2003b. Characterizing worker exposure to bitumen during hot mix paving and asphalt mixing operations. Am. Ind. Hyg. Assoc. J. 63 (3), 293–299.

Butler, M.A., Burr, G., Dankovic, D., Lunsford, R.A., Miller, A., Nguyen, M., et al., 2000. Hazard Review: Health Effects of Occupational Exposure to Asphalt. National Institute of Occupational Safety and Health, Cincinnati, OH.

Cerniglia, C.E., 1992. Biodegradation of polycyclic aromatic hydrocarbons. Biodegradation 3, 351–368.

Claydon, M.F., Christian, F., Eyres, A.R., Guelfo, G., Lentge, H., Mitchell, J.G., et al., 1984. Review of Bitumen Fume Exposures and Guidance on Measurement Report No. 6184. International Study Group for Conservation of Clean Air and Water-Europe, The Hague, Netherlands.

Cooper, D.G., Zajic, J.E., 1980. Surface-active compounds from microorganisms. Adv. Appl. Microbiol. 26, 229–253.

Corbett, L.W., 1975. Reaction variables in the air blowing of asphalt. Ind. Eng. Chem. Process Des. Dev. 14 (2), 181–187.

Corbett, L.W., Petrossi, U., 1978. Differences in distillation and solvent separated asphalt residua. Ind. Eng. Chem. Prod. Res. Dev. 17 (4), 342–346.

D'Andrea, A., Bonora, V., Drago, D., 2004. Asphalt concrete with bottom ash: environmental aspects. In: Proceedings. International Conference of Restoration, Recycling and Rejuvenation Technology for Engineering and Architecture Application held in Cesena, Italy, June 7–11, pp. 56–63.

Daugherty, D.R., Pietrzak, R.F., Fuhrmann, M., Columbo, P., 1988. An Experimental Survey of the Factors that Affect Leaching from Low-Level Radioactive Waste Forms Report No. BNL-52125. Brookhaven National Laboratory, Upton, NY.

Ekstrom, L., Kriech, A., Bowen, C., Johnson, S., Breuer, D., 2001. International studies to compare methods for personal sampling of bitumen fumes. J. Environ. Monit. 3, 439–445.

Fedorak, P.M., Semple, K.M., Vazquez-Duhalt, R., Westlake, D.W.S., 1993. Chloroperoxidase mediated modifications of petroporphyrins and asphaltenes. Enzyme Microb. Technol. 15, 429–437.

Foght, J.M., Gutnick, D.L., Westlake, D.W.S., 1989. Effect of emulsan on biodegradation of crude oil by pure and mixed bacterial cultures. Appl. Environ. Microbiol., 36–42. January.

Foght, J.M., Fedorak, P.M., Westlake, D.W.S., 1990. Mineralization of [^{14}C] Hexadecane and [^{14}C] phenanthrene in crude oil: specificity among bacterial isolates. Can. J. Microbiol. 36, 169–175.

Frontera-Suau, R., Bost, F., McDonald, T., Morris, P.J., 2002. Aerobic biodegradation of hopanes and other biomarkers by crude oil degrading enrichment cultures. Environ. Sci. Technol. 36, 4585–4592.

Fuhrmann, M., Pietrazak, R.F., Franz, E.M., Heiser III, J.H., Columbo, P., 1989. Optimization of the Factors that Accelerate Leaching Report No. BNL-52204. Brookhaven National Laboratory, Upton, NY.

Gamble, J.F., Nicolich, M.J., Barone, N.J., Vincent, W.J., 1999. Exposure-response of asphalt fumes with changes in pulmonary function and symptoms. Scand. J. Work Environ. Health 25 (3), 186–206.

Garcia-Arellano, H., Buenrostro-Gonzalez, E., Vazquez-Duhalt, R., 2004. Biocatalytic transformation of petroporphyrins by chemical modified cytochrome c. Biotechnol. Bioeng. 85, 790–798.

Gary, J.G., Handwerk, G.E., Kaiser, M.J., 2007. Petroleum Refining: Technology and Economics, fifth ed. CRC Press, Taylor & Francis Group, Boca Raton, FL.

Gibson, D.T., Subramanian, V., 1984. Microbial Degradation of Aromatic Hydrocarbons. In: Gibson, D.T. (Ed.), Microbial Degradation of Organic Compounds McGraw–Hill, New York, NY, pp. 181–252.

Goldman, S., Shabtai, Y., Rubinovitz, C., Rosenberg, E., Gutnick, D.L., 1982. Emulsan in *Acinetobacter calcoaceticus* RAG-1: distribution of cell-free and cell-associated cross-reacting material. Appl. Environ. Microbiol. 44, 165–170.

Greenwood, P.F., Wibrow, S., George, S.J., Tibbett, M., 2008. Sequential hydrocarbon biodegradation in a soil from arid coastal australia, treated with oil under laboratory-controlled conditions. Org. Geochem. 39, 1336–1346.

Guiliano, M., Boukir, A., Doumenq, P., Mille, G., 2000. Supercritical fluid extraction of BAL 150 crude oil asphaltenes. Energy Fuels 14, 89–94.
Gutnick, D.L., 1987. The emulsan polymer: perspectives on a microbial capsule as an industrial product. Biopolymers 26, 223–240.
Hao, R., Lu, A., 2009. Biodegradation of heavy oils by halophilic bacterium. Prog. Nat. Sci. 19, 997–1001.
Harvey, R., 1997. Polycyclic Aromatic Hydrocarbons. Wiley-VCH, New York, NY.
Heikkilä, P., Riala, R., Hämeilä, M., Nykyri, E., Pfäffli, P., 2002. Occupational exposure to bitumen during road paving. Am. Ind. Hyg. Assoc. J. 63, 156–165.
Hofrichter, M., Bublitz, F., Fritsche, W., 1997. Fungal attack on coal: I. Modification of hard coal by fungi. Fuel Proc. Technol. 52, 43–53.
Hoiberg, A.J., 1964. Bituminous Materials: Asphalts, Tars, and Pitches. John Wiley and Sons, New York, NY.
Hsu, C.S., Robinson, P.R., 2006. Practical Advances in Petroleum Processing, vols. 1 and 2. Springer, New York, NY.
Irwin, R.J., Van Mouwerik, M., Stevens, L., Seese, M.D., Basham, W., 1997. Environmental Contaminants Encyclopedia. National Park Service, Water Resources Division, Fort Collins, CO.
Iturbe, R., Flores, C., Castro, A., Torres, L.G., 2007. Sub-soil contamination due to oil spills in zones surrounding oil pipeline-pump stations and oil pipeline right-of-ways in Southwest-Mexico. Environ. Monit. Assess. 133, 387–398.
Kanaly, R.A., Harayama, S., 2000. Biodegradation of high-molecular weight polycyclic aromatic hydrocarbons by bacteria. J. Bacteriol. 182, 2059–2067.
Kennepohl, G.J.A., 2008. Asphalt Pavements and the Environment Proceedings. ISAP International Symposium on Asphalt Pavements and Environments. International Society for Asphalt Pavements, Zürich, Switzerland, August.
Kim, J.-S., Crowley, D.E., 2007. Microbial diversity in natural asphalts of the rancho La Brea Tar Pits. Appl. Environ. Microbiol. 73, 4579–4591.
Kim, I.S., Park, J., Kim, K.W., 2001. Enhanced biodegradation of polycyclic aromatic hydrocarbons using nonionic surfactants in soil slurry. Appl. Geochem. 16, 1419–1428.
Kitto, A.M., Pirbazari, M., Badriyha, B.N., Ravindran, V., Synolakis, C.E., 1997. Emissions of volatile and semi-volatile organic compounds and particulate matter from hot asphalts. Environ. Technol. 18, 121–138.
Kriech, A.J., Kurek, J.T., Osborn, L.A., Blackburn, G.R., 1997. Extraction and analysis of asphalt pavement core samples: detection of coal tar-derived species using chemical and biological methods. Polycyclic Aromat. Compd. 16, 3–40.
Kriech, A.J., Kurek, J.T., Wissel, H.L., Osborn, L.V., Blackburn, G.R., 2002. Evaluation of worker exposure to asphalt paving fumes using traditional and nontraditional techniques. J. Am. Ind. Hyg. Assoc. 63 (5), 628–635.
Lacotte, D.J., Mille, G., Acquaviva, M., Bertrand, J.C., 1996. Arabian light 150 asphaltene biotransformation with *n*-alkanes as co-substrate. Chemosphere 32, 1755–1761.
Lange, C.R., Stroup-Gardiner, M., 2007. Temperature-dependent chemical-specific emission rates of aromatics and polyaromatic hydrocarbons (PAHs) in bitumen fume. J. Occup. Environ. Hyg. 4 (S1), 72–76.
Lavania, M., Cheema, S., Sarma, P.M., Mandal, A.K., Lal, B., 2012. Biodegradation of asphalt by *Garciaella petrolearia* TERIG02 for viscosity reduction of heavy oil. Biodegradation 23, 15–24.
Legret, M., Odie, L., Demare, D., Jullien, A., 2005. Leaching of heavy metals and polycyclic aromatic hydrocarbons from reclaimed asphalt pavement. Water Res. 39 (15), 3675–3685.

Luey, J., Li, S.W., 1993. Determination of the Biodegradation Rate of Asphalt for the Hanford Grout Vaults Report. Contract DE-AC06-76RLO 1830. Pacific Northwest Laboratory Richland, US Department of Energy, Richland, WA, April.

Machackova, J., Wittlingerova, Z., Vlk, K., Zima, J., Ales, L., 2008. Comparison of two methods for assessment of in situ jet fuel remediation efficiency. Water Air Soil Pollut. 187, 181–194.

Mahler, B.J., LaVista, J., 2015. Coal-Tar-Sealant Runoff Causes Toxicity and DNA Damage. USGS Newsroom. United States Geological Survey, Reston, VA, <http://www.usgs.gov/newsroom/article.asp?ID=4190&from=news_side#.VS6JEfnF9kZ> (accessed 15.04.15).

Mogollon, L., Rodriguez, R., Larrota, W., Ortiz, C., Torres, R., 1998. Biocatalytic removal of nickel and vanadium from petroporphyrins and asphaltenes. Appl. Biochem. Biotechnol. 70–72, 765–777.

Morais, E.B., de Tauk-Tornisielo, S.M., 2009. Biodegradation of oil refinery residues using mixed-culture of microorganisms isolated from land farming. Braz. Arch. Biol. Technol. 52 (6), 1571–1578.

Moretti, L., Di Mascio, P., D'Andrea, A., 2013. Environmental impact assessment of road asphalt pavements. Mod. Appl. Sci. 7 (11), 1–11.

Moschopedis, S.E., Speight, J.G., 1973. The oxidation of petroleum fractions. Fuel 52, 83.

Moschopedis, S.E., Speight, J.G., 1975. Oxidation of bitumen. Fuel 54, 210.

Moschopedis, S.E., Speight, J.G., 1977. The effect of air blowing on the properties and constitution of natural bitumen. J. Mater. Sci. 12, 990.

Moschopedis, S.E., Speight, J.G., 1978. The influence of metal salts on bitumen oxidation. Fuel 57, 235.

Mundt, D.J., Marano, K.M., Nunes, A.P., Adams, R.C., 2009. A review of changes in composition of hot mix asphalt in the United States. J. Occup. Environ. Hyg. 6 (11), 714–725.

Naranjo, L., Urbina, H., De Sisto, A., León, V., 2007. Isolation of autochthonous non-white rot fungi with potential for enzymatic upgrading of Venezuelan extra-heavy crude oil. Biocatal. Biotransform. 25, 1–9.

Okoh, A.I., 2006. Biodegradation alternative in the cleanup of petroleum hydrocarbon pollutants. Biotechnol. Mol. Biol. Rev. 1 (2), 38–50.

Ourisson, G., Albrecht, P., Rohmer, M., 1979. The Hopanoids. Paleochemistry and biochemistry of a group of natural products. Pure Appl. Chem. 51, 709–729.

Parkash, S., 2003. Refining Processes Handbook. Gulf Professional Publishing, Elsevier, Amsterdam, Netherlands.

Pendrys, J.P., 1989. Biodegradation of asphalt cement-20 by aerobic bacteria. Appl. Environ. Microbiol. 55 (6), 1357–1362.

Petersen, J.C., 1986. Quantitative Functional Group Analysis of Asphalts Using Differential Infrared Spectrometry and Selective Chemical Reactions—Theory and Application Transportation Research Record 1096. Transportation Research Board, National Research Council., pp. 1–11.

Petersen, J.C., Plancher, H., Ensley, E.K., Venable, R.L., Miyake, G., 1982. Chemistry of Asphalt-Aggregate Interactions: Relationship with Pavement Moisture Damage Prediction Test Transportation Research Record No. 843. Transportation Research Board, National Research Council, Washington, DC, pp. 95–104.

Phillips, U.A., Traxler, R.W., 1963. Microbial degradation of asphalt. Appl. Microbiol. 11, 235–238.

Pineda, F.G., Mesta-Howard, A.M., Boll-Argüello, G., Lira-Galeana, C., 2004. A Microbial consortium isolated from a crude oil sample that uses asphaltenes as a carbon and energy source. Biodegradation 15, 145–151.

Rabus, R., Kube, M., Heider, J., Beck, A., Heitmann, K., Widdel, F., et al., 2005. The genome sequence of an anaerobic aromatic-degrading denitrifying bacterium, Strain Ebn1. Arch. Microbiol. 183, 27–36.
Ramos, J.L., Duque, E., Gallegos, M.T., Godyoy, P., Ramos-Gonzalez, M.I., Rojas, A., et al., 2002. Mechanisms of solvent tolerance in gram-negative bacteria. Annu. Rev. Microbiol. 56, 743–768.
Ray, D.L., Guzzo, L., 1990. Trashing The Planet: How Science Can Help Us Deal With Acid Rain, Depletion of The Ozone, and Nuclear Waste (Among Other Things). Regnery Gateway, Washington, DC.
Rontani, J.F., Bosser-Joulak, F., Rambeloarisoa, E., Bertrand, J.C., Giusti, G., Faure, R., 1985. Analytical study of asyhart crude oil biodegradation. Chemosphere 14, 1413–1422.
Rosenberg, E., Rubinovitz, C., Gottlieb, A., Rosenhak, S., Ron, E.Z., 1988. Production of Biodispersan by *Acinetobacter calcoaceticus* A2. Appl. Environ. Microbiol. 54, 317–322.
Roy, T.A., Kriech, A.J., Mackerer, C.R., 2007. Percutaneous absorption of polycyclic aromatic compounds from bitumen fume condensate. J. Occup. Environ. Hyg. 4 (1), 137–143.
Sar, N., Rosenberg, E., 1983. Emulsifier production by *Acinetobacter calcoaeticus* strains. Curr. Microbiol. 9, 309–314.
Shoham, Y., Rosenberg, E., 1983. Enzymatic depolymerization of emulsan. J. Bacteriol. 156 (1), 61–167.
Shoham, Y., Rosenberg, M., Rosenberg, E., 1983. Bacterial degradation of emulsan. Appl. Environ. Microbiol. 46, 573–579.
Speight, J.G., 1992. Asphalt. In: Kirk–Othmer's Encyclopedia of Chemical Technology, fourth ed., vol. 3, pp. 689–724.
Speight, J.G., 2005. Environmental Analysis and Technology for the Refining Industry. John Wiley & Sons Inc., Hoboken, NJ.
Speight, J.G., 2013. The Chemistry and Technology of Coal, third ed. CRC Press, Taylor & Francis Group, Boca Raton, FL.
Speight, J.G., 2014. The Chemistry and Technology of Petroleum, fifth ed. CRC Press, Taylor & Francis Group, Boca Raton, FL.
Speight, J.G., 2015. Handbook of Petroleum Product Analysis, second ed. John Wiley & Sons Inc., Hoboken, NJ.
Speight, J.G., Arjoon, K.K., 2012. Bioremediation of Petroleum and Petroleum Products. Scrivener Publishing, Salem, MA.
Speight, J.G., Ozum, B., 2002. Petroleum Refining Processes. Marcel Dekker Inc., New York, NY.
Testa, S.M., 1995. Chemical aspects of cold-mix asphalt incorporating contaminated soil. J. Soil Contam. 4 (2), 1–17.
Thouand, G., Bauda, P., Oudot, J., Kirsch, G., Sutton, C., Vidalie, J.F., 1999. Laboratory evaluation of crude oil biodegradation with commercial or natural microbial inocula. Can. J. Microbiol. 45, 106–115.
Traxler, R.W., Proteau, P.R., Traxler, R.N., 1965. Action of microorganisms on bituminous materials. I. Effect of bacteria on asphalt viscosity. Appl. Microbiol. 13 (6), 838–841.
Uribe-Alvarez, C., Ayala, M., Perezgasga, L., Naranjo, L., Urbina, H., Vazquez-Duhalt, R., 2011. First evidence of mineralization of petroleum asphaltenes by a strain of *Neosartorya fischeri*. Microbiol. Biotechnol. 4 (5), 663–672.
US EPA, 1996. Study of Secleted Petroleum Refining Residuals. Office of Solid Waste Management, Environmetnal Protection Agency, Washington, DC.
Van Der Meer, J.R., 1994. Genetic adaptation of bacteria to chlorinated aromatic compounds. FEMS Microbiol. Rev. 15, 239–249.
Van Der Meer, J.R., Devos, W.M., Harayama, S., Zehnder, A.J.B., 1992. Molecular mechanisms of genetic adaptation to xenobiotic compounds. Microbiol. Rev. 56, 677–694.

Venkateswaran, K., Hoaki, T., Kato, M., Maruyama, T., 1995. Microbial degradation of resins fractionated from Arabian light crude oil. Can. J. Microbiol. 41, 418–424.

WHO, 2004. Asphalt (Bitumen) Concise International Chemical Assessment Document No. 59. World Health Organization, Geneva, Switzerland.

Whyte, L.G., Bourbonnière, L., Greer, C.W., 1997. Biodegradation of petroleum hydrocarbons by psychrotrophic *Pseudomonas* strains possessing both alkane (*alk*) and naphthalene (*nah*) catabolic pathways. Appl. Environ. Microbiol. 63 (9), 3719–3723.

Winter, K., 2006. Leaching Characteristics of Recycled Asphalt Pavement: RAP may be Suitable as a Fill Material. VDOT Research Library. Virginia Department of Transportation, Charlottesville, VA.

Wolfe, D.L., Armentrout, D., Arends, C.B., Baker, H.M., Petersen, J.C., 1986. Crude Source Effects on the Chemical, Morphological, and Viscoelastic Properties of Styrene/Butadiene Latex Modified Asphalt Cements Transportation Research Record No. 1096. Transportation Research Board, National Research Council, Washington, DC, pp. 12–21.

Wyndham, R.C., Costerton, J.W., 1981. *In vitro* microbial degradation of bituminous hydrocarbons and *in situ* colonization of bitumen surfaces within the athabasca oil sands deposit. Appl. Environ. Microbiol. 41, 791–800.

Zosim, Z., Gutnick, D.L., Rosenberg, E., 1987. Effect of protein content on the surface activity and viscosity of emulsan. Colloid Polymer Sci. 265, 442–447.

Zrafi-Nouira, I., Guermazi, S., Chouari, R., Safi, N.M.D., Pelletier, E., Backhrouf, A., et al., 2009. Molecular diversity analysis and bacterial population dynamics of an adapted seawater microbiota during the degradation of tunisian zarzantine oil. Biodegradation 20, 467–486.

APPENDIX 1

ASTM International Paving Standards and Road Standards

The road and paving standards developed by ASTM International provide the specifications and test methods pertinent to the material, physical, mechanical, performance, and application requirements of road surfaces and pavements. These geotechnical surfaces are laid down on specified areas intended to sustain either or both foot and vehicular traffic. The types of paving covered by these standards are typically made of asphalt, concrete, and bituminous materials. These road and paving standards allow geotechnical engineering firms and construction companies to examine and evaluate paving materials to ensure strength and durability toward safe application and use.

AGGREGATE SPECIFICATIONS

D242 Standard Specification for Mineral Filler for Bituminous Paving Mixtures

D448 Standard Classification for Sizes of Aggregate for Road and Bridge Construction

D692 Standard Specification for Coarse Aggregate for Bituminous Paving Mixtures

D1073 Standard Specification for Fine Aggregate for Bituminous Paving Mixtures

D1139 Standard Specification for Aggregate for Single or Multiple Bituminous Surface Treatments

D2940 Standard Specification for Graded Aggregate Material for Bases or Subbases for Highways or Airports

D5106 Standard Specification for Steel Slag Aggregates for Bituminous Paving Mixtures

D6155 Standard Specification for Nontraditional Coarse Aggregates for Bituminous Paving Mixtures

AGGREGATE TESTS

D546 Standard Test Method for Sieve Analysis of Mineral Filler for Bituminous Paving Mixtures

D2419 Standard Test Method for Sand Equivalent Value of Soils and Fine Aggregate

D3042 Standard Test Method for Insoluble Residue in Carbonate Aggregates

D3319 Standard Practice for Accelerated Polishing of Aggregates Using the British Wheel

D3744 Standard Test Method for Aggregate Durability Index

D4469 Standard Practice for Calculating Percent Asphalt Absorption by the Aggregate in an Asphalt Pavement Mixture

D4791 Standard Test Method for Flat Particles, Elongated Particles, or Flat and Elongated Particles in Coarse Aggregate

D4792 Standard Test Method for Potential Expansion of Aggregates from Hydration Reactions

D5148 Standard Test Method for Centrifuge Kerosine Equivalent

D5444 Standard Test Method for Mechanical Size Analysis of Extracted Aggregate

D5821 Standard Test Method for Determining the Percentage of Fractured Particles in Coarse Aggregate

D6928 Standard Test Method for Resistance of Coarse Aggregate to Degradation by Abrasion in the Micro-Deval Apparatus

D7172 Standard Test Method for Determining the Relative Density (Specific Gravity) and Absorption of Fine Aggregates Using Infrared

D7370 Standard Test Method for Determination of Relative Density and Absorption of Fine, Coarse, and Blended Aggregate Using Combined Vacuum Saturation and Rapid Submersion

D7428 Standard Test Method for Resistance of Fine Aggregate to Degradation by Abrasion in the Micro-Deval Apparatus

ANALYSIS OF ASPHALT MIXTURES

D1461 Standard Test Method for Moisture or Volatile Distillates in Bituminous Paving Mixtures

D1856 Standard Test Method for Recovery of Asphalt from Solution by Abson Method

D2172 Standard Test Methods for Quantitative Extraction of Bitumen from Bituminous Paving Mixtures
D4125 Standard Test Methods for Asphalt Content of Bituminous Mixtures by the Nuclear Method
D4887 Standard Practice for Preparation of Viscosity Blends for Hot Recycled Bituminous Materials
D5404 Standard Practice for Recovery of Asphalt from Solution Using the Rotary Evaporator
D6307 Standard Test Method for Asphalt Content of Hot-Mix Asphalt by Ignition Method
D6390 Standard Test Method for Determination of Draindown Characteristics in Uncompacted Asphalt Mixtures
D7906 Standard Practice for Recovery of Asphalt from Solution Using Toluene and the Rotary Evaporator

ASPHALT SPECIFICATIONS

D946 Standard Specification for Penetration-Graded Asphalt Cement for Use in Pavement Construction
D2026 Standard Specification for Cutback Asphalt (Slow-Curing Type)
D2027 Standard Specification for Cutback Asphalt (Medium-Curing Type)
D2028 Standard Specification for Cutback Asphalt (Rapid-Curing Type)
D2399 Standard Practice for Selection of Cutback Asphalts
D3141 Standard Specification for Asphalt for Undersealing Portland-Cement Concrete Pavements
D3381 Standard Specification for Viscosity-Graded Asphalt Cement for Use in Pavement Construction
D4311 Standard Practice for Determining Asphalt Volume Correction to a Base Temperature
D4552 Standard Practice for Classifying Hot-Mix Recycling Agents
D5710 Standard Specification for Trinidad Lake Modified Asphalt
D6114 Standard Specification for Asphalt-Rubber Binder
D6154 Standard Specification for Chemically Modified Asphalt Cement for Use in Pavement Construction
D6373 Standard Specification for Performance Graded Asphalt Binder
D6626 Standard Specification for Performance Graded Trinidad Lake Modified Asphalt Binder

ASPHALT SURFACE TREATMENTS

D1369 Standard Practice for Quantities of Materials for Bituminous Surface Treatments

D2995 Standard Practice for Estimating Application Rate of Bituminous Distributors

D3910 Standard Practices for Design, Testing, and Construction of Slurry Seal

D5360 Standard Practice for Design and Construction of Bituminous Surface Treatments

D5624 Standard Practice for Determining the Transverse-Aggregate Spread Rate for Surface Treatment Applications

D6372 Standard Practice for Design, Testing, and Construction of Micro-Surfacing

D7000 Standard Test Method for Sweep Test of Bituminous Emulsion Surface Treatment Samples

D7564 Standard Practice for Construction of Asphalt-Rubber Cape Seal

BRIDGES AND STRUCTURES

D3633 Standard Test Method for Electrical Resistivity of Membrane-Pavement Systems

D3963 Standard Specification for Fabrication and Jobsite Handling of Epoxy-Coated Steel Reinforcing Bars

D4014 Standard Specification for Plain and Steel-Laminated Elastomeric Bearings for Bridges

D4071 Standard Practice for Use of Portland Cement Concrete Bridge Deck Water Barrier Membrane Systems

D4580 Standard Practice for Measuring Delaminations in Concrete Bridge Decks by Sounding

D4788 Standard Test Method for Detecting Delaminations in Bridge Decks Using Infrared Thermography

D5977 Standard Specification for High Load Rotational Spherical Bearings for Bridges and Structures

D6087 Standard Test Method for Evaluating Asphalt-Covered Concrete Bridge Decks Using Ground Penetrating Radar

D6153 Standard Specification for Materials for Bridge Deck Waterproofing Membrane Systems

D6297 Standard Specification for Asphaltic Plug Joints for Bridges

CALCIUM AND SODIUM CHLORIDES AND OTHER DEICING MATERIALS

D98 Standard Specification for Calcium Chloride

D345 Standard Test Method for Sampling and Testing Calcium Chloride for Roads and Structural Applications

D632 Standard Specification for Sodium Chloride

COLD MIX ASPHALT

D4215 Standard Specification for Cold-Mixed, Cold-Laid Bituminous Paving Mixtures

D6627 Standard Test Method for Determination of a Volatile Distillate Fraction of Cold Asphalt Mixtures

D6704 Standard Test Method for Determining the Workability of Asphalt Cold Mix Patching Material

D7196 Standard Test Method for Raveling Test of Cold Mixed Emulsified Asphalt Samples

D7229 Standard Test Method for Preparation and Determination of the Bulk Specific Gravity of Dense-Graded Cold Mix Asphalt (CMA) Specimens by Means of the Superpave Gyratory Compactor

DURABILITY AND DISTILLATION TESTS

D243 Standard Test Method for Residue of Specified Penetration

D402 Standard Test Method for Distillation of Cutback Asphalt

D1754 Standard Test Method for Effects of Heat and Air on Asphaltic Materials (Thin-Film Oven Test)

D2872 Standard Test Method for Effect of Heat and Air on a Moving Film of Asphalt (Rolling Thin-Film Oven Test)

D6521 Standard Practice for Accelerated Aging of Asphalt Binder Using a Pressurized Aging Vessel (PAV)

EFFECT OF WATER AND OTHER ELEMENTS ON ASPHALT COATED AGGREGATES

D1075 Standard Test Method for Effect of Water on Compressive Strength of Compacted Bituminous Mixtures

D3625 Standard Practice for Effect of Water on Bituminous-Coated Aggregate Using Boiling Water

D4867 Standard Test Method for Effect of Moisture on Asphalt Concrete Paving Mixtures

D7870 Standard Practice for Moisture Conditioning Compacted Asphalt Mixture Specimens by Using Hydrostatic Pore Pressure

EMULSIFIED ASPHALT SPECIFICATIONS

D977 Standard Specification for Emulsified Asphalt

D2397 Standard Specification for Cationic Emulsified Asphalt

D3628 Standard Practice for Selection and Use of Emulsified Asphalts

D5505 Standard Practice for Classifying Emulsified Recycling Agents

EMULSIFIED ASPHALT TEST

D244 Standard Test Methods and Practices for Emulsified Asphalts

D6929 Standard Practice for Freezing of Emulsified Asphalts

D6930 Standard Test Method for Settlement and Storage Stability of Emulsified Asphalts

D6933 Standard Test Method for Oversized Particles in Emulsified Asphalts (Sieve Test)

D6934 Standard Test Method for Residue by Evaporation of Emulsified Asphalt

D6935 Standard Test Method for Determining Cement Mixing of Emulsified Asphalt

D6936 Standard Test Method for Determining Demulsibility of Emulsified Asphalt

D6937 Standard Test Method for Determining Density of Emulsified Asphalt

D6997 Standard Test Method for Distillation of Emulsified Asphalt

D6998 Standard Practice for Evaluating Aggregate Coating Using Emulsified Asphalts

D6999 Standard Practice for Miscibility of Emulsified Asphalts

D7226 Standard Test Method for Determining the Viscosity of Emulsified Asphalts Using a Rotational Paddle Viscometer

D7402 Standard Practice for Identifying Cationic Emulsified Asphalts

D7403 Standard Test Method for Determination of Residue of Emulsified Asphalt by Low Temperature Vacuum Distillation

D7404 Standard Test Method for Determination of Emulsified Asphalt Residue by Moisture Analyzer

D7496 Standard Test Method for Viscosity of Emulsified Asphalt by Saybolt Furol Viscometer

D7497 Standard Practice for Recovering Residue from Emulsified Asphalt Using Low Temperature Evaporative Technique

FIELD METHODS FOR MEASURING TIRE PAVEMENT FRICTION

E274 Standard Test Method for Skid Resistance of Paved Surfaces Using a Full-Scale Tire

E445 Standard Test Method for Stopping Distance on Paved Surfaces Using a Passenger Vehicle Equipped with Full-Scale Tires

E556 Standard Test Method for Calibrating a Wheel Force or Torque Transducer Using a Calibration Platform (User Level)

E670 Standard Test Method for Side Force Friction on Paved Surfaces Using the Mu-Meter

E1337 Standard Test Method for Determining Longitudinal Peak Braking Coefficient of Paved Surfaces Using Standard Reference Test Tire

E1859 Standard Test Method for Friction Coefficient Measurements Between Tire and Pavement Using a Variable Slip Technique

E1890 Standard Guide for Validating New Area Reference Skid Measurement Systems and Equipment

E1911 Standard Test Method for Measuring Paved Surface Frictional Properties Using the Dynamic Friction Tester

E1960 Standard Practice for Calculating International Friction Index of a Pavement Surface

E2100 Standard Practice for Calculating the International Runway Friction Index

E2101 Standard Test Method for Measuring the Frictional Properties of Winter Contaminated Pavement Surfaces Using an Averaging-Type Spot Measuring Decelerometer

E2340 Standard Test Method for Measuring the Skid Resistance of Pavements and Other Trafficked Surfaces Using a Continuous Reading, Fixed-Slip Technique

E2341 Standard Test Method for Determining the Stopping Distance Number by Initial Speed and Stopping Distance at Traffic Incident Sites

E2666 Standard Practice for Correlations of Mu Values of Continuous Friction Measurement Equipment to Determine Maintenance Levels for Use at Airports

E2793 Standard Guide for the Evaluation, Calibration, and Correlation of E274 Friction Measurement Systems and Equipment

E28833 Standard Guide for Standard Guide for the Evaluation, and Calibration, Continuous Friction Measurement Equipment (CFME)

FORMED IN-PLACE SEALANTS FOR JOINTS AND CRACKS IN PAVEMENTS

D1985 Standard Practice for Preparing Concrete Blocks for Testing Sealants, for Joints and Cracks

D5078 Standard Specification for Crack Filler, Hot-Applied, for Asphalt Concrete and Portland Cement Concrete Pavements

D5167 Standard Practice for Melting of Hot-Applied Joint and Crack Sealant and Filler for Evaluation

D5249 Standard Specification for Backer Material for Use with Cold- and Hot-Applied Joint Sealants in Portland-Cement Concrete and Asphalt Joints

D5329 Standard Test Methods for Sealants and Fillers, Hot-Applied, for Joints and Cracks in Asphaltic and Portland Cement Concrete Pavements

D5893 Standard Specification for Cold Applied, Single Component, Chemically Curing Silicone Joint Sealant for Portland Cement Concrete Pavements

D6690 Standard Specification for Joint and Crack Sealants, Hot Applied, for Concrete and Asphalt Pavements

FUNDAMENTAL/MECHANISTIC TESTS

D3387 Standard Test Method for Compaction and Shear Properties of Bituminous Mixtures by Means of the US Corps of Engineers Gyratory Testing Machine (GTM)

D6931 Standard Test Method for Indirect Tensile (IDT) Strength of Bituminous Mixtures

D7312 Standard Test Method for Determining the Permanent Shear Strain and Complex Shear Modulus of Asphalt Mixtures Using the Superpave Shear Tester (SST)

D7313 Standard Test Method for Determining Fracture Energy of Asphalt-Aggregate Mixtures Using the Disk-Shaped Compact Tension Geometry

D7369 Standard Test Method for Determining the Resilient Modulus of Bituminous Mixtures by Indirect Tension Test

D7460 Standard Test Method for Determining Fatigue Failure of Compacted Asphalt Concrete Subjected to Repeated Flexural Bending

D7552 Standard Test Method for Determining the Complex Shear Modulus (G*) of Bituminous Mixtures Using Dynamic Shear Rheometer

HIGHWAY TRAFFIC CONTROL MATERIALS

D4280 Standard Specification for Extended Life Type, Nonplowable, Raised Retroreflective Pavement Markers

D4383 Standard Specification for Plowable, Raised Retroreflective Pavement Markers

D4505 Standard Specification for Preformed Retroreflective Pavement Marking Tape for Extended Service Life

D4592 Standard Specification for Preformed Retroreflective Pavement Marking Tape for Limited Service Life

D4956 Standard Specification for Retroreflective Sheeting for Traffic Control

D6628 Standard Specification for Color of Pavement Marking Materials

D7585 Standard Practice for Evaluating Retroreflective Pavement Markings Using Portable Hand-Operated Instruments

D7942 Standard Specification for Thermoplastic Pavement Markings in Non-Snow Plow Areas

MECHANICAL TESTS OF ASPHALT MIXTURES

D1074 Standard Test Method for Compressive Strength of Bituminous Mixtures

D1560 Standard Test Methods for Resistance to Deformation and Cohesion of Bituminous Mixtures by Means of Hveem Apparatus

D1561 Standard Practice for Preparation of Bituminous Mixture Test Specimens by Means of California Kneading Compactor

D5581 Standard Test Method for Resistance to Plastic Flow of Bituminous Mixtures Using Marshall Apparatus (6-inch Diameter Specimen)

D6925 Standard Test Method for Preparation and Determination of the Relative Density of Asphalt Mix Specimens by Means of the Superpave Gyratory Compactor

D6926 Standard Practice for Preparation of Bituminous Specimens Using Marshall Apparatus

D6927 Standard Test Method for Marshall Stability and Flow of Bituminous Mixtures

D7115 Standard Test Method for Measurement of Superpave Gyratory Compactor (SGC) Internal Angle of Gyration Using Simulated Loading

METHODOLOGY FOR ANALYZING PAVEMENT ROUGHNESS

E1170 Standard Practices for Simulating Vehicular Response to Longitudinal Profiles of Traveled Surfaces

E1489 Standard Practice for Computing Ride Number of Roads from Longitudinal Profile Measurements Made by an Inertial Profile Measuring Device

E1926 Standard Practice for Computing International Roughness Index of Roads from Longitudinal Profile Measurements

E1927 Standard Guide for Conducting Subjective Pavement Ride Quality Ratings

E2034 Standard Practices for Simulating Truck Response to Longitudinal Profiles of Vehicular Traveled Surfaces

E2955 Standard Practice for Simulating Profilograph Response to Longitudinal Profiles of Traveled Surfaces

METHODS FOR MEASURING PROFILE AND ROUGHNESS

E950 Standard Test Method for Measuring the Longitudinal Profile of Traveled Surfaces with an Accelerometer Established Inertial Profiling Reference

E1082 Standard Test Method for Measurement of Vehicular Response to Traveled Surface Roughness

E1215 Standard Specification for Trailers Used for Measuring Vehicular Response to Road Roughness

E1274 Standard Test Method for Measuring Pavement Roughness Using a Profilograph

E1364 Standard Test Method for Measuring Road Roughness by Static Level Method
E1448 Standard Practice for Calibration of Systems Used for Measuring Vehicular Response to Pavement Roughness
E1703 Standard Test Method for Measuring Rut-Depth of Pavement Surfaces Using a Straightedge
E2133 Standard Test Method for Using a Rolling Inclinometer to Measure Longitudinal and Transverse Profiles of a Traveled Surface
E2560 Standard Specification for Data Format for Pavement Profile

METHODS OF SAMPLING

D75 Standard Practice for Sampling Aggregates
D140 Standard Practice for Sampling Bituminous Materials
D979 Standard Practice for Sampling Bituminous Paving Mixtures
D3665 Standard Practice for Random Sampling of Construction Materials
D5361 Standard Practice for Sampling Compacted Bituminous Mixtures for Laboratory Testing

MISCELLANEOUS ASPHALT TESTS

D70 Standard Test Method for Density of Semi-Solid Bituminous Materials (Pycnometer Method)
D2042 Standard Test Method for Solubility of Asphalt Materials in Trichloroethylene
D3142 Standard Test Method for Specific Gravity, API Gravity, or Density of Cutback Asphalts by Hydrometer Method
D3143 Standard Test Method for Flash Point of Cutback Asphalt with Tag Open-Cup Apparatus
D3279 Standard Test Method for *n*-Heptane Insolubles
D3289 Standard Test Method for Density of Semi-Solid and Solid Bituminous Materials (Nickel Crucible Method)
D4124 Standard Test Method for Separation of Asphalt into Four Fractions
D5546 Standard Test Method for Solubility of Asphalt Binders in Toluene by Centrifuge
D6608 Standard Practice for the Identification of Trinidad Lake Asphalt in Asphalt Mixes
D6703 Standard Test Method for Automated Heithaus Titrimetry

D7173 Standard Practice for Determining the Separation Tendency of Polymer from Polymer Modified Asphalt
D7553 Standard Test Method for Solubility of Asphalt Materials in *N*-Propyl Bromide

PAVEMENT MANAGEMENT AND DATA NEEDS

D5340 Standard Test Method for Airport Pavement Condition Index Surveys
D6433 Standard Practice for Roads and Parking Lots Pavement Condition Index Surveys
E1166 Standard Guide for Network Level Pavement Management
E1656 Standard Guide for Classification of Automated Pavement Condition Survey Equipment
E1777 Standard Guide for Prioritization of Data Needs for Pavement Management
E1778 Standard Terminology Relating to Pavement Distress
E1889 Standard Guide for Pavement Management Implementation
E2840 Standard Practice for Pavement Condition Index Surveys for Interlocking Concrete Roads and Parking Lots

PAVEMENT TESTING AND EVALUATION

D1195 Standard Test Method for Repetitive Static Plate Load Tests of Soils and Flexible Pavement Components, for Use in Evaluation and Design of Airport and Highway Pavements
D1196 Standard Test Method for Nonrepetitive Static Plate Load Tests of Soils and Flexible Pavement Components, for Use in Evaluation and Design of Airport and Highway Pavements
D4602 Standard Guide for Nondestructive Testing of Pavements Using Cyclic-Loading Dynamic Deflection Equipment
D4694 Standard Test Method for Deflections with a Falling-Weight-Type Impulse Load Device
D4695 Standard Guide for General Pavement Deflection Measurements
D4748 Standard Test Method for Determining the Thickness of Bound Pavement Layers Using Short-Pulse Radar
D5858 Standard Guide for Calculating *In Situ* Equivalent Elastic Moduli of Pavement Materials Using Layered Elastic Theory

D6951 Standard Test Method for Use of the Dynamic Cone Penetrometer in Shallow Pavement Applications

D7228 Standard Test Method for Prediction of Asphalt-Bound Pavement Layer Temperatures

E2583 Standard Test Method for Measuring Deflections with a Light Weight Deflectometer (LWD)

E2835 Standard Test Method for Measuring Deflections Using a Portable Impulse Plate Load Test Device

PLANT-MIXED ASPHALT SURFACES AND BASES

D2489 Standard Practice for Estimating Degree of Particle Coating of Bituminous-Aggregate Mixtures

D6932 Standard Guide for Materials and Construction of Open-Graded Friction Course Plant Mixtures

D7064 Standard Practice for Open-Graded Friction Course (OGFC) Mix Design

PREFORMED JOINT FILLERS, SEALERS, AND SEALING SYSTEMS

D517 Standard Specification for Asphalt Plank

D545 Standard Test Methods for Preformed Expansion Joint Fillers for Concrete Construction (Nonextruding and Resilient Types)

D994 Standard Specification for Preformed Expansion Joint Filler for Concrete (Bituminous Type)

D1751 Standard Specification for Preformed Expansion Joint Filler for Concrete Paving and Structural Construction (Nonextruding and Resilient Bituminous Types)

D1752 Standard Specification for Preformed Sponge Rubber Cork and Recycled PVC Expansion Joint Fillers for Concrete Paving and Structural Construction

D2628 Standard Specification for Preformed Polychloroprene Elastomeric Joint Seals for Concrete Pavements

D2835 Standard Specification for Lubricant for Installation of Preformed Compression Seals in Concrete Pavements

D3542 Standard Specification for Preformed Polychloroprene Elastomeric Joint Seals for Bridges

D4070 Standard Specification for Adhesive Lubricant for Installation of Preformed Elastomeric Bridge Compression Seals in Concrete Structures

D5973 Standard Specification for Elastomeric Strip Seals with Steel Locking Edge Rails Used in Expansion Joint Sealing

D6924 Standard Specification for Preformed Thermoplastic Vulcanizate Elastomeric Joint Seals for Bridges

D7174 Standard Specification for Preformed Closed-Cell Polyolefin Expansion Joint Fillers for Concrete Paving and Structural Construction

QUALITY CONTROL, INSPECTION, AND TESTING AGENCIES

D3666 Standard Specification for Minimum Requirements for Agencies Testing and Inspecting Road and Paving Materials

D4561 Standard Practice for Quality Control Systems for Organizations Producing and Applying Bituminous Paving Materials

D7495 Standard Specification for Minimum Requirements for Accreditation Bodies That Accredit Agencies Testing and Inspecting Road and Paving Materials

RHEOLOGICAL TESTS

D5 Standard Test Method for Penetration of Bituminous Materials

D113 Standard Test Method for Ductility of Bituminous Materials

D139 Standard Test Method for Float Test for Bituminous Materials

D2170 Standard Test Method for Kinematic Viscosity of Asphalts (Bitumens)

D2171 Standard Test Method for Viscosity of Asphalts by Vacuum Capillary Viscometer

D2493 Standard Viscosity-Temperature Chart for Asphalts

D4957 Standard Test Method for Apparent Viscosity of Asphalt Emulsion Residues and Non-Newtonian Bitumens by Vacuum Capillary Viscometer

D5801 Standard Test Method for Toughness and Tenacity of Bituminous Materials

D6084 Standard Test Method for Elastic Recovery of Bituminous Materials by Ductilometer

D6648 Standard Test Method for Determining the Flexural Creep Stiffness of Asphalt Binder Using the Bending Beam Rheometer (BBR)

D6723 Standard Test Method for Determining the Fracture Properties of Asphalt Binder in Direct Tension (DT)

D6816 Standard Practice for Determining Low-Temperature Performance Grade (PG) of Asphalt Binders

D7175 Standard Test Method for Determining the Rheological Properties of Asphalt Binder Using a Dynamic Shear Rheometer

D7405 Standard Test Method for Multiple Stress Creep and Recovery (MSCR) of Asphalt Binder Using a Dynamic Shear Rheometer

D7643 Standard Practice for Determining the Continuous Grading Temperatures and Continuous Grades for PG Graded Asphalt Binders

D7741 Standard Test Method for Measurement of Apparent Viscosity of Asphalt-Rubber or Other Asphalt Binders by Using a Rotational Hand Held Viscometer

SPECIFIC GRAVITY AND DENSITY OF ASPHALT MIXTURES

D1188 Standard Test Method for Bulk Specific Gravity and Density of Compacted Bituminous Mixtures Using Coated Samples

D2041 Standard Test Method for Theoretical Maximum Specific Gravity and Density of Bituminous Paving Mixtures

D2726 Standard Test Method for Bulk Specific Gravity and Density of Non-Absorptive Compacted Bituminous Mixtures

D2950 Standard Test Method for Density of Bituminous Concrete in Place by Nuclear Methods

D3203 Standard Test Method for Percent Air Voids in Compacted Dense and Open Bituminous Paving Mixtures

D3549 Standard Test Method for Thickness or Height of Compacted Bituminous Paving Mixture Specimens

D6752 Standard Test Method for Bulk Specific Gravity and Density of Compacted Bituminous Mixtures Using Automatic Vacuum Sealing Method

D6857 Standard Test Method for Maximum Specific Gravity and Density of Bituminous Paving Mixtures Using Automatic Vacuum Sealing Method

D6995 Standard Test Method for Determining Field VMA Based on the Maximum Specific Gravity of the Mix (Gmm)

D7063 Standard Test Method for Effective Porosity and Effective Air Voids of Compacted Bituminous Paving Mixture Samples

D7113 Standard Test Method for Density of Bituminous Paving Mixtures in Place by the Electromagnetic Surface Contact Methods

D7227 Standard Practice for Rapid Drying of Compacted Asphalt Specimens Using Vacuum Drying Apparatus

SPECIFICATIONS AND TEST FOR TAR AND TAR PRODUCTS

D20 Standard Test Method for Distillation of Road Tars

D490 Standard Specification for Road Tar

D633 Standard Volume Correction Table for Road Tar

D1665 Standard Test Method for Engler Specific Viscosity of Tar Products

D2728) Standard Practice for Paving Uses and Application Temperatures for Road Tars

STATISTICAL PROCEDURES AND EVALUATION OF DATA

D4460 Standard Practice for Calculating Precision Limits Where Values Are Calculated from Other Test Methods

D6607 Standard Practice for Inclusion of Precision Statement Variation in Specification Limits

SURFACE CHARACTERISTICS RELATED TO TIRE PAVEMENT SLIP RESISTANCE

E303 Standard Test Method for Measuring Surface Frictional Properties Using the British Pendulum Tester

E660 Standard Practice for Accelerated Polishing of Aggregates or Pavement Surfaces Using a Small-Wheel, Circular Track Polishing Machine

E1845 Standard Practice for Calculating Pavement Macrotexture Mean Profile Depth

E2157 Standard Test Method for Measuring Pavement Macrotexture Properties Using the Circular Track Meter

E2380 Standard Test Method for Measuring Pavement Texture Drainage Using an Outflow Meter

TERMINOLOGY

D8 Standard Terminology Relating to Materials for Roads and Pavements

E867 Standard Terminology Relating to Vehicle-Pavement Systems

TIRE AND SLIDER CHARACTERISTICS

E501 Standard Specification for Standard Rib Tire for Pavement Skid-Resistance Tests

E524 Standard Specification for Standard Smooth Tire for Pavement Skid-Resistance Tests

E155 Standard Specification for Special Purpose, Smooth-Tread Tire, Operated on Fixed Braking Slip Continuous Friction Measuring Equipment

E1844 Standard Specification for A Size 10 and 4–5 Smooth-Tread Friction Test Tire

TRAFFIC MONITORING

E1318 Standard Specification for Highway Weigh-in-Motion (WIM) Systems with User Requirements and Test Methods

E1957 Standard Practice for Installing Using Pneumatic Tubes with Roadway Traffic Counters and Classifiers

E2259 Standard Guide for Archiving and Retrieving ITS-Generated Data

E2300 Standard Specification for Highway Traffic Monitoring Devices

E2415 Standard Practice for Installing Piezoelectric Highway Traffic Sensors

E2467 Standard Practice for Developing Axle Count Adjustment Factors

E2468 Standard Practice for Metadata to Support Archived Data Management Systems

E2532 Standard Test Methods for Evaluating Performance of Highway Traffic Monitoring Devices

E2561 Standard Practice for the Installation of Inductive Loop Detectors

E2667 Standard Practice for Acquiring Intersection Turning Movement Traffic Data

E2759 Standard Practice for Highway Traffic Monitoring Truth-in-Data

APPENDIX 2

List of ASTM International and AASHTO Test Methods Related to Asphalt Testing (Listed by Subject)

The road and paving standards developed by ASTM International and AASHTO also provide the specifications and test methods pertinent to the types of paving, which are covered by these standards are typically made of asphalt, concrete, and bituminous materials. These road and paving standards allow engineering firms and construction companies to design paving materials and pavement to ensure strength and durability as well as safe application and use.

- Abrasion Resistance: ASTM D4886; Test Method for Abrasion Resistance of Geotextiles (Sand paper/Sliding Block Method)
- Accelerated Weathering: ASTM D4798; Practice for Accelerated Weathering Test Conditions and Procedures for Bituminous Materials (Xenon-Arc Method)
- Accelerated Weathering: ASTM D4799; Practice for Accelerated Weathering Test Conditions and Procedures for Bituminous Materials (Fluorescent UV, Water Spray, and Condensation Method)
- Accelerated Weathering: ASTM D5329, Section 15; Pertaining to Accelerated Weathering
- Accelerated Weathering: ASTM G151; Pertaining to Accelerated Weathering; Practice for Exposing Nonmetallic Materials in Accelerated Test Devices That Use Laboratory Light Sources
- Accelerated Weathering: ASTM G153; Pertaining to Accelerated Weathering; Practice for Operating Enclosed Carbon Arc Light Apparatus for Exposure of Nonmetallic Materials
- Accelerated Weathering: ASTM G154; Pertaining to Accelerated Weathering; Practice for Operating Fluorescent Light Apparatus for UV Exposure on Nonmetallic Materials Striping Paints (Traffic Marking)

- Accelerated Weathering: ASTM G154; Pertaining to Accelerated Weathering; Practice for Operating Fluorescent Light Apparatus for UV Exposure on Nonmetallic Materials
- Accelerated Weathering: ASTM G155; Pertaining to Accelerated Weathering; Practice for Operating Xenon Arc Light Apparatus for Exposure of Nonmetallic Materials
- Accelerated Weathering: ASTM G23; Pertaining to Accelerated Weathering; Practice for Operating Light-Exposure Apparatus (Carbon-Arc Type) with and without Water for Exposure of Nonmetallic Materials
- Accelerated Weathering: PRI; Accelerated Weathering by APWS, a unique weathering system that allows for aging of complete pavement specimens.
- Accelerated Weathering: PRI TM-039; Accelerated Weathering Test Conditions and Procedures for Pavement Material
- Accelerated Weathering/Aging: AASHTO R28; Practice for Accelerated Aging of Asphalt Binder Using a Pressurized Aging Vessel (PAV)
- Accelerated Weathering/Aging: ASTM D6521; Practice for Accelerated Aging of Asphalt Binder Using a Pressurized Aging Vessel (PAV)
- Accelerated Weathering—Pressure Aging Method: AASHTO R28; Practice for Accelerated Aging of Asphalt Binder Using a Pressurized Aging Vessel (PAV)
- Accelerated Weathering—Pressure Aging Method: ASTM D6521; Practice for Accelerated Aging of Asphalt Binder Using a Pressurized Aging Vessel (PAV)
- Accelerated Weathering—UV/Condensate): ASTM D4799; Practice for Accelerated Weathering Test Conditions and Procedures for Bituminous Materials (Fluorescent UV, Water Spray, and Condensation Method)
- Accelerated Weathering—Xenon Arc: ASTM D4798; Practice for Accelerated Weathering Test Conditions and Procedures for Bituminous Materials (Xenon-Arc Method)
- Acetone Extract: ASTM D297; Test Method for Rubber Products Chemical Analysis, para 14.1
- Acid/Base #: ASTM D664; Acid Number of Petroleum Products by Potentiometric Titration

- Acid/Base #: ASTM D664; Acid Number of Petroleum Products by Potentiometric Titration
- Adherent Coating: ASTM D5711; Test Method for Determining the Adherent Coating on Coarse Aggregates
- Adhesion to Damp, Wet, or Underwater Surfaces: ASTM D6511, Section 17; Test Method for Solvent Bearing Bituminous Compounds, Wet Surface Adhesion
- Aggregate (Sand) Content in Bituminous Emulsions: ASTM D2939, Section 28; Aggregate (Sand) Content in Bituminous Emulsions
- Aggregate (Sand) Content in Bituminous Emulsions: ASTM D2939, Section 28; Aggregate (Sand) Content in Bituminous Emulsions
- Aggregate Gradation—Mixture: AASHTO T30; Mechanical Analysis of Extracted Aggregate
- Aggregate Properties (SE, Soundness LA Abrasion): ASTM C88, ASTM C131, ASTM D2419—See Aggregate Section
- Aging: ASTM D573; Test Method for Rubber—Deterioration in an Air Oven
- Aging (also see Deterioration in the following): ASTM D5721; Test Method for Air-Oven Aging of Polyolefin Geomembranes
- Air Void Determination: AASHTO T269; Percent Air Voids in Compacted Dense and Open Bituminous Paving Mixtures
- Air Void Determination: ASTM D3203; Test Method for Percent Air Voids in Compacted Dense and Open Bituminous Paving Mixtures
- Air Voids—Hot Mix/Paving Mixtures: ASTM D3203; Test Method for Percent Air Voids in Compacted Dense and Open Bituminous Paving Mixtures
- Air Voids—Mixture: ASTM D3203; Test Method for Percent Air Voids in Compacted Dense and Open Bituminous Paving Mixtures
- Aluminum Content: ASTM D6511, Section 14; Test Method for Solvent Bearing Bituminous Compounds, Aluminum Content
- Aniline Point: ASTM D611; Aniline Point and Mixed Aniline Point of Petroleum Products and Hydrocarbon Solvents
- Anti-Stripping Additive in Asphalt: ID DOT IT-99; Detection of Anti-Stripping Additive in Asphalt
- API Gravity: ASTM D1250; Guide for Use of Petroleum Measurements Tables
- API Gravity: ASTM D287; API Gravity of Crude Petroleum and Petroleum Products (Hydrometer Method)

- Application Behavior—Brush: ASTM D2939, Section 17; Brush-Application Behavior
- Application Behavior—Spray: ASTM D2939, Section 18; Spray-Application Behavior
- Aromatic/Aliphatic Ratio: ASTM D6805; Test Method for Infrared (IR) Procedure for Determination of Aromatic/Aliphatic Ration of Bituminous Emulsions
- Ash Content: AASHTO T111; Inorganic Matter of Ash in Bituminous Materials
- Ash Content: ASTM D128; Test Method for Analysis of Lubricating Grease
- Ash Content: ASTM D2939, Section 10; Ash Content
- Ash Content: ASTM D2939, Section 10; Ash Content
- Ash Content: ASTM D297; Test Method for Rubber Products Chemical Analysis, para 14–51
- Ash Content: ASTM D482; Ash from Petroleum Products
- Ash Content: ASTM D6511, Section 9; Test Method for Solvent Bearing Bituminous Compounds, Ash Content
- Ash, %: ASTM D482; Ash from Petroleum Products
- Asphalt (Resid/VTB) from Crude via Distillation: ASTM D5236; Distillation of Heavy Hydrocarbon Mixture (Vacuum Potstill Method)
- Asphalt/Bitumen Content: ASTM D4; Test Method for Bitumen Content
- Asphalt Cement Content (from Pavement Mixtures): AASHTO T308; Asphalt Content of Hot Mix Asphalt by Ignition Oven
- Asphalt Cement Content (from Pavement Mixtures): ASTM D6307; Asphalt Content of Hot Mix Asphalt by Ignition Oven
- Asphalt Compatibility: ASTM D5329, Section 14; Asphalt Compatibility
- Asphalt Composition (SARA Analysis with Iatroscan): ASTM D4124; Separation of Asphalt into Four Fractions (asphaltenes, saturates, naphthene aromatics, and polar aromatics)
- Asphalt Content (Extraction and Recovery): ASTM D6847; Test Method for Quantitative Extraction and Recovery of Asphalt Binder from Asphalt Mixtures
- Asphalt Content (Extraction): AASHTO T164; Quantitative Extraction of Asphalt Binder from Bituminous Paving Mixtures
- Asphalt Content (Extraction): ASTM D2172; Quantitative Extraction of Asphalt Binder from Hot-Mix Asphalt (HMA)

- Asphalt Content (Extraction): CA DOT T 362; Method of Determining Asphalt Content of Bituminous Mixtures by Vacuum Extraction
- Asphalt Content (Ignition Oven): AASHTO T308; Asphalt Content of Hot Mix Asphalt by Ignition Oven
- Asphalt Content (Ignition Oven): ASTM D6307; Asphalt Content of Hot Mix Asphalt by Ignition Oven
- Asphalt Content—Mixtures: ASTM D2172; Quantitative Extraction of Asphalt Binder from Hot-Mix Asphalt (HMA)
- Asphalt Content—Mixtures: ASTM D545, Section 7.5; Asphalt Content; Test Methods for Preformed Expansion Joint Fillers for Concrete Construction
- Asphaltene Content,%: ASTM D3279; *n*-Heptane Insolubles
- Asphalt Grade Verification AASHTO M320
- Asphalt Pavement Analyzer (APA) Rut Test: AASHTO TP63; Determining Rutting Susceptibility of Asphalt Paving Mixtures Using the Asphalt Pavement Analyzer (APA)
- Asphalt Recovery, Abson Method: AASHTO T170; Recovery of Asphalt from Solution by Abson Recovery
- Asphalt Recovery, Abson Method: ASTM D1856; Recovery of Asphalt from Solution by Abson Recovery
- Asphalt Recovery, Rotovap: ASTM D5404; Recovery of Asphalt from Solution Using the Rotavapor Evaporator
- Asphalt Retention: ASTM D6140; Test Method for Determining the Asphalt Retention of Paving Fabrics
- Auto Ignition: ASTM E 659; Auto-Ignition Temperature of Liquid Chemicals
- Base/Neutrals/Acids: EPA 1311/8270; Test Method for Determining TCLP Bases, Neutrals, and Acids
- Bases, Neutrals, and Acids: EPA 1311/8260; Test Method for Determining TCLP Volatiles
- Beam Fatigue: AASHTO TP9; Standard Test Method for Determining the Creep Compliance and Strength of Hot Mix Asphalt (HMA) Using the Indirect Tensile Test Device
- Behavior at 60°C (140°F): ASTM D6511, Section 12; Test Method for Solvent Bearing Bituminous Compounds, Behavior at 60°C (140°F)
- Binder Content—Mixture: ASTM D2172; Quantitative Extraction of Asphalt Binder from Hot-Mix Asphalt (HMA)
- Binder Content—Mixture: ASTM D6307; Asphalt Content of Hot Mix Asphalt by Ignition Oven

- Bitumen Content: ASTM D4; Test Method for Bitumen Content
- Boiling in Hydrochloric Acid: ASTM D545, Section 7.4; Boiling in Hydrochloric Acid; Test Methods for Preformed Expansion Joint Fillers for Concrete Construction
- Bond Strength: AASHTO T250; Test Method for Bond Strength of Thermoplastic Traffic Marking Materials
- Bond Strength: ASTM D4796; Test Method for Bond Strength of Thermoplastic Traffic Marking Materials
- Bond, Fuel-Immersed: ASTM D5329, Section 11; Bond Fuel-Immersed
- Bond, Non-Immersed: ASTM D5329, Section 9; Bond Non-Immersed
- Breaking Strength and Elongation: ASTM D4632; Test Method for Grab Breaking Load and Elongation of Geotextiles
- Bulk Gravity—Mixture: ASTM D2726; Test Method for Bulk Gravity of Compacted Bituminous Mixtures Using Saturated Surface-Dry Specimens
- Bulk Specific Gravity: AASHTO T166; Test Method for Bulk Specific Gravity of Non-Absorptive Compacted Bituminous Mixtures Using Saturated Surface-Dry Specimens
- Bulk Specific Gravity: AASHTO T331; Bulk Specific Gravity and Density of Compacted Asphalt Mixtures Using Automatic Vacuum Sealing Method
- Bulk Specific Gravity: ASTM D2726; Bulk Gravity of Compacted Bituminous Mixtures Using Saturated Surface-Dry Specimens
- Bulk Specific Gravity: ASTM D6752; Test Method for Bulk Specific Gravity and Density of Compacted Bituminous Mixtures Using Automatic Vacuum Sealing Method
- Carbon Black Content: ASTM D297; Test Method for Rubber Products Chemical Analysis, para 38–39
- Carbon Black Dispersion: ASTM D5596; Test Method for Microscopic Evaluation of Dispersion of Carbon Black in Polyolefin Geosynthetics
- Carbon, Hydrogen, Nitrogen, and Oxygen Content: ASTM D5291; Instrumental Determination of C, H, N, and O in Petroleum Products and Lubricants
- Carbonyl Index: ASTM E 1252; General Techniques for Obtaining Infrared Spectra for Qualitative Analysis
- Cement Mixing: AASHTO T59, Section 12; Determining Cement Mixing of Emulsified Asphalt

- Cement Mixing: ASTM D6935; Determining Cement Mixing of Emulsified Asphalt
- Characterization: ASTM D2007; Test Method for Characteristic Groups in Rubber Extender and Processing Oils and Other Petroleum Derived Oils by Clay-Gel Absorption Chromatographic Method
- Chemical Resistance—Geogrids: ASTM D6213; Practice for Tests to Evaluate Chemical Resistance of Geogrids to Liquids
- Chemical Resistance—Geomembranes: ASTM D5747; Practice for Tests to Evaluate Chemical Resistance of Geomembranes to Liquids
- Chemical Resistance—Geonets: ASTM D6388; Practice for Tests to Evaluate Chemical Resistance of Geonets to Liquids
- Chemical Resistance—Geotextiles: ASTM D6389; Practice for Tests to Evaluate Chemical Resistance of Geotextiles to Liquids
- Chloride Content: ASTM D4294
- Chloride Ion Penetration Resistance: AASHTO T259; Method of Test for Resistance of Concrete to Chloride Ion Penetration
- Chloride Ion Penetration Resistance: AASHTO T277; Method of Test for Electrical Indication of Concrete's Ability to Resist Chloride Ion Penetration
- Chloride Ion Penetration Resistance: ASTM C1202; Test Method for Electrical Indication of Concrete's Ability to Resist Chloride Ion Penetration
- Chlorides (Organic): ASTM D4929; Determination of Organic Chloride Content in Crude Oil
- Clay Content: AASHTO T176; Plastic Fines in Graded Aggregates and Soils by Use of Sand Equivalent Test
- Clay Lumps and Friable particles: AASHTO T112; Clay Lumps and Friable Particles in Aggregates
- Clay Lumps and Friable particles: ASTM C142; Test Method for Clay Lumps and Friable Particles in Aggregates
- Coarse Aggregate Angularity: PA DOT 621; Determination of the Percentage of Fractured Particles in Coarse Aggregate
- Coating and Stripping—Mixture: ASTM D1664; Coating and Stripping of Bitumen—Aggregate Mixtures
- Coating and Water Resistance: AASHTO T59, Section 22; Ash Content
- Coating and Water Resistance: ASTM D244, Sections 22–29; Coating Ability and Water Resistance

- Color: AASHTO T250, Section 7; Reflectance, Color, and Yellowness Index
- Color: ASTM D4960; Test Method for Evaluation of Color for Thermoplastic Traffic Marking Materials
- Compatibility: ASTM D5976; Specification for Type I Polymer Modified Asphalt Cement for Use in Pavement Construction
- Compression: ASTM D575; Test Method for Rubber Properties in Compression
- Compression,: ASTM D545, Section 7.2; Recovery; Test Methods for Preformed Expansion Joint Fillers for Concrete Construction
- Compressive Strength: AASHTO T167; Compressive Strength of Bituminous Mixtures
- Compressive Strength: ASTM D1074; Test Method of Compressive Strength of Bituminous Mixtures
- Compressive Strength—Effects of Water: AASHTO T165; Effect of Water on Compressive Strength of Compacted Bituminous Mixtures
- Compressive Strength—Effects of Water: ASTM D1075; Effect of Water on Compressive Strength of Compacted Bituminous Mixtures
- Cone Penetration, Fuel-Immersed: ASTM D217; Test Method for Cone Penetration of Lubricating Greases
- Cone Penetration, Fuel-Immersed: ASTM D5329, Section 7; Test Method for Cone Penetration of Lubricating Greases, Fuel-Immersed
- Cone Penetration, Non-Immersed: ASTM D217; Test Method for Cone Penetration of Lubricating Greases
- Cone Penetration, Non-Immersed: ASTM D5329, Section 6; Test Method for Cone Penetration of Lubricating Greases, Non-Immersed
- Conradson Carbon Residue (Coking Value): ASTM D189; Conradson Carbon Residue of Petroleum Products
- Consistency: ASTM D562; Consistency of Paints Measuring Krebs Units Viscosity Using a Stormer-Type Viscometer
- Consistency: ASTM D6511, Section 11; Test Method for Solvent Bearing Bituminous Compounds, Consistency (ASTM D562)
- Consistency (Saybolt Furol): AASHTO T59, Section 17; AASHTO T72; Saybolt Viscosity
- Consistency (Saybolt Furol): ASTM D244, Sections 17–21; ASTM D88; Saybolt Viscosity
- Constant Weight: ASTM D1037; Test Method for Evaluating the Properties of Wood-Based Fiber and Particle Panel Materials

- Contaminate Fractions: ASTM D7213; Boiling Range Distribution of Petroleum Distillates in the Boiling Range from 100°C to 615°C by Gas Chromatography
- Corrosion Evaluation: ASTM D1654; Test Method for Evaluation of Painted or Coated Specimens Subjected to Corrosive Environments
- Corrosion Resistance: ASTM D5968; Evaluation of Corrosiveness of Diesel Engine Oil
- Corrosion Resistance (Copper): ASTM D4048; Detection of Copper Corrosion from Lubricating Greases
- Corrosion Resistance (Copper): ASTM D4048; Detection of Copper Corrosion from Lubricating Greases
- Crack Resistance: ASTM D5397; Test Method for Evaluation of Stress Crack Resistance of Polyolefin Geomembranes Using Notched Constant Tensile Load
- Cracking Resistance: AASHTO M 248; Test Method for Cracking Resistance (on Concrete Block), para 4.3.3
- Creep Compliance: AASHTO T322; Determining the Creep Compliance and Strength of Hot-Mix Asphalt (HMA) Using the Indirect Tensile Test Device
- Creep Stiffness: AASHTO T313; Determining the Flexural Creep Stiffness of Asphalt Binders Using the Bending Beam Rheometer (BBR)
- Creep Stiffness: ASTM D6648; Determining the Flexural Creep Stiffness of Asphalt Binder Using the Bending Beam Rheometer (BBR)
- Critical Cracking Temperature: ASTM D6816; Determining Low-Temperature Performance Grade (PG) of Asphalt Binders
- Crude Oil Compatibility: ASTM D6703; Test Method for Automated Heithaus Titrimetry
- Crush Resistance: ASTM D1213; Test Method for Crushing Resistance
- Dampproofing: ASTM D449. 2014; Standard Specification for Asphalt Used in Dampproofing and Waterproofing
- Degradation (LA Abrasion) Aggregates: ASTM C131; Test Method for Resistance to Degradation of Small-Size Coarse Aggregate by Abrasion and Impact in the Los Angeles Machine
- Degradation (LA Abrasion)Aggregates: ASTM C131; Test Method for Resistance to Degradation of Small-Size Coarse Aggregate by Abrasion and Impact in the Los Angeles Machine

- Demulsibility: AASHTO T59, Section 10; Determining Demulsibility of Emulsified Asphalt
- Demulsibility: ASTM D6936; Determining Demulsibility of Emulsified Asphalt
- Density: AASHTO T228; Density of Semi-Solid Bituminous Materials (Pycnometer Method)
- Density: AASHTO T250, Section 13; Test Method for Density and SG (Relative Density) of Plastics by Displacement
- Density: ASTM D1475; Test Method for Density of Paint, Varnish, Lacquer, and Related Products
- Density: ASTM D545, Section 7.7; Density; Test Methods for Preformed Expansion Joint Fillers for Concrete Construction
- Density: ASTM D70; Density of Semi-Solid Bituminous Materials (Pycnometer Method)
- Density: ASTM D792; Test Method for Density and Specific Gravity (Relative Density) of Plastics by Displacement
- Density/Unit Weight: ASTM D6937; Determining Density of Emulsified Asphalts
- Density, Absorption and Voids: ASTM C642; Test Method for Density, Absorption, and Voids in Hardened Concrete
- Density, Relative (SG) or API: AASHTO T227; Test Method for Density, Relative Density (SG) or API Gravity of Crude Petroleum and Liquid Petroleum Products by Hydrometer Method
- Density, Relative (SG) or API: ASTM D1298; Test Method for Density, Relative Density (Specific Gravity) or API Gravity of Crude Petroleum and Liquid Petroleum Products by Hydrometer Method
- Deterioration: ASTM D5970; Test Method for Deterioration of Geotextiles from Outdoor Exposure
- Deterioration (Aging): ASTM D4355; Test Method for Deterioration of Geotextiles from Exposure to Ultraviolet Light and Water (Xenon-Arc Type Apparatus)
- Direct Flame Test: ASTM D2939, Section 20; Direct Flame Test
- Direct Flame Test: ASTM D2939, Section 20; Direct Flame Test
- Direct Tension: AASHTO T314; Determining Fracture Properties of Asphalt Binders in Direct Tension (DTT)
- Direct Tension: ASTM D6723; Determining Fracture Properties of Asphalt Binder in Direct Tension (DTT)
- Direct Tension: ASTM D6723; Determining Fracture Properties of Asphalt Binder in Direct Tension (DTT)

- Distillation: AASHTO T59, Section 5; Distillation of Emulsified Asphalt
- Distillation: AASHTO T78; Test Method for Distillation of Cutback Asphaltic (Bituminous) Products
- Distillation: ASTM D20; Distillation of Road Tars
- Distillation: ASTM D2892; Test Method for Distillation of Crude Petroleum (15-Theoretical Plate Column)
- Distillation: ASTM D402; Test Method for Distillation of Cutback Asphaltic (Bituminous) Products
- Distillation: ASTM D6997; Distillation of Emulsified Asphalt
- Distillation (low temperature vacuum): AASHTO T59, Section 64
- Distillation (low temperature vacuum): ASTM D244, Sections 64–69; Residue by Low-Temperature Vacuum Distillation
- Distillation (SimDis): ASTM D5307; Determination of Boiling Range Distribution of Crude Petroleum by Gas Chromatography
- Distillation (Steam): ASTM D255; Steam Distillation of Bituminous Protective Coatings
- Distillation of Petroleum Products: ASTM D86; Distillation of Petroleum Products at Atmospheric Pressure
- Distillation of Road Tars: ASTM D20; Distillation of Road Tars
- Draindown: ASTM D6390; Test Method for Determination of Draindown Characteristics in Uncompacted Asphalt Mixtures
- Draindown—Hot Mix/Paving Mixtures: ASTM D6390; Test Method for Determination of Draindown Characteristics in Uncompacted Asphalt Mixtures
- Drying: ASTM D4061; Test Method for Retroflectance of Horizontal Coating (Doctor Blade Application)
- Drying (Curing): ASTM D1640; Test Method for Drying, Curing or Film Formation of Organic Coatings at Room Temperature
- Drying Time: ASTM D2939, Section 13; Drying Time
- Drying Time: ASTM D2939, Section 13; Test Method for Determining Firm Set
- Drying Time: ASTM D711; Test Method for No-Pick-Up Time of Traffic Paint
- Drying Time: ASTM T248; Test Method for Drying Time
- Ductility: AASHTO T51; Test Method for Ductility of Bituminous Materials
- Ductility: ASTM D113; Ductility of Bituminous Materials
- Durability Index: AASHTO T210; Aggregate Durability Index

- Durability Index: ASTM D3744; Aggregate Durability Index
- Dynamic Modulus: AASHTO TP62; Determining Dynamic Modulus of Hot-Mix Asphalt Concrete Mixtures
- Dynamic Modulus—Hot Mix/Paving Mixtures: AASHTO TP62; Determining Dynamic Modulus of Hot-Mix Asphalt Concrete Mixtures
- Dynamic Modulus—Hot Mix/Paving Mixtures: ASTM D3497; Test Method for Dynamic Modulus of Asphalt Mixtures
- Dynamic Shear: AASHTO T315; Determining the Rheological Properties of Asphalt Binders Using a Dynamic Shear Rheometer
- Dynamic Shear: ASTM D7175; Determining Properties of Asphalt Binder Using a Dynamic Shear Rheometer (DSR)
- Dynamic Shear: ASTM D7175; Determining Properties of Asphalt Binder Using a Dynamic Shear Rheometer (DSR)
- Dynamic Shear and Phase Angle: AASHTO T315; Determining the Rheological Properties of Asphalt Binders Using a Dynamic Shear Rheometer
- Dynamic Shear and Phase Angle: ASTM D7175; Determining Properties of Asphalt Binder Using a Dynamic Shear Rheometer (DSR)
- Dynamic Shear Rheometer: AASHTO T315; Determining the Rheological Properties of Asphalt Binders Using a Dynamic Shear Rheometer
- Dynamic Shear Rheometer: ASTM D7175; Determining Properties of Asphalt Binder Using a Dynamic Shear Rheometer (DSR)
- EDSI (inorganics): ASTM E1508; Quantitative Analysis by Energy-Dispersive Spectroscopy
- Effect of Moisture on Paving Mixtures: ASTM D4867; Test Method for Determining Effect of Moisture on Asphalt Concrete Paving Mixtures
- Effect of Water on Coated Aggregates: ASTM D3625; Practice for Effect of Water on Bituminous-Coated Aggregate Using Boiling Water
- Effective Porosity: ASTM D7063; Test Method for Effective Porosity and Effective Air Voids of Compacted Bituminous Paving Mixtures Using Marshall Apparatus
- Effects of Exposure (Aging): ASTM D7238; Test Method for Effect of Exposure of Unreinforced Polyolefin Geomembranes Using Fluorescent UV Condensation Apparatus

- Effects of Temperature: ASTM D4594; Test Method for Effects of Temperature on Stability of Geotextiles
- Elastic Recovery: AASHTO T301; Elastic Recovery of Asphalt Materials by Means of a Ductilometer
- Elastic Recovery: ASTM D6084; Elastic Recovery Test of Bituminous Materials by Means of a Ductilometer
- Elemental Analysis of Lubricant and Additive Components: ASTM D4927M; Elemental Analysis of Lubricant and Additive Components, Barium, Calcium, Phosphorus, Sulfur, and Zinc by Wavelength Dispersive X-Ray Fluorescence
- Emulsified Asphalt: ASTM D977; Standard Specification for Emulsified Asphalt. Annual Book of Standards.
- Emulsified Asphalt Properties (Residue from Distillation and Pen of Residue): ASTM D244 and ASTM D2397—See Section on Emulsified Asphalt
- Emulsified Asphalt: ASTM D2397; Standard Specification for Cationic Emulsified Asphalt. Annual Book of Standards.
- Evaluation and Characterization of Additives and Modifiers: PRI Additives and Modifiers; PRI can evaluate, assess, and characterize additives and modifiers, expanding to the basic tenants outlined in AASHTO R15; Practice for Asphalt Additives and Modifiers. We maintain an inventory of asphalts, binders, and aggregates to be used in the additive and modifier projects to ensure continuity. Often successful evaluation projects result in intellectual property via patents for our clients
- Expansion: ASTM D4792; Test Method for Potential Expansion of Aggregate from Hydration Reactions
- Expansion in Boiling Water: ASTM D545, Section 7.1; Expansion in Boiling Water; Test Methods for Preformed Expansion Joint Fillers for Concrete Construction
- Extraction of Bitumen (Centrifuge): AASHTO T164, Section Method A; Quantitative Extraction of Bitumen from Bituminous Paving Mixtures
- Extraction of Bitumen (Centrifuge): ASTM D2172, Section Method A; Quantitative Extraction of Bitumen from Bituminous Paving Mixtures
- Extraction of Bitumen (Reflux): AASHTO T164, Section Method B; Quantitative Extraction of Bitumen from Bituminous Paving Mixtures

- Extraction of Bitumen (Reflux): ASTM D2172, Section Method B; Quantitative Extraction of Bitumen from Bituminous Paving Mixtures
- Extrusion: ASTM D545, Section 7.3; Extrusion; Test Methods for Preformed Expansion Joint Fillers for Concrete Construction
- Fatigue Life: AASHTO T321; Fatigue Life of Compacted Hot Mix Asphalt (HMA) Subjected to Repeated Flexural Bending
- Fiber Type Identification: ASTM D276; Test Method for Identification of Fibers in Textiles
- Field (Aggregate) Coating Test: AASHTO T59, Sections 43–48; Field Coating Test of Emulsified Asphalts
- Field (Aggregate) Coating Test: ASTM D244, Sections 52–57; Field Coat Testing on Emulsified Asphalts
- Filtering Efficiency: ASTM D5141; Test Method for Determining Filtering Efficiency and Flow Rate for Silt Fence Application of a Geotextile Using Site Specific Soils
- Fine Aggregate Angularity—Aggregates: AASHTO T304; Test Method for Uncompacted Void Content of Fine Aggregates (Fine Aggregate Angularity)
- Fine Aggregate Angularity—Aggregates: AASHTO TP33; Test Method for Uncompacted Void Content of Fine Aggregate
- Flammability: ASTM D6668; Test Method for Discrimination Between Flammability Ratings $F=0$ and $F=1$
- Flash and Fire Points: AASHTO T48; Test Method for Flash and Fire Points by Cleveland Open Cup
- Flash Point: ASTM D92; Flash and Fire Points by Cleveland Open Cup
- Flash Point: ASTM D93; Flash Point by Pensky-Martens Closed Cup Tester
- Flash Point (COC): AASHTO T48; Flash and Fire Points by Cleveland Open Cup Tester
- Flash Point, Pensky-Martens: AASHTO T48; Flash and Fire Points by Cleveland Open Cup Tester
- Flash Point, Pensky-Martens: ASTM D93; Flash Point by Pensky-Martens Closed Cup Tester
- Flash Point, Tag Open Cup: AASHTO T79; Flash Point with Tag Open Cup Apparatus for Use with Materials Having a Flash Less Than 93.3°C (Cutback)
- Flash Point, Tag Open Cup: ASTM D3143; Flash Point of Cutback Asphalt with Tag Open Cup Apparatus

- Flat and Elongated Particles—Aggregates: ASTM D4791; Test Method for Flat Particles, Elongated Particles, or Flat and Elongated particles in Coarse Aggregates
- Flexibility: ASTM D2939, Section 16; Flexibility
- Flexibility: ASTM D3111; Test Method for Flexibility of Hot-Melt Adhesive by Mandrel Bend Test Method
- Flexibility*: ASTM D5329, Section 18; Flexibility
- Float: ASTM D139; Float Test for Bituminous Materials
- Flow: ASTM D5329, Section 8; Flow
- Flowability: AASHTO T250, Section 9; Test Method for Flowability
- Force Ductility and Ratio: AASHTO T300; Force Ductility Test of Bituminous Materials
- Fractured Faces—Aggregates: ASTM D5821; Test Method for Determining Percentage of Fractured particles in Coarse Aggregate
- Freezing: AASHTO T59, Section 16; Test Method for Determining Damage from Freezing
- Gel Permeation Chromatography: ASTM E 1252; General Techniques for Obtaining Infrared Spectra for Qualitative Analysis
- Gradation (Extracted Aggregates): AASHTO T30; Mechanical Analysis of Extracted Aggregates (requires extraction by AASHTO T164/ ASTM D2172 Quantitative Extraction of Bitumen…)
- Graded Asphalt Cement: ASTM D 3381; Graded Asphalt Cement for Use in Pavement Construction—Tables 1–3; Viscosity
- Grade Verification AASHTO M320
- Hamburg Loaded Wheel: AASHTO T324; Hamburg Wheel—Track Testing of Compacted Hot-Mix Asphalt (HMA)
- Hardness: ASTM D2240; Test Method for Rubber Property—Durometer Hardness
- Hardness: ASTM D3363; Test Method for Film Hardness by Pencil Test
- Heat Loss (Oven Loss): ASTM D6; Loss on Heating of Oil and Asphaltic
- Heat of Combustion/Heating Value: ASTM D240; Heat for Combustion of Liquid Hydrogen Fuels by Bomb Calorimeter
- Hydraulic Conductivity: ASTM D5567; Test Method for Hydraulic Conductivity Ratio (HCR) Testing of Soil/Geotextile Systems
- Identification of Cationic Rapid Setting Emulsions: AASHTO T59, Sections 27–35; Identification Test for Rapid Setting Cationic Emulsified Asphalt

- Identification of Cationic Rapid Setting Emulsions: ASTM D244, Sections 38–45; Identification Test for Rapid Setting Cationic Emulsified Asphalt
- Identification of Cationic Slow Setting Emulsions: AASHTO T59, Sections 36–42; Identification of Cationic Slow Set Emulsions
- Identification of Cationic Slow Setting Emulsions: ASTM D244, Sections 46–51; Identification of Cationic Slow Setting Emulsions
- Identification of TLA: ASTM D6608; Practice for Identification of Trinidad Lake Asphalt in Asphalt Mixes
- Impact Resistance: AASHTO T250, Section 11; Test Method for Impact Resistance
- Impact Resistance: ASTM D256; Test Method for Determining Izod Pendulum Impact Resistance of Notched Specimens
- Index Flux: ASTM D5587; Test method for Tearing Strength of Fabrics by the Trapezoid Procedure
- Indirect Tensile Strength: ASTM D6931; Test Method for Indirect Tensile (IDT) Strength of Bituminous Mixtures
- Infrared Emittance: ASTM C1371; Test Method for Determination of Emittance of Materials Near Room Temperature
- Insolubles in Carbonate Aggregates: ASTM D3042; Test Method for Insoluble Residue in Carbonate Aggregates
- Kerosene Equivalent: ASTM D5148; Test Method for Centrifuge Kerosene Equivalent
- Lightweight Pieces in Aggregate: AASHTO T113; Lightweight Pieces in Aggregate
- Lightweight Pieces in Aggregate: ASTM C123; Test Method for Lightweight Particles in Aggregate
- Lime: AASHTO T219; Testing Lime for Chemical Constituents and Particle Size
- Lime: AASHTO T232; Determination of Lime Content in Lime-Treated Soils by Titration
- Liquid and Plastic Limits (PI)—Aggregates: ASTM D4318; Test Method for Liquid Limit, Plastic Limit, and Plasticity Index of Soils
- Liquid Limit: AASHTO T89; Determining the Liquid Limits of Soils
- Los Angeles Abrasion (Resistance to Degradation): AASHTO T96; Resistance to Degradation of Small-Size Coarse Aggregate by Abrasion and Impact in the Los Angeles Machine

- Los Angeles Abrasion (Resistance to Degradation): ASTM C131; Test Method for Resistance to Degradation of Small-Size Coarse Aggregate by Abrasion and Impact in the Los Angeles Machine
- Lottman (Modified): AASHTO T283; Test Method for Resistance of Compacted Bituminous Mixture to Moisture Induced Damage
- Marshall Stability and Flow: AASHTO T245; Resistance to Plastic Flow of Bituminous Mixtures Using Marshall Apparatus
- Mass: ASTM D5261; Test Method for Measuring Mass per Unit Area of Geotextiles
- Mechanical Analysis of Extracted Aggregates: AASHTO T30; Mechanical Analysis of Extracted Aggregates (requires extraction by AASHTO T164/ASTM D2172 Quantitative Extraction of Bitumen…)
- Melt Flow Rate: ASTM D1238; Melt Flow Rate of Thermoplastics by Extrusion Plastometer
- Melting: ASTM D5167, Section 15; Practice for Melting of Hot-Applied Joint and Crack Sealant and Filler for Evaluation Required for Individual and/or Specification Tests
- Metals and Metals Sample Preparation: ASTM D5708; Determination of Nickel, Vanadium, and Iron in Crude Oils and Residual Fuels by Inductively Coupled Plasma (ICP) Atomic Emission Spectroscopy
- Metals (Ni, V, Fe): ASTM D5708; Determination of Nickel, Vanadium, and Iron in Crude Oils and Residual Fuels by Inductively Coupled Plasma (ICP) Atomic Emission Spectroscopy
- Micro-Deval: AASHTO T327; Resistance of Coarse Aggregate to Degradation by Abrasion in the Micro-Deval Apparatus
- Mineral Matter: ASTM D6511, Section 19; Test Method for Solvent Bearing Bituminous Compounds, Mineral Matter (ASTM D4)
- Mineral Stabilizers and Bitumen: ASTM D6511, Section 18; Test Method for Solvent Bearing Bituminous Compounds, Mineral Stabilizers (ASTM D4)
- Miscibility with Water: AASHTO T59, Section 15; Miscibility of Emulsified Asphalts
- Miscibility with Water: ASTM D6999; Miscibility of Emulsified Asphalts
- Moisture (Effect of)—Hot Mix/Paving Mixtures: ASTM D4867; Test Method for Determining Effect of Moisture on Asphalt Concrete Paving Mixtures

- Moisture—Aggregates: ASTM C566; Test Method for Total Moisture Content of Aggregate by Drying
- Moisture—Aggregates: ASTM D1864; Moisture in Mineral Aggregate Used on Built-Up-Roofs
- Moisture Content: AASHTO T265; Laboratory Determination of Moisture Content of Soils
- Moisture Content: AASHTO T329; Moisture Content of Hot-Mix (HMA) by Oven Method
- Moisture Content: ASTM C566; Test Method for Total Moisture Content of Aggregate by Drying
- Moisture Content: ASTM D2216; Laboratory Determination of Water (Moisture) Content of Soil and Rock by Mass
- Moisture or Volatile Content: AASHTO T110; Moisture of Volatile Distillates in Bituminous Paving Mixtures
- Moisture or Volatile Content: ASTM D1461; Moisture of Volatile Distillates in Bituminous Paving Mixtures
- Moisture or Volatile Distillates—Mixture: ASTM D1461; Moisture of Volatile Distillates in Bituminous Paving Mixtures
- Moisture Resistance—Hot Mix/Paving Mixtures: AASHTO T283; Test Method for Resistance of Compacted Bituminous Mixture to Moisture Induced Damage
- Moisture Resistance—Hot Mix/Paving Mixtures: ASTM D1075; Effect of Water on Compressive Strength of Compacted Bituminous Mixtures
- Multiple Stress Creep and Recovery: AASHTO TP70; Multiple Stress Creep Recovery Test of Asphalt Binders Using a Dynamic Shear Rheometer (MSCR)
- *n*-Heptane/*n*-Pentane Insolubles: ASTM D3279; *n*-Heptane Insolubles (also other solvents)
- Natural Rubber: ASTM D297; Test Method for Rubber Products Chemical Analysis, para 11–13
- Nitrogen: ASTM D4629; Trace Nitrogen in Liquid Petroleum Hydrocarbons by Syringe/Inlet Oxidative Combustion and Chemiluminescence
- Nonvolatile Content: ASTM D6511, Section 7; Test Method for Solvent Bearing Bituminous Compounds, Nonvolatile
- Oil Distillate by Micro Distillation: AASHTO T59; Distillation of Petroleum Products at Atmospheric Pressure

- Oil Distillate by Micro Distillation: ASTM D86; Distillation of Petroleum Products at Atmospheric Pressure
- Oil Swell: ASTM D471; Test Method for Rubber Property—Effect of Liquids
- Oliensis Test ASTM D1370
- Opening Size: ASTM D4751; Test Method for Determining Apparent Opening Size of a Geotextile
- Ozone Resistance: ASTM D1149; Test Method for Rubber Deterioration-Surface Ozone Cracking in a Chamber
- Particle Charge: ASTM D244, Sections 11–16; Particle Charge
- Particle Coating—Hot Mix/Paving Mixtures: AASHTO T195; Practice for Estimating Degree of Particle Coating of Bituminous-Aggregate Mixtures
- Particle Coating—Hot Mix/Paving Mixtures: ASTM D2489; Practice for Estimating Degree of Particle Coating of Bituminous-Aggregate Mixtures
- Particle Coating—Hot Mix/Paving Mixtures: ASTM D2489; Practice for Estimating Degree of Particle Coating of Bituminous-Aggregate Mixtures
- Particle Size by Hydrometer: ASTM D422; Test Method for Particle-Size Analysis of Soils
- Penetration: AASHTO T49; Test Method for Penetration of Bituminous Materials
- Penetration: ASTM D5; Test Method for Penetration of Bituminous Materials
- Penetration, Cone: ASTM D217; Test Method for Cone Penetration of Lubricating Greases
- Penetration-Graded Asphalt: ASTM D946; Specification for Penetration-Graded Asphalt Cement for Use in Pavement Construction
- Penetration-Graded Cement: ASTM D946; Specification for Penetration-Graded Cement for Use in Pavements Construction (individual DOTs may have additional requirements)
- Penetration-Graded Cement: AASHTO M20; Specification for Penetration-Graded Cement for Use in Pavements Construction (individual DOTs may have additional requirements)
- Performance Strength: ASTM D4885; Test Method for Determining Performance Strength of Geomembranes by Wide Strip Tensile Method

- Permanent Shear Strain and Stiffness: AASHTO T320; Determining the Permanent Shear Strain and Stiffness of Asphalt Mixtures Using the Superpave Shear Tester (SST)
- Permeability: AASHTO T215; Permeability of Granular Soils (Constant Head)
- Permeability: ASTM D2434; Permeability of Granular Soils (Constant Head)
- Permeability (Impermeability): ASTM D5084; Test Method for Measurement of Hydraulic Conductivity of Saturated Porous Materials Using a Flexible Wall Permeometer
- Permeability, Water: ASTM D4491; Test Method for Water Permeability of Geotextiles by Permittivity
- pH: AASHTO T200; Test Method for pH of Aqueous Solutions with the Glass Electrode
- pH: ASTM E70; Test Method for pH of Aqueous Solutions with Glass Electrode
- Pigment Content: ASTM D2371; Test Method for Pigment Content of Solvent-Reducible Paints
- Pliability at 0°C (32°F): ASTM D6511, Section 13; Test Method for Solvent Bearing Bituminous Compounds, Pliability
- Polishing: AASHTO T279; Accelerated Polishing of Aggregates Using British Wheel
- Polishing: ASTM D3319; Accelerated Polishing of Aggregates Using British Wheel
- Polymer Content: AASHTO T302; Polymer Content of Polymer-Modified Emulsified Asphalt and Residue and Asphalt Binders (calibration curve required)
- Pour Point: ASTM D97; Pour Point of Petroleum Products
- Proctor Compaction: AASHTO T180; Moisture-Density Relations of Soils Using a 4.54-kg (10-lb) Rammer and a 457-mm (18-in) Drop
- Proctor Compaction: ASTM D1557; Test Method for Laboratory Compaction Characteristics of Soil Using Modified Effort
- Puncture Strength: ASTM D6241; Test Method for Static Puncture Strength of Geotextiles and Geotextile Related Products Using a 50-mm Probe
- Quantitative Extraction of Asphalt Binder: AASHTO T164, Section Method A; Quantitative Extraction of Bitumen from Bituminous Paving Mixtures

- Quantitative Extraction of Asphalt Binder: AASHTO T319; Quantitative Extraction and Recovery of Asphalt Binder from Asphalt Mixtures
- Quantitative Extraction of Asphalt Binder: ASTM D2172; Quantitative Extraction of Asphalt Binder from Hot-Mix Asphalt (HMA)
- Quinoline Insolubles: ASTM D2318; Quinoline-Insolubles Content of Tar and Pitch
- RAP (recovered binder viscosity): AASHTO T202; Test Method for Viscosity of Asphalts by Vacuum Capillary Viscometer
- RAP (recovered binder viscosity): ASTM D2171; Viscosity of Asphalts by Vacuum Capillary Viscometer
- RAP (Recovery): AASHTO T170; Recovery of Asphalt from Solution by Abson Recovery
- RAP (Recovery): ASTM D1856; Recovery of Asphalt from Solution by Abson Recovery
- RAP or Millings Recovered Binder Viscosity: AASHTO T164; Extraction + ASTM D5404; Recovery + ASTM D2171; Absolute Viscosity
- RAP or Millings Recovered Binder Viscosity: ASTM D2172; Extraction + ASTM D5404; Recovery + ASTM D2171; Absolute Viscosity
- Raveling—Mixtures: ASTM D7196; Test Method for Raveling Test of Cold Mixed Bituminous Emulsion Samples
- Recovery: ASTM D545, Section 7.2; Recovery; Test Methods for Preformed Expansion Joint Fillers for Concrete Construction
- Redispersibility: ASTM D4866, Section 6.1; Test Method for Redispersibility; Specification for Coal Tar Pitch Emulsion Pavement Sealer Mix Formulations
- Reflectance: AASHTO T250, Section 7; Reflectance, Color, and Yellowness Index
- Reflectance: ASTM E1349; Test Method for Reflectance Factor and Color by Spectrophotometry Using Bidirectional Geometry
- Reflectance of Aluminum Coatings: ASTM D6511, Section 15; Test Method for Solvent Bearing Bituminous Compounds, Reflectance
- Repeated Shear: AASHTO T320; Determining the Permanent Shear Strain and Stiffness of Asphalt Mixtures Using the Superpave Shear Tester (SST)
- Residue by Evaporation: AASHTO T59, Section 7; Residue by Evaporation of Emulsified Asphalt

- Residue by Evaporation: ASTM D2939, Section 8; Residue by Evaporation
- Residue by Evaporation: ASTM D6934; Residue by Evaporation of Emulsified Asphalt
- Residue Penetration: ASTM D243; Residue of Specified Penetration
- Resilience: ASTM D5329; Test Methods for Sealants and Fillers, Hot Applied, for Joints and Cracks in Asphaltic and Portland Cement Concrete Pavements
- Resilience*: ASTM D5; Test Method for Penetration of Bituminous Materials
- Resilience*: ASTM D5329, Section 12; Penetration of Bituminous Materials
- Resilience, Oven-Aged*: ASTM D5; Test Method for Penetration of Bituminous Materials
- Resilience, Oven-Aged*: ASTM D5329, Section 13; Penetration of Bituminous Materials
- Resilient Modulus: ASTM D4123; Indirect Tension Test for Resilient Modulus of Bituminous Mixtures
- Resistance to Freezing: AASHTO T161; Method of Test for Resistance of Concrete to Rapid Freezing and Thawing
- Resistance to Freezing: ASTM D2939, Section 6; Resistance to Freezing
- Resistance to Heat: ASTM D2939, Section 14; Resistance to Heat
- Resistance to Impact: ASTM D2939, Section 26; Resistance to Impact
- Resistance to Impact After Accelerated Weathering: ASTM D2939, Section 27; Resistance to Impact After Accelerated Weathering
- Resistance to Kerosene: ASTM D2939, Section 25; Resistance to Kerosene
- Resistance to Kerosene: ASTM D4866, Section 6.4; Resistance to Kerosene (ASTM D4866 (legacy), Specification for Coal Tar Pitch Emulsion Pavement Sealer Mix Formulations)
- Resistance to Plastic Flow—Mixture: AASHTO T245; Resistance to Plastic Flow of Bituminous Mixtures Using Marshall Apparatus
- Resistance to Volatilization: ASTM D2939, Section 23; Resistance to Volatilization
- Resistance to Water: ASTM D2939, Section 15; Resistance to Water
- Rolling Thin Film Oven Test (RTFOT): AASHTO T240; Test Method for Effect of Heat and Air on a Moving Film of Asphalt (RTFOT Method)

- Rolling Thin Film Oven Test (RTFOT): ASTM D2872; Test Method for Effect of Heat and Air on a Moving Film of Asphalt (RTFOT Method)
- Roofing Asphalt: ASTM D312. 2014; Standard Specification for Asphalt Used in Roofing
- Rotational (Brookfield) Viscosity: AASHTO T316; Viscosity Determination of Asphalt Binders Using Rotational Viscometer
- Rotational (Brookfield) Viscosity: ASTM D2196 (Modified); Rheological Properties of Non-Newtonian Material by Rotational (Brookfield Type) Viscometer
- Rotational (Brookfield) Viscosity: ASTM D4402; Viscosity Determinations of Unfilled Asphalts Using the Brookfield Thermosel Apparatus
- Rotational (Brookfield) Viscosity: ASTM D5801; Toughness and Tenacity of Bituminous Materials
- Roundness: ASTM D1155; Test Method for Roundness of Glass Spheres
- Rubber Hydrocarbon Content: ASTM D297; Test Method for Rubber Products Chemical Analysis, para 24
- Rusting: ASTM D610; Test Method for Evaluating Degree of Rusting on Painted Steel Surface
- Rutting Resistance: AASHTO TP63; Determining Rutting Susceptibility of Asphalt Paving Mixtures Using the Asphalt Pavement Analyzer (APA)
- Salt Fog (Corrosion Resistance): ASTM D5894; Practice for Cyclic Fog Salt For/UV Exposure of Painted Metal
- Sample Preparation: ASTM D4887; Practice for Preparation of Viscosity Blends for Hot Recycled Bituminous Materials
- Sample Preparation and Bulk Specific Gravity: ASTM D7229; Preparation and Determination of Bulk Specific Gravity of Dense-Graded Cold Mix Asphalt (CMA) Specimens by Means of Superpave Gyratory Compactor
- Sample Preparation—Hot Mix/Paving Mixtures: AASHTO T312; Preparing and Determining Density of Hot-Mix Asphalt (HMA) Specimens by Means of Superpave Gyrator Compactor
- Sample Preparation—Hot Mix/Paving Mixtures: ASTM D5167, Section 15; Practice for Melting of Hot-Applied Joint and Crack Sealant and Filler for Evaluation Required for Individual and/or Specification Tests

- Sample Preparation—Hot Mix/Paving Mixtures: ASTM D6926; Preparation of Bituminous Specimens by Marshall Apparatus
- Sample Size Reduction: AASHTO T248; Reducing Samples of Aggregate to Testing Size
- Sample Size Reduction: AASHTO T328; Reducing Samples of Hot-Mix Asphalt to Testing Size
- Sample Size Reduction: ASTM C702; Practice for Reducing Samples of Aggregate to Testing Size
- Sampling: ASTM D 140. 2014. Standard Practice for Sampling Bituminous Materials
- Sampling: ASTM D4354; Practice for Sampling of Geosynthetics for Testing
- Sand Equivalency—Aggregates: AASHTO T176; Plastic Fines in Graded Aggregates and Solid by Use of Sand Equivalent Test
- Sand Equivalency—Aggregates: ASTM D2419; Test Method for Sand Equivalent Value of Soils and Fine Aggregates
- SARA Analysis by Iatroscan: ASTM D4124; Separation of Asphalt into Four Fractions (asphaltenes, saturates, naphthene aromatics, and polar aromatics)
- SBS Content: AASHTO T302; Polymer Content of Polymer-Modified Emulsified Asphalt and Residue and Asphalt Binders (calibration curve required)
- Scaling Resistance: ASTM C672; Test Method for Scaling Resistance of Concrete Surfaces Exposed to Deicing Chemicals
- Sediment by Extraction: ASTM D473; Sediment in Crude Oils and Fuels by the Extraction Method
- Separation ("cigar tube test"): ASTM D7173; Determining Separation Tendency of Polymer from Polymer Modified Asphalt
- Settlement: AASHTO T59, Section 18; Settlement and Storage Stability of Emulsified Asphalts, Asphalt Emulsions: Anionic, Cationic, Polymer Modified, Mineral Colloid, and Settlement: ASTM D6930; Settlement and Storage Stability of Emulsified Asphalts
- Sieve Analysis (Glass Beads): ASTM D1214; Test Method for Sieve Analysis of Glass Spheres
- Sieve Analysis (Gradation)—Aggregates: AASHTO T27; Sieve Analysis of Fine and Coarse Aggregates
- Sieve Analysis (Gradation)—Aggregates: ASTM C136; Test Method for Sieve Analysis of Fine and Coarse Aggregates

- Sieve Analysis (Granular Mineral—Roofing): ASTM D451; Sieve Analysis of Granular Mineral Surfacing for Asphalt Roofing Products
- Sieve Analysis (Mineral Filler): AASHTO T37; Sieve Analysis of Mineral Filler for Bituminous Paving Materials
- Sieve Analysis (Mineral Filler): ASTM D546; Sieve Analysis of Mineral Filler for Paving Materials
- Sieve Analysis (Surfacing- Roofing): ASTM D452; Sieve Analysis of Surfacing for Asphalt Roofing Products
- Sieve Test: AASHTO T59, Section 13; Oversized Particles in Emulsified Asphalts (Sieve Test)
- Sieve Test: ASTM D6933; Oversized Particles in Emulsified Asphalts (Sieve Test)
- Softening Point: AASHTO T53; Softening Point of Bitumen (Ring and Ball Apparatus)
- Softening Point: ASTM D36; Softening Point of Bitumen (Ring and Ball Apparatus)
- Softening Point: ASTM D2319; Standard Test Method for Softening Point of Pitch (Cube-in-Air Method).
- Softening Point: ASTM D3104; Standard Test Method for Softening Point of Pitches (Mettler Softening Point Method).
- Softening Point (Legacy): ASTM D3461; Test Method for Softening Point of Asphalt and Pitch (Mettler Cup and Ball Method)
- Softening Point*: AASHTO T53; Softening Point of Bitumen (Ring and Ball Apparatus)
- Softening Point*: ASTM D36; Softening Point of Bitumen (Ring and Ball Apparatus)
- Softening Point: ASTM D61; Standard Test Method for Softening Point of Pitches (Cube-in-Water Method)
- Softening Points: AASHTO M 248; Test Method for Softening Point After Heating, para 4.3.5 (ASTM D36)
- Solar Reflectance: ASTM C1549; Test Method for Determination of Solar Reflectance Near Ambient Temperature Using a Portable Solar Reflectometer
- Solar Reflectance Index (SRI): ASTM E1980; Practice for Calculating Solar Reflectance Index for Horizontal and Low Slope Opaque Surfaces
- Solids: ASTM D1644; Test Method for Non-Volatiles of Varnishes
- Solubility: AASHTO T44; Solubility of Bituminous Materials

- Solubility: ASTM D2042; Solubility of Asphalt Material in Trichloroethylene (also other solvents)
- Solubility: ASTM D6511, Section 8; Test Method for Solvent Bearing Bituminous Compounds, Solubility (ASTM D4)
- Solubility in CS2: ASTM D2939, Section 24; Solubility of Residue in Trichloroethylene
- Solubility in TCE: AASHTO T44; Solubility of Asphalt Materials in Trichloroethylene
- Solubility in TCE: ASTM D2042; Solubility of Asphalt Material in Trichloroethylene (also other solvents)
- Solubility in TCE: ASTM D2939, Section 24; Solubility of Residue in Trichloroethylene
- Solubility in Toluene by Centrifuge: ASTM D5546; Test Method for Solubility of Polymer-Modified Asphalt Materials in 1,1,1-Trichloroethane
- Solubility*: ASTM D5329, Section 17; Solubility
- Soundness (Freezing and Thawing): AASHTO T103; Soundness of Aggregates by Freezing and Thawing
- Soundness (Magnesium or Sodium): AASHTO T104; Soundness of Aggregates by Use of Sodium Sulfate or Magnesium Sulfate
- Soundness (Magnesium or Sodium): ASTM C88; Test Method for Soundness of Aggregates by Use of Sodium Sulfate or Magnesium Sulfate
- Specific Gravity: AASHTO T228; Density of Semi-Solid Bituminous Materials (pycnometer method)
- Specific Gravity: ASTM D3289; Density of Semi-Solid and Solid Bituminous Materials (Ni Crucible Method)
- Specific Gravity: ASTM D70; Density of Semi-Solid Bituminous Materials (pycnometer method)
- Specific Gravity and Absorption—Fine Aggregates: ASTM C128; Test Method for Density, Relative Density (Specific Gravity), and Absorption of Fine Aggregates
- Specific Gravity and Absorption—Coarse-Aggregates: ASTM C127; Test Method for Density, Relative Density (Specific Gravity), and Absorption of Coarse Aggregates
- Specific Gravity and Absorption—Fine Aggregates: ASTM C128; Test Method for Density, Relative Density (Specific Gravity), and Absorption of Fine Aggregates

- Specific Gravity and Absorption of Coarse Aggregate: AASHTO T85; Specific Gravity and Absorption of Coarse Aggregates
- Specific Gravity and Absorption of Coarse Aggregate: ASTM C127; Test Method for Density, Relative Density (Specific Gravity), and Absorption of Coarse Aggregates
- Specific Gravity and Absorption of Fine Aggregate: AASHTO T84; Specific Gravity and Absorption of Fine Aggregates
- Specific Gravity and Absorption of Fine Aggregate: ASTM C128; Test Method for Density, Relative Density (Specific Gravity), and Absorption of Fine Aggregates
- Specific Gravity and Absorption-Coarse Aggregates: ASTM C127; Test Method for Density, Relative Density (Specific Gravity), and Absorption of Coarse Aggregates
- Specific Gravity and Absorption- Fine Aggregates: ASTM D2489; Practice for Estimating Degree of Particle Coating of Bituminous-Aggregate Mixtures
- Specific Gravity and Absorption-Coarse Aggregates: ASTM C127; Test Method for Density, Relative Density (Specific Gravity), and Absorption of Coarse Aggregates
- Specific Gravity (Soil): AASHTO T100; Test Method for Specific Gravity of Soil Solids by Water Pycnometer
- Specific Gravity (Soil): ASTM D854; Test Method for Specific Gravity of Soil Solids by Water Pycnometer
- Specific Gravity and Absorption of Fine Aggregate: ASTM C127; Test Method for Density, Relative Density (Specific Gravity), and Absorption of Coarse Aggregates
- Specification Conformance: ASTM D4759; Test Method for Determining Specification Conformance of Geosynthetics
- Spot Test: AASHTO T102; Spot Test of Asphaltic Materials
- Stain Index: ASTM D2746; Staining Tendency of Asphalt
- Storage Life: AASHTO M248; Test Method for Assessing Storage Life (1 year), para 4.4
- Storage Stability: AASHTO T59, Section 18; Settlement and Storage Stability of Emulsified Asphalts
- Storage Stability: ASTM D6930; Settlement and Storage Stability of Emulsified Asphalts
- Sulfur: ASTM D2622; Sulfur in Petroleum Products by Wavelength Dispersive X-Ray Fluorescence Spectroscopy

- Sulfur: ASTM D4294; Sulfur in Petroleum and Petroleum Products by Energy Dispersive X-Ray Fluorescence Spectroscopy
- Sweep Test: ASTM D7000; Sweep Test of Bituminous Emulsion Surface Treatment Samples
- Tar Acids: ASTM D453; Test Method for Tar Acids in Creosote-Coal Tar Solutions
- Tear Strength, Internally Reinforces: ASTM D5884; Test Method for Determining Tearing Strength of Internally Reinforced Geomembranes
- Tear Strength, Trapezoid: ASTM D4533; Test Method for Trapezoid Tearing Strength of Geotextiles
- Tensile Adhesion*: ASTM D5329, Section 16; Tensile Adhesion
- Tensile and Elongation: ASTM D412; Test Method for Vulcanized Rubber and Thermoplastic Elastomers-Tension
- Tensile Strength: ASTM D412; Test Method for Vulcanized Rubber and Thermoplastic Elastomers-Tension
- Tensile Strength: ASTM D4595; Test Method for Tensile Properties of Geotextiles by Wide-Width Strip Method
- Tensile Strength, Polyethylene, and Polypropylene: ASTM D6693; Test Method for Determining Tensile Properties on Non-reinforced Polyethylene and Flexible Polypropylene Geomembranes
- Tension Creep: ASTM D5262; Test Method for Evaluating Unconfined Tension Creep and Creep Rupture Behavior of Geosynthetics
- Theoretical Max. Gravity—Hot Mix/Paving Mixtures: ASTM D2041; Theoretical Maximum Specific Gravity and Density of Bituminous Paving Mixtures
- Theoretical Maximum Gravity: AASHTO T209; Theoretical Maximum Specific Gravity and Density of Bituminous Paving Mixtures
- Theoretical Maximum Gravity: ASTM D2041; Theoretical Maximum Specific Gravity and Density of Bituminous Paving Mixtures
- Thickness: ASTM D5199; Test Method for Measuring the Normal Thickness of Geosynthetics
- Thickness and Height—Hot Mix/Paving Mixtures: ASTM D3549; Test Method for Determining Thickness or Height of Compacted Bituminous Paving Mixtures Specimens
- Thickness, Core: ASTM D5994; Test Method for Measuring Core Thickness of Textured Geomembranes

- Thin Film Oven Test (TFOT): AASHTO T179; Test Method for Effects of Heat and Air on Asphaltic Materials (Thin Film Oven Method)
- Thin Film Oven Test (TFOT): ASTM D95; Test Method for Water in Petroleum Products and Bituminous Materials by Distillation
- Titanium Dioxide Content: AASHTO T250, Section 8; Titanium Dioxide Determination
- Toughness and Tenacity: ASTM D5801; Toughness and Tenacity of Bituminous Materials
- Uncompacted Void Content—Aggregates: ASTM C1252; Test Method for Uncompacted Void Content of Fine Aggregate (as influenced by Particle Shape, Surface Texture, and Grading)
- Uniformity: ASTM D2939, Section 5; Uniformity
- Uniformity: ASTM D6511, Section 5; Test Method for Solvent Bearing Bituminous Compounds, Uniformity
- Unit Weight: AASHTO T19; Bulk Density (Unit Weight) and Voids in Aggregate
- Unit Weight: AASHTO T228; Density of Semi-Solid Bituminous Materials (Pycnometer Method)
- Unit Weight: ASTM C29; Test Method for Bulk Density (Unit Weight) and Voids in Aggregate
- Unit Weight: ASTM D70; Density of Semi-Solid Bituminous Materials (pycnometer method)
- Unit Weight and Voids—Aggregates: ASTM C29; Test Method for Bulk Density (Unit Weight) and Voids in Aggregate
- Vacuum Distillation (Cut Points): ASTM D1160; Distillation of Petroleum Products at Reduced Pressure
- Viscosity: ASTM D1200; Test Method for Viscosity by Furol Viscosity Cup
- Viscosity: ASTM D2170; Test Method for Kinematic Viscosity of Asphalts
- Viscosity—Asphalt Binders: AASHTO T201; Test Method for Kinematic Viscosity of Asphalts
- Viscosity—Asphalt Binders: ASTM D1084; Test Method for Viscosity of Adhesives
- Viscosity—Asphalt Binders: ASTM D1200; Test Method for Viscosity by Furol Viscosity Cup
- Viscosity—Asphalt Binders: ASTM D3794; Guide for Testing Coil Coatings

- Viscosity—Asphalt Binders: ASTM D445; Kinematic Viscosity of Transparent and Opaque Liquids (and calculation of dynamic viscosity)
- Viscosity—Modified Asphalt Binders: ASTM D3381; ASTM D3381 + Toughness and Tenacity, Spot Test, and Others
- Viscosity, Absolute: AASHTO T202; Test Method for Viscosity of Asphalts by Vacuum Capillary Viscometer

VIEW TESTING AND SPECIFICATION PAGE

- Viscosity, Absolute: ASTM D2171; Viscosity of Asphalts by Vacuum Capillary Viscometer
- Viscosity, Absolute/Dynamic: AASHTO T202; Test Method for Viscosity of Asphalts by Vacuum Capillary Viscometer
- Viscosity, Absolute/Dynamic: ASTM D2171; Viscosity of Asphalts by Vacuum Capillary Viscometer
- Viscosity-Graded Cement: AASHTO M226; Specification for Viscosity-Graded Cement for Use in Pavement Construction
- Viscosity-Graded Cement: ASTM D3381; Specification for Viscosity-Graded Cement for Use in Pavement Construction
- Viscosity, Kinematic: AASHTO T201; Test Method for Kinematic Viscosity of Asphalts
- Viscosity, Kinematic: ASTM D2170; Test Method for Kinematic Viscosity of Asphalts
- Viscosity, Rotational: AASHTO T316; Viscosity Determination of Asphalt Binders using Rotational Viscometer

ASPHALT BINDERS: CONVENTIONAL PROPERTIES

- Viscosity, Rotational: ASTM D2196 (Modified); Rheological Properties of Non-Newtonian Material by Rotational (Brookfield type) Viscometer
- Viscosity, Rotational: ASTM D4402; Viscosity Determinations of Unfilled Asphalts Using the Brookfield Thermosel Apparatus
- Viscosity, Rotational: ASTM D7226; Test Method for Determining the Viscosity of Emulsified Asphalts Using a Rotational Paddle Viscometer
- Viscosity, Saybolt: AASHTO T59, Section 9; Saybolt Viscosity
- Viscosity, Saybolt: AASHTO T72; Test Method for Saybolt Furol Viscosity

- Viscosity, Saybolt: ASTM D88; Saybolt Furol Viscosity of Bituminous Materials
- Viscosity, Saybolt Furol: AASHTO T72; Saybolt Furol Viscosity of Bituminous Materials at High Temperatures
- Viscosity, Saybolt Furol: ASTM D88; Saybolt Furol Viscosity of Bituminous Materials
- Viscosity, Specific (Engler): ASTM D1665; Test Method for Engler Specific Viscosity of Tar Products
- Viscosity, Stormer: ASTM D562; Consistency of Paints Measuring Krebs Units Viscosity Using a Stormer-Type Viscometer
- Voids in Fine Aggregate: AASHTO T304; Test Method for Uncompacted Void Content of Fine Aggregates (Fine Aggregate Angularity)
- Volatile Content: ASTM D2369; Test Methods for Volatile Content of Coatings
- Volatile Distillation—Mixtures: ASTM D6627; Test Method for Determination of a Volatile Distillate Fraction of Cold Asphalt Mixtures
- Volatile Organic Content (VOC): ASTM D6511, Section 20; Test Method for Solvent Bearing Bituminous Compounds, Ash Content (EPA 24A—using ASTM D2369, D4017, D1475)
- Volatiles: ASTM D2939, Section 9; Volatiles
- Water Absorption: ASTM D545, Section 7.6; Water Absorption; Test Methods for Preformed Expansion Joint Fillers for Concrete Construction
- Water Content: AASHTO T55; Test Method for Water in Petroleum Products and Bituminous Materials by Distillation
- Water Content: ASTM D4377; Water in Crude Oils by Potentiometric Karl Fischer Titration
- Water Content: ASTM D95; Test Method for Water in Petroleum Products and Bituminous Materials by Distillation
- Waterproofing: ASTM D449. 2014; Standard Specification for Asphalt Used in Dampproofing and Waterproofing
- Water Resistance: ASTM D1735; Practice for Testing Water Resistance of
- Water Resistance: ASTM D2247; Practice for Testing Water Resistance of Coatings in 100% RH
- Wear Resistance (Dry and/or Wet): ASTM D2486; Determination of Relative Wear Resistance of Bituminous Pavement Sealers in Both Unconditioned and Conditioned (Water) Environments

- Wear Resistance (Dry and/or Wet): PRI TM 035; Determination of Relative Wear Resistance of Bituminous Pavement Sealers in Both Unconditioned and Conditioned (Water) Environments
- Weight per Gallon: AASHTO T59, Section 10; Determining Density of Emulsified Asphalt
- Weight per Gallon: ASTM D2939, Section 7; Weight per Gallon
- Weight per Gallon: ASTM D6511, Section 6; Test Method for Solvent Bearing Bituminous Compounds, Weight per Gallon (Cup Method)
- Weight per Gallon: ASTM D6937; Determining Density of Emulsified Asphalts
- Weights and Measures: ASTM D3549; Test Method for Thickness or Height of Compacted Bituminous Paving Mixture Specimens
- Wet Film Continuity: ASTM D2939, Section 22; Wet Film Continuity
- Wet Film Continuity: ASTM D2939, Section 22; Wet Film Continuity
- Wet Flow: ASTM D2939, Section 19; Wet Flow
- Workability—Mixtures: ASTM D6704; Test Method for Determining the Workability of Asphalt Cold Mix Patching Material

GLOSSARY

The following list represents a selection of definitions that are commonly used in reference to asphalt science and technology. Also included are definitions used in refining operations (processes, equipment, and products) that will also be of use to the reader. In this Glossary and throughout the text, the preference has been to use North American terminology; readers from Europe and other countries will need to substitute the word "asphalt" by the word "bitumen" when necessary.

AASHTO American Association of State Highway and Transportation Officials; an Organization of highway engineers from the 50 States that develops guides and standards.

ABN separation a method of fractionation by which petroleum is separated into acidic, basic, and neutral constituents.

Absolute permeability ability of a rock to conduct a fluid when only one fluid is present in the pores of the rock.

Absolute viscosity a term used interchangeably with viscosity to distinguish it from either kinematic viscosity or commercial viscosity. Absolute viscosity is the ratio of shear stress to shear rate—the internal resistance of a fluid to flow. The common unit of absolute viscosity is the poise. Absolute viscosity divided by fluid density equals kinematic viscosity. It is occasionally referred to as dynamic viscosity. Absolute viscosity and kinematic viscosity are expressed in fundamental units. Commercial viscosity such as Saybolt viscosity is expressed in arbitrary units of time, usually seconds.

Absorber see Absorption tower.

Absorption the assimilation of one material into another; in petroleum refining, the use of an absorptive liquid to selectively remove components from a process stream.

Absorption gasoline gasoline extracted from natural gas or refinery gas by contacting the absorbed gas with an oil and subsequently distilling the gasoline from the higher-boiling components.

Absorption oil oil used to separate the heavier components from a vapor mixture by absorption of the heavier components during intimate contacting of the oil and vapor; used to recover natural gasoline from wet gas.

Absorption plant a plant for recovering the condensable portion of natural or refinery gas, by absorbing the higher-boiling hydrocarbons in an absorption oil, followed by separation and fractionation of the absorbed material.

Absorption tower a tower or column which promotes contact between a rising gas and a falling liquid so that part of the gas may be dissolved in the liquid.

Acetone-benzol process a dewaxing process in which acetone and benzol (benzene or aromatic naphtha) are used as solvents.

Acid catalyst a catalyst having acidic character; the aluminas are examples of such catalysts.

Acid deposition acid rain; a form of pollution depletion in which pollutants, such as nitrogen oxides and sulfur oxides, are transferred from the atmosphere to soil or water; often referred to as atmospheric self-cleaning. The pollutants usually arise from the use of fossil fuels.

Acidity the capacity of an acid to neutralize a base such as a hydroxyl ion (OH^-).

Acid-modified asphalt asphalt bitumen modified by the addition of inorganic acids, typically phosphoric, or polyphosphoric acid.

Acid number a measure of the reactivity of petroleum with a caustic solution and given in terms of milligrams of potassium hydroxide that are neutralized by one gram of petroleum.

Acid rain the precipitation phenomenon that incorporates anthropogenic acids and other acidic chemicals from the atmosphere to the land and water (see Acid deposition).

Acid sludge the residue left after treating petroleum oil with sulfuric acid for the removal of impurities; a black, viscous substance containing the spent acid and impurities.

Acid treating a process in which unfinished petroleum products, such as gasoline, kerosene, and lubricating-oil stocks, are contacted with sulfuric acid to improve their color, odor, and other properties.

Additive a material added to another (usually in small amounts) in order to enhance desirable properties or to suppress undesirable properties.

Additive level the total percentage of all additives in asphalt (expressed in % of weight or % of volume).

Adhesion the process of forming a chemical bond between the asphalt binder and the aggregate.

Adhesion (active) the formation of a chemical bond between the asphalt binder and the aggregate in the presence of water.

Adsorption transfer of a substance from a solution to the surface of a solid resulting in relatively high concentration of the substance at the place of contact; see also Chromatographic adsorption.

Adsorption gasoline natural gasoline obtained an the adsorption process from wet gas.

Aggregate any hard, inert, mineral material used for mixing in graduated fragments; includes sand, gravel, crushed stone, or slag.

Aggregate pop-outs generally small, shallow pits or depressions in the surface of either concrete or hot-mix asphalt pavements that result mostly from the weathering of individual aggregate particles (in whole or in part); may also be the direct result of the expansion of absorbed moisture due to freezing; in others, this may be caused by chemical changes of an aggregate component; typical materials that may result in aggregate pop-outs include chert, soft fine-grained limestone, shale, pyrite, or coal.

Air blowing the process by which compressed air is blown into a bitumen feedstock typically at 230–260°C (445–500°F), sometimes in the presence of catalysts (typically ferric chloride, phosphoric acid, or phosphorus pentoxide); this process results in complex reactions which raise the softening point and viscosity of the asphalt bitumen. See oxidized asphalt.

Air-blown asphalt asphalt produced by blowing air through residua at elevated temperatures. See oxidized asphalt.

Air-blown asphalts asphalt products produced by air blowing. See oxidized asphalts.

Airlift Thermofor catalytic cracking a moving-bed continuous catalytic process for conversion of heavy gas oils into lighter products; the catalyst is moved by a stream of air.

Air pollution the discharge of toxic gases and particulate matter introduced into the atmosphere, principally as a result of human activity.

Air rectification the process of mildly blowing (mildly oxidizing) asphalt to improve the properties for use. See Air-rectified asphalt.

Air-rectified asphalt (synonym semi-blown asphalt) an asphalt that has been subjected to mild oxidation with the goal of producing an asphalt meeting paving-grade asphalt specifications; air-rectified asphalts are used in paving applications as well as roofing applications and also for some industrial applications.

Air-refined asphalt penetration-grade asphalt produced by partial blowing; archaic term, no longer in use.

Air sweetening a process in which air or oxygen is used to oxidize lead mercaptides to disulfides instead of using elemental sulfur.

Air toxics hazardous air pollutants.

Air voids air spaces in the aggregate.

Albertite a black, brittle, natural hydrocarbon possessing a conchoidal fracture and a specific gravity of approximately 1.1.

Alicyclic hydrocarbon a compound containing carbon and hydrogen only, which has a cyclic structure (e.g., cyclohexane); also collectively called naphthenes.

Aliphatic hydrocarbon a compound containing carbon and hydrogen only which has an open-chain structure (e.g., as ethane, butane, octane, butene) or a cyclic structure (e.g., cyclohexane).

Aliquot that quantity of material of proper size for measurement of the property of interest; test portions may be taken from the gross sample directly, but often preliminary operations such as mixing or further reduction in particle size are necessary.

Alkali any substance having basic (as opposed to acidic) properties. In a restricted sense it is applied to the hydroxides of ammonium, lithium, potassium, and sodium.

Alkaline a high pH usually of an aqueous solution; aqueous solutions of sodium hydroxide, sodium orthosilicate, and sodium carbonate are typical alkaline materials.

Alkalinity the capacity of a base to neutralize the hydrogen ion (H^+).

Alkali treatment see Caustic wash.

Alkali wash see Caustic wash.

Alkanes hydrocarbons that contain only single carbon–hydrogen bonds. The chemical name indicates the number of carbon atoms and ends with the suffix "ane."

Alkenes hydrocarbons that contain carbon–carbon double bonds. The chemical name indicates the number of carbon atoms and ends with the suffix "ene."

Alkylate the product of an alkylation process.

Alkylate bottoms residua from fractionation of alkylate; the alkylate product which boils higher than the aviation gasoline range; sometimes called heavy alkylate or alkylate polymer.

Alkylation in the petroleum industry, a process by which an olefin (e.g., ethylene) is combined with a branched-chain hydrocarbon (e.g., *iso*-butane); alkylation may be accomplished as a thermal or as a catalytic reaction.

Alkyl groups a group of carbon and hydrogen atoms that branch from the main carbon chain or ring in a hydrocarbon molecule. The simplest alkyl group, a methyl group, is a carbon atom attached to three hydrogen atoms.

Alligator cracking cracking which is characterized by interconnected or interlaced cracks in the wheel path, forming a series of small polygons.

Alpha-scission the rupture of the aromatic carbon–aliphatic carbon bond that joins an alkyl group to an aromatic ring.

Alumina (Al_2O_3) used in separation methods as an adsorbent and in refining as a catalyst.

American Society for Testing and Materials (ASTM) the official organization in the United States for designing standard tests for petroleum and other industrial products.

Amine washing a method of gas cleaning whereby acidic impurities such as hydrogen sulfide and carbon dioxide are removed from the gas stream by washing with an amine (usually an alkanolamine).

Amphoteric possession of the quality of reacting either as an acid or as a base.

Analytical equivalence the acceptability of the results obtained from different laboratories; a range of acceptable results.

Analyte the chemical for which a sample is tested, or analyzed.

Antibody a molecule having chemically reactive sites specific for certain other molecules.

Aniline point the temperature, usually expressed in °F, above which equal volumes of a petroleum product are completely miscible; a qualitative indication of the relative proportions of paraffins in a petroleum product which are miscible with aniline only at higher temperatures; a high aniline point indicates low aromatics.

Anticline structural configuration of a package of folding rocks in which the rocks are tilted in different directions from the crest.

Antiknock resistance to detonation or pinging in spark-ignition engines.

Antiknock agent a chemical compound such as tetraethyl lead which, when added in small amount to the fuel charge of an internal-combustion engine, tends to lessen knocking.

Antistripping agent an additive used in an asphaltic binder to overcome the natural affinity of an aggregate for water instead of asphalt.

API gravity a measure of the *lightness* or *heaviness* of petroleum which is related to density and specific gravity.

$$^{\circ}\text{API} = \left(141.5/\text{sp gr} @ 60^{\circ}\text{F}\right) - 131.5$$

Apparent bulk density the density of a catalyst as measured; usually loosely compacted in a container.

Apparent viscosity the viscosity of a fluid, or several fluids flowing simultaneously, measured in a porous medium (rock), and subject to both viscosity and permeability effects; also called effective viscosity.

Aquifer a subsurface rock interval that will produce water.

Aromatic hydrocarbon a hydrocarbon characterized by the presence of an aromatic ring or condensed aromatic rings; benzene and substituted benzene, naphthalene and substituted naphthalene, phenanthrene and substituted phenanthrene, as well as the higher condensed ring systems; compounds that are distinct from those of aliphatic compounds or alicyclic compounds.

Aromatization the conversion of non-aromatic hydrocarbons to aromatic hydrocarbons by: (i) rearrangement of aliphatic (non-cyclic) hydrocarbons into aromatic ring structures; and (ii) dehydrogenation of alicyclic hydrocarbons (naphthenes).

Arosorb process a process for the separation of aromatics from non-aromatics by adsorption on a gel from which they are recovered by desorption.

Asphalt the non-volatile product obtained by distillation and treatment of an asphaltic crude oil; a manufactured product.

Asphalt a mixture of asphalt and mineral materials used as a paving material that is typically produced at temperatures in the range of 140–160°C (280–320°F); in North America the term asphalt is synonymous with asphalt as used in Europe and also the term hot mix asphalt; in some parts of the world the term is synonymous with the binder–aggregate mix.

Asphalt (air-rectified) asphalt that has been subjected to mild oxidation with the goal of producing an asphalt meeting paving-grade asphalt specifications; air-rectified asphalts are used in paving applications as well as roofing applications and also for some industrial applications.

Asphalt base course a foundation course consisting of mineral aggregate, bound together with asphaltic material.

Asphalt binder the term used in the United States and some other countries for the organic binder.

Asphalt binder course an intermediate course between a base course and an asphalt surface course; the binder course is usually a coarse-graded aggregate asphalt concrete containing little or no mineral matter passing through a No. 200 sieve.

Asphalt blocks small size blocks (typically, 50 lb, 20 kg) of bonding asphalt for being melted in kettles.

Asphalt cement asphalt especially prepared as to quality and consistency for direct use in the manufacture of bituminous pavements; the term used in the United States and some other countries for the asphalt binder; asphalt that is refined to meet specifications for paving, industrial, and special purposes; penetration is usually between 40 and 300 dmm.

Asphalt cold mix asphalt mixtures made using cutback asphalts or asphalt emulsions, which can be placed at ambient temperatures.

Asphalt concrete high-quality, thoroughly controlled hot mixture of asphalt cement and well-graded high-quality aggregate thoroughly compacted into a uniform dense mass.

Asphalt emulsion an emulsion of asphalt cement in water containing a small amount of emulsifying agent; a mixture of two normally immiscible components (asphalt and water) and an emulsifying agent (usually a surfactant); asphalt emulsions are utilized in paving, roofing, and waterproofing operations; called *emulsified asphalts* in North America.

Asphalt enamel (asphalt paint) an external coating for protecting steel pipes; the term can also be used for asphalt paints (formulated cutback asphalts or asphalt emulsions).

Asphaltene (asphaltene constituents) the brown to black powdery material produced by treatment of petroleum, petroleum residua, or bituminous materials with a low-boiling liquid hydrocarbon, e.g., pentane or heptane; soluble in benzene (and other aromatic solvents), carbon disulfide, and chloroform (or other chlorinated hydrocarbon solvents).

Asphaltene association factor the number of individual asphaltene species which associate in non-polar solvents as measured by molecular weight methods; the molecular weight of asphaltenes in toluene divided by the molecular weight in a polar non-associating solvent, such as dichlorobenzene, pyridine, or nitrobenzene.

Asphaltene constituent highly polar aromatic materials which have a high viscosity or stiffness at ambient temperatures and are responsible for the overall stiffness of asphalts; the asphaltene fraction can be precipitated with n-heptane and are sometimes referred to as *n*-heptane insolubles.

Asphaltene fraction (petroleum) a complex combination of hydrocarbonaceous species obtained as a complex solid black product by the separation of petroleum residues by means of a special treatment of a low-boiling hydrocarbon, such as *n*-heptane cut; also contains vanadium and nickel.

Asphalt flux oil used to reduce the consistency or viscosity of hard asphalt to the point required for use.

Asphalt fume the gases and vapors emitted from heated asphalt, and the aerosols and mists resulting from the condensation of vapors after volatilization from heated asphalt.

Asphalt grading terminology there are currently three main grading systems employed world-wide for identifying and specifying asphalt used in road construction; these systems are penetration, viscosity, and performance graded; each system has test methods that are unique to that system, similar asphalts are used across all grading systems; the particular system used within a given country or region is generally a result of historical practices or governmental stipulations.

Asphaltic pyrobitumen see Asphaltoid.

Asphaltic road oil a thick, fluid solution of asphalt; usually a residual oil. see also Non-asphaltic road oil.

Asphalt intermediate course a course between a base course and asphalt surface course; sometimes called binder course.

Asphaltite a variety of naturally occurring, dark brown to black, solid, non-volatile bituminous material that is differentiated from bitumen primarily by a high content of material insoluble in *n*-pentane (asphaltene) or other liquid hydrocarbons.

Asphalt joint filler an asphaltic product used for filling cracks and joints in pavement and other structures.

Asphalt millings the fine particles (generally from dust to less than an inch or so) of bitumen and inorganic material that are produced by the mechanical grinding of bituminous concrete surfaces; sources include highway departments and local contractors hired to conduct road improvements; quantities of asphalt millings from a particular operation can vary greatly and quality will vary depending on the original asphalt character, age, weather, and other site-specific conditions; millings contain approximately 5–7% asphalt, which is used as a binder for the quarry materials (stone, rock, sand, silt) that make up the load-bearing portion of a bituminous concrete surface; produced by grinding a bituminous concrete-paved surface.

Asphaltoid a group of brown to black, solid bituminous materials of which the members are differentiated from asphaltites by their infusibility and low solubility in carbon disulfide.

Asphalt macadam a type of asphalt mix with a high stone content and containing 3–5% (w/w) of asphalt.

Asphalt mastic a void-less asphalt mixture with asphalt as a binder containing filler and/or limestone fine aggregates and/or natural asphalt and/or sand.

Asphalt mixes (mixtures) mixtures of graded mineral aggregates (sized stone fractions, sands, and fillers) with a controlled amount of penetration-graded asphalt.

Asphalt overlay one or more courses of asphalt construction on an existing pavement; generally includes a leveling course to correct the contour of the old pavement, followed by a uniform course or courses to provide needed thickness.

Asphalt (oxidized) obtained by blowing air through a heated residuum, or raffinate from a deasphalting process with or without a catalyst; a complex black solid obtained by blowing air through a heated residuum, or raffinate from a deasphalting process with or without a catalyst.

Asphalt paint a cutback asphalt made to treat bare metal or concrete or wood surfaces giving a bond between the surface and an enamel or a bituminous membrane or bonding asphalt; a specialized cutback asphalt product that contains relatively small amounts of other materials that are not native to asphalt or to the diluents typically used in cutback products, such as lampblack, aluminum flakes, and mineral pigments; used as a protective coating in waterproofing operations and other similar applications.

Asphalt pavement pavement consisting of a surface course of mineral aggregate coated and cemented together with asphalt cement on supporting courses such as asphalt bases; crushed stone, slag, or gravel.

Asphalt (petroleum-derived) a dark brown to black cement-like residuum obtained from the distillation of suitable crude oils; the distillation processes may involve one or more of the following: atmospheric distillation, vacuum distillation, steam distillation; further processing of distillation residuum may be needed to yield a material whose physical properties are suitable for commercial applications; the additional processes can involve air oxidation, solvent stripping or blending of residua of different stiffness characteristics.

Asphalt primer a liquid asphaltic material of low viscosity which, upon application to a non-bituminous surface to waterproof the surface and prepare it for further construction; cutback asphalt made to treat bare metal surfaces giving a bond between the metal and an enamel.

Asphalt roofing felt a sheet material, impregnated with asphalt, generally supplied in rolls and used in roof construction.

Asphalt soil stabilization (soil treatment) treatment of naturally occurring non-plastic or moderately plastic soils with liquid asphalt at normal temperatures; after mixing, aeration and compaction provide water-resistant base and sub-base courses with improved load-bearing qualities.

Asphalt surface treatments applications of asphaltic materials to any type of road or pavement surface, with or without a cover of mineral aggregate that produces an increase in thickness of less than 1 inch.

Asphaltum see Asphalt.

Associated molecular weight the molecular weight of asphaltene constituents in an associating (non-polar) solvent, such as toluene.

ASTM American Society for Testing and Materials; a national organization of users and producers of materials that establishes standards.

Atmospheric distillation distillation at atmospheric pressure

Atmospheric equivalent boiling point (AEBP) a mathematical method of estimating the boiling point at atmospheric pressure of non-volatile fractions of petroleum.

Atmospheric residuum a residuum obtained by distillation of a crude oil under atmospheric pressure and which boils above 350°C (660°F).

Attainment area a geographical area that meets NAAQS criteria for air pollutants (see also Non-attainment area).

Attapulgus clay see Fuller's earth.

Autofining a catalytic process for desulfurizing distillates.

Average particle size the weighted average particle diameter of a catalyst.

Aviation gasoline any of the special grades of gasoline suitable for use in certain airplane engines.

Aviation turbine fuel see Jet fuel.

Back mixing the phenomenon observed when a catalyst travels at a slower rate in the riser pipe than the vapors.

BACT best available control technology.

Baghouse a filter system for the removal of particulate matter from gas streams; so called because of the similarity of the filters to coal bags.

Bari-Sol process a dewaxing process which employs a mixture of ethylene dichloride and benzol as the solvent.

Barrel the unit of measurement of liquids in the petroleum industry; equivalent to 42 US standard gallons or 33.6 imperial gallons.

Base a generic term for material installed prior to asphalt paving. May be a crushed stone product or asphalt product; the base material provides the load-bearing characteristics of the finished pavement and may vary from 3–4″ for a residential driveway to 18″ or more for parking areas or roadways; the correct type and amount of base material must be determined and specified prior to paving. Lack of adequate base material is a primary cause of pavement failure.

Base course the layer of material immediately beneath the surface or intermediate course; may be composed of crushed stone; crushed slag; crushed or uncrushed gravel and sand; or combinations of these materials; may also be bound with asphalt (asphalt base course).

Basement the foot or base of a sedimentary sequence composed of igneous or metamorphic rocks.

Base number the quantity of acid, expressed in milligrams of potassium hydroxide per gram of sample that is required to titrate a sample to a specified end-point.

Base oils Petroleum-derived products consisting of complex mixtures of straight and branch-chained paraffinic, naphthenic (cycloparaffin), and aromatic hydrocarbons, with carbon numbers of 15 or more and boiling-points in the range of 300–600°C (570–1110°F); depending on climatic conditions base oils can be used to reduce the low stiffness of asphalt to resist low-temperature cracking of pavements.

Base stock a primary refined petroleum fraction into which other oils and additives are added (blended) to produce the finished product.

Basic nitrogen nitrogen (in petroleum) which occurs in pyridine form

Basic sediment and water (bs&w, bsw) the material which collects in the bottom of storage tanks, usually composed of oil, water, and foreign matter; also called bottoms, bottom settlings.

Basin a geological receptacle in which a sedimentary column is deposited that shares a common tectonic history at various stratigraphic levels.

Battery a series of stills or other refinery equipment operated as a unit.

Baumé gravity the specific gravity of liquids expressed as degrees on the Baum (°Bé) scale; for liquids lighter than water:

$$\text{Sp gr } 60°\text{F} = 140/(130 + °\text{BJ})$$

For liquids heavier than water:

$$\text{Sp gr } 60°\text{F} = 145/(145 - °\text{BJ})$$

Bauxite mineral matter used as a treating agent; hydrated aluminum oxide formed by the chemical weathering of igneous rock.

Bbl see Barrel.

Bell cap a hemispherical or triangular cover placed over the riser in a (distillation) tower to direct the vapors through the liquid layer on the tray; see Bubble cap.

Bender process a chemical treating process using lead sulfide catalyst for sweetening light distillates by which mercaptans are converted to disulfides by oxidation.

Bending beam rheometer an instrument used to determine the low-temperature stiffness properties of asphalts that have been laboratory-aged to simulate extended aging of the asphalt in asphalt pavements; data are used as part of the performance-graded asphalt specification.

Bentonite montmorillonite (a magnesium–aluminum silicate); used as a treating agent.

Benzene a colorless aromatic liquid hydrocarbon (C_6H_6).

Benzin a refined light naphtha used for extraction purposes.

Benzine an obsolete term for light petroleum distillates covering the gasoline and naphtha range; see Ligroine.

Benzol the general term which refers to commercial or technical (not necessarily pure) benzene; also the term used for aromatic naphtha.

Beta-scission the rupture of a carbon–carbon bond; two bonds removed from an aromatic ring.

Billion 1×10^9

Binder material serving to adhere to aggregate and ensure cohesion of the mixture; a more general term used to identify asphalt plus potential modifiers used to produce asphalt mixes; the term binder reflects that some asphalt mixes may utilize modified asphalts.

Binder course a transitional layer of asphaltic paving between the base and the surface course.

Biocide any chemical capable of killing bacteria and biorganisms.

Biogenic material derived from bacterial or vegetation sources.

Biological lipid any biological fluid that is miscible with a non-polar solvent. These materials include waxes, essential oils, chlorophyll, etc.

Biological oxidation the oxidative consumption of organic matter by bacteria by which the organic matter is converted into gases.

Biomass biological organic matter.

Biopolymer a high-molecular-weight carbohydrate produced by bacteria.

Bitumen a semi-solid to solid hydrocarbonaceous material found filling pores and crevices of sandstone, limestone, or argillaceous sediments.

Bituminous containing bitumen or constituting the source of bitumen; of or related to asphalt and bitumen.

Bituminous rock see Bituminous sand.

Bituminous sand a formation in which the bituminous material (see Bitumen) is found as a filling in veins and fissures in fractured rock or impregnating relatively shallow sand, sandstone, and limestone strata; a sandstone reservoir (often referred to as an oil sand deposit) that is impregnated with a heavy, viscous dark-colored tar-like material that cannot be retrieved through a well by conventional production techniques.

Black acid(s) a mixture of the sulfonates found in acid sludge which are insoluble in naphtha, benzene, and carbon tetrachloride; very soluble in water but insoluble in 30% sulfuric acid; in the dry, oil-free state, the sodium soaps are black powders.

Black oil any of the dark-colored oils; a term now often applied to heavy oil.

Black soap see Black acid.

Black strap the black material (mainly lead sulfide) formed in the treatment of sour light oils with doctor solution and found at the interface between the oil and the solution.

Blacktop common term for asphalt but should not be used in requesting any specifications or work as the term is widely used with various meanings in different areas; sometimes blacktop is used to refer to a penetration pavement or hot oil treatment. See fog seal.

Bleeding asphalt (flushed asphalt) caused by the upward movement of asphalt to the pavement surface.

Blended asphalt blends of two or more asphalts with different physical characteristics or blends of asphalt(s) and high-boiling-range petroleum fractions (e.g., vacuum gas oil) in order to achieve desired physical properties.

Blown asphalt the asphalt prepared by air blowing a residuum or an asphalt.

Blowing still (oxidizer, asphalt-blowing unit) equipment used to air-blow asphalt.

Bogging a condition that occurs in a coking reactor when the conversion to coke and light ends is too slow causing the coke particles to agglomerate.

Boiling point a characteristic physical property of a liquid at which the vapor pressure is equal to that of the atmosphere and the liquid is converted to a gas.

Boiling range the range of temperature, usually determined at atmospheric pressure in standard laboratory apparatus, over which the distillation of an oil commences, proceeds, and finishes.

Bonding asphalt oxidized asphalt or polymer-modified asphalt used for roofing.

Borrow suitable material from sources outside the roadway prism used primarily for embankments.

Bottled gas usually butane or propane, or butane–propane mixtures, liquefied and stored under pressure for domestic use; see also Liquefied petroleum gas.

Bottoms the liquid which collects in the bottom of a vessel (tower bottoms, tank bottoms) during distillation; also the deposit or sediment formed during storage of petroleum or a petroleum product; see also Residuum and Basic sediment and water.

Bright stock refined, high-viscosity lubricating oils usually made from residual stocks by processes such as a combination of acid treatment or solvent extraction with dewaxing or clay finishing.

Briquette See briquetting.

Briquetting the process by which fine materials (e.g., coal dusts, metal tailings) are mixed with asphalt (or other) binder to form conveniently handled blocks or pellets.

British thermal unit see Btu.

Bromine number the number of grams of bromine absorbed by 100 g of oil which indicates the percentage of double bonds in the material.

Brown acid oil-soluble petroleum sulfonates found in acid sludge which can be recovered by extraction with naphtha solvent. Brown-acid sulfonates are somewhat similar to mahogany sulfonates but are more water-soluble. In the dry, oil-free state, the sodium soaps are light-colored powders.

Brown soap see Brown acid.

Brønsted acid a chemical species which can act as a source of protons.

Brønsted base a chemical species which can accept protons.

BS&W see Basic sediment and water.

BTEX benzene, toluene, ethylbenzene, and the xylene isomers.

Btu (British thermal unit) the energy required to raise the temperature of one pound of water one degree Fahrenheit.

BTU see Btu.

Bubble cap an inverted cup with a notched or slotted periphery to disperse the vapor in small bubbles beneath the surface of the liquid on the bubble plate in a distillation tower.

Bubble plate a tray in a distillation tower.

Bubble point the temperature at which incipient vaporization of a liquid in a liquid mixture occurs, corresponding with the equilibrium point of 0% vaporization or 100% condensation.

Bubble tower a fractionating tower so constructed that the vapors rising pass up through layers of condensate on a series of plates or trays (see Bubble plate); the vapor passes from one plate to the next above by bubbling under one or more caps (see Bubble cap) and out through the liquid on the plate where the less volatile portions of vapor condense in bubbling through the liquid on the plate, overflow to the next lower plate, and ultimately back into the reboiler thereby effecting fractionation.

Bubble tray a circular, perforated plate having the internal diameter of a bubble tower, set at specified distances in a tower to collect the various fractions produced during distillation.

Buckley-Leverett method a theoretical method of determining frontal advance rates and saturations from a fractional flow curve.

Built up roofing (BUR) (in North America) a continuous roofing membrane consisting of plies of saturated organic (e.g., cellulose) felts or coated inorganic (e.g., glass fiber) felts, assembled in place with alternate layers of asphalt or coal tar pitch, and surfaced with mineral aggregate, a granule surfaced sheet, or a roof coating; in Europe: a continuous roofing membrane consisting of plies of coated inorganic (e.g., glass fiber) felts, assembled in place with alternate layers of asphalt, and surfaced with mineral aggregate, a granule surfaced sheet, or a roof coating.

Built up roofing asphalt oxidized asphalt used in the construction of low slope built up roofing (bur) systems; this material is called built-up roofing asphalt (BURA) in North America.

Bumping the knocking against the walls of a still occurring during distillation of petroleum or a petroleum product which usually contains water.

Bunker C oil see No. 6 Fuel oil.

Burner fuel oil any petroleum liquid suitable for combustion.

Burning oil an illuminating oil, such as kerosene (kerosine) suitable for burning in a wick lamp.

Burning point see Fire point.

Burning-quality index an empirical numerical indication of the likely burning performance of a furnace or heater oil; derived from the distillation profile and the API gravity, and generally recognizing the factors of paraffin character and volatility.

Burton process an older thermal cracking process in which oil was cracked in a pressure still and any condensation of the products of cracking also took place under pressure.

Butane dehydrogenation a process for removing hydrogen from butane to produce butenes and, on occasion, butadiene.

Butane vapor-phase isomerization a process for isomerizing n-butane to *iso*-butane using aluminum chloride catalyst on a granular alumina support and with hydrogen chloride as a promoter.

C_1, C_2, C_3, C_4, C_5 fractions a common way of representing fractions containing a preponderance of hydrocarbons having 1, 2, 3, 4, or 5 carbon atoms, respectively, and without reference to hydrocarbon type.

CAA Clean Air Act; this act is the foundation of air regulations in the United States.

Calcining heating a metal oxide or an ore to decompose carbonates, hydrates, or other compounds, often in a controlled atmosphere.

Capillary action the rise or movement of water in the voids of a soil caused by capillary forces.

Capillary forces interfacial forces between immiscible fluid phases, resulting in pressure differences between the two phases.

Capillary number N_c, the ratio of viscous forces to capillary forces, and equal to viscosity times velocity divided by interfacial tension.

Capillary pressure a force per area unit resulting from the surface forces to the interface between two fluids.

Carbene the pentane- or heptane-insoluble material that is insoluble in benzene or toluene but which is soluble in carbon disulfide (or pyridine).

Carboid the pentane- or heptane-insoluble material that is insoluble in benzene or toluene and which is also insoluble in carbon disulfide (or pyridine).

Carbonate washing processing using a mild alkali (e.g., potassium carbonate) process for emission control by the removal of acid gases from gas streams.

Carbon-forming propensity see Carbon residue.

Carbonization the conversion of an organic compound into char or coke by heat in the substantial absence of air; often used in reference to the destructive distillation (with simultaneous removal of distillate) of coal.

Carbon–oxygen log information about the relative abundance of elements such as carbon, oxygen, silicon, and calcium in a formation; usually derived from pulsed neutron equipment.

Carbon rejection upgrading processes in which coke is produced, e.g., coking.

Carbon residue the amount of carbonaceous residue remaining after thermal decomposition of petroleum, a petroleum fraction, or a petroleum product in a limited amount of air; also called the coke- or *carbon-forming propensity*; often prefixed by the terms Conradson or Ramsbottom in reference to the inventor of the respective tests.

Cascade tray a fractionating device consisting of a series of parallel troughs arranged in stair-step fashion in which liquid from the tray above enters the uppermost trough and liquid thrown from this trough by vapor rising from the tray below impinges against a plate and a perforated baffle and liquid passing through the baffle enters the next layer of the troughs.

Casinghead gas natural gas which issues from the casinghead (the mouth or opening) of an oil well.

Casinghead gasoline the liquid hydrocarbon product extracted from casinghead gas by one of three methods: compression, absorption, or refrigeration; see also Natural gasoline.

Catagenesis the alteration of organic matter during the formation of petroleum that may involve temperatures in the range 50°C (120°F) to 200°C (390°F); see also Diagenesis and Metagenesis.

Catalyst a chemical agent which, when added to a reaction (process), will enhance the conversion of a feedstock without being consumed in the process.

Catalyst selectivity the relative activity of a catalyst with respect to a particular compound in a mixture, or the relative rate in competing reactions of a single reactant.

Catalyst stripping the introduction of steam, at a point where spent catalyst leaves the reactor, in order to strip, i.e., remove, deposits retained on the catalyst.

Catalytic activity the ratio of the space velocity of the catalyst under test to the space velocity required for the standard catalyst to give the same conversion as the catalyst being tested; usually multiplied by 100 before being reported.

Catalytic air-blown asphalt oxidized asphalts produced using catalysts in air blowing

Catalytic cracking the conversion of high-boiling feedstocks into lower-boiling products by means of a catalyst which may be used in a fixed bed or fluid bed.

Catalytic reforming rearranging hydrocarbon molecules in a gasoline-boiling-range feedstock to produce other hydrocarbons having a higher antiknock quality; isomerization of paraffins, cyclization of paraffins to naphthenes, dehydrocyclization of paraffins to aromatics.

Cat cracking see Catalytic cracking.

Catforming a process for reforming naphtha using a platinum–silica–alumina catalyst which permits relatively high space velocities and results in the production of high-purity hydrogen.

Caustic consumption the amount of caustic lost from reacting chemically with the minerals in the rock, the oil, and the brine.

Cement-treated base consists of specified soil aggregates and Portland cement concrete mixed in a pug mill and deposited on the subgrade to the specified thickness.

Caustic wash the process of treating a product with a solution of caustic soda to remove minor impurities; often used in reference to the solution itself.

Ceresin a hard, brittle wax obtained by purifying ozokerite; see Microcrystralline wax and Ozokerite.

Cetane index an approximation of the cetane number calculated from the density and mid-boiling point temperature; see also Diesel index.

Cetane number a number indicating the ignition quality of diesel fuel; a high cetane number represents a short ignition delay time; the ignition quality of diesel fuel can also be estimated from the following formula:

$$\text{Diesel index} = (\text{aniline point } (^\circ\text{F}) \times \text{API gravity})100$$

CFR Code of Federal Regulations; Title 40 (40 CFR) contains the regulations for protection of the environment.

Characterization factor the UOP characterization factor K, defined as the ratio of the cube root of the molal average boiling point, T_B, in degrees Rankine (°R = °F + 460), to the specific gravity at 60°F/60°F:

$$K = (T_B)^{1/3} / \text{sp gr}$$

The value ranges from 12.5 for paraffin stocks to 10.0 for the highly aromatic stocks; also called the Watson characterization factor.

Cheesebox still an early type of vertical cylindrical still designed with a vapor dome.

Chelating agents complex-forming agents having the ability to solubilize heavy metals.

Chemical octane number the octane number added to gasoline by refinery processes or by the use of octane number improvers such as tetraethyl lead.

Chemical waste any solid, liquid, or gaseous material discharged from a process and that may pose substantial hazards to human health and environment.

Chlorex process a process for extracting lubricating-oil stocks in which the solvent used is Chlorex (β- β-dichlorodiethyl ether).

Chromatographic adsorption selective adsorption on materials such as activated carbon, alumina, or silica gel; liquid or gaseous mixtures of hydrocarbons are passed through the adsorbent in a stream of diluent, and certain components are preferentially adsorbed.

Chromatographic separation the separation of different species of compounds according to their size and interaction with the rock as they flow through a porous medium.

Chromatography a method of separation based on selective adsorption; see also Chromatographic adsorption.

Clarified oil the heavy oil which has been taken from the bottom of a fractionator in a catalytic cracking process and from which residual catalyst has been removed.

Clarifier equipment for removing the color or cloudiness of an oil or water by separating the foreign material through mechanical or chemical means; may involve centrifugal action, filtration, heating, or treatment with acid or alkali.

Clay silicate minerals that also usually contain aluminum and have particle sizes less than 0.002 micron; used in separation methods as an adsorbent and in refining as a catalyst.

Clay contact process see Contact filtration.

Clay refining a treating process in which vaporized gasoline or other light petroleum product is passed through a bed of granular clay such as fuller's earth.

Clay regeneration a process in which spent coarse-grained adsorbent clays from percolation processes are cleaned for reuse by deoiling them with naphtha, steaming out the excess naphtha, and then roasting in a stream of air to remove carbonaceous matter.

Clay treating see Gray clay treating.

Clay wash light oil, such as kerosene (kerosine) or naphtha, used to clean fuller's earth after it has been used in a filter.

Clastic composed of pieces of pre-existing rock.

Cleanup a preparatory step following extraction of a sample media designed to remove components that may interfere with subsequent analytical measurements.

Cloud point the temperature at which paraffin wax or other solid substances begin to crystallize or separate from the solution, imparting a cloudy appearance to the oil when the oil is chilled under prescribed conditions.

CMA cold mix asphalt.

Coal an organic rock.

Coalescence the union of two or more droplets to form a larger droplet and, ultimately, a continuous phase.

Coal tar the specific name for the tar produced from coal; a dark brown to black, highly aromatic material manufactured during the high-temperature carbonization of bituminous coals which differs from asphalt substantially in composition and physical characteristics; previously used in the roofing and paving industries as an alternative to asphalt.

Coal tar pitch the specific name for the pitch produced from coal; a black or dark-brown cementitious solid that is obtained as a non-volatile residue in the partial evaporation or fractional distillation of coal tar; has been used in the past in roofing as an alternative to asphalt.

Coarse aggregate aggregate particles retained on a no. 8 sieve.

Coarse graded aggregate an aggregate having a continuous grading in size of particles from coarse through fine with a predominance of coarse sizes.

Coating asphalt air-blown or oxidized or polymer-modified asphalt used to manufacture roofing membranes or shingles.

Coke a gray to black solid carbonaceous material produced from petroleum during thermal processing; characterized by having a high carbon content (95% + by weight) and a honeycomb type of appearance and is insoluble in organic solvents.

Coke drum a vessel in which coke is formed and which can be isolated from the process for cleaning.

Coke number used, particularly in Great Britain, to report the results of the Ramsbottom carbon residue test, which is also referred to as a coke test.

Coker the processing unit in which coking takes place.

Coking a process for the thermal conversion of petroleum in which gaseous, liquid, and solid (coke) products are formed.

Cold adhesive bituminous cutback used as a glue for application at ambient temperature of polymer-modified asphalt membranes.

Cold-applied roofing asphalt asphalt roofing products that are applied at ambient temperatures at the work place without any heating (e.g., peel-and-stick asphalt membrane or membranes applied with the use of a cold adhesive).

Cold pressing the process of separating wax from oil by first chilling (to help form wax crystals) and then filtering under pressure in a plate and frame press.

Cold settling processing for the removal of wax from high-viscosity stocks, wherein a naphtha solution of the waxy oil is chilled and the wax crystallizes out of the solution.

Colloid mills high-speed shearing devices in which hot asphalt can be dispersed using a surfactant in an aqueous solution to produce an asphalt emulsion

Colored mineral granules naturally colored or factory-colored minerals used as light surface protection for asphalt membranes or asphalt shingles.

Color stability the resistance of a petroleum product to color change due to light, aging, etc.

Combustible liquid a liquid with a flash point in excess of 37.8°C (100°F) but below 93.3°C (200°F).

Compaction the densification of crushed stone base, subgrade soil, or asphaltic material by means of vibration or rolling.

Composition the general chemical make-up of petroleum.

Composition map a means of illustrating the chemical make-up of petroleum using chemical and/or physical property data.

Con Carbon see Carbon residue.

Condensate a mixture of light hydrocarbon liquids obtained by condensation of hydrocarbon vapors: predominately butane, propane, and pentane with some heavier hydrocarbons and relatively little methane or ethane; see also Natural gas liquids.

Conductivity a measure of the ease of flow through a fracture, perforation, or pipe.

Conradson carbon residue see Carbon residue.

Contact filtration a process in which finely divided adsorbent clay is used to remove color bodies from petroleum products.

Contaminant a substance that causes deviation from the normal composition of an environment.

Contingent resource the amounts of hydrocarbons estimated at a given date that are potentially recoverable from known accumulations but are not considered commercially recoverable under the economic evaluation conditions corresponding to that date.

Continuous contact coking a thermal conversion process in which petroleum-wetted coke particles move downward into the reactor in which cracking, coking, and drying take place to produce coke, gas, gasoline, and gas oil.

Continuous contact filtration a process to finish lubricants, waxes, or special oils after acid treating, solvent extraction, or distillation.

Conversion the thermal treatment of petroleum which results in the formation of new products by the alteration of the original constituents.

Conversion cost the cost of changing a production well to an injection well, or some other change in the function of an oilfield installation.

Conversion factor the percentage of feedstock converted to light ends, gasoline, other liquid fuels, and coke,

Copper sweetening processes involving the oxidation of mercaptans to disulfides by oxygen in the presence of cupric chloride.

Core a cylindrical rock sample taken from a formation when drilling, used to determine the rock's permeability, porosity, hydrocarbon saturation, and other productivity-associated properties.

Co-surfactant a chemical compound, typically alcohol that enhances the effectiveness of a surfactant.

Course, asphalt base a foundation course consisting of mineral aggregate, bound together with asphalt material.

Course, asphalt surface the top course of an asphalt pavement, sometimes called an asphalt wearing course.

Cp (centipoise) a unit of viscosity.

Cracked residua residua that have been subjected to temperatures above 350°C (660°F) during the distillation process.

Crack filler a material that is placed in a pavement crack or joint to fill but not necessarily seal the void created by the crack or joint.

Crack sealant a material that has adhesive and cohesive properties to seal cracks, joints, or other narrow openings (less than 1 1/2″ wide) in pavements against the entrance or passage of water or other debris

Cracking the thermal processes by which the constituents of petroleum are converted to lower-molecular-weight products.

Cracking activity see Catalytic activity.

Cracking coil equipment used for cracking heavy petroleum products consisting of a coil of heavy pipe running through a furnace so that the oil passing through it is subject to high temperature.

Cracking-residue asphalt (thermal asphalt) archaic term, used only infrequently.

Cracking still the combined equipment-furnace, reaction chamber, fractionator for the thermal conversion of heavier feedstocks to lighter products.

Cracking temperature the temperature (350°C; 660°F) at which the rate of thermal decomposition of petroleum constituents becomes significant.

Criteria air pollutants air pollutants or classes of pollutants regulated by the Environmental Protection Agency; the air pollutants are (including VOCs): ozone, carbon monoxide, particulate matter, nitrogen oxides, sulfur dioxide, and lead.

Crocodile cracking see Alligator cracking.

Cross-linking combining of two or polymer molecules by use of a chemical that mutually bonds with a part of the chemical structure of the polymer molecules.

Crude assay a procedure for determining the general distillation characteristics (e.g., distillation profile) and other quality information of crude oil.

Crude oil see Petroleum.

Crude petroleum a naturally occurring mixture, consisting predominantly of hydrocarbons but also containing sulfur, nitrogen, or oxygen derivatives of hydrocarbons, which can be removed from the earth in a liquid state.

Crude scale wax the wax product from the first sweating of the slack wax.

Crude still distillation equipment in which crude oil is separated into various products.

Crushed stone the product resulting from the artificial crushing of rocks, boulders, or large cobblestones with the particles resulting from the crushing operation having all faces fractured.

Crusher run aggregates that have received little or no screening after initial crushing operations; generally more economical than screened aggregates.

Cryogenic plant a processing plant capable of producing liquid natural gas products, including ethane, at very low operating temperatures.

Cryogenics the study, production, and use of low temperatures.

Cumene a colorless liquid [$C_6H_5CH(CH_3)_2$] used as an aviation gasoline blending component and as an intermediate in the manufacture of chemicals.

Cut the portion of the roadway formed by excavation below the surface of the earth.

Cutback the term applied to the products from blending heavier feedstocks or products with lighter oils to bring the heavier materials to the desired specifications.

Cutback asphalt asphalt liquefied by the addition of a volatile liquid such as naphtha or kerosene which, after application and on exposure to the atmosphere, evaporates leaving the asphalt; asphalt whose viscosity has been reduced by the addition of a cutback solvent derived from petroleum.

Cutback solvent (petroleum) relatively volatile petroleum solvent used in the manufacture of cutback asphalt; typically white spirit (Stoddard solvent) and kerosene are the petroleum-derived solvents employed.

Cut point the boiling-temperature division between distillation fractions of petroleum.

Cutting oil an oil to lubricate and cool metal-cutting tools; also called cutting fluid, cutting lubricant.

Cycle stock the product taken from some later stage of a process and recharged (recycled) to the process at some earlier stage.

Cyclic constituents (naphthene aromatics, naphthenes) compounds with aromatic and naphthenic nuclei with side chain constituents; viscous liquids and that represent the major proportion of the dispersion medium for the asphaltenes and adsorbed resins in asphalt; often constitute 30–60% (w/w) of the total asphalt.

Cyclic steam injection the alternating injection of steam and production of oil with condensed steam from the same well or wells.

Cyclization the process by which an open-chain hydrocarbon structure is converted to a ring structure, e.g., hexane to benzene.

Cyclone a device for extracting dust from industrial waste gases. It is in the form of an inverted cone into which the contaminated gas enters tangential from the top; the gas is propelled down a helical pathway, and the dust particles are deposited by means of centrifugal force onto the wall of the scrubber.

Deactivation reduction in catalyst activity by the deposition of contaminants (e.g., coke, metals) during a process.

Dealkylation the removal of an alkyl group from aromatic compounds.

Deasphalted oil typically the soluble material after the insoluble asphaltic constituents have been removed; commonly, but often incorrectly, used in place of *deasphaltened oil*; see Deasphalting.

Deasphaltened oil the fraction of petroleum after only the asphaltene constituents have been removed.

Deasphaltening removal of a solid powdery asphaltene fraction from petroleum by the addition of the low-boiling liquid hydrocarbons such as *n*-pentane or *n*-heptane under ambient conditions.

Deasphalting the removal of the asphaltene fraction from petroleum by the addition of a low-boiling hydrocarbon liquid such as *n*-pentane or *n*-heptane; more correctly the removal asphalt (tacky, semi-solid) from petroleum (as occurs in a refinery asphalt plant) by the addition of liquid propane or liquid butane under pressure.

Debutanization distillation to separate butane and lighter components from higher-boiling components.

Decant oil the highest boiling product from a catalytic cracker; also referred to as slurry oil, clarified oil, or bottoms.

Decarbonizing a thermal conversion process designed to maximize coker gas-oil production and minimize coke and gasoline yields; operated at essentially lower temperatures and pressures than delayed coking.

Decoking removal of petroleum coke from equipment such as coking drums; hydraulic decoking uses high-velocity water streams.

Decolorizing removal of suspended, colloidal, and dissolved impurities from liquid petroleum products by filtering, adsorption, chemical treatment, distillation, bleaching, etc.

Deep lift asphalt pavement a pavement in which the asphalt base course is placed in one or more lifts of 4 or more inches compacted thickness.

De-ethanization distillation to separate ethane and lighter components from propane and higher-boiling components; also called de-ethanation.

Degradation the loss of desirable physical properties of EOR fluids, e.g., the loss of viscosity of polymer solutions.

Dehydrating agents substances capable of removing water (drying) or the elements of water from another substance.

Dehydrocyclization any process by which both dehydrogenation and cyclization reactions occur.

Dehydrogenation the removal of hydrogen from a chemical compound; for example, the removal of two hydrogen atoms from butane to make butene(s) as well as the removal of additional hydrogen to produce butadiene.

Delayed coking a coking process in which the thermal reactions are allowed to proceed to completion to produce gaseous, liquid, and solid (coke) products.

Delimitation an exploration activity that increases or decreases reserves by means of drilling delimiting wells.

Demethanization the process of distillation in which methane is separated from the higher-boiling components; also called demethanation.

Dense-graded aggregate a mineral aggregate uniformly graded from the maximum size down to and including sufficient mineral dust to reduce the void space in the compacted aggregate to exceedingly small dimensions approximating the size of voids in the dust itself.

Density the mass (or weight) of a unit volume of any substance at a specified temperature; see also Specific gravity.

Deoiling reduction in quantity of liquid oil entrained in solid wax by draining (sweating) or by a selective solvent; see MEK deoiling.

Depentanizer a fractionating column for the removal of pentane and lighter fractions from a mixture of hydrocarbons.

Depropanization distillation in which lighter components are separated from butanes and higher-boiling material; also called depropanation.

Desalting removal of mineral salts (mostly chlorides) from crude oils.

Design thickness the total pavement structure thickness above the subgrade.

Desorption the reverse process of adsorption whereby adsorbed matter is removed from the adsorbent; also used as the reverse of absorption.

Desulfurization the removal of sulfur or sulfur compounds from a feedstock.

Detergent oil lubricating oil possessing special sludge-dispersing properties for use in internal-combustion engines.

Dewaxing see Solvent dewaxing.

Dew point pressure the pressure at which the first drop of liquid is formed, when it goes from the vapor phase to the two-phase region.

Devolatilized fuel smokeless fuel; coke that has been reheated to remove all of the volatile material.

Diagenesis the concurrent and consecutive chemical reactions which commence the alteration of organic matter (at temperatures up to 50°C (120°F)) and ultimately result in the formation of petroleum from the marine sediment; see also Catagenesis and Metagenesis.

Diagenetic rock rock formed by conversion through pressure or chemical reaction; from a rock, e.g., sandstone is a diagenetic.

Diesel fuel fuel used for internal combustion in diesel engines; usually that fraction which distills after kerosene.

Diesel cycle a repeated succession of operations representing the idealized working behavior of the fluids in a diesel engine.

Diesel index an approximation of the cetane number of diesel fuel calculated from the density and aniline point.

Diesel knock the result of a delayed period of ignition and the accumulation of diesel fuel in the engine.

Differential-strain analysis measurement of thermal stress relaxation in a recently cut well.

Discovered resource The volume of hydrocarbons tested through wells drilled.

Discovery the incorporation of reserves attributable to drilling exploratory wells that test hydrocarbon-producing formations.

Distillation a process for separating liquids with different boiling points.

Distillation curve see Distillation profile.

Distillation loss the difference, in a laboratory distillation, between the volume of liquid originally introduced into the distilling flask and the sum of the residue and the condensate recovered.

Distillation range the difference between the temperature at the initial boiling point and at the end point, as obtained by the distillation test.

Distillation profile the distillation characteristics of petroleum or petroleum products showing the temperature and the per cent distilled.

Doctor solution a solution of sodium plumbite used to treat gasoline or other light petroleum distillates to remove mercaptan sulfur; see also Doctor test.

Doctor sweetening a process for sweetening gasoline, solvents, and kerosene by converting mercaptans to disulfides using sodium plumbite and sulfur.

Doctor test a test used for the detection of compounds in light petroleum distillates which react with sodium plumbite; see also Doctor solution.

Dome a geological structure with a semi-spherical shape or relief.

Domestic heating oil see No. 2 Fuel Oil.

Donor solvent process a conversion process in which hydrogen donor solvent is used in place of or to augment hydrogen.

Downcomer a means of conveying liquid from one tray to the next below in a bubble tray column.

Drainage structures and facilities for collecting and carrying away water.

Drainage radius the distance from which fluids flow to the well, that is, the distance reached by the influence of disturbances caused by pressure drops.

Drill stem test (*formation test*) the conventional formation test method.

Drum-mixer an asphalt mixing device in which mixtures of mineral aggregate and asphalt are heated and combined continuously in a rotating drum.

Dry gas equivalent to liquid (DGEL) the volume of crude oil that, because of its heat rate, is equivalent to the volume of dry gas.

Dry gas natural gas containing negligible amounts of hydrocarbons heavier than methane. Dry gas is also obtained from the processing complexes.

Drying removal of a solvent or water from a chemical substance; also referred to as the removal of solvent from a liquid or suspension.

Dropping point the temperature at which grease passes from a semisolid to a liquid state under prescribed conditions.

Dry gas a gas which does not contain fractions that may easily condense under normal atmospheric conditions.

Dry point the temperature at which the last drop of petroleum fluid evaporates in a distillation test.

Dualayer distillate process a process for removing mercaptans and oxygenated compounds from distillate fuel oils and similar products, using a combination of treatment with concentrated caustic solution and electrical precipitation of the impurities.

Dualayer gasoline process a process for extracting mercaptans and other objectionable acidic compounds from petroleum distillates; see also Dualayer solution.

Dualayer solution a solution which consists of concentrated potassium or sodium hydroxide containing a solubilizer; see also Dualayer gasoline process.

Dubbs cracking an older continuous, liquid-phase thermal cracking process formerly used.

Durability testing See weathering test.

Dynamic shear rheometer a testing device used to determine the stiffness of asphalts over a range of temperatures and test frequencies: typically a standard amount of asphalt (25 mm in diameter by 1 mm in thickness) tested between two flat plates (25 mm in diameter)—an oscillatory stress or strain of known value is applied to the asphalt sample and the resultant strain or stress is measured; the data are used to calculate the stiffness of the asphalt and are part of the specification within the performance-graded system of specifications.

Ebullated bed a process in which the catalyst bed is in a suspended state in the reactor by means of a feedstock recirculation pump which pumps the feedstock upwards at sufficient speed to expand the catalyst bed at approximately 35% above the settled level.

Economic limit the point at which the revenues obtained from the sale of hydrocarbons match the costs incurred in its exploitation.

Economic reserves the accumulated production that is obtained from a production forecast in which economic criteria are applied.

Edeleanu process a process for refining oils at low temperature with liquid sulfur dioxide (SO_2), or with liquid sulfur dioxide and benzene; applicable to the recovery of aromatic concentrates from naphtha and heavier petroleum distillates.

Effective porosity a fraction that is obtained by dividing the total volume of communicated pores by the total rock volume.

Effective viscosity see Apparent viscosity.

Effluent any contaminating substance, usually a liquid, which enters the environment via a domestic industrial, agricultural, or sewage plant outlet.

Elastomer a polymeric substance (natural or synthetic) which when stretched to a length that is less than its point of rupture and released will recover substantially to its originally length; examples are vulcanized natural rubber, styrene butadiene latex rubber, styrene butadiene styrene block copolymer.

Electric desalting a continuous process to remove inorganic salts and other impurities from crude oil by settling out in an electrostatic field.

Electrical precipitation a process using an electrical field to improve the separation of hydrocarbon reagent dispersions. May be used in chemical treating processes on a wide variety of refinery stocks.

Electrofining a process for contacting a light hydrocarbon stream with a treating agent (acid, caustic, doctor, etc.), then assisting the action of separation of the chemical phase from the hydrocarbon phase by an electrostatic field.

Electrolytic mercaptan process a process in which aqueous caustic solution is used to extract mercaptans from refinery streams.

Electrostatic precipitators devices used to trap fine dust particles (usually in the size range 30–60 microns) that operate on the principle of imparting an electric charge to particles in an incoming air stream and which are then collected on an oppositely charged plate across a high-voltage field.

Eluate the solutes, or analytes, moved through a chromatographic column (see *elution*).

Eluent solvent used to elute sample.

Elution a process whereby a solute is moved through a chromatographic column by a solvent (liquid or gas) or eluent.

Embankment a structure of soil, soil aggregate, or broken rock between the embankment foundation and the subgrade.

Emission control the use of gas cleaning processes to reduce emissions.

Emission standard the maximum amount of a specific pollutant permitted to be discharged from a particular source in a given environment.

Emulsified asphalt an emulsion of asphalt cement and water that contains a small amount of an emulsifying agent, a heterogeneous system containing two normally immiscible phases (asphalt and water), in which the water forms the continuous phase of the emulsion and minute globules of asphalt form the discontinuous phase; emulsified asphalt may be either anionic, electronegatively charged asphalt globules or cationic, electropositively charged asphalt globules, depending upon the emulsifying agent. See asphalt emulsion.

Emulsion a dispersion of very small drops of one liquid in an immiscible liquid, such as oil in water.

Emulsion breaking the settling or aggregation of colloidal-sized emulsions from suspension in a liquid medium.

End-of-pipe emission control the use of specific emission control processes to clean gases after production of the gases.

Energy the capacity of a body or system to do work, measured in joules (SI units); also the output of fuel sources.

Energy from biomass the production of energy from biomass.

Engler distillation a standard test for determining the volatility characteristics of a gasoline by measuring the per cent distilled at various specified temperatures.

Entrained bed a bed of solid particles suspended in a fluid (liquid or gas) at such a rate that some of the solid is carried over (entrained) by the fluid.

EPA Environmental Protection Agency.

Equiviscous (equi-viscous) temperature (EVT) the temperature at which asphalt has a viscosity that is optimum for application in built-up roofing (BUR) systems.

Erosion removal and transportation of soil by the action of water or wind.

ESAL equivalent single axel load.

Ester a compound formed by the reaction between an organic acid and an alcohol. Ethoxylated alcohols (i.e., alcohols having ethylene oxide functional groups attached to the alcohol molecule).

Ethanol see Ethyl alcohol.

Ethyl alcohol (ethanol or grain alcohol) an inflammable organic compound (C_2H_5OH) formed during fermentation of sugars; used as an intoxicant and as a fuel.

Evaporation a process for concentrating non-volatile solids in a solution by boiling off the liquid portion of the waste stream.

Evaporites sedimentary formations consisting primarily of salt, anhydrite, or gypsum, as a result of evaporation in coastal waters.

Expanding clays clays that expand or swell on contact with water, e.g., montmorillonite.

Exploratory well a well that is drilled without detailed knowledge of the underlying rock structure in order to find hydrocarbons whose exploitation is economically profitable.

Explosive limits the limits of percentage composition of mixtures of gases and air within which an explosion takes place when the mixture is ignited.

Extract the portion of a sample preferentially dissolved by the solvent and recovered by physically separating the solvent.

Extractive distillation the separation of different components of mixtures which have similar vapor pressures by flowing a relatively high-boiling solvent, which is selective for one of the components in the feed, down a distillation column as the distillation proceeds; the selective solvent scrubs the soluble component from the vapor.

Extra heavy oil crude oil with relatively high fractions of heavy components, high specific gravity (low API density), and high viscosity but mobile at reservoir conditions; thermal recovery methods are the most common form of commercially exploiting this kind of oil.

Fabric filters filters made from fabric materials and used for removing particulate matter from gas streams (see Baghouse).

Facies one or more layers of rock that differs from other layers in composition, age or content.

Fat oil the bottom or enriched oil drawn from the absorber as opposed to lean oil.

Faujasite a naturally occurring silica-alumina (SiO_2–Al_2O_3) mineral.

Fault a fractured surface of geological strata along which there has been differential movement.

FCC fluid catalytic cracking.

FCCU fluid catalytic cracking unit.

Feedstock petroleum as it is fed to the refinery; a refinery product that is used as the raw material for another process; the term is also generally applied to raw materials used in other industrial processes.

Ferrocyanide process a regenerative chemical treatment for mercaptan removal using caustic-sodium ferrocyanide reagent.

Field-scale the application of EOR processes to a significant portion of a field.

Filler (paving) fine mineral matter employed to give body to a bituminous binder or to fill the voids of a sand.

Filler (roofing) fine mineral matter, typically limestone, or slate dust mixed with asphalt prior to being applied as a coating in the manufacture of roofing shingles and other roofing products.

Filtration the use of an impassable barrier to collect solids but which allows liquids to pass.

Fine aggregate aggregate particles passing a no. 8 sieve.

Fine graded aggregate an aggregate having a continuous grading in sizes of particles from coarse through fine with a predominance of fine sizes.

Fingering the formation of finger-shaped irregularities at the leading edge of a displacing fluid in a porous medium which move out ahead of the main body of fluid.

Fire point the lowest temperature at which, under specified conditions in standardized apparatus, a petroleum product vaporizes sufficiently rapidly to form above its surface an air–vapor mixture which burns continuously when ignited by a small flame.

First contact miscibility see miscibility.

Fischer-Tropsch process a process for synthesizing hydrocarbons and oxygenated chemicals from a mixture of hydrogen and carbon monoxide.

Five-spot an arrangement or pattern of wells with four injection wells at the comers of a square and a producing well in the center of the square.

Fixed bed a stationary bed (of catalyst) to accomplish a process (see Fluid bed).

Flammability range the range of temperature over which a chemical is flammable.

Flammable a substance that will burn readily.

Flammable liquid a liquid having a flash point below 37.8°C (100°F).

Flammable solid a solid that can ignite from friction or from heat remaining from its manufacture, or which may cause a serious hazard if ignited.

Flash point the lowest temperature to which the product must be heated under specified conditions to give off sufficient vapor to form a mixture with air that can be ignited momentarily by a flame.

Flexible pavement a pavement (roadway) constructed from asphalt in contrast to a pavement constructed from Portland cement concrete; roadways; a road surfacing made from layers of asphalt mixes which maintains intimate contact with, and distributes loads to, the subgrade and depends on aggregate interlock, particle friction, and cohesion for stability.

Flexicoking a modification of the fluid coking process insofar as the process also includes a gasifier adjoining the burner/regenerator to convert excess coke to a clean fuel gas.

Flocculation threshold the point at which constituents of a solution (e.g., asphaltene constituents or coke precursors) will separate from the solution as a separate (solid) phase.

Floc point the temperature at which wax or solids separate as a definite floc.

Flue gases the gaseous products of the combustion process mostly comprised of carbon dioxide, nitrogen, and water vapor; gas from the combustion of fuel, the heating value of which has been substantially spent and which is, therefore, discarded to the flue or stack.

Fluid bed use of an agitated bed of inert granular material to accomplish a process in which the agitated bed resembles the motion of a fluid; a bed (of catalyst) that is agitated by an upward passing gas in such a manner that the particles of the bed simulate the movement of a fluid and has the characteristics associated with a true liquid. See Fixed bed.

Fluid catalytic cracking cracking in the presence of a fluidized bed of catalyst.

Fluid coking a continuous fluidized solids process that cracks feed thermally over heated coke particles in a reactor vessel to gas, liquid products, and coke.

Fluidized bed combustion a process used to burn low-quality solid fuels in a bed of small particles suspended by a gas stream (usually air that will lift the particles but not blow them out of the vessel). Rapid burning removes some of the offensive by-products of combustion from the gases and vapors that result from the combustion process.

Fluid saturation the portion of the pore space occupied by a specific fluid; oil, gas and water may exist.

Flushed asphalt see Bleeding asphalt.

Flux a term referring to a raw material from which oxidized asphalt is made; typically soft asphalt is used, although asphalt of higher viscosity can be included within the definition of flux; also called *flux oil*, which is a relatively non-volatile fluid used in the manufacture of non-fluxed asphalt.

Fluxed asphalt (petroleum) asphalt where the viscosity has been reduced by the addition of a flux oil derived from petroleum; typically gas oil of various distillation ranges is employed as the flux oil; fluxed asphalt differs from cutback asphalt which is also reduced-viscosity asphalt in that the flux oil has negligible volatility at ambient temperatures compared to the petroleum solvents used to produce cutback asphalts.

Flux oil (petroleum) high flashpoint hydrocarbon oils (generally paraffinic) added to a roofing flux prior to oxidizing; the purpose of a flux oil is to enable manufacture of oxidized asphalt with higher penetration values at a given softening point than would be possible without incorporation of the flux oil; can also refer to flux or flux oil; relatively non-volatile fluid used in the manufacture of fluxed asphalt, it also refers to the diluent used in the manufacture of oxidized asphalt.

Fly ash particulate matter produced from mineral matter in coal that is converted during combustion to finely divided inorganic material and which emerges from the combustor in the gases.

Fog seal a light application of liquid asphalt without mineral aggregate cover; slow-setting asphalt emulsion diluted with water is the preferred type.

Foots oil the oil sweated out of slack wax; named from the fact that the oil goes to the foot, or bottom, of the pan during the sweating operation.

Formation an interval of rock with distinguishable geologic characteristics.

Formation resistance factor the ratio between the resistance of rock saturated 100% with brine divided by the resistance of the saturating water.

Fossil fuel resources a gaseous, liquid, or solid fuel material formed in the ground by chemical and physical changes (diagenesis) in plant and animal residues over geological time; natural gas, petroleum, coal, and oil shale.

Fractional composition the composition of petroleum as determined by fractionation (separation) methods.

Fractional distillation the separation of the components of a liquid mixture by vaporizing and collecting the fractions, or cuts, which condense in different temperature ranges.

Fractional flow the ratio of the volumetric flow rate of one fluid phase to the total fluid volumetric flow rate within a volume of rock.

Fractional flow curve the relationship between the fractional flow of one fluid and its saturator during simultaneous flow of fluids through rock.

Fracture a natural or man-made crack in rock.

Fractionating column a column arranged to separate various fractions of petroleum by a single distillation and which may be tapped at different points along its length to separate various fractions in the order of their boiling points.

Fractionation the separation of petroleum into the constituent fractions using solvent or adsorbent methods; chemical agents such as sulfuric acid may also be used.

Frasch process a process formerly used for removing sulfur by distilling oil in the presence of copper oxide.

Free water (groundwater) water that is free to move through a soil mass under the influence of gravity.

French drain a trench loosely backfilled with stones, the largest being placed on the bottom with the size decreasing toward the top.

Fuel oil also called heating oil is a distillate product that covers a wide range of properties; see also No. 1 to No. 4 Fuel oils.

Full-depth asphalt pavement an asphalt pavement in which asphalt mixtures are employed for all courses above the subgrade or improved subgrade; a full-depth asphalt pavement is laid directly on the prepared subgrade.

Fuller's earth a clay which has high adsorptive capacity for removing color from oils; attapulgus clay is a widely used fuller's earth.

Fume-suppressing BUR asphalt proprietary BUR asphalt product which contains small amounts of polymer (added during manufacture or at the jobsite) that forms a layer on the surface of the heated asphalt, lowering the rate of fume generation; also known as *low-fuming asphalt*.

Functional group the portion of a molecule that is characteristic of a family of compounds and determines the properties of these compounds.

Furfural extraction a single-solvent process in which furfural is used to remove aromatic, naphthene, olefin, and unstable hydrocarbons from a lubricating-oil charge stock.

Furnace oil a distillate fuel primarily intended for use in domestic heating equipment.

Gap-graded asphalt mix a mix that is similar in function to dense-graded asphalt mixes insofar as they provide dense impervious layers when properly compacted; this type of mix contains aggregate that ranges in size from coarse to fine, with some intermediate sizes missing or present in small amounts.

Gas compressibility ratio (*Z*) the ratio between an actual gas volume and an ideal gas volume. This is an adimensional amount that usually varies between 0.7 and 1.2.

Gas lift an artificial production system that is used to raise the well fluid by injecting gas down the well through tubing or through the tubing-casing annulus.

Gas oil a liquid petroleum distillate with a viscosity and boiling-range between those of kerosene and lubricating oil.

Gas–oil ratio ratio of the number of cubic feet of gas measured at atmospheric (standard) conditions to barrels of produced oil measured at stocktank conditions.

Gas–oil sulfonate sulfonate made from a specific refinery stream, in this case the gas–oil stream.

Gasoline fuel for the internal combustion engine that is commonly, but improperly, referred to simply as gas.

Gaseous pollutants gases released into the atmosphere that act as primary or secondary pollutants.

Gasohol a term for motor vehicle fuel comprising between 80–90% unleaded gasoline and 10–20% ethanol (see also Ethyl alcohol).

Gas oil a petroleum distillate with a viscosity and boiling range between those of kerosine and lubricating oil.

Gas reversion a combination of thermal cracking or reforming of naphtha with thermal polymerization or alkylation of hydrocarbon gases carried out in the same reaction zone.

Geological province a region of large dimensions characterized by similar geological and development histories.

Gilsonite an asphaltite that is >90% bitumen; a natural hydrocarbonaceous material found in the Uintah basin in north-eastern Utah, USA.

Girbotol process a continuous, regenerative process to separate hydrogen sulfide, carbon dioxide, and other acid impurities from natural gas, refinery gas, etc., using mono-, di-, or triethanolamine as the reagent.

Glance pitch an asphaltite.

Glass mat (felt mat) a non-woven mat made with short glass fibers adhered together with a resin and suitable for coating and impregnation with asphalt for roofing products.

Glycol-amine gas treating a continuous, regenerative process to simultaneously dehydrate and remove acid gases from natural gas or refinery gas.

Graben a dip or depression formed by tectonic processes, limited by normal-type faults.

Grahamite an asphaltite.

Gravel a coarse granular material (usually larger than 1/4 inch in diameter) resulting from the natural erosion and disintegration of rock; crushed gravel is the result of artificial crushing with most fragments having at least one face resulting from fracture.

Gravitational segregation a driving mechanism in which the fluids tend to separate according to their specific gravities.

Gravity see API gravity.

Gravity drainage the movement of fluids that results from the force of gravity.

Gravity segregation partial separation of fluids caused by the gravity force acting on differences in density.

Gray clay treating a fixed-bed, usually fuller's earth, vapor-phase treating process to selectively polymerize unsaturated gum-forming constituents (diolefins) in thermally cracked gasoline.

Grain alcohol see Ethyl alcohol.

Gravimetric gravimetric methods weigh a residue.

Gravity drainage the movement of oil in a reservoir that results from the force of gravity.

Gravity segregation partial separation of fluids in a reservoir caused by the gravity force acting on differences in density.

Greenhouse effect warming of the earth due to entrapment of the sun's energy by the atmosphere.

Greenhouse gases gases that contribute to the greenhouse effect.

Guard bed a bed of an adsorbent (such as, for example, bauxite) that protects a catalyst bed by adsorbing species detrimental to the catalyst.

Gulf HDS process a fixed-bed process for the catalytic hydrocracking of heavy stocks to lower-boiling distillates with accompanying desulfurization.

Gulfining a catalytic hydrogen treating process for cracked and straight-run distillates and fuel oils, to reduce sulfur content; improve carbon residue, color, and general stability; and effect a slight increase in gravity.

Gum an insoluble tacky semi-solid material formed as a result of the storage instability and/or the thermal instability of petroleum and petroleum products.

Handling efficiency shrinkage factor a fraction of natural gas that is derived from considering self-consumption and the lack of capacity to handle such; obtained from the gas-handling statistics of the final period in the area corresponding to the field being studied.

HAP(s) hazardous air pollutant(s).

Hard asphalt asphalt possessing low penetration value and high softening-point; used in the manufacture of high-modulus asphalt mixtures.

Hardness the concentration of calcium and magnesium in brine.

HCPV hydrocarbon pore volume.

Heating oil see Fuel oil.

Heat value the amount of heat released per unit of mass or per unit of volume when a substance is completely burned. The heat power of solid and liquid fuels is expressed in calories per gram or in Btu per pound. For gases, this parameter is generally expressed in kilocalories per cubic meter or in Btu per cubic foot.

Heavy ends the highest boiling portion of a petroleum fraction; see also Light ends.

Heavy fuel oil fuel oil having a high density and viscosity; generally residual fuel oil such as No. 5 and No 6. fuel oil

Heavy oil typically petroleum having an API gravity of less than 20°.

Heavy petroleum see Heavy oil.

Heteroatom compounds chemical compounds which contain nitrogen and/or oxygen and/or sulfur and/or metals bound within their molecular structure(s).

Heterogeneity lack of uniformity in properties such as permeability.

HF alkylation an alkylation process whereby olefins (C_3, C_4, C_5) are combined with *iso*-butane in the presence of hydrofluoric acid catalyst.

HMA hot mix asphalt.

Horst rock of the earth's crust rising between two faults; it is the opposite of a graben.

Hortonsphere a spherical pressure-type tank used to store a volatile liquid which prevents the excessive evaporation loss that occurs when such products are placed in conventional storage tanks.

Hot-applied roofing application of roofing membranes with hot bonding asphalt as a glue by mopping, pouring, or with mechanical spreaders (pour & roll technique) this is also called hot bonding roofing.

Hot bonding roofing See hot-applied roofing.

Hot filtration test a test for the stability of a petroleum product.

Hot mix asphalt a mixture of asphalt and mineral materials used as a paving material that is typically produced at temperatures in the range of 140–160°C (280–320°F); in Europe, the term is synonymous with asphalt; differentiation between warm mix asphalt and hot mix asphalt is not often obvious. See warm mix asphalt.

Hot mix asphalt concrete abbreviation of the proper name for what is commonly referred to by names such as *asphalt*, *hot-mix*, and *blacktop*; this term should always be used in specifying asphalt pavement work to avoid any confusion or misinterpretation of the material desired; produced in many different grades from coarse base mixes to specialized mixes for surfacing and repair. In most instances the grades are specified according to state department of transportation guidelines.

Hot spot an area of a vessel or line wall appreciably above normal operating temperature, usually as a result of the deterioration of an internal insulating liner which exposes the line or vessel shell to the temperature of its contents.

Hot welding roofing See torching

Houdresid catalytic cracking a continuous moving-bed process for catalytically cracking reduced crude oil to produce high-octane gasoline and light distillate fuels.

Houdriflow catalytic cracking a continuous moving-bed catalytic cracking process employing an integrated single vessel for the reactor and regenerator kiln.

Houdriforming a continuous catalytic reforming process for producing aromatic concentrates and high-octane gasoline from low-octane straight naphtha.

Houdry butane dehydrogenation a catalytic process for dehydrogenating light hydrocarbons to their corresponding mono- or diolefins.

Houdry fixed-bed catalytic cracking a cyclic regenerable process for cracking of distillates.

Houdry hydrocracking a catalytic process combining cracking and desulfurization in the presence of hydrogen.

Huff-and-puff a cyclic EOR method in which steam or gas is injected into a production well; after a short shut-in period, oil and the injected fluid are produced through the same well.

Hydration the association of molecules of water with a substance.

Hydrocarbon compounds chemical compounds containing only carbon and hydrogen.

Hydrocarbons organic compounds containing only hydrogen and carbon.

Hydrolysis a chemical reaction in which water reacts with another substance to form one or more new substances.

Hydroconversion a term often applied to hydrocracking

Hydrocracking a catalytic high-pressure high-temperature process for the conversion of petroleum feedstocks in the presence of fresh and recycled hydrogen; carbon–carbon bonds are cleaved in addition to the removal of heteroatomic species.

Hydrocracking catalyst a catalyst used for hydrocracking which typically contains separate hydrogenation and cracking functions.

Hydrodenitrogenation the removal of nitrogen by hydrotreating.

Hydrodesulfurization the removal of sulfur by hydrotreating.

Hydrofining a fixed-bed catalytic process to desulfurize and hydrogenate a wide range of charge stocks from gases through waxes.

Hydroforming a process in which naphtha is passed over a catalyst at elevated temperatures and moderate pressures, in the presence of added hydrogen or hydrogen-containing gases, to form high-octane motor fuel or aromatics.

Hydrogen blistering blistering of steel caused by trapped molecular hydrogen formed as atomic hydrogen during corrosion of steel by hydrogen sulfide.

Hydrogen addition an upgrading process in the presence of hydrogen, e.g., hydrocracking; see Hydrogenation.

Hydrogenation the chemical addition of hydrogen to a material. In non-destructive hydrogenation, hydrogen is added to a molecule only if, and where, unsaturation with respect to hydrogen exists.

Hydrogen transfer the transfer of inherent hydrogen within the feedstock constituents and products during processing.

Hydroprocessing a term often equally applied to hydrotreating and to hydrocracking; also often collectively applied to both.

Hydrostatic pressure the pressure in a liquid under static conditions; the product of the unit weight of the liquid and the difference in elevation between the given points and the free water elevation.

Hydrotreating the removal of heteroatomic (nitrogen, oxygen, and sulfur) species by treatment of a feedstock or product at relatively low temperatures in the presence of hydrogen.

Hydrovisbreaking a non-catalytic process, conducted under similar conditions to visbreaking, which involves treatment with hydrogen to reduce the viscosity of the feedstock and produce more stable products than is possible with visbreaking.

Hydropyrolysis a short residence time high-temperature process using hydrogen.

Hyperforming a catalytic hydrogenation process for improving the octane number of naphtha through removal of sulfur and nitrogen compounds.

Hypochlorite sweetening the oxidation of mercaptans in a sour stock by agitation with aqueous, alkaline hypochlorite solution; used where avoidance of free-sulfur addition is desired, because of stringent copper strip requirements and minimum expense is not the primary object.

Ignitability characteristic of liquids whose vapors are likely to ignite in the presence of ignition source; also characteristic of non-liquids that may catch fire from friction or contact with water and that burn vigorously.

Illuminating oil oil used for lighting purposes.

Immiscibility the inability of two or more fluids to have complete mutual solubility; they co-exist as separate phases.

Immiscible two or more fluids that do not have complete mutual solubility and co-exist as separate phases.

Improved subgrade any course or courses of select or improved material between the foundation soil and the sub-base is usually referred to as the improved subgrade; the improved subgrade can be made up of two or more courses of different-quality materials.

Impurities and plant liquefiables shrinkage factor the fraction obtained by considering the non-hydrocarbon gas impurities (sulfur, carbon dioxide, nitrogen compounds, etc.) contained in the sour gas, in addition to shrinkage caused by the generation of liquids in a gas processing plant.

Impurities shrinkage factor the fraction that results from considering the non-hydrocarbon gas impurities (sulfur, carbon dioxide, nitrogen compounds, etc.) contained in the sour gas. It is obtained from the operation statistics of the last annual period of the gas processing complex (GPC) that processes the production of the field analyzed.

Incompatibility the *immiscibility* of petroleum products and also of different crude oils which is often reflected in the formation of a separate phase after mixing and/or storage.

Infill drilling drilling additional wells within an established pattern.

Infrared spectroscopy an analytical technique that quantifies the vibration (stretching and bending) that occurs when a molecule absorbs (heat) energy in the infrared region of the electromagnetic spectrum.

Inhibitor a substance, the presence of which, in small amounts, in a petroleum product prevents or retards undesirable chemical changes from taking place in the product, or in the condition of the equipment in which the product is used.

Inhibitor sweetening a treating process to sweeten gasoline of low mercaptan content, using a phenylenediamine type of inhibitor, air, and caustic.

Initial boiling point the recorded temperature when the first drop of liquid falls from the end of the condenser.

Initial vapor pressure the vapor pressure of a liquid of a specified temperature and 0% evaporated.

Instability the inability of a petroleum product to exist for periods of time without change to the product.

Integrity maintenance of a slug or bank at its preferred composition without too much dispersion or mixing.

Interface the thin surface area separating two immiscible fluids that are in contact with each other.

Interfacial film a thin layer of material at the interface between two fluids which differs in composition from the bulk fluids.

Interfacial tension the strength of the film separating two immiscible fluids, e.g., oil and water or microemulsion and oil; measured in dynes (force) per centimeter or millidynes per centimeter.

Interfacial viscosity the viscosity of the interfacial film between two immiscible liquids.

Interference testing a type of pressure transient test in which pressure is measured over time in a closed-in well while nearby wells are produced; flow and communication between wells can sometimes be deduced from an interference test.

Interphase mass transfer the net transfer of chemical compounds between two or more phases.

Iodine number a measure of the iodine absorption by oil under standard conditions; used to indicate the quantity of unsaturated compounds present; also called iodine value.

Ion exchange a means of removing cations or anions from solution onto a solid resin.

Ion exchange capacity a measure of the capacity of a mineral to exchange ions in amount of material per unit weight of solid.

Ions chemical substances possessing positive or negative charges in solution.

Isocracking a hydrocracking process for conversion of hydrocarbons which operates at relatively low temperatures and pressures in the presence of hydrogen and a catalyst to produce more valuable, lower-boiling products.

Isoforming a process in which olefinic naphtha is contacted with an alumina catalyst at high temperature and low pressure to produce isomers of higher octane number.

Iso-Kel process a fixed-bed, vapor-phase isomerization process using a precious metal catalyst and external hydrogen.

Isomate process a continuous, non-regenerative process for isomerizing C_5–C_8 normal paraffin hydrocarbons, using aluminum chloride-hydrocarbon catalyst with anhydrous hydrochloric acid as a promoter.

Isomerate process a fixed-bed isomerization process to convert pentane, heptane, and heptane to high-octane blending stocks.

Isomerization the conversion of a *normal* (straight-chain) paraffin hydrocarbon into an *iso* (branched-chain) paraffin hydrocarbon having the same atomic composition.

Isopach a line on a map designating points of equal formation thickness.

Iso-plus Houdriforming a combination process using a conventional Houdriformer operated at moderate severity, in conjunction with one of three possible alternatives, including the use of an aromatic recovery unit or a thermal reformer; see Houdriforming.

Jet fuel fuel meeting the required properties for use in jet engines and aircraft turbine engines.

Joint reflection cracking reflection cracks arising from underlying Portland cement concrete pavement joint movement.

Kaolinite a clay mineral formed by hydrothermal activity at the time of rock formation or by chemical weathering of rock with high feldspar content; usually associated with intrusive granite rock with high feldspar content.

Kata-condensed aromatic compounds Compounds based on linear condensed aromatic hydrocarbon systems, e.g., anthracene and naphthacene (tetracene).

Kauri butanol number A measurement of solvent strength for hydrocarbon solvents; the higher the kauri-butanol (KB) value, the stronger the solvency; the test method (ASTM D1133) is based on the principle that kauri resin is readily soluble in butyl alcohol but not in hydrocarbon solvents and the resin solution will tolerate only a certain amount of dilution and is reflected as a cloudiness when the resin starts to come out of solution; solvents such as toluene can be added in a greater amount (and thus have a higher KB value) than weaker solvents like hexane.

Kerogen a complex carbonaceous (organic) material that occurs in sedimentary rock and shale; generally insoluble in common organic solvents.

Kerosene (kerosine) a fraction of petroleum that was initially sought as an illuminant in lamps; a precursor to diesel fuel; a petroleum distillate consisting of hydrocarbons with carbon numbers predominantly in the range of C_9 through C_{16} and boiling in the range of 150–190°C (300–550°F).

K-factor see Characterization factor.

Kinematic viscosity the ratio of viscosity to density, both measured at the same temperature.

Knock the noise associated with self-ignition of a portion of the fuel–air mixture ahead of the advancing flame front.

LAER lowest achievable emission rate; the required emission rate in non-attainment permits.

Lake asphalt most common form of bitumen, occurring in Trinidad (Republic of Trinidad and Tobago).

Lamp burning a test of burning oils in which the oil is burned in a standard lamp under specified conditions in order to observe the steadiness of the flame, the degree of encrustation of the wick, and the rate of consumption of the kerosene.

Lamp oil see Kerosene.

Laydown the portion of the asphalt paving process where the hot asphalt is actually placed or laid down by the paving machine.

Leaded gasoline gasoline containing tetraethyl lead or other organometallic lead antiknock compounds.

Lean gas the residual gas from the absorber after the condensable gasoline has been removed from the wet gas.

Lean oil absorption oil from which gasoline fractions have been removed; oil leaving the stripper in a natural-gasoline plant.

Leveling course an asphalt–aggregate mixture of variable thickness used to eliminate regularities in the contour of an existing surface before superimposed treatment or construction.

Lewis acid a chemical species which can accept an electron pair from a base.

Lewis base a chemical species which can donate an electron pair.

Light ends the lower-boiling components of a mixture of hydrocarbons; see also Heavy ends, Light hydrocarbons.

Light hydrocarbons hydrocarbons with molecular weights less than that of heptane (C_7H_{16}).

Light oil the products distilled or processed from crude oil up to, but not including, the first lubricating-oil distillate.

Light petroleum petroleum having an API gravity greater than 20°.

Ligroine (Ligroin) a saturated petroleum naphtha boiling in the range of 20–135°C (68–275°F) and suitable for general use as a solvent; also called benzine or petroleum ether.

Limolite fine-grain sedimentary rock that is transported by water. The granulometrics range from fine sand to clay.

Linde copper sweetening a process for treating gasoline and distillates with a slurry of clay and cupric chloride.

Liquid petrolatum see White oil.

Liquefied petroleum gas propane, butane, or mixtures thereof, gaseous at atmospheric temperature and pressure, held in the liquid state by pressure to facilitate storage, transport, and handling.

Liquid asphalt an asphalt material having a soft or fluid consistency that is beyond the range of measurement by the normal penetration test, the limit of which is 300 maximum; includes cutback asphalt and emulsified asphalt.

Liquid chromatography a chromatographic technique that employs a liquid mobile phase.

Liquid/liquid extraction an extraction technique in which one liquid is shaken with or contacted by an extraction solvent to transfer molecules of interest into the solvent phase.

Liquid sulfur dioxide–benzene process a mixed-solvent process for treating lubricating-oil stocks to improve viscosity index; also used for dewaxing.

Lithology the geological characteristics of rock.

Live steam steam coming directly from a boiler before being utilized for power or heat.

Liver the intermediate layer of dark-colored, oily material, insoluble in weak acid and in oil, which is formed when acid sludge is hydrolyzed.

Long-term mixture conditioning a laboratory aging procedure used to simulate the effects of the aging of hot mix asphalt that occurs over the service life of a pavement.

Lorenz coefficient a permeability heterogeneity factor.

Loss on heating a common industrial asphalt test which measures the weight loss after exposing a small asphalt sample to 163°C (325°F) for 5 h.

Lower-phase microemulsion a microemulsion phase containing a high concentration of water that, when viewed in a test tube, resides near the bottom with oil phase on top.

Low-slope roofing roofing products designed for a roof slope of less than or equal to 14 degrees.

Lube see Lubricating oil.

Lube cut a fraction of crude oil of suitable boiling range and viscosity to yield lubricating oil when completely refined; also referred to as lube oil distillates or lube stock.

Lubricating oil a fluid lubricant used to reduce friction between bearing surfaces.

MACT maximum achievable control technology. Applies to major sources of hazardous air pollutants.

Mahogany acids oil-soluble sulfonic acids formed by the action of sulfuric acid on petroleum distillates. They may be converted to their sodium soaps (mahogany soaps) and extracted from the oil with alcohol for use in the manufacture of soluble oils, rust preventives, and special greases. The calcium and barium soaps of these acids are used as detergent additives in motor oils; see also Brown acids and Sulfonic acids.

Major source a source that has a potential to emit for a regulated pollutant that is at or greater than an emission threshold set by regulations.

Maltenes relatively low-molecular-weight soluble (in n-heptane) fraction of asphalt; believed to dissolve, or disperse the asphaltenes in the colloidal structure of asphalt; the fraction of petroleum that is soluble in, for example, pentane or heptane; deasphaltened oil; also the term arbitrarily assigned to the pentane-soluble portion of petroleum that is relatively high boiling (>300°C, 760 mm) (see also Petrolenes).

Map cracking large pattern cracking.

Marine engine oil oil used as crankcase oil in marine engines.

Marine gasoline fuel for motors in marine service.

Marine sediment the organic biomass from which petroleum is derived.

Marsh an area of spongy waterlogged ground with large numbers of surface water pools. Marshes usually result from: (i) an impermeable underlying bedrock; (ii) surface deposits of glacial boulder clay; (iii) a basin-like topography from which natural drainage is poor; (iv) very heavy rainfall in conjunction with a correspondingly low evaporation rate; (v) low-lying land, particularly at estuarine sites at or below sea level.

Mass spectrometer an analytical technique that *fractures* organic compounds into characteristic "fragments" based on functional groups that have a specific mass-to-charge ratio.

Mastic asphalt voidless asphalt mixture with asphalt as a binder in which the volume of the filler and binder exceeds the volume of remaining voids.

Mat, asphalt a term used to describe the fresh asphalt surface behind the paving machine; most commonly used to refer to the asphalt during the laydown and compaction phase of construction.

Materials any substances specified for use in the construction of the project and its appurtenances.

Mayonnaise low-temperature sludge; a black, brown, or gray deposit having a soft, mayonnaise-like consistency; not recommended as a food additive!

MCL maximum contaminant level as dictated by regulations.

Medicinal oil highly refined, colorless, tasteless, and odorless petroleum oil used as a medicine in the nature of an internal lubricant; sometimes called liquid paraffin.

Membrane technology gas separation processes utilizing membranes that permit different components of a gas to diffuse through the membrane at significantly different rates.

MDL See Method detection limit.

Medium curing asphalt liquid asphalt composed of asphalt cement and a kerosene-type diluent of medium volatility.

MEK (methyl ethyl ketone) a colorless liquid ($CH_3COCH_2CH_3$) used as a solvent; as a chemical intermediate; and in the manufacture of lacquers, celluloid, and varnish removers.

MEK deoiling a wax-deoiling process in which the solvent is generally a mixture of methyl ethyl ketone and toluene.

MEK dewaxing a continuous solvent dewaxing process in which the solvent is generally a mixture of methyl ethyl ketone and toluene.

Membrane a factory-made flexible layer of asphalt with internal or external incorporation of one or more carriers, supplied in roll form ready for use.

Metagenesis the alteration of organic matter during the formation of petroleum that may involve temperatures above 200°C (390°F). See also catagenesis and diagenesis.

Metamorphic rocks rocks resulting from the transformation that commonly takes place at great depths due to pressure and temperature. The original rocks may be sedimentary, igneous, or metamorphic.

Methanol see Methyl alcohol.

Method Detection Limit the smallest quantity or concentration of a substance that the instrument can measure.

Methyl t-butyl ether an ether added to gasoline to improve its octane rating and to decrease gaseous emissions; see Oxygenate.

Mercapsol process a regenerative process for extracting mercaptans, utilizing aqueous sodium (or potassium) hydroxide containing mixed cresols as solubility promoters.

Mercaptans organic compounds having the general formula R-SH.

Metagenesis the alteration of organic matter during the formation of petroleum that may involve temperatures above 200°C (390°F); see also Catagenesis and Diagenesis.

Methyl alcohol (methanol; wood alcohol) a colorless, volatile, inflammable, and poisonous alcohol (CH_3OH) traditionally formed by destructive distillation of wood or, more recently, as a result of synthetic distillation in chemical plants.

Methyl ethyl ketone see MEK.

Mica a complex aluminum silicate mineral that is transparent, tough, flexible, and elastic.

Micellar fluid (surfactant slug) an aqueous mixture of surfactants, co-surfactants, salts, and hydrocarbons. The term micellar is derived from the word micelle, which is a submicroscopic aggregate of surfactant molecules and associated fluid.

Micelle the structural entity by which asphaltene constituents are dispersed in petroleum.

Microcarbon residue the carbon residue determined using a themogravimetric method. See also Carbon residue.

Microcrystalline wax wax extracted from certain petroleum residua and having a finer and less apparent crystalline structure than paraffin wax.

Microemulsion a stable, finely dispersed mixture of oil, water, and chemicals (surfactants and alcohols).

Microorganisms animals or plants of microscopic size, such as bacteria.

Microscopic displacement efficiency the efficiency with which an oil displacement process removes the oil from individual pores in the rock.

Mid-boiling point the temperature at which approximately 50% of a material has distilled under specific conditions.

Middle distillate distillate boiling between the kerosene and lubricating oil fractions.

Middle-phase microemulsion a microemulsion phase containing a high concentration of both oil and water that, when viewed in a test tube, resides in the middle with the oil phase above it and the water phase below it.

Migration (primary) the movement of hydrocarbons (oil and natural gas) from mature, organic-rich source rocks to a point where the oil and gas can collect as droplets or as a continuous phase of liquid hydrocarbon.

Migration (secondary) the movement of the hydrocarbons as a single, continuous fluid phase through water-saturated rocks, fractures, or faults followed by accumulation of the oil and gas in sediments (traps) from which further migration is prevented.

Mineral aggregate a combination of stone fractions and filler.

Mineral dust the portion of the fine aggregate passing a No. 200 sieve.

Mineral filler a finely divided mineral product at least 65% of which will pass a No. 200 sieve; pulverized limestone is the most common manufactured filler, although other stone dust, hydrated lime, Portland cement, and certain natural deposits of finely divided mineral matter are also used.

Mineral hydrocarbons petroleum hydrocarbons, considered *mineral* because they come from the earth rather than from plants or animals.

Mineral oil the older term for petroleum; the term was introduced in the nineteenth century as a means of differentiating petroleum (rock oil) from whale oil which, at the time, was the predominant illuminant for oil lamps.

Minerals naturally occurring inorganic solids with well-defined crystalline structures.

Mineral seal oil a distillate fraction boiling between kerosene and gas oil.

Mineral wax yellow to dark brown, solid substances that occur naturally and are composed largely of paraffins; usually found associated with considerable mineral matter, as a filling in veins and fissures or as an interstitial material in porous rocks.

Minimum miscibility pressure (MMP) see Miscibility.

Miscibility an equilibrium condition, achieved after mixing two or more fluids, which is characterized by the absence of interfaces between the fluids: (i) *first-contact miscibility:* miscibility in the usual sense, whereby two fluids can be mixed in all proportions without any interfaces forming. Example: At room temperature and pressure, ethyl alcohol and water are first-contact miscible. (ii) *Multiple-contact miscibility (dynamic miscibility):* miscibility that is developed by repeated enrichment of one fluid phase with components from a second fluid phase with which it comes into contact. (iii) *Minimum miscibility* pressure: the minimum pressure above which two fluids become miscible at a given temperature, or can become miscible, by dynamic processes.

Mitigation identification, evaluation, and cessation of potential impacts of a process product or by-product.

Mixed-phase cracking the thermal decomposition of higher-boiling hydrocarbons to gasoline components.

Modified asphalt bituminous binder whose rheological properties have been modified during manufacture by the use of one or more chemical agents.

Modified naphtha insolubles (MNI) an insoluble fraction obtained by adding naphtha to petroleum; usually the naphtha is modified by adding paraffin constituents; the fraction might be equated to asphaltenes *if* the naphtha is equivalent to n-heptane, but usually it is not.

Molecular sieve a synthetic zeolite mineral having pores of uniform size; it is capable of separating molecules, on the basis of their size, structure, or both, by absorption or sieving.

Motor Octane Method a test for determining the knock rating of fuels for use in spark-ignition engines; see also Research Octane Method.

Moving-bed catalytic cracking a cracking process in which the catalyst is continuously cycled between the reactor and the regenerator.

MSDS Material safety data sheet.

MTBE see Methyl t-butyl ether.

MTV material transfer vehicle.

NAAQS National Ambient Air Quality Standards; standards exist for the pollutants known as the criteria air pollutants: nitrogen oxides (NO_x), sulfur oxides (SO_x), lead, ozone, particulate matter, less than 10 microns in diameter, and carbon monoxide (CO).

Naft pre-Christian era (Greek) term for naphtha.

Napalm a thickened gasoline used as an incendiary medium that adheres to the surface it strikes.

Naphtha a generic term applied to refined, partly refined, or unrefined petroleum products and liquid products of natural gas, the majority of which distills below 240°C (464°F); the volatile fraction of petroleum which is used as a solvent or as a precursor to gasoline.

Naphthenes cycloparaffins.

Native asphalt (natural asphalt) naturally occurring mixture of bitumen and mineral matter formed by oil seepages in the earth's crust; includes Trinidad lake asphalt, rock asphalt, Gilsonite, Selenice, and others. See bitumen.

Natural asphalt asphalt occurring in nature that has been derived from petroleum by natural processes of evaporation of volatile fractions leaving the asphalt fractions; native asphalts of most importance are found in the Trinidad and Bermudez lake deposits; asphalt from these sources is called Lake Asphalt; also (more correctly) called bitumen.

Natural gasoline a mixture of liquid hydrocarbons extracted from natural gas suitable for blending with refinery gasoline.

Natural gasoline plant a plant for the extraction of fluid hydrocarbon, such as gasoline and liquefied petroleum gas, from natural gas.

NESHAP National Emissions Standards for Hazardous Air Pollutants; emission standards for specific source categories that emit or have the potential to emit one or more hazardous air pollutants; the standards are modeled on the best practices and most effective emission reduction methodologies in use at the affected facilities.

Neutralization a process for reducing the acidity or alkalinity of a waste stream by mixing acids and bases to produce a neutral solution; also known as pH adjustment.

Neutral oil a distillate lubricating oil with viscosity usually not above 200 s at 100°F.

Neutralization number the weight, in milligrams, of potassium hydroxide needed to neutralize the acid in 1 g of oil; an indication of the acidity of an oil.

Non-asphaltic road oil any of the non-hardening petroleum distillates or residual oils used as dust layers. They have sufficiently low viscosity to be applied without heating and, together with asphaltic road oils, are sometimes referred to as dust palliatives.

Non-attainment area a geographical area that does not meet NAAQS criteria for air pollutants (see also Attainment area).

Non-ionic surfactant a surfactant molecule containing no ionic charge.

Non-Newtonian a fluid that exhibits a change of viscosity with flow rate.

Normal fault the result of the downward displacement of one of the strata from the horizontal. The angle is generally between 25° and 60° and it is recognized by the absence of part of the stratigraphic column.

NOx oxides of nitrogen.

Nuclear magnetic resonance spectroscopy an analytical procedure that permits the identification of complex molecules based on the magnetic properties of the atoms they contain.

Number 1 Fuel oil (No. 1 Fuel oil) very similar to kerosene and is used in burners where vaporization before burning is usually required and a clean flame is specified.

Number 2 Fuel oil (No. 2 Fuel oil) also called domestic heating oil; has properties similar to diesel fuel and heavy jet fuel; used in burners where complete vaporization is not required before burning.

Number 4 Fuel oil (No. 4 Fuel oil) a light industrial heating oil and is used where preheating is not required for handling or burning; there are two grades of No. 4 fuel oil, differing in safety (flash point) and flow (viscosity) properties.

Number 5 Fuel oil (No. 5 Fuel oil) a heavy industrial fuel oil which requires preheating before burning.

Number 6 Fuel oil (No. 6 Fuel oil) a heavy fuel oil and is more commonly known as Bunker C oil when it is used to fuel ocean-going vessels; preheating is always required for burning this oil.

Octane barrel yield a measure used to evaluate fluid catalytic cracking processes; defined as (RON + MON)/2 times the gasoline yield, where RON is the research octane number and MON is the motor octane number.

Oils that portion of the maltenes that is not adsorbed by a surface-active material such as clay or alumina.

Oil sand see Tar sand.

Oil shale a fine-grained impervious sedimentary rock which contains an organic material called kerogen.

Olefin synonymous with *alkene*.

OOIP see Oil originally in place.

Open-graded aggregate an aggregate containing little or no mineral filler or in which the void spaces in the compacted aggregate are relatively large.

Optimum salinity the salinity at which a middle-phase microemulsion containing equal concentrations of oil and water results from the mixture of a micellar fluid (surfactant slug) with oil.

Organic sedimentary rocks rocks containing organic material such as residues of plant and animal remains/decay.

Overhead that portion of the feedstock which is vaporized and removed during distillation.

Oxidation a process which can be used for the treatment of a variety of inorganic and organic substances; a chemical reaction that occurs between binder and oxygen that changes the stiffness of the binder.

Oxidized asphalt asphalt in which the rheological properties have been substantially modified by reaction with air at elevated temperatures; often referred to as *blown asphalt*.

Oxidized asphalt membrane a roofing asphalt product typically made by coating a glass fiber or polyester mat with a mixture of oxidized asphalt and mineral filler, and then packaging the finished product in rolls; in North America these products may be made with a mineral granule surface and are called *roll roofing*:

Oxidizer See blowing still.

Ozokerite (Ozocerite) a naturally occurring wax; when refined also known as ceresin.

Oxygenate an oxygen-containing compound that is blended into gasoline to improve its octane number and to decrease gaseous emissions.

Oxygenated gasoline gasoline with added ethers or alcohols, formulated according to the Federal Clean Air Act to reduce carbon monoxide emissions during winter months.

Oxygen scavenger a chemical which reacts with oxygen in injection water, used to prevent degradation of polymer.

PAH, PAC polycyclic aromatic hydrocarbons is the collective name for a large group of several hundred chemicals that have a characteristic structure of two or more fused aromatic rings; a class of organic compounds and also a sub-group of the larger family of chemicals: polycyclic aromatic compounds (PAC): also called polynuclear aromatic compounds (PNAs).

Pale oil a lubricating oil or a process oil refined until its color, by transmitted light, is straw to pale yellow.

Paraffinum liquidum see Liquid petrolatum.

Paraffin wax the colorless, translucent, highly crystalline material obtained from the light lubricating fractions of paraffin crude oils (wax distillates).

Particle density the density of solid particles.

Particulate matter (particulates) particles in the atmosphere or on a gas stream that may be organic or inorganic and originate from a wide variety of sources and processes.

Particle size distribution the particle size distribution (of a catalyst sample) expressed as a percent of the whole.

Pavement structure (combination or composite) all courses of selected material placed on the foundation or subgrade soil, other than any layers or courses constructed in grading operations; when the asphalt pavement is on an old Portland cement concrete base or other rigid-type base, the pavement structure is referred to as a combination or composite-type pavement structure.

Paving asphalt asphalt used to coat mineral aggregate, mainly used in the construction and maintenance of paved surfaces and hydraulic works.

Paving machine (paver) the equipment used to place the asphalt or concrete materials in their finished position; these machines should be the appropriate size for the project and must be able to place a consistently smooth, even finish at the proper depth and provide initial compaction; a machine designed for placement of a uniform asphalt mat onto a road surface prior to roller compaction.

PCC Portland cement concrete.

Penetration-grade asphalt asphalt classified by the depth to which a standard needle will penetrate the asphalt sample under specified test conditions.

Penetration index indication of the thermal susceptibility of a bituminous binder; calculated from the values of penetration and the softening point.

Penetration test specification test to measure the hardness of asphalt under specified conditions. In which the indentation of asphalt in tenths of a millimeter at 25°C (77°F) is measured using a standard needle with a loading of 100 g and 5 s duration.

Penex process a continuous, non-regenerative process for isomerization of C_5 and/or C_6 fractions in the presence of hydrogen (from reforming) and a platinum catalyst.

Pentafining a pentane isomerization process using a regenerable platinum catalyst on a silica-alumina support and requiring outside hydrogen.

Pepper sludge the fine particles of sludge produced in acid treating which may remain in suspension.

Percolation the movement of free water through soil.

Performance-graded asphalt asphalt classified based on the research results of the US Strategic Highway Research Program (SHRP); performance-graded specifications are based on the stiffness of the asphalt at the high- and low-temperature environment in which the asphalt will be expected to perform within pavement; currently performance-graded asphalts are most widely utilized in the United States and Canada.

Peri-condensed aromatic compounds Compounds based on angular condensed aromatic hydrocarbon systems, e.g., phenanthrene, chrysene, picene, etc.

Permeability the ease of flow of the water through the rock; a measure of the rate or volume of flow of water through soil.

Petrol a term commonly used in some countries for gasoline.

Petrolatum a semisolid product, ranging from white to yellow in color, produced during refining of residual stocks; see Petroleum jelly.

Petrolenes the term applied to that part of the pentane-soluble or heptane-soluble material that is low boiling (<300°C, <570°F, 760 mm) and can be distilled without thermal decomposition (see also Maltenes).

Petroleum (crude oil) a naturally occurring mixture of gaseous, liquid, and solid hydrocarbon compounds usually found trapped deep underground beneath impermeable cap rock and above a lower dome of sedimentary rock such as shale; most petroleum reservoirs occur in sedimentary rocks of marine, deltaic, or estuarine origin.

Petroleum asphalt asphalt refined from crude petroleum. See Asphalt.

Petroleum ether see Ligroine.

Petroleum jelly a translucent, yellowish to amber or white, hydrocarbon substance (melting point: 38–54°C) having almost no odor or taste, derived from petroleum and used principally in medicine and pharmacy as a protective dressing and as a substitute for fats in ointments and cosmetics; also used in many types of polishes and in lubricating greases, rust preventives, and modeling clay; obtained by dewaxing heavy lubricating-oil stocks.

Petroleum pitch the non-volatile residue from the distillation of thermal cracked or steam-cracked residuum and/or catalytically cracked clarified oil with a softening point from 40 to 180°C (104 to 355°F); composed primarily of a complex combination of three or more membered condensed ring aromatic hydrocarbons.

Petroleum refinery see Refinery.

Petroleum refining a complex sequence of events that result in the production of a variety of products.

Petroleum resins a complex combination of organic compounds, predominantly hydrocarbons, obtained as a fraction of the extract from solvent extraction of residuum; predominantly of high-molecular-weight compounds with high carbon-to-hydrogen ratios.

Petroporphyrins see Porphyrins.

Phase a separate fluid that co-exists with other fluids; gas, oil, water, and other stable fluids such as microemulsions are all called phases in EOR research.

Phase behavior the tendency of a fluid system to form phases as a result of changing temperature, pressure, or the bulk composition of the fluids or of individual fluid phases.

Phase separation the formation of a separate phase that is usually the prelude to coke formation during a thermal process; the formation of a separate phase as a result of the instability/incompatibility of petroleum and petroleum products.

pH adjustment neutralization.

Phosphoric acid polymerization a process using a phosphoric acid catalyst to convert propene, butene, or both, to gasoline or petrochemical polymers.

Photoionization a gas chromatographic detection system that utilizes a *detector (PID)* ultraviolet lamp as an ionization source for analyte detection. It is usually used as a selective detector by changing the photon energy of the ionization source.

Pilot project a project that is being executed in a small representative sector of a project where tests performed are similar to those that will be implemented throughout the project; the purpose is to gather information and/or obtain results that could be used to generalize an exploitation strategy in the field.

PINA analysis a method of analysis for paraffins, *iso*-paraffins, naphthenes, and aromatics.

PIONA analysis a method of analysis for paraffins, *iso*-paraffins, olefins, naphthenes, and aromatics.

Pipe still a still in which heat is applied to the oil while being pumped through a coil or pipe arranged in a suitable firebox.

Pipestill gas the most volatile fraction that contains most of the gases that are generally dissolved in the crude. Also known as pipestill light ends.

Pipestill light ends see *Pipestill gas*.

Pitch the non-volatile, brown to black, semi-solid to solid viscous product from the destructive distillation of many bituminous or other organic materials, especially coal.

Plant liquefiables shrinkage factor the fraction arising from considering the liquefiables obtained in transportation to the processing complexes.

Plant liquids natural gas liquids recovered in gas processing complexes, mainly consisting of ethane, propane, and butane.

Plant mix a mixture produced in an asphalt mixing plant, that consists of mineral aggregate uniformly coated with asphalt cement or liquid asphalt.

Plastomer a polymer type which exhibits stiffness and strength but does not recover substantially when deformed; examples of this type of polymer used in asphalts are ethylene vinyl acetate, ethylene methacrylate, polyethylene, and atactic polypropylene.

Platforming a reforming process using a platinum-containing catalyst on an alumina base.

PNA a polynuclear aromatic compound.

PNA analysis a method of analysis for paraffins, naphthenes, and aromatics.

Polar aromatics resins; the constituents of petroleum that are predominantly aromatic in character and contain polar (nitrogen, oxygen, and sulfur) functions in their molecular structure(s).

Polished aggregate the result of use of a soft aggregate (such as limestone) included in an asphalt surface mix, which becomes polished (smoothed) by traffic and the aggregate loses its ability to squeeze out the water, and the pavement can become slippery.

Pollutant a chemical (or chemicals) introduced into the land water and air systems that is (are) not indigenous to these systems; also an indigenous chemical (or chemicals) introduced into the land water and air systems in amounts greater than the natural abundance.

Pollution the introduction into the land water and air systems of a chemical or chemicals that are not indigenous to these systems or the introduction into the land water and air systems of indigenous chemicals in greater-than-natural amounts.

Polycyclic aromatic hydrocarbons (PAHs) polycyclic aromatic hydrocarbons are a suite of compounds comprised of two or more condensed aromatic rings. They are found in many petroleum mixtures, and they are predominantly introduced to the environment through natural and anthropogenic combustion processes.

Polyforming a process charging both C_3 and C_4 gases with naphtha or gas oil under thermal conditions to produce gasoline.

Polymer gasoline the product of polymerization of gaseous hydrocarbons to hydrocarbons boiling in the gasoline range.

Polymerization the combination of two olefin molecules to form a higher-molecular-weight paraffin.

Polymer-modified asphalt modified asphalt/asphalt in which the modifier used is one or more organic polymers.

Polymer-modified asphalt membrane a factory-made flexible layer of straight run and/ or oxidized asphalt modified with elastomeric or plastomeric polymers with internal or external incorporation of one or more carriers, supplied in roll form ready for use.

Polymer stability the ability of a polymer to resist degradation and maintain its original properties.

Polynuclear aromatic compound an aromatic compound having two or more fused benzene rings, e.g., naphthalene, phenanthrene.

Polysulfide treating a chemical treatment used to remove elemental sulfur from refinery liquids by contacting them with a non-regenerable solution of sodium polysulfide.

PONA analysis a method of analysis for paraffins (P), olefins (O), naphthenes (N), and aromatics (A).

Pop-outs generally small, shallow pits or depressions in the surface of either concrete or hot mix asphalt pavements that result mostly from the weathering of individual aggregate particles (in whole or in part); may also be the direct result of the expansion of absorbed moisture due to freezing; in others, this may be caused by chemical changes of an aggregate component; typical materials that may result in aggregate. Pore diameter: the average pore size of a solid material, e.g., catalyst.

Pore space a small hole in rock that contains fluid or fluids; a 4-inch cube of rock may contain millions of inter-connected pore spaces.

Pore volume the total volume of all pores and fractures in a rock.

Porosity the percentage of rock volume available to contain water or other fluid.

Porphyrins organometallic constituents of petroleum that contain vanadium or nickel; the degradation products of chlorophyll that became included in the protopetroleum.

Portland cement concrete a composite material that consists essentially of Portland cement and water as a binding medium in which is mixed coarse and fine particles of crushed stone.

Positive bias a result that is incorrect and too high.

Possible reserves reserves where there is an even greater degree of uncertainty but about which there is some information.

Potential reserves reserves based upon geological information about the types of sediments where such resources are likely to occur and they are considered to represent an educated guess.

Pour point the lowest temperature at which oil will pour or flow when it is chilled without disturbance under definite conditions.

Powerforming a fixed-bed naphtha-reforming process using a regenerable platinum catalyst.

Power-law exponent an exponent used to model the degree of viscosity change of some non-Newtonian liquids.

Precipitation number the number of milliliters of precipitate formed when 10 ml of lubricating oil is mixed with 90 ml of petroleum naphtha of a definite quality and centrifuged under definitely prescribed conditions.

Primary structure the chemical sequence of atoms in a molecule.

Prime coat an application of low-viscosity liquid asphalt to an absorbent surface; used to prepare an untreated base for an asphalt surface.

Probable reserves mineral reserves that are nearly certain but about which a slight doubt exists.

Propane-precipitated asphalt (propane asphalt) see solvent precipitation and solvent asphalt.

Propane deasphalting solvent deasphalting using propane as the solvent.

Propane decarbonizing a solvent extraction process used to recover catalytic cracking feed from heavy fuel residues.

Propane dewaxing a process for dewaxing lubricating oils in which propane serves as solvent.

Propane fractionation a continuous extraction process employing liquid propane as the solvent; a variant of propane deasphalting.

Prospective resource the amount of hydrocarbons evaluated at a given date of accumulations not yet discovered but which have been inferred and are estimated as recoverable.

Protopetroleum a generic term used to indicate the initial product formed when changes have occurred to the precursors of petroleum.

PSD prevention of significant deterioration.

PTE potential to emit; the maximum capacity of a source to emit a pollutant, given its physical or operation design, and considering certain controls and limitations.

Pug mill mixer used to combine stone materials and asphalt in an asphalt-mixing plant; the mixing is effected by high-speed stirring with paddle blades at elevated temperatures.

Pyrobitumen see Asphaltoid.

Pyrolysis exposure of a feedstock to high temperatures in an oxygen-poor environment.

Pyrophoric substances that catch fire spontaneously in air without an ignition source.

Quadrillion 1×10^{15}

Quench the sudden cooling of hot material discharging from a thermal reactor.

RACT Reasonably Available Control Technology standards; implemented in areas of non-attainment to reduce emissions of volatile organic compounds and nitrogen oxides.

Raffinate the portion of the oil which remains undissolved in a solvent refining process; the part of a liquid remaining after the more soluble components have been extracted by a solvent.

Raffinate (petroleum, residual oil decarbonization) a complex combination of hydrocarbons obtained as the solvent-insoluble fraction from C_5–C_7 solvent decarbonization of a residual oil; consists predominantly of aromatic hydrocarbons having carbon numbers predominantly higher than C_{34} and boiling above approximately 495°C (923°F).

Ramsbottom carbon residue see Carbon residue.

Rapid curing asphalt liquid asphalt composed of asphalt cement and naphtha or gasoline-type diluent of high volatility.

Raveling the progressive disintegration of an asphalt mix layer from the surface downward as a result of the dislodgement of aggregate particles.

Raw materials minerals extracted from the earth prior to any refining or treating.

Reclaimed asphalt pavement removed and/or reprocessed pavement materials containing asphalt and aggregates.

Recycle ratio the volume of recycle stock per volume of fresh feed; often expressed as the volume of recycle divided by the total charge.

Recycle stock the portion of a feedstock which has passed through a refining process and is recirculated through the process.

Recycling the use or reuse of chemical waste as an effective substitute for commercial products or as an ingredient or feedstock in an industrial process.

Reduced crude a residual product remaining after the removal, by distillation or other means, of an appreciable quantity of the more volatile components of crude oil.

Refinery a series of integrated unit processes by which petroleum can be converted to a slate of useful (salable) products; a facility composed of a group of separation and chemical engineering unit processes used for refining crude oil into different oil products.

Refinery gas a gas (or a gaseous mixture) produced as a result of refining operations.

Refining the processes by which petroleum is distilled and/or converted by application of physical and chemical processes to form a variety of products.

Reflective cracking cracks in an asphalt overlay pavement caused by cracks in the existing pavement *reflecting* up through the overlay; specialized techniques and materials such multi-membrane paving fabrics help reduce this problem. See also Joint reflective cracking.

Reformate the liquid product of a reforming process.

Reformed gasoline gasoline made by a reforming process.

Reforming the conversion of hydrocarbons with low octane numbers into hydrocarbons having higher octane numbers; e.g., the conversion of a n-paraffin into a iso-paraffin.

Reformulated gasoline (RFG) gasoline designed to mitigate smog production and to improve air quality by limiting the emission levels of certain chemical compounds such as benzene and other aromatic derivatives; often contains oxygenates.

Reid vapor pressure a measure of the volatility of liquid fuels, especially gasoline.

Regeneration the reactivation of a catalyst by burning off the coke deposits.

Regenerator a reactor for catalyst reactivation.

Regression a geological term used to define the elevation of one part of the continent over sea level, as a result of the ascent of the continent or the lowering of the sea level.

Rehabilitation the renewal of an existing surface or pavement structure by repair, recycling, or overlay techniques.

Relative permeability the permeability of rock to gas, oil, or water, when any two or more are present, expressed as a fraction of the sir phase permeability of the rock.

Renewable energy sources solar, wind, and other non-fossil fuel energy sources.

Rerunning the distillation of oil which has already been distilled.

Research Octane Method a test for determining the knock rating, in terms of octane numbers, of fuels for use in spark-ignition engines; see also Motor Octane Method.

Reserve replacement rate a rate that indicates the amount of hydrocarbons replaced or incorporated by new discoveries compared with what has been produced in a given period. It is the coefficient that arises from dividing the new discoveries by production during the period of analysis. It is generally referred to in annual terms and is expressed as a percentage.

Reserve–production ratio the result of dividing the remaining reserve at a given date by the production in a period. This indicator assumes constant production, hydrocarbon prices, and extraction costs, without variation over time, in addition to the non-existence of new discoveries in the future.

Reserves well-identified resources that can be profitably extracted and utilized with existing technology.

Residual asphalt see Straight-run asphalt.

Residual fuel oil obtained by blending the residual product(s) from various refining processes with suitable diluent(s) (usually middle distillates) to obtain the required fuel oil grades.

Residual resistance factor the reduction in permeability of rock to water caused by the adsorption of polymer.

Residuum (resid; *pl*: residua) the residue obtained from petroleum after non-destructive distillation has removed all the volatile materials from crude oil, e.g. an atmospheric (345°C, 650°F+) residuum.

Residuum (dewaxed heavy paraffinic, vacuum) a complex combination of hydrocarbons obtained as the residue from the molecular distillation of a dewaxed heavy paraffinic distillate; consists of hydrocarbonaceous molecular species having carbon numbers predominantly greater than 80 and boiling above approximately 450°C (840°F).

Residuum (distillation, residue hydrogenation) a complex combination of hydrocarbons obtained as a residue from the distillation of crude oil under vacuum; consists predominantly of hydrocarbonaceous molecular species having carbon numbers predominantly in the range above 50 and boiling in the range above approximately 360°C (680°F).

Residues (hydrodesulfurized vacuum) obtained by treating a vacuum residuum with hydrogen in the presence of a catalyst under conditions primarily to remove organic sulfur compounds; of hydrocarbonaceous molecular species having carbon numbers predominantly greater than 34 and boiling approximately above 495°C (923°F).

Residues (hydrodesulfurized vacuum) a complex combination of hydrocarbons obtained by treating a vacuum residuum with hydrogen in the presence of a catalyst under conditions primarily to remove organic sulfur compounds; consists of hydrocarbonaceous molecular species having carbon numbers predominantly greater than 34 and boiling approximately above 495°C (923°F).

Residues (thermally cracked vacuum) a complex combination of hydrocarbons obtained from the vacuum distillation of the products from a thermal cracking process; consists predominantly of hydrocarbonaceous molecular species having carbon numbers predominantly greater than 34 and boiling above approximately 495°C (925°F).

Residues (petroleum, vacuum) residuum from the vacuum distillation of the residuum from atmospheric distillation of crude oil. It consists of hydrocarbons having carbon numbers predominantly greater than 34 and boiling above approximately 495°C (925°F).

Residues (vacuum) a complex residuum from the vacuum distillation of the residuum from atmospheric distillation of a crude oil; consists of hydrocarbon having carbon numbers predominantly greater than 34 and boiling above approximately 495°C (923°F).

Residuum (vacuum, residue hydrogenation) a complex combination of hydrocarbons obtained as a residue from the distillation of crude oil under vacuum; consists predominantly of hydrocarbons having carbon numbers predominantly in the range above 50 and boiling in the range above approximately 500°C (930°F).

Resin constituents (polar aromatics) a fraction of relatively high molecular weight present in the maltene (non-asphaltene) fraction; dispersing agents (often referred to as peptizers) for the asphaltene constituents; separated using solvent precipitation and adsorption chromatography.

Resins that portion of the maltenes that is adsorbed by a surface-active material such as clay or alumina; the fraction of deasphaltened oil that is insoluble in liquid propane but soluble in *n*-heptane.

Resistance factor a measure of resistance to flow of a polymer solution relative to the resistance to flow of water.

Resource the total amount of a commodity (usually a mineral but can include non-minerals such as water and petroleum) that has been estimated to be ultimately available.

Resurfacing existing surfaces may be improved by resurfacing (or overlaying) with a plant mix asphalt mat of varying thicknesses; may be considered in two categories: (i) overlays to provide smooth, skid-resistant, and water-resistant surfaces or to make improvements in grade and/or cross-section; and (ii) overlays to strengthen existing pavements to handle heavier loads or increased traffic; sometimes called overlays.

Retention the loss of chemical components due to adsorption onto the rock's surface, precipitation, or to trapping within the rock.

Retention time the time it takes for an eluate to move through a chromatographic system and reach the detector. Retention times are reproducible and can therefore be compared to a standard for analyte identification.

Reverse fault the result of compression forces where one of the strata is displaced upwards from the horizontal.

Rexforming a process combining Platforming with aromatics extraction, wherein low-octane raffinate is recycled to the Platformer.

Rich oil absorption oil containing dissolved natural gasoline fractions.

Rigid pavement a pavement structure that distributes loads to the subgrade, having as one course a Portland cement concrete slab of relatively high bending resistance.

Riser the part of the bubble-plate assembly which channels the vapor and causes it to flow downward to escape through the liquid; also the vertical pipe where fluid catalytic cracking reactions occur.

Road (roadway) a general term denoting a public way for purposes of vehicular travel, including the entire area within the right-of-way.

Roadbed the graded portion of a highway within the top and side slopes, prepared as a foundation for the pavement structure and shoulders.

Road oil term sometimes used for very soft vacuum residue or harder asphalt that have flux oil added, or cutbacks that have been produced using petroleum with a boiling point greater than 225°C (435°F) added to reduce the viscosity; generally used to produce asphalt paving mixes for use on very low-volume roads in moderate to cold climates.

Rock from which crushed stone, sand, and gravel are made; the rock most suitable for making good aggregates.

Rock asphalt Naturally occurring bituminous material: usually a combination of bitumen and limestone; found in south-eastern France, Sicily, and elsewhere in the world; bitumen which occurs in formations that have a limiting ratio of bitumen-to-rock matrix.

Rock matrix the granular structure of a rock or porous medium.

Roll roofing See oxidized asphalt membrane or polymer-modified membrane.

Rolling thin film oven test (RTFOT) a common paving asphalt test which subjects a thin film of asphalt on the inside of a rolling glass jar to 163°C (325°F) for 85 min.

Roofing flux low-viscosity, high-flashpoint, generally paraffinic residue of vacuum distillation of an appropriate petroleum crude oil used as a feedstock in the manufacture of oxidized asphalt used in roofing applications.

Roofing asphalt asphalt used for manufacture of roofing systems or roofing products, such as asphalt shingles, polymer-modified membranes, saturated felt under-layer, and roofing adhesives.

Roofing cement a material made by adding filler and fibers to either an asphalt emulsion or cutback asphalt to make an adhesive used for maintenance and in applying flashings on a new roof; depending on the performance characteristics sought for particular cements, the asphalt used in the formulation may be oxidized or straight-run.

Roofing felt a sheet material, impregnated with asphalt, generally supplied in rolls and used in roof construction. See asphalt roofing felt.

Roofing kettle a vessel used to heat binders such as oxidized asphalt for use in the construction of built-up roofing and some polymer-modified asphalt roof systems.

Roofing shingles a steep-slope roofing product; typically made by coating a glass mat with filled coating asphalt and then surfacing with colored mineral granules.

Rotary drum dryer a device in an asphalt-mixing plant used to dry and heat stone materials.

Routing enlargement of pavement cracks using a specialized machine which provides a uniform-width reservoir for the sealant; proper choice of bit size will result in the proper depth to width ratio (depth/width); this procedure greatly increases the effectiveness and durability of crack sealing.

Salinity the concentration of salt in water.

Sand a course granular mineral mainly comprising quartz grains that is derived from the chemical and physical weathering of rocks rich in quartz, notably sandstone and granite.

Sandstone a sedimentary rock formed by compaction and cementation of sand grains; can be classified according to the mineral composition of the sand and cement.

SARA analysis a method of fractionation by which petroleum is separated into saturates, aromatics, resins, and asphaltene fractions.

SARA separation see SARA analysis.

Saturant asphalt asphalt that is used to saturate organic felt to make roofing felt or to make organic-based shingles; can be straight-run or oxidized asphalt.

Saturates paraffins and cycloparaffins (naphthenes); predominantly straight and branched-chain aliphatic hydrocarbons present in asphalt, together with alkyl naphthenes and some alkyl aromatics; can be equivalent to 5–20% (w/w) of the asphalt.

Saturation the ratio of the volume of a single fluid in the pores to pore volume, expressed as a percent and applied to water, oil, or gas separately; the sum of the saturations of each fluid in a pore volume is 100%.

Saturation pressure the pressure at which the first gas bubble is formed, when it goes from the liquid phase to the two-phase region.

Saybolt Furol viscosity the time, in seconds (Saybolt Furol Seconds, SFS), for 60 ml of fluid to flow through a capillary tube in a Saybolt Furol viscometer at specified temperatures between 70°F and 210°F; the method is appropriate for high-viscosity oils such as transmission, gear, and heavy fuel oils.

Saybolt Universal viscosity the time, in seconds (Saybolt Universal Seconds, SUS), for 60 ml of fluid to flow through a capillary tube in a Saybolt Universal viscometer at a given temperature.

Scale wax the paraffin derived by removing the greater part of the oil from slack wax by sweating or solvent deoiling.

Screed leveling device at the rear of a paving machine.

Screen factor a simple measure of the viscoelastic properties of polymer solutions.

Scrubber a device that uses water and chemicals to clean air pollutants from combustion exhaust.

Scrubbing purifying a gas by washing with water or chemicals; less frequently, the removal of entrained materials.

Seal coat (sealcoat) a thin asphalt surface treatment used to waterproof and improve the texture of an asphalt-wearing surface; depending on the purpose, seal coats may or may not be covered with aggregate; the main types of seal coats are aggregate seals, fog seals, emulsion slurry seals, and sand seals.

Seal coating application of a sealant (usually coal-tar emulsion or asphalt emulsion type) to preserve, protect, and beautify asphalt pavements. Generally used on low-traffic streets or off-street locations. There are many different seal-coating products available from low-end "do-it-yourself" grades sold in buckets to commercial products. Commercial products are sold in bulk in concentrated form. Water, silica sand (or other suitable aggregates), and specialized additives are then added in precise ratios (mix design) by the contractor in an agitated mixing tank. Care should be taken to choose the appropriate product, mix design, and application method for the specific project and location.

Secondary pollutants a pollutant (chemical species) produced by interaction of a primary pollutant with another chemical or by dissociation of a primary pollutant or by other effects within a particular ecosystem.

Secondary structure the ordering of the atoms of a molecule in space relative to each other.

Sediment an insoluble solid formed as a result of the storage instability and/or the thermal instability of petroleum and petroleum products.

Sedimentary formed by or from deposits of sediments, especially from sand grains or silts transported from their source and deposited in water, as sandstone and shale; or from calcareous remains of organisms, as limestone.

Sedimentary strata typically consist of mixtures of clay, silt, sand, organic matter, and various minerals; formed by or from deposits of sediments, especially from sand grains or silts transported from their source and deposited in water, such as sandstone and shale; or from calcareous remains of organisms, such as limestone.

Seismic section a seismic profile that uses the reflection of seismic waves to determine the geological subsurface.

Selective solvent a solvent which, at certain temperatures and ratios, will preferentially dissolve more of one component of a mixture than of another and thereby permit partial separation.

Select material suitable material obtained from roadway cuts, borrow areas, or commercial sources and designated or reserved for use as foundation for the sub-base, for sub-base material, shoulder surfacing, or other specific purposes.

Selenice bitumen (natural asphalt) from Albania.

Self-adhesive asphalt membrane roofing or waterproofing polymer-modified asphalt membrane applied at ambient temperature with the peel and stick method.

Semi-blown asphalt see air-rectified asphalt.

Separation process an upgrading process in which the constituents of petroleum are separated, usually without thermal decomposition, e.g. distillation and deasphalting.

Separator-Nobel dewaxing a solvent (tricholoethylene) dewaxing process.

Separatory funnel glassware shaped like a funnel with a stoppered rounded top and a valve at the tapered bottom, used for liquid/liquid separations.

Shear mechanical deformation or distortion, or partial destruction of a polymer molecule as it flows at a high rate.

Shear rate a measure of the rate of deformation of a liquid under mechanical stress.

Shear-thinning the characteristic of a fluid whose viscosity decreases as the shear rate increases.

Shell fluid catalytic cracking a two-stage fluid catalytic cracking process in which the catalyst is regenerated.

Shell still a still formerly used in which the oil was charged into a closed, cylindrical shell and the heat required for distillation was applied to the outside of the bottom from a firebox.

Short-term mixture conditioning a laboratory procedure used to simulate the effects of the aging of hot mix asphalt and binder absorption that occurs during the pre-compaction phase of the construction process.

Shoulder the portion of the roadway contiguous with the traveled way for accommodation of stopped vehicles, for emergency use, and for lateral support of base and surface courses.

Sidestream a liquid stream taken from any one of the intermediate plates of a bubble tower.

Sidestream stripper a device used to perform further distillation on a liquid stream from any one of the plates of a bubble tower, usually by the use of steam.

Single-well tracer a technique for determining residual oil saturation by injecting an ester, allowing it to hydrolyze; following dissolution of some of the reaction product in residual oil the injected solutions produced back and analyzed.

Slack wax the soft, oily crude wax obtained from the pressing of paraffin distillate or wax distillate.

Slag the air-cooled, non-metallic byproduct of a blast furnace operation, consisting essentially of silicates and alumino-silicates of lime and other bases, that is developed simultaneously with iron in a blast furnace.

Slime a name used for petroleum in ancient texts.

Slim tube testing laboratory procedure for the determination of minimum miscibility pressure using long, small-diameter, sand-packed, oil-saturated, stainless steel tube.

Slow-curing asphalt liquid asphalt composed of asphalt cement and oils of low volatility.

Sludge a semi-solid to solid product which results from the storage instability and/or the thermal instability of petroleum and petroleum products.

Slurry hydroconversion process a process in which the feedstock is contacted with hydrogen under pressure in the presence of a catalytic coke-inhibiting additive.

Slurry phase reactors tanks into which wastes, nutrients, and microorganisms are placed.

Slurry seal a mixture of slow-setting emulsified asphalt, fine aggregate, and mineral filler with water added to produce slurry consistency.

Slurry seal process a seal-coating process generally used on runways, streets, and roadways; in this process the coating is manufactured by the application equipment as it is being applied; a closely specified blend of graded asphalt emulsion, additives, and aggregate slurry seal is generally classified as Type I, II, or III depending on the size of aggregate used—a large aggregate slurry seal with additional polymers may also be referred to as micro-surfacing; used infrequently on parking areas due to the potential for tracking in hot weather.

Smoke point a measure of the burning cleanliness of jet fuel and kerosine.

Sodium hydroxide treatment see Caustic wash.

Sodium plumbite a solution prepared from a mixture of sodium hydroxide, lead oxide, and distilled water; used in making the doctor test for light oils such as gasoline and kerosine.

Soft-applied roofing asphalt roofing products that are applied by heating the asphalt membrane sufficiently with a torch or hot air welder to ensure good adhesion to the substrate.

Softening-point a specification test measuring the temperature, measured in °C, at which material under standardized test conditions attains a specific consistency.

Soil aggregate natural or prepared mixtures consisting predominantly of hard, durable particles or fragments of stone, slag, gravel, or sand, that contain some soil-clay or stone dust conforming to specified requirements.

Soil cement base a mixture of the natural subgrade material and Portland cement in the correct amounts; after thorough mixing, the proper amount of water is added, and the material is compacted to the required thickness.

Soil support the ability of the roadbed material, or subgrade soil, to support the traffic loads transmitted through a flexible pavement structure.

Solubility parameter a measure of the solvent power and polarity of a solvent.

Solutizer-steam regenerative process a chemical treating process for extracting mercaptans from gasoline or naphtha, using solutizers (potassium isobutyrate, potassium alkyl phenolate) in strong potassium hydroxide solution.

Solvent a liquid in which certain kinds of molecules dissolve. While they typically are liquids with low boiling points, they may include high-boiling liquids, supercritical fluids, or gases.

Solvent asphalt the asphalt produced by solvent extraction of residua or by light hydrocarbon (propane) treatment of a residuum or an asphaltic crude oil.

Solvent deasphalting a process for removing asphaltic and resinous materials from reduced crude oils, lubricating-oil stocks, gas oils, or middle distillates through the extraction or precipitant action of low-molecular-weight hydrocarbon solvents; see also Propane deasphalting.

Solvent decarbonizing see Propane decarbonizing.

Solvent deresining see Solvent deasphalting.

Solvent dewaxing a process for removing wax from oils by means of solvents, usually by chilling a mixture of solvent and waxy oil, filtration or by centrifuging the wax which precipitates, and solvent recovery.

Solvent extraction a process for separating liquids by mixing the stream with a solvent that is immiscible with part of the waste but that will extract certain components of the waste stream.

Solvent extracts aromatic by-products (extracts) obtained from the refining of base oils.

Solvent precipitation the process by which a hard product, propane-precipitated asphalt, is separated from a vacuum residue by solvent precipitation (usually with propane); propane-precipitated asphalt is an asphalt by this definition; in the United States, this process is called "solvent deasphalting" and the product, solvent-refined asphalt.

Solvent-refined asphalt term used in the United States for propane-precipitated asphalt, also referred to as propane deasphalter pitch or propane deasphalter asphalt.

Solvent naphtha a refined naphtha of restricted boiling range used as a solvent; also called petroleum naphtha; petroleum spirits.

Solvent refining see Solvent extraction.

Sonication a physical technique employing ultrasound to intensely vibrate a sample media in extracting solvent and to maximize solvent–analyte interactions.

Sonic log a well log based on the time required for sound to travel through rock, useful in determining porosity.

Sour crude oil crude oil containing an abnormally large amount of sulfur compounds; see also Sweet crude oil.

SOx oxides of sulfur.

Soxhlet extraction an extraction technique for solids in which the sample is repeatedly contacted with solvent over several hours, increasing extraction efficiency.

Specific gravity the mass (or weight) of a unit volume of any substance at a specified temperature compared to the mass of an equal volume of pure water at a standard temperature; see also Density.

Spent catalyst catalyst that has lost much of its activity due to the deposition of coke and metals.

Spontaneous ignition ignition of a fuel, such as coal, under normal atmospheric conditions; usually induced by climatic conditions.

Stabilization the removal of volatile constituents from a higher boiling fraction or product (stripping); the production of a product which, to all intents and purposes, does not undergo any further reaction when exposed to the air.

Stabilizer a fractionating tower for removing light hydrocarbons from an oil to reduce vapor pressure particularly applied to gasoline.

Standard conditions the reference amounts for pressure and temperature—in the English system, they are 14.73 pounds per square inch for the pressure and 60°F for temperature.

Standpipe the pipe by which catalyst is conveyed between the reactor and the regenerator.

Stationary phase in chromatography, the porous solid or liquid phase through which an introduced sample passes. The different affinities the stationary phase has for a sample allow the components in the sample to be separated, or resolved.

Steam cracking a conversion process in which the feedstock is treated with superheated steam.

Steam distillation distillation in which vaporization of the volatile constituents is effected at a lower temperature by introduction of steam (open steam) directly into the charge.

Steam drive injection (steam injection) EOR process in which steam is continuously injected into one set of wells (injection wells) or other injection source to effect oil displacement toward and production from a second set of wells (production wells); steam stimulation of production wells is *direct steam stimulation* whereas steam drive by steam injection to increase production from other wells is *indirect steam stimulation*.

Steam-refined asphalt vacuum residue that has been subjected to steam stripping.

Steam stimulation injection of steam into a well and the subsequent production of oil from the same well.

Steam stripping injection of steam into a residue which aids vacuum distillation.

Steep-slope roofing roofing products designed for a roof slope of more than 14 degrees.

Stimulation the process of acidifying or fracturing carried out to expand existing ducts or to create new ones in the source rock formation.

Stone mastic asphalt, stone matrix asphalt (SMA) referred to as stone mastic asphalt in Europe or stone matrix asphalt in the United States; a graded asphalt mixture with asphalt as a binder, composed of a coarse crushed aggregate skeleton bound with a mastic mortar.

Stone-matrix asphalt mix a mix composed of high-quality coarse aggregate, combined with a large amount of mastic composed of a high-performance asphalt binder, mineral filler, and a small amount of fibrous material.

Storage stability (or storage instability) the ability (inability) of a liquid to remain in storage over extended periods of time without appreciable deterioration as measured by gum formation and the depositions of insoluble material (sediment).

Straight-reduced asphalts vacuum residues used as asphalts; steam stripping may have been used in their production; straight reduced asphalts refer to asphalt produced to a specific target grade without blending with other asphalt grades to achieve the desired result.

Straight-run asphalt asphalt produced by the distillation of asphaltic crude oil; asphalt that is similar to straight-reduced asphalts and steam-refined asphalt.

Straight-run products obtained from a distillation unit and used without further treatment.

Strata layers including the solid iron-rich inner core, molten outer core, mantle, and crust of the earth.

Stratigraphy the sub-discipline of geology that studies the origin, composition, distribution, and succession of rock strata.

Straw oil pale paraffin oil of straw color used for many process applications.

Stripping a means of separating volatile components from less volatile ones in a liquid mixture by the partitioning of the more volatile materials to a gas phase of air or steam (stabilization).

Structural nose a term used in structural geology to define a geometric form protruding from a main body.

Sub-base the course in the asphalt pavement structure immediately below the base course is the sub-base. If the subgrade soil is of adequate quality, it may serve as the sub-base.

Subdrain a structure placed beneath the ground surface to collect and carry away underground water.

Subgrade the uppermost material placed in embankments or unmoved from cuts in the normal grading of the roadbed: the foundation for the asphalt pavement structure; sometimes called basement soil or foundation soil.

Subgrade stabilization modification of roadbed soils by admixing with stabilizing or chemical agents that will increase load-bearing capacity, firmness, and resistance to weathering or displacement.

Subsurface drainage removal of free water from various structural components of the pavement or the surrounding soil.

Sulfonic acids acids obtained from petroleum or a petroleum product with strong sulfuric acid.

Sulphur extended asphalt a hot mixed asphalt in which part of the bituminous binder is replaced by elemental sulfur, typically at levels between 20 and 40% (w/w) of the original asphalt content.

Sulfuric acid alkylation an alkylation process in which olefins (C_3, C_4, and C_5) combine with *iso*-butane in the presence of a catalyst (sulfuric acid) to form branched chain hydrocarbons used especially in gasoline blending stock.

Supercritical fluid an extraction method where the extraction fluid is present at a pressure and temperature above its critical point.

Super-light oil oil with a specific gravity typically higher than 38° API.

Surface active material a chemical compound, molecule, or aggregate of molecules with physical properties that cause it to adsorb at the interface between *two* immiscible liquids, resulting in a reduction of interfacial tension or the formation of a microemulsion.

Surface course one or more layers of a pavement structure designed to accommodate the traffic load, the top layer of which resists skidding, traffic abrasion, and the disintegrating effects of climate; sometimes called the wearing course.

Surface dressing process used to seal road surfaces; a thin film of asphalt, cutback asphalt, or asphalt emulsions is spread, covered with a single or double layer of chippings, and then rolled.

Surface treatment may include surface dressing and other techniques, such as spraying with minor amounts of asphalt emulsion to bind surfaces together.

Surfactant a type of chemical, characterized as one that reduces interfacial resistance to mixing between oil and water or changes the degree to which water wets the rock.

Suspensoid catalytic cracking a non-regenerative cracking process in which cracking stock is mixed with slurry of catalyst (usually clay) and cycle oil and passed through the coils of a heater.

SW-846 an EPA multi-volume publication entitled *Test Methods for Evaluating Solid Waste, Physical/Chemical Methods*; the official compendium of analytical and sampling methods that have been evaluated and approved for use in complying with the RCRA regulations and that functions primarily as a guidance document setting forth acceptable, although not required, methods for the regulated and regulatory communities to use in responding to RCRA-related sampling and analysis requirements. SW-846 changes over time as new information and data are developed.

Sweated wax a crude wax freed from oil by having been passed through a sweater.

Sweating the separation of paraffin oil and low-melting wax from paraffin wax.

Sweet crude oil crude oil containing little sulfur; see also Sour crude oil.

Sweetening the process by which petroleum products are improved in odor and color by oxidizing or removing the sulfur-containing and unsaturated compounds.

Swelling increase in the volume of crude oil caused by absorption of EOR fluids, especially carbon dioxide. Also increase in volume of clays when exposed to brine.

Swept zone the volume of rock that is effectively swept by injected fluids.

Synthetic crude oil (syncrude) a hydrocarbon product produced by the conversion of coal, oil shale, or tar sand bitumen that resembles conventional crude oil; can be refined in a petroleum refinery.

Tack coat asphalt oil, usually emulsion type, applied to existing pavement during repairs or overlay paving to create a bond between the old and new asphalt.

Tar the volatile, brown to black, oily, viscous product from the destructive distillation of many bituminous or other organic materials, especially coal; a name used for petroleum in ancient texts.

Target analyte target analytes are compounds that are required analytes in US EPA analytical methods. BTEX and PAHs are examples of petroleum-related compounds that are target analytes in US EPA Methods.

Tar sand see Bituminous sand.

Technical reserves the accumulative production derived from a production forecast in which economic criteria are not applied.

Terminal a facility outside a refinery where asphalt is held for intermediate storage prior to delivery to (or collection by) customers.

Tertiary structure the three-dimensional structure of a molecule.

Tetraethyl lead (TEL) an organic compound of lead, $Pb(CH_3)_4$, which, when added in small amounts, increases the antiknock quality of gasoline.

Thermal coke the carbonaceous residue formed as a result of a non-catalytic thermal process; the Conradson carbon residue; the Ramsbottom carbon residue.

Thermal cracking a process which decomposes, rearranges, or combines hydrocarbon molecules by the application of heat, without the aid of catalysts.

Thermally cracked asphalts also known as residues (petroleum); asphalt produced by thermal cracking.

Thermal polymerization a thermal process to convert light hydrocarbon gases into liquid fuels.

Thermal process any refining process which utilizes heat, without the aid of a catalyst.

Thermal reforming a process using heat (but no catalyst) to effect molecular rearrangement of low-octane naphtha into gasoline of higher antiknock quality.

Thermal stability (thermal instability) the ability (inability) of a liquid to withstand relatively high temperatures for short periods of time without the formation of carbonaceous deposits (sediment or coke).

Thermofor catalytic cracking a continuous, moving-bed catalytic cracking process.

Thermofor catalytic reforming a reforming process in which the synthetic, bead-type catalyst of coprecipitated chromia (Cr_2O_3) and alumina (Al_2O_3) flows down through the reactor concurrent with the feedstock.

Thermofor continuous percolation a continuous clay treating process to stabilize and decolorize lubricants or waxes.

Thief zone any geologic stratum not intended to receive injected fluids in which significant amounts of injected fluids are lost; fluids may reach the thief zone due to an improper completion or a faulty cement job.

Thin-layer chromatography (TLC) a chromatographic technique employing a porous medium of glass coated with a stationary phase. An extract is spotted near the bottom of the medium and placed in a chamber with solvent (mobile phase). The solvent moves up the medium and separates the components of the extract, based on affinities for the medium and solvent.

Topped crude petroleum that has had volatile constituents removed up to a certain temperature, e.g., 250°C+ (480°F+) topped crude; not always the same as a residuum.

Topping the distillation of crude oil to remove light fractions only.

Topping plant a stand-alone distillation plant; usually found in terminals and used to remove distillate materials added to asphalts for transportation purposes.

Total petroleum hydrocarbons (TPH) the family of several hundred chemical compounds that originally come from petroleum.

Torching application of a roofing membrane with a propane gas flame, used for melting the side of the roofing membrane, without addition of hot bonding asphalt; also called hot welding roofing.

Total thickness the thickness from the top of the formation of interest down to a vertical boundary determined by a water level or by a change of formation.

Tower equipment for increasing the degree of separation obtained during the distillation of oil in a still.

TPH E gas chromatographic test for TPH extractable organic compounds.

TPH V gas chromatographic test for TPH volatile organic compounds.

TPH-D(DRO) gas chromatographic test for TPH diesel-range organics.

TPH-G(GRO) gas chromatographic test for TPH gasoline-range organics.

Trace element those elements that occur at very low levels in a given system.

Trinidad Lake asphalt bitumen (natural asphalt) obtained from the Trinidad Lake, island of Trinidad, Republic of Trinidad and Tobago.

Transgression a geological term used to define the immersion of one part of the continent under sea level as a result of a descent of the continent or an elevation of the sea level.

Transport liquefiables shrinkage factor the fraction obtained by considering the liquefiables obtained in transportation to the processing complexes.

Trap a sediment in which oil and gas accumulate from which further migration is prevented.

Treatment any method, technique, or process that changes the physical and/or chemical character of petroleum.

Triaxial borehole seismic survey a technique for detecting the orientation of hydraulically induced fractures, wherein a tool holding three mutually seismic detectors is clamped in the borehole during fracturing; fracture orientation is deduced through analysis of the detected microseismic perpendicular events that are generated by the fracturing process.

Trickle hydrodesulfurization a fixed-bed process for desulfurizing middle distillates.

Trillion 1×10^{12}

True boiling point (True boiling range) the boiling point (boiling range) of a crude oil fraction or a crude oil product under standard conditions of temperature and pressure.

Tube-and-tank cracking a older liquid-phase thermal cracking process.

Ultimate analysis elemental composition.

Ultrafining a fixed-bed catalytic hydrogenation process to desulfurize naphtha and upgrade distillates by essentially removing sulfur, nitrogen, and other materials.

Ultraforming a low-pressure naphtha-reforming process employing onstream regeneration of a platinum-on-alumina catalyst and producing high yields of hydrogen and high-octane-number reformate.

Unassociated molecular weight the molecular weight of asphaltenes in an non-associating (polar) solvent, such as dichlorobenzene, pyridine, or nitrobenzene.

Unconformity a surface of erosion that separates younger strata from older rocks.

Underdrain a perforated or porous-walled pipe placed with suitable pervious backfill beneath the ground surface to collect and carry away underground water.

Underlayment factory-made flexible sheets of asphalt (oxidized or modified) which are used as underlay to coverings of sloping roofs (e.g., tiles, slates, shingles).

Unifining a fixed-bed catalytic process to desulfurize and hydrogenate refinery distillates.

Unisol process a chemical process for extracting mercaptan sulfur and certain nitrogen compounds from sour gasoline or distillates using regenerable aqueous solutions of sodium or potassium hydroxide containing methanol.

Universal viscosity see Saybolt Universal viscosity.

Unresolved complex the thousands of compounds that a gas chromatograph *mixture (UCM)* is unable to fully separate.

Unstable usually refers to a petroleum product that has more volatile constituents present or refers to the presence of olefin and other unsaturated constituents.

UOP alkylation a process using hydrofluoric acid (which can be regenerated) as a catalyst to unite olefins with *iso*-butane.

UOP copper sweetening a fixed-bed process for sweetening gasoline by converting mercaptans to disulfides by contact with ammonium chloride and copper sulfate in a bed.
UOP fluid catalytic cracking a fluid process of using a reactor-over-regenerator design.
Upgrading the conversion of petroleum to value-added salable products.
Upper-phase microemulsion a microemulsion phase containing a high concentration of oil that, when viewed in a test tube, resides on top of a water phase.
Urea dewaxing a continuous dewaxing process for producing low-pour-point oils, and using urea which forms a solid complex (adduct) with the straight-chain wax paraffins in the stock; the complex is readily separated by filtration.
Vacuum distillation distillation under reduced pressure.
Vacuum residuum a residuum obtained by distillation of a crude oil under vacuum (reduced pressure); that portion of petroleum which boils above a selected temperature such as 510°C (950°F) or 565°C (1050°F).
Vapor-phase cracking a high-temperature, low-pressure conversion process.
Vapor-phase hydrodesulfurization a fixed-bed process for desulfurization and hydrogenation of naphtha.
Vertical seismic profiling a method of conducting seismic surveys in the bore hole for detailed subsurface information.
VGC (viscosity-gravity constant) an index of the chemical composition of crude oil defined by the general relation between specific gravity, sg, at 60° F and Saybolt Universal viscosity, SUV, at 100° F:

$$a = 10\,\mathrm{sg} - 1.0752\log(\mathrm{SUV} - 38) / 10\,\mathrm{sg} - \log(\mathrm{SUV} - 38)$$

The constant, *a*, is low for the paraffin crude oils and high for the naphthenic crude oils.
VI (Viscosity index) an arbitrary scale used to show the magnitude of viscosity changes in lubricating oils with changes in temperature.
Visbreaking a relatively mild thermal cracking operation mainly used to reduce the viscosity and pour point of vacuum residues for subsequent use in heavy fuel oils; the process converts a proportion of the residue feedstock to distillate product, e.g. gas oil; a process for reducing the viscosity of heavy feedstocks by controlled thermal decomposition.
Viscosity a measure of the ability of a liquid to flow or a measure of its resistance to flow; the force required to move a plane surface of area 1 m^2 over another parallel plane surface 1 m away at a rate of 1 m/s when both surfaces are immersed in the fluid.
Viscosity-graded asphalt asphalt which is graded and specified by the viscosity at a standard temperature, which is typically 60°C (140°F).
Viscosity-gravity constant see VGC.
Viscosity index see VI.
VFA voids filled with asphalt.
VMA voids in mineral aggregate.
VOC (VOCs) volatile organic compound(s); volatile organic compounds are regulated because they are precursors to ozone; carbon-containing gases and vapors from incomplete gasoline combustion and from the evaporation of solvents.
Voids air spaces in the aggregate.
Volatile compounds a relative term that may mean (i) any compound that will purge, (ii) any compound that will elute before the solvent peak (usually those <C6), or (iii) any compound that will not evaporate during a solvent removal step.

VTM percent voids in the total mix.

Warm mix asphalt asphalt mixtures produced at lower temperatures as compared to those typically associated with rolled asphalt pavement; produced and placed at temperatures typically 10–40°C (50–100°F) lower than conventional rolled asphalt; the differentiation between warm mix asphalt and hot mix asphalt is not often obvious.

Watson characterization factor see Characterization factor.

Wax see Mineral wax and Paraffin wax.

Wax distillate a neutral distillate containing a high percentage of crystallizable paraffin wax, obtained on the distillation of paraffin or mixed-base crude, and on reducing neutral lubricating stocks.

Wax fractionation a continuous process for producing waxes of low oil content from wax concentrates; see also MEK deoiling.

Wax manufacturing a process for producing oil-free waxes.

Wearing course the top course of asphalt pavements, also called the surface course.

Weathered crude oil crude oil which, due to natural causes during storage and handling, has lost an appreciable quantity of its more volatile components; also indicates uptake of oxygen.

Weathering test various accelerated durability tests have been developed for oxidized asphalts used in roofing applications; the most prevalent is the xenon arc accelerated weathering test, where thin oxidized asphalt films are applied to aluminum panels and then subjected to light, heat, and water sprays in several combinations of time and temperature.

Well abandonment the final activity in the operation of a well when it is permanently closed under safety and environment preservation conditions.

Wellbore (well bore) the hole in the earth comprising a well.

Wet gas gas containing a relatively high proportion of hydrocarbons which are recoverable as liquids; see also Lean gas.

Wet scrubbers devices in which a counter-current spray liquid is used to remove impurities and particulate matter from a gas stream.

Wettability the relative degree to which a fluid will spread on (or coat) a solid surface in the presence of other immiscible fluids.

Wettability number a measure of the degree to which a rock is water-wet or oil-wet, based on capillary pressure curves.

Wettability reversal the reversal of the preferred fluid wettability of a rock, e.g., from water-wet to oil-wet, or vice versa.

White oil a generic name applied to highly refined, colorless hydrocarbon oils of low volatility, and covering a wide range of viscosity.

White spirit a distillate petroleum product free of rancid or objectionable odors, boiling-range 150–200°C (300–390°F); sometimes described as Stoddard solvent.

WMA warm mix asphalt.

Wobbe Index (or Wobbe Number) the calorific value of a gas divided by the specific gravity.

Wood alcohol see Methyl alcohol.

Zeolite a crystalline aluminosilicate used as a catalyst and having a particular chemical and physical structure.

INDEX

Note: Page numbers followed by "*f*" and "*t*" refer to figures and tables, respectively.

H

I

J

K

L

M

U

V

W

X

Z

Kolibri 360 GmbH
Ferdinand-Jühlke-Straße 7
99095 Erfurt, Deutschland
produktsicherheit@kolibri360.de